BIRKHÄUSER

Trends in Mathematics

Trends in Mathematics is a series devoted to the publication of volumes arising from conferences and lecture series focusing on a particular topic from any area of mathematics. Its aim is to make current developments available to the community as rapidly as possible without compromise to quality and to archive these for reference.

Proposals for volumes can be sent to the Mathematics Editor at either

Birkhäuser / Springer Basel AG
P.O. Box 133
CH-4010 Basel
Switzerland

or

Birkhauser Boston
233 Spring Street
New York, NY 10013
USA

Material submitted for publication must be screened and prepared as follows:

All contributions should undergo a reviewing process similar to that carried out by journals and be checked for correct use of language which, as a rule, is English. Articles without proofs, or which do not contain any significantly new results, should be rejected. High quality survey papers, however, are welcome.

We expect the organizers to deliver manuscripts in a form that is essentially ready for direct reproduction. Any version of TeX is acceptable, but the entire collection of files must be in one particular dialect of TeX and unified according to simple instructions available from Birkhäuser.

Furthermore, in order to guarantee the timely appearance of the proceedings it is essential that the final version of the entire material be submitted no later than one year after the conference. The total number of pages should not exceed 350. The first-mentioned author of each article will receive 25 free offprints. To the participants of the congress the book will be offered at a special rate.

Advances in Ring Theory

Dinh Van Huynh
Sergio R. López-Permouth
Editors

Birkhäuser

Editors:

Dinh Van Huynh
Sergio R. López-Permouth
Department of Mathematics
Ohio University
321 Morton Hall
Athens, OH 45701
USA
e-mail: huynh@math.ohiou.edu
lopez@math.ohiou.edu

2000 Mathematics Subject Classification 06, 11-17, 22

Library of Congress Control Number: 2010921344

Bibliographic information published by Die Deutsche Bibliothek. Die Deutsche Bibliothek lists this publication in the Deutsche Nationalbibliografie; detailed bibliographic data is available in the Internet at http://dnb.ddb.de

ISBN 978-3-0346-0285-3

P.O. Box 133, CH-4010 Basel, Switzerland
Part of Springer Science+Business Media
Printed on acid-free paper produced from chlorine-free pulp. TCF ∞
Cover Design: Alexander Faust, Basel, Switzerland
Printed in Germany

ISBN 978-3-0346-0285-3
e-ISBN 978-3-0346-0286-0

9 8 7 6 5 4 3 2 1

www.birkhauser.ch

Contents

S.K. Jain

Preface

The International Conference on Algebra and its Applications held in Athens, Ohio, June 18–21, 2008 and sponsored by the Ohio University Center for Ring Theory and its Applications (CRA) had as its central purpose to honor Surender K. Jain, the Center's retiring first director, on the dual occasion of his 70th birthday and of his retirement from Ohio University. With this volume we celebrate the contributions to Algebra of our distinguished colleague. One of Surender's main attributes has been the way in which he radiates enthusiasm about mathematical research; his eagerness to pursue mathematical problems is contagious; we hope that reading this excellent collection of scholarly writings will have a similar effect on our readers and that you will be inspired to continue the pursuit of Ring Theory as well as Algebra and its Applications.

As with previous installments of CRA conferences, the underlying principle behind the meeting was to bring together specialists on the various areas of Algebra in order to promote communication and cross pollination between them. In particular, a common philosophy of our conferences through the years has been to bring algebraists who focus on the theoretical aspects of our field with those others who embrace applications of Algebra in diverse areas. Clearly, as a reflection of the interests of the organizers, the applications we emphasized were largely within the realm of Coding Theory. The philosophy behind the organization of the conference has undoubtedly impacted this Proceedings volume.

For the most part, the contributors delivered related talks at the conference itself. However, there are also a couple of contributions in this volume from authors who could not be present at the conference but wanted to participate and honor Dr. Jain on this occasion. All papers were subject to a strict process of refereeing and, in fact, not all submissions were accepted for publication.

We would like to take this opportunity to thank all the anonymous referees who delivered their verdicts about the submitted papers within a very tight schedule; they also provided valuable feedback on many of the papers that appear here in final form. Likewise, we wish to express our deep appreciation to Sylvia Lotrovsky and Thomas Hempfling of Birkhäuser for their diligent efforts to bring this volume to completion.

Advances in Ring Theory
Trends in Mathematics, 1–17

Applications of Cogalois Theory to Elementary Field Arithmetic

Toma Albu

Dedicated to S.K. Jain on his 70th birthday

Abstract. The aim of this expository paper is to present those basic concepts and facts of Cogalois theory which will be used for obtaining in a natural and easy way some interesting results in elementary field arithmetic.

Mathematics Subject Classification (2000). Primary 12-06, 12E30, 11-06, 11A99; Secondary 12F05, 12F10, 12F99, 12Y05.

Keywords. Cogalois theory, elementary field arithmetic, field extension, Galois extension, radical extension, Kneser extension, Cogalois extension, G-Cogalois extension.

1. Introduction

A standard, very concrete, and not so hard exercise in any undergraduate abstract algebra course anyone of us has encountered is the following one:

Consider the field extension $\mathbb{Q} \subseteq \mathbb{Q}(\sqrt{2}, \sqrt[3]{5})$.

(a) *Calculate the* degree $[\mathbb{Q}(\sqrt{2}, \sqrt[3]{5}) : \mathbb{Q}]$ *of this extension.*

(b) *Find a* primitive element *of this extension.*

Surely, it is natural to ask what about the same questions when we replace the very particular radicals $\sqrt{2}$ and $\sqrt[3]{5}$ by arbitrary finitely many radicals of positive integers. More precisely, we have the following

Problem. *Consider the field extension*

$$\mathbb{Q} \subseteq \mathbb{Q}(\sqrt[n_1]{a_1}, \dots, \sqrt[n_r]{a_r}),$$

The author gratefully acknowledges partial financial support from the grant ID-PCE 1190/2008 awarded by the Consiliul Naţional al Cercetării Ştiinţifice în Învăţământul Superior (CNCSIS), România.

where $r, n_1, \ldots, n_r, a_1, \ldots, a_r$ *are positive integers, and where* $\sqrt[n_i]{a_i}$ *is the positive real* n_i*th root of* a_i *for each* i, $1 \leqslant i \leqslant r$.

(a) *Calculate the* degree $[\mathbb{Q}(\sqrt[n_1]{a_1}, \ldots, \sqrt[n_r]{a_r}) : \mathbb{Q}]$ *of this extension.*
(b) *Find a* primitive element *of this extension.*

More than twenty years ago we first thought about this challenging problem. A first attempt to solve it, even in a more general case, was the introduction and investigation of the so-called *Kummer extensions with few roots of unity*, see Albu [1]. After that, we discovered, little by little, the fundamental papers of Kneser [25] and Greither and Harrison [20] and got more and more involved in their topic, which lead to what is nowadays called *Cogalois theory*. There are at least two reasons for presenting this material to *ring and module theorists*:

- firstly, to make a *propaganda* of this pretty nice and equally new theory in field theory by providing a gentle and as short as possible introduction to a general audience and readership of its basic notions and results, and
- secondly, we want to show how this theory has nice applications in solving some interesting and nontrivial problems of elementary *field arithmetic*, including that mentioned above concerning the computation of the degree and finding a (canonical) primitive element of field extensions like $\mathbb{Q} \subseteq \mathbb{Q}(\sqrt[n_1]{a_1}, \ldots, \sqrt[n_r]{a_r})$.

2. Notation and terminology

By $\mathbb{N}$ we denote the set $\{0, 1, 2, \ldots\}$ of all natural numbers, by $\mathbb{N}^*$ the set $\mathbb{N} \setminus \{0\}$ of all strictly positive natural numbers, and by $\mathbb{Q}$ (resp. $\mathbb{R}$, $\mathbb{C}$) the field of all rational (resp. real, complex) numbers. For any $\varnothing \neq A \subseteq \mathbb{C}$ (resp. $\varnothing \neq X \subseteq \mathbb{R}$) we denote $A^* = A \setminus \{0\}$ (resp. $X_+ = \{\, x \in X \mid x \geqslant 0 \,\}$). If $a \in \mathbb{R}_+^*$ and $n \in \mathbb{N}^*$, then the unique positive real root of the equation $x^n - a = 0$ will be denoted by $\sqrt[n]{a}$. For any set M, $|M|$ will denote the cardinal number of M.

A *field extension* is a pair (F, E) of fields, where F is a subfield of E (or E is an overfield of F), and in this case we shall write E/F. Very often, instead of "field extension" we shall use the shorter term "extension". If E is an overfield of a field F, we will also say that E is an extension of F. By an *intermediate field* of an extension E/F we mean any subfield K of E with $F \subseteq K$, and the set of all intermediate fields of E/F is a complete lattice that will be denoted by $\mathbb{I}(E/F)$.

Throughout this paper F always denotes a field, $\mathrm{Char}(F)$ its characteristic, $e(F)$ its characteristic exponent (that is, $e(F) = 1$ if F has characteristic 0, and $e(F) = p$ if F has characteristic $p > 0$), and Ω a fixed algebraically closed field containing F as a subfield. Any considered overfield of F is supposed to be a subfield of Ω.

For an arbitrary nonempty subset S of Ω and a number $n \in \mathbb{N}^*$ we denote throughout this paper:

$$\begin{aligned} S^* &= S \setminus \{0\}, \\ S^n &= \{\, x^n \mid x \in S \,\}, \\ \mu_n(S) &= \{\, x \in S \mid x^n = 1 \,\}. \end{aligned}$$

By a *primitive nth root of unity* we mean any generator of the cyclic group $\mu_n(\Omega)$; ζ_n will always denote such an element.

For an arbitrary group G, the notation $H \leqslant G$ means that H is a subgroup of G. The lattice of all subgroups of G will be denoted by $\mathbb{L}(G)$. For any subset M of G, $\langle M \rangle$ will denote the subgroup of G generated by M.

For a field extension E/F we shall denote by $[E : F]$ the *degree*, and by $\mathrm{Gal}\,(E/F)$ the *Galois group* of E/F. For any subgroup Δ of $\mathrm{Gal}\,(E/F)$, $\mathrm{Fix}\,(\Delta)$ will denote the fixed field of Δ. If E/F is an extension and $A \subseteq E$, then $F[A]$ will denote the smallest subring of E containing both A and F as subsets. We also denote by $F(A)$ the smallest subfield of E containing both A and F as subsets, called the subfield of E obtained by adjoining to F the set A. For all other undefined terms and notation concerning basic field theory the reader is referred to Bourbaki [17], Karpilovsky [24], and/or Lang [26].

3. What is Cogalois theory?

Cogalois theory, a fairly new area in field theory, investigates field extensions, finite or not, that possess a so-called *Cogalois correspondence*. The subject is somewhat dual to the very classical *Galois theory* dealing with field extensions possessing a *Galois correspondence*.

In what follows we are intending to briefly explain the meaning of such extensions. An interesting but difficult problem in field theory is to describe in a satisfactory manner the set $\mathbb{I}(E/F)$ of all intermediate fields of a given field extension E/F, which, in general is a complicated-to-conceive, potentially infinite set of hard-to-describe-and-identify objects. This is a very particular case of a more general problem in mathematics: *Describe in a satisfactory manner the collection* $\underline{\mathrm{Sub}}(X)$ *of all subobjects of a given object* X *of a category* $\mathcal{C}$. For instance, if G is a group, then an important problem in group theory is to describe the set $\mathbb{L}(G)$ of all subgroups of G. Observe that for any field F we may consider the category $\mathcal{E}_F$ of all field extensions of F. If E is any object of $\mathcal{E}_F$, i.e., a field extension E/F, then the set $\mathbb{I}(E/F)$ of all subfields of E containing F, i.e., of all intermediate fields of E/F, is precisely the set $\underline{\mathrm{Sub}}(E)$ of all subobjects of E in $\mathcal{E}_F$.

Another important problem in field theory is to calculate the *degree* of a given field extension E/F.

Answers to these two problems are given for particular field extensions by *Galois theory* invented by E. Galois (1811–1832) and by *Kummer theory* invented by E. Kummer (1810–1873). Let us briefly recall the solutions offered by these two theories in answering the two problems presented above.

The fundamental theorem of finite Galois theory (FTFGT). *If E/F is a finite Galois extension with Galois group Γ, then the canonical map*

$$\alpha : \mathbb{I}(E/F) \longrightarrow \mathbb{L}(\Gamma),\ \ \alpha(K) = \mathrm{Gal}(E/K),$$

is a lattice anti-isomorphism, i.e., a bijective order-reversing map. Moreover, $[E : F] = |\Gamma|$.

We say that such an E/F is an *extension with Γ-Galois correspondence.*

In this way, the lattice $\mathbb{I}(E/F)$ of all subobjects of an object $E \in \mathcal{E}_F$, which has the additional property that is a finite Galois extension of F, can be described by the lattice of all subobjects of the object $\mathrm{Gal}\,(E/F)$ in the category $\mathcal{G}_f$ of all finite groups. In principle, this category is more suitable than the category $\mathcal{E}_F$ of all field extensions of F, since the set of all subgroups of a finite group is a far more benign object. Thus, many questions concerning a field are best studied by transforming them into group theoretical questions in the group of automorphisms of the field.

Note that for an *infinite* Galois extension E/F the FTFGT fails. In this case the Galois group $\mathrm{Gal}\,(E/F)$ is in fact a *profinite group*, that is, a projective limit of finite groups, or equivalently, a Hausdorff, compact, totally disconnected topological group; its topology is the so called *Krull topology*. The description of $\mathbb{I}(E/F)$ is given by

The fundamental theorem of infinite Galois theory (FTIGT). *If E/F is an arbitrary Galois extension with Galois group Γ, then the canonical map*

$$\alpha : \mathbb{I}(E/F) \longrightarrow \overline{\mathbb{L}}(\Gamma),\ \ \alpha(K) = \mathrm{Gal}(E/K),$$

is a lattice anti-isomorphism, where $\overline{\mathbb{L}}(\Gamma)$ denotes the lattice of all closed subgroups of the group Γ endowed with the Krull topology.

Observe that the lattice $\overline{\mathbb{L}}(\Gamma)$ is nothing else than the lattice of all subobjects of Γ in the category of all profinite groups.

However, the Galois group of a given Galois field extension E/F, finite or not, is in general difficult to be concretely described; so, it will be desirable to impose additional conditions on E/F such that the lattice $\mathbb{I}(E/F)$ be isomorphic (or anti-isomorphic) to the lattice $\mathbb{L}(\Delta)$ of all subgroups of some other group Δ, easily computable and appearing explicitly in the data of the given Galois extension E/F. A class of such Galois extensions is that of *classical Kummer extensions.* We recall their definition below.

Definition. A field extension E/F is said to be a *classical n-Kummer extension*, with n a given positive integer, if the following three conditions are satisfied:

(1) $\gcd(n, e(F)) = 1$,
(2) $\zeta_n \in F$,
(3) $E = F(\{\sqrt[n]{a_i}\,|\, i \in I\,\})$,

where I is an arbitrary set, finite or not, $a_i \in F^*$, and $\sqrt[n]{a_i}$ is a certain root in Ω of the polynomial $X^n - a_i$, $i \in I$.

Note that the extension E/F is finite if and only if the set I in the definition above can be chosen to be finite.

For a classical n-Kummer extension E/F we denote by

$$\mathrm{Kum}(E/F) := F^*\langle\{\sqrt[n]{a_i} \mid i \in I\,\rangle/F^*$$

the so-called *Kummer group* of E/F. The next result is a part of the so-called *Kummer theory*.

The fundamental theorem of Kummer theory (FTKT). *Let E/F be a classical n-Kummer extension with Kummer group Δ. Then there exists a canonical lattice isomorphism*

$$\mathbb{I}(E/F) \xrightarrow{\sim} \mathbb{L}(\Delta).$$

Observe that the Kummer group Δ of a classical n-Kummer extension E/F is intrinsically given with the extension E/F and easily manageable as well. This group is isomorphic, but not canonically, with the *character group* $\widehat{\Gamma}$ of the Galois group Γ of E/F; in particular, it follows that for E/F finite, the group Δ is isomorphic with Γ, and in particular it has exactly $[E : F]$ elements. Consequently, if E/F is a finite classical n-Kummer extension, say $E = F(\sqrt[n]{a_1}, \ldots, \sqrt[n]{a_r})$, then

$$[\,F(\sqrt[n]{a_1}, \ldots, \sqrt[n]{a_r}) : F\,] = |F^*\langle\,\sqrt[n]{a_1}, \ldots, \sqrt[n]{a_r}\,\rangle/F^*|.$$

Note also that any classical n-Kummer extension E/F is a Galois extension with an Abelian Galois group of exponent a divisor of n (this means that $\sigma^n = 1_E$ for all $\sigma \in \mathrm{Gal}(E/F)$), and conversely, any Galois extension E/F such that $\gcd(n, e(F)) = 1$, $\zeta_n \in F$ for some $n \in \mathbb{N}^*$, and such that the Galois group of E/F is an Abelian group of exponent a divisor of n, is a classical n-Kummer extension.

On the other hand, there exists a fairly large class of field extensions which are not necessarily Galois, but enjoy a property similar to that in FTKT or is dual to that in FTFGT. Namely, these are the extensions E/F for which there exists a canonical lattice isomorphism (and *not* a lattice anti-isomorphism as in the Galois case) between $\mathbb{I}(E/F)$ and $\mathbb{L}(\Delta)$, where Δ is a certain group canonically associated with the extension E/F. We call the members of this class *extensions with Δ-Cogalois correspondence*. Their prototype is the field extension

$$\mathbb{Q}\,(\sqrt[n_1]{a_1}, \ldots, \sqrt[n_r]{a_r})/\mathbb{Q}\,,$$

where $r, n_1, \ldots, n_r, a_1, \ldots, a_r$ are positive integers, and where $\sqrt[n_i]{a_i}$ is the positive real n_ith root of a_i for each i, $1 \leqslant i \leqslant r$. For such an extension, the associated group Δ is the factor group $\mathbb{Q}^*\langle\sqrt[n_1]{a_1}, \ldots, \sqrt[n_r]{a_r}\,\rangle/\,\mathbb{Q}^*$. Note that the finite classical n-Kummer extensions have a privileged position: they are at the same time extensions with Galois and with Cogalois correspondences, and the two groups appearing in this setting are isomorphic.

After 1930 there were attempts to weaken the condition $\zeta_n \in F$ in the definition of a Kummer extension in order to effectively compute the degree of particular finite radical extensions, i.e., of extensions of type $F\,(\sqrt[n_1]{a_1}\,,\ldots,\sqrt[n_r]{a_r}\,)/F$, where F was mainly an algebraic number field. All these attempts finally lead to what nowadays is called *Cogalois theory*, also spelled *co-Galois theory*.

The main precursors of Cogalois theory, in chronological order, are H. Hasse (1930), A. Besicovitch (1940) [16] , L.J. Mordell (1953) [27], C.L. Siegel (1972) [29], M. Kneser (1975) [25] whose paper brilliantly superseded all the previous work done in computing the degree of finite radical extensions, A. Schinzel (1975) [28], D. Gay, W.Y. Vélez (1978) [19], etc.

In our opinion, Cogalois theory was born in 1986, with birthplace *Journal of Pure and Applied Algebra* [20], and having C. Greither and D.K. Harrison as parents. In that paper [20], the *Cogalois extensions* have been introduced and investigated for the first time in the literature, and other classes of finite field extensions possessing a Cogalois correspondence, including the so-called *neat presentations* have been considered.

Besides the Cogalois extensions introduced by Greither and Harrison [20] in 1986, new basic classes of finite radical field extensions the Cogalois theory deals with, namely the *G-Kneser extensions*, *strongly G-Kneser extensions*, and *G-Cogalois extensions* were introduced and investigated in 1995 by T. Albu and F. Nicolae [9]. Note that the frame of G-Cogalois extensions permits a simple and unified manner to study the classical Kummer extensions, the Kummer extensions with few roots of unity, the Cogalois extensions, and the neat presentations. In 2001 an *infinite Cogalois theory* investigating infinite radical extensions has been developed by T. Albu and M. Ţena, in 2003 appeared the author's monograph "Cogalois theory" [7], and in 2005 the infinite Cogalois theory has been generalized to arbitrary profinite groups by T. Albu and Ş.A. Basarab [8], leading to a so-called *abstract Cogalois theory* for arbitrary profinite groups.

Roughly speaking, Cogalois theory investigates *radical extensions*, finite or not, i.e., extensions of type E/F with $E = F(\{\sqrt[n_i]{a_i}\,|\, i \in I\,\})$, $n_i \in \mathbb{N}^*$, $a_i \in F^*$, $i \in I$, I an arbitrary set, finite or not, such that there exists a lattice isomorphism

$$\mathbb{I}(E/F) \xrightarrow{\sim} \mathbb{L}(\Delta),$$

where Δ is a group canonically associated with the given extension E/F. Mostly, $\Delta = F^*\langle\{\sqrt[n_i]{a_i}\,|\, i \in I\,\rangle/F^*$.

4. Basic concepts and results of Cogalois theory

In this section we will briefly present some of the basic notions and facts of Cogalois theory, namely those of *G-radical extension*, *G-Kneser extension*, *Cogalois extension*, *strongly G-Kneser extension*, and *G-Cogalois extension*.

G-Radical extensions

The notion of *radical extension* is rather basic and well known in Galois theory. However, our terminology used in the previous section is somewhat different from that commonly used in Galois theory (see, e.g., Kaplansky [23], Karpilovsky [24], Lang [26]), but they agree for simple extensions. Note that radical extensions have been called *coseparable* by Greither and Harrison [20]. As explained above, by a radical extension we mean a field extension E/F such that E is obtained by adjoining to the base field F an arbitrary set of "radicals" over F, i.e., of elements $x \in E$ such that $x^n = a \in F$ for some $n \in \mathbb{N}^*$. Such an x is denoted by $\sqrt[n]{a}$ and is called an *nth radical* of a.

We reformulate below this notion using the following notation applicable to any extension E/F:

$$T(E/F) := \{\, x \in E^* \mid x^n \in F^* \text{ for some } n \in \mathbb{N}^* \,\}.$$

Observe that for every element in $x \in T(E/F)$ there exists an $n \in \mathbb{N}^*$ such that $x^n = a \in F$, so x is an *nth radical* of a. Thus, $T(E/F)$ is precisely the set of all "radicals" belonging to E of elements of F^*. This observation suggests the following

Definition 4.1. An extension E/F is said to be *radical* (resp. G-radical) if there exists a set A with $A \subseteq T(E/F)$ (resp. a group G with $F^* \leqslant G \leqslant T(E/F)$) such that $E = F(A)$ (resp. $E = F(G)$).

Observe that any radical extension E/F is G-radical for some G; indeed, if $E = F(A)$ for some $A \subseteq T(E/F)$, then just take as G the subgroup $G = F^*\langle\, A \,\rangle$ of the multiplicative group E^* of E generated by F^* and A.

G-Kneser extensions

The basic concept of G-Kneser extension has been introduced by Albu and Nicolae [9] for finite extensions and by Albu and Ţena [13] for infinite extensions.

Definition 4.2. A finite extension E/F is said to be *G-Kneser* if it is a G-radical extension such that $|G/F^*| = [E : F]$ (only the inequality $|G/F^*| \leqslant [E : F]$ is sufficient). The extension E/F is called *Kneser* if it is G-Kneser for some group G.

Note that a finite G-radical extension E/F is G-Kneser if and only if there exists a set of representatives of the quotient group G/F^* which is linearly independent over F if and only if every set of representatives of G/F^* is a vector space basis of E over F. This implies an easy procedure to exhibit vector space bases for such extensions: first, list all the elements, with no repetition, of the quotient group G/F^* and then take representatives of the cosets from this list.

The Kneser criterion

We present now a crucial result which characterizes separable G-Kneser extensions E/F according to whether or not certain roots of unity belonging to G are in F. Originally, it has been established by Kneser [25] only for finite extensions. The general case has been proved by Albu and Ţena [13] using the fact that the property of an arbitrary G-radical extension being G-Kneser is of finite character.

Theorem 4.3. (The Kneser criterion). *An arbitrary separable G-radical extension E/F is G-Kneser if and only if $\zeta_p \in G \Longrightarrow \zeta_p \in F$ for every odd prime p and $1 \pm \zeta_4 \in G \Longrightarrow \zeta_4 \in F$.*

Note that the *separability* condition cannot be dropped from the Kneser criterion.

The Kneser criterion is a very powerful tool in Cogalois theory. We only mention a few of applications:

- in proving the Greither-Harrison criterion (see Theorem 4.5);
- in investigating G-Cogalois extensions (see Section 4);
- in elementary field arithmetic (see Section 6);
- in Gröbner bases (see Subsection 7.1);
- in classical algebraic number theory (see Subsection 7.2).

Cogalois extensions

Remember that for any extension E/F we use the following notation throughout this paper:

$$T(E/F) := \{\, x \in E^* \mid x^n \in F^* \text{ for some } n \in \mathbb{N}^* \,\}.$$

Since $F^* \leqslant T(E/F)$, it makes sense to consider the quotient group $T(E/F)/F^*$, which is nothing else than the torsion group $t(E^*/F^*)$ of the quotient group E^*/F^*. This group, playing a major role in Cogalois theory, is somewhat dual to the Galois group of E/F, which explains the terminology below.

Definition 4.4. The *Cogalois group* of an arbitrary field extension E/F, denoted by $Cog\,(E/F)$, is the quotient group $T(E/F)/F^*$. The extension E/F is said to be *Cogalois* if it is $T(E/F)$-Kneser.

Clearly, a finite extension E/F is Cogalois if and only if it is radical, i.e., $E = F(T(E/F))$, and $|\mathrm{Cog}\,(E/F)| = [\,E : F\,]$ (only the inequality $|\mathrm{Cog}\,(E/F)| \leqslant [\,E : F\,]$ is sufficient).

Observe that, in contrast to the fact that the Galois group $\mathrm{Gal}(E/F)$ of an arbitrary extension E/F is in general not Abelian, the Cogalois group $\mathrm{Cog}(E/F)$ of any extension E/F is always a torsion Abelian group.

The computation of the Cogalois group of an extension is not an easy task. For quadratic extensions of $\mathbb{Q}$ we have a complete description of such groups (see Albu, Nicolae, and Ţena [12]). Note also that a nice result due to Greither and Harrison [20] says that the Cogalois group of any extension E/F of algebraic number fields is finite.

The term of "Cogalois extension" appeared for the first time in the literature in 1986 in the fundamental paper of Greither and Harrison [20], where the Cogalois extensions were introduced as follows: a finite extension E/F is called *conormal* (resp. *coseparable*) if $|\mathrm{Cog}(E/F)| \leqslant [\,E : F\,]$ (resp. if E/F is radical), and is called *Cogalois* if it is both conormal and coseparable. So, the Greither and Harrison's terminology for finite Cogalois extensions has been chosen to agree with the dual of the following well-known characterization: an extension, finite or not, is Galois if and only if it is both *normal* and *separable*.

A basic concept in the theory of radical extensions is that of *purity*: we say that an extension E/F is *pure* if $\mu_p(E) \subseteq F$ for every p, p odd prime or 4. This concept is somewhat related to that used in group theory: a subgroup H of an Abelian multiplicative group G is called *pure* if $G^n \cap H = H^n$ for every $n \in \mathbb{N}^*$.

The next result, characterizing Cogalois extensions in terms of purity is due to Greither and Harrison [20] for finite extensions, and to Albu and Ţena [13] for arbitrary extensions. The original proof in [20] involves the machinery of the cohomology of groups. A very short and simple proof, based only on the Kneser criterion is due to Albu and Ţena [13].

Theorem 4.5. (The Greither–Harrison criterion). *An arbitrary extension E/F is Cogalois if and only if it is radical, separable, and pure.*

Corollary 4.6. *Any G-radical extension E/F with E a subfield of $\mathbb{R}$ is Cogalois, and* $\mathrm{Cog}\,(E/F) = G/F^*$.

Proof. Clearly E/F is pure, so by the Greither–Harrison criterion, it is Cogalois. Now, by the Kneser criterion, it is also G-Kneser. This implies that $G = T(E/F)$ (see Albu [7] for more details), so $\mathrm{Cog}\,(E/F) = G/F^*$. □

Galois and Cogalois connections

Let E/F be an arbitrary field extension, and denote by Γ the Galois group $\mathrm{Gal}(E/F)$ of E/F. Then, it is easily seen that the maps

$$\alpha : \mathbb{I}(E/F) \longrightarrow \mathbb{L}(\Gamma),\ \ \alpha(K) = \mathrm{Gal}(E/K),$$

and

$$\beta : \mathbb{L}(\Gamma) \longrightarrow \mathbb{I}(E/F),\ \ \beta(\Delta) = \mathrm{Fix}(\Delta),$$

yield a *Galois connection* between the lattice $\mathbb{I}(E/F)$ of all intermediate fields of the extension E/F and the lattice $\mathbb{L}(\Gamma)$ of all subgroups of Γ. We call it the *standard Galois connection associated with the extension E/F*.

Recall that a *Galois connection* between the posets $(X, \leqslant)$ and $(Y, \leqslant)$ is a pair of order-reversing maps $\alpha : X \longrightarrow Y$ and $\beta : Y \longrightarrow X$ such that $x \leqslant (\beta \circ \alpha)(x), \forall\, x \in X,$ and $y \leqslant (\alpha \circ \beta)(y), \forall\, y \in Y$.

If the maps α and β are both order-preserving instead of order-reversing, we obtain a *Cogalois connection* between X and Y. More precisely, a *Cogalois connection* between the posets $(X, \leqslant)$ and $(Y, \leqslant)$ is a pair of order-preserving maps $\alpha : X \longrightarrow Y$ and $\beta : Y \longrightarrow X$ such that $(\beta \circ \alpha)(x) \leqslant x, \forall\, x \in X,$ and $y \leqslant (\alpha \circ \beta)(y), \forall\, y \in Y$.

The prototype of a Cogalois connection is that canonically associated with any radical extension. Let E/F be an arbitrary G-radical extension. Then, the maps

$$\varphi : \mathbb{I}(E/F) \longrightarrow \mathbb{L}(G/F^*),\ \varphi(K) = (K \cap G)/F^*,$$

and

$$\psi : \mathbb{L}(G/F^*) \longrightarrow \mathbb{I}(E/F),\ \psi(H/F^*) = F(H),$$

establish a Cogalois connection between the lattices $\mathbb{I}(E/F)$ and $\mathbb{L}(G/F^*)$, called the *standard Cogalois connection associated with the extension* E/F. Notice that, in contrast with the standard Galois connection which is associated with any extension, the standard Cogalois connection is associated only with radical extensions.

The considerations above naturally lead us to define the following dual concepts. An extension E/F with Galois group Γ is said to be an *extension with* Γ*-Galois correspondence* if the standard Galois connection associated with E/F yields a lattice anti-isomorphism between the lattices $\mathbb{I}(E/F)$ and $\mathbb{L}(\Gamma)$. Dually, a G-radical extension E/F is said to be an *extension with* G/F^**-Cogalois correspondence* if the standard Cogalois connection associated with E/F yields a lattice isomorphism between the lattices $\mathbb{I}(E/F)$ and $\mathbb{L}(G/F^*)$.

The next result (see Albu [7]) shows that the finite extensions with Γ-Galois correspondence are precisely the Galois extensions.

Proposition 4.7. *A finite extension E/F with Galois group Γ is Galois if and only if it is an extension with Γ-Galois correspondence, in other words, the maps α and β from the standard Galois connection associated with E/F are lattice anti-isomorphisms, inverse to one another, between the lattices $\mathbb{I}(E/F)$ and $\mathbb{L}(\Gamma)$.*

Strongly G-Kneser extensions

Similarly to the fact that a subextension of a normal extension is not necessarily normal, a subextension of a Kneser extension is not necessarily Kneser, So, it makes sense to consider the extensions that inherit the property of being Kneser, which will be called *strongly Kneser*.

Definition 4.8. An extension E/F is said to be *strongly G-Kneser* if it is a G-radical extension such that, for every intermediate field K of E/F, the extension E/K is K^*G-Kneser, or equivalently, the extension K/F is $K^* \cap G$-Kneser. The extension E/F is called *strongly Kneser* if it is strongly G-Kneser for some G.

The next result gives a characterization of G-Kneser extensions E/F which are extensions with G/F^*-Cogalois correspondence, and is somewhat dual to the corresponding result in Proposition 4.7 for Galois extensions.

Theorem 4.9. *The following assertions are equivalent for an arbitrary G-radical extension E/F.*

(1) *E/F is strongly G-Kneser.*
(2) *E/F is G-Kneser with G/F^*-Cogalois correspondence, i.e., the maps*

$$\varphi : \mathbb{I}(E/F) \longrightarrow \mathbb{L}(G/F^*) \quad \text{and} \quad \psi : \mathbb{L}(G/F^*) \longrightarrow \mathbb{I}(E/F)$$

defined above are isomorphisms of lattices, inverse to one another.

G-Cogalois extensions

An intrinsic characterization of strongly G-Kneser extension is available for separable extensions. Such extensions deserve a special name.

Definition 4.10. An extension E/F is called *G-Cogalois* if it is a separable strongly G-Kneser extension.

G-Cogalois extensions play in Cogalois theory the same role as that of Galois extensions in Galois theory. These extensions can be nicely characterized within the class of G-radical extensions by means of a certain sort of local "purity", called n-purity.

We say that an extension E/F is n-pure for some $n \in \mathbb{N}^*$ if $\mu_p(E) \subseteq F$ for all p, p odd prime or 4, with $p \mid n$. Recall that the *exponent* $\exp(T)$ of a finite multiplicative group T is the least number $n \in \mathbb{N}^*$ with the property that $T^n = \{e\}$.

Theorem 4.11. (The n-purity criterion [9]). *A finite separable G-radical extension E/F with $\exp(G/F^*) = n$ is G-Cogalois if and only if it is n-pure.*

The n-purity criterion is a powerful tool in Cogalois theory. Note that for infinite extensions a similar criterion for G-Cogalois extensions, namely the $\mathcal{P}_G$-*purity criterion*, has been established by Albu [3].

The next result is due to Albu and Nicolae [9] for finite extensions and to Albu and Ţena [13] for infinite extensions.

Theorem 4.12. *Let E/F be an extension which is simultaneously G-Cogalois and H-Cogalois. Then $G = H$.*

In view of Theorem 4.12, the group G of any G-Cogalois extension, finite or not, is uniquely determined. So, it makes sense to introduce the following concept.

Definition 4.13. If E/F is a G-Cogalois extension, then the group G/F^* is called the Kneser group of the extension E/F and is denoted by $\mathrm{Kne}(E/F)$.

Observe that for any G-Cogalois extension E/F one has $\mathrm{Kne}(E/F) \leqslant \mathrm{Cog}(E/F)$.

5. Examples of G-Cogalois extensions

The n-purity criterion for finite extensions or the $\mathcal{P}_G$-purity criterion for infinite extensions immediately provide the following large classes of G-Cogalois extensions:

- $\mathbb{Q}\,(\sqrt[n_1]{a_1}, \ldots, \sqrt[n_r]{a_r})/\mathbb{Q}$, with
$$\mathrm{Kne}\,(\mathbb{Q}\,(\sqrt[n_1]{a_1}, \ldots, \sqrt[n_r]{a_r})/\mathbb{Q}) = \mathbb{Q}^*\langle \sqrt[n_1]{a_1}, \ldots, \sqrt[n_r]{a_r}\rangle / \mathbb{Q}.$$
- *Cogalois extensions* E/F, with $\mathrm{Kne}(E/F) = \mathrm{Cog}\,(E/F)$.

- *Classical n-Kummer extensions* E/F, $E = F(\{\sqrt[n]{a_i} \mid i \in I\})$, with
$$\mathrm{Kne}(E/F) = F^*\langle\{\sqrt[n]{a_i} \mid i \in I\}\rangle/F^*.$$
The Kneser and the Galois groups of such extensions E/F are related by a (non-canonical) isomorphism
$$\mathrm{Kne}(E/F) \simeq \mathrm{Hom}_c(\mathrm{Gal}(E/F), \mu_n(F)),$$
where the subscript "$_c$" means "continuous". In particular, if E/F is a finite classical n-Kummer extension, then $\mathrm{Kne}(E/F) \simeq \mathrm{Gal}(E/F)$. Note that the whole classical Kummer theory can be immediately deduced from Cogalois theory using an infinite variant of the n-purity criterion.
- Various *generalizations* of classical n-Kummer extensions, including *generalized n-Kummer extensions*, *n-Kummer extensions with few roots of unity*, and *quasi-n-Kummer extensions*, have been introduced and investigated by Albu [1] and Albu and Nicolae [9] for finite extensions, and by Albu and Ţena [13] for infinite extensions. All of these are extensions E/F with $E = F(\{\sqrt[n]{a_i} \mid i \in I\})$, $\gcd(n, e(F)) = 1$, and where the condition $\zeta_n \in F$ in the definition of a classical n-Kummer extension (see Section 3, before FTKT) is replaced by the condition $\mu_n(E) \subseteq F$ for generalized n-Kummer extensions, by the condition $\mu_n(E) \subseteq \{-1, 1\}$ for n-Kummer extensions with few roots of unity, and by the condition $\zeta_p \in F$ for every p, p odd prime or 4, with $p \mid n$ for quasi-n-Kummer extensions.

  A theory of these generalizations of classical n-Kummer extensions can be developed using the properties of G-Cogalois extensions, and it turns out that this theory is very similar to the classical Kummer theory. Since, in general, they are not Galois extensions, no other approach (e.g., via Galois theory, as in the case of classical n-Kummer extensions) is applicable.

6. Applications to elementary field arithmetic

In this section we present interesting applications of Cogalois theory to completely solve some very concrete and natural questions in elementary field arithmetic. Many of them, to the best of our knowledge, cannot be solved without involving the machinery of Cogalois theory, e.g., 6.3, 6.4, 6.8, etc. Note also that most of these applications hold in more general cases, and not only for finite real radical extensions of $\mathbb{Q}$ as they appear in 6.1–6.5 (see Albu [7]).

If not indicated otherwise, $r, n_1, \dots, n_r$ will denote in this section elements of $\mathbb{N}^*$, $a_1, \dots, a_r$ elements of $\mathbb{Q}_+^*$, and $\sqrt[n_i]{a_i}$ the positive real n_ith root of a_i, $1 \leqslant i \leqslant r$.

6.1. Effective degree computation:
$[\mathbb{Q}(\sqrt[n_1]{a_1}, \dots, \sqrt[n_r]{a_r}) : \mathbb{Q}] = |\mathbb{Q}^*\langle\sqrt[n_1]{a_1}, \dots, \sqrt[n_r]{a_r}\rangle/\mathbb{Q}^*|$.

Proof. This follows at once from the Kneser criterion since $\mathbb{Q}(\sqrt[n_1]{a_1}, \dots, \sqrt[n_r]{a_r})/\mathbb{Q}$ is a $\mathbb{Q}^*\langle\sqrt[n_1]{a_1}, \dots, \sqrt[n_r]{a_r}\rangle$-Kneser extension. □

6.2. Exhibiting extension basis:

A vector space basis for the extension $\mathbb{Q}(\sqrt[n_1]{a_1}, \ldots, \sqrt[n_r]{a_r})/\mathbb{Q}$ is easily obtained as soon as we have listed, with no repetition, all the elements of its Kneser group $\mathbb{Q}^*\langle \sqrt[n_1]{a_1}, \ldots, \sqrt[n_r]{a_r} \rangle/\mathbb{Q}^*$. Then any set of representatives of the cosets from this list is a basis of the extension, as this has been justified in Section 4 just after Definition 4.2.

We illustrate this with the following concrete extension $\mathbb{Q}(\sqrt[4]{20}, \sqrt[6]{500})/\mathbb{Q}$. Denote for simplicity

$$E = \mathbb{Q}(\sqrt[4]{20}, \sqrt[6]{500}),\ G = \mathbb{Q}^*\langle \sqrt[4]{20}, \sqrt[6]{500} \rangle,\ a = \sqrt[6]{500},\ b = \sqrt[4]{20},$$

and for every $x \in G$ let $\widehat{x}$ denote its coset $x\mathbb{Q}^*$ in the quotient group $G/\mathbb{Q}^*$.

We are going now to explicitly describe the Kneser group $G/\mathbb{Q}^*$ of $E/\mathbb{Q}$. Since $\operatorname{ord}(\widehat{a}) = 6$, $\operatorname{ord}(\widehat{b}) = 4$, and $\widehat{b^2} = \widehat{a^3} = \widehat{\sqrt{5}}$, we have

$$\begin{aligned} G/\mathbb{Q}^* &= \mathbb{Q}^*\langle a, b \rangle/\mathbb{Q}^* = \langle \widehat{a}, \widehat{b} \rangle = \{ \widehat{a^i} \cdot \widehat{b^j} \mid 0 \leqslant i \leqslant 5,\ 0 \leqslant j \leqslant 1 \} \\ &= \{ \widehat{1}, \widehat{a}, \widehat{a^2}, \widehat{a^3}, \widehat{a^4}, \widehat{a^5}, \widehat{b}, \widehat{a} \cdot \widehat{b}, \widehat{a^2} \cdot \widehat{b}, \widehat{a^3} \cdot \widehat{b}, \widehat{a^4} \cdot \widehat{b}, \widehat{a^5} \cdot \widehat{b} \}. \end{aligned}$$

Since $\widehat{b} \notin \langle \widehat{a} \rangle$, we have $|\langle \widehat{a}, \widehat{b} \rangle| = 12$. Thus $[E : \mathbb{Q}] = 12$, and, as explained above, a basis of the extension $E/\mathbb{Q}$ is the set

$$\{ \sqrt[6]{500}^{\,i} \cdot \sqrt[4]{20}^{\,j} \mid 0 \leqslant i \leqslant 5,\ 0 \leqslant j \leqslant 1 \}.$$

Observe that $G/\mathbb{Q}^* = \langle \widehat{a}, \widehat{b} \rangle = \langle \widehat{ab} \rangle = \langle\, \widehat{\sqrt[12]{2000000000}} \,\rangle$, so it is a cyclic group of order 12. It follows that another basis of the extension $E/\mathbb{Q}^*$ is the set $\{ \sqrt[12]{2000000000}^{\,i} \mid 0 \leqslant i \leqslant 11 \}$.

6.3. Finding all intermediate fields:

All the intermediate fields of the G-Cogalois extension $\mathbb{Q}(\sqrt[n_1]{a_1}, \ldots, \sqrt[n_r]{a_r})/\mathbb{Q}$, that is to say, all the subfields of the field $\mathbb{Q}(\sqrt[n_1]{a_1}, \ldots, \sqrt[n_r]{a_r})$, are, by Theorem 4.9, exactly $\mathbb{Q}(H)$, where $\mathbb{Q}^* \leqslant H \leqslant \mathbb{Q}^*\langle \sqrt[n_1]{a_1}, \ldots, \sqrt[n_r]{a_r} \rangle$. So, knowing all the subgroups of its Kneser group $\mathbb{Q}^*\langle \sqrt[n_1]{a_1}, \ldots, \sqrt[n_r]{a_r} \rangle/\mathbb{Q}^*$ we can completely describe all the subfields of $\mathbb{Q}(\sqrt[n_1]{a_1}, \ldots, \sqrt[n_r]{a_r})$.

Consider the concrete example $E = \mathbb{Q}(\sqrt[4]{20}, \sqrt[6]{500})$ in 6.2. We know that $\operatorname{Kne}(E/\mathbb{Q}^*)$ is a cyclic group of order 12 generated by $\widehat{c}$, where $c = \sqrt[12]{2000000000}$, so its subgroups are precisely the following: $\langle \widehat{c} \rangle$, $\langle \widehat{c^2} \rangle$, $\langle \widehat{c^3} \rangle$, $\langle \widehat{c^4} \rangle$, $\langle \widehat{c^6} \rangle$, $\langle \widehat{c^{12}} \rangle$. Consequently, all the subfields of E are:

$$\mathbb{Q},\ \mathbb{Q}(c),\ \mathbb{Q}(c^2),\ \mathbb{Q}(c^3),\ \mathbb{Q}(c^4),\ \mathbb{Q}(c^6),$$

where $c = \sqrt[12]{2000000000}$.

Note that for every positive divisor d of $[E : \mathbb{Q}] = 12$, there exists a unique subfield K of E with $[K : \mathbb{Q}] = d$, in other words, the extension $E/\mathbb{Q}$ has the so-called *unique subfield property* (USP), and this property holds because its Kneser group is cyclic (see Subsection 6.10).

6.4. Primitive element:

$\mathbb{Q}(\sqrt[n_1]{a_1}, \dots, \sqrt[n_r]{a_r}) = \mathbb{Q}(\sqrt[n_1]{a_1} + \dots + \sqrt[n_r]{a_r})$.

Proof. By Section 5, $\mathbb{Q}(\sqrt[n_1]{a_1}, \dots, \sqrt[n_r]{a_r})/\mathbb{Q}$ is a $\mathbb{Q}^*\langle \sqrt[n_1]{a_1}, \dots, \sqrt[n_r]{a_r} \rangle$-Cogalois extension; apply now the n-purity criterion to deduce that $\sqrt[n_1]{a_1} + \dots + \sqrt[n_r]{a_r}$ is a primitive element of the extension $\mathbb{Q}(\sqrt[n_1]{a_1}, \dots, \sqrt[n_r]{a_r})/\mathbb{Q}$ (see Albu and Nicolae [10] for more details). □

6.5. When is a sum of radicals of positive rational numbers a rational number?

Answer: $\sqrt[n_1]{a_1} + \dots + \sqrt[n_r]{a_r} \in \mathbb{Q} \iff \sqrt[n_i]{a_i} \in \mathbb{Q}$ for all i, $1 \leqslant i \leqslant r$.

Proof. If $\sqrt[n_1]{a_1} + \dots + \sqrt[n_r]{a_r} \in \mathbb{Q}$, then

$$\mathbb{Q}(\sqrt[n_1]{a_1}, \dots, \sqrt[n_r]{a_r}) = \mathbb{Q}(\sqrt[n_1]{a_1} + \dots + \sqrt[n_r]{a_r}) = \mathbb{Q}$$

by 6.4, and consequently $\sqrt[n_1]{a_1}, \dots, \sqrt[n_r]{a_r} \in \mathbb{Q}$. □

6.6. When can a positive algebraic number α be written as a finite sum of real numbers of type $\pm \sqrt[n_i]{a_i}$, $1 \leqslant i \leqslant r$?

Answer: *An algebraic number $\alpha \in \mathbb{R}_+^*$ has the property above if and only if the extension $\mathbb{Q}(\alpha)/\mathbb{Q}$ is radical, or Kneser, or Cogalois.*

Proof. Assume that α can be written as a finite sum of real numbers of type $\pm \sqrt[n_i]{a_i}$, $1 \leqslant i \leqslant r$, $r, n_i \in \mathbb{N}^*$, $a_i \in \mathbb{Q}_+^*$. Then $\mathbb{Q}(\alpha)$ is a subfield of the field $\mathbb{Q}(\sqrt[n_1]{a_1}, \dots, \sqrt[n_r]{a_r}) \subseteq \mathbb{R}$, so it is a pure extension. Being clearly separable and radical, it is Cogalois by the Greither–Harrison criterion (Theorem 4.5), and so is also its subextension $\mathbb{Q}(\alpha)/\mathbb{Q}$.

Now assume that the finite extension $\mathbb{Q}(\alpha)/\mathbb{Q}$ is radical. Again by the Greither–Harrison criterion, it is also Cogalois. According to a result of Greither and Harrison [20] mentioned in Section 4 after Definition 4.4, the Cogalois group $\mathrm{Cog}(\mathbb{Q}(\alpha)/\mathbb{Q}) = T(\mathbb{Q}(\alpha)/\mathbb{Q})/\mathbb{Q}^*$ of the extension $\mathbb{Q}(\alpha)/\mathbb{Q}$ is finite. Let $\{x_1, \dots, x_r\}$ be a set of representatives of this finite group. Observe that $x_i \equiv -x_i \pmod{\mathbb{Q}^*}$, so we may assume that $x_i > 0$ for all i, $1 \leqslant i \leqslant r$. Then $\mathbb{Q}(\alpha) = \mathbb{Q}(x_1, \dots, x_r)$, and for every i, $1 \leqslant i \leqslant r$, there exists $n_i \in \mathbb{N}^*$ such that $x_i^{n_i} = a_i \in \mathbb{Q}_+^*$, and so $\mathbb{Q}(\alpha) = \mathbb{Q}(\sqrt[n_1]{a_1}, \dots, \sqrt[n_r]{a_r})$ as desired. □

6.7. When can a positive superposed radical not be decomposed into a finite sum of real numbers of type $\pm \sqrt[n_i]{a_i}$, $1 \leqslant i \leqslant r$?

Answer: By 6.6, a superposed radical $\alpha = \sqrt[n_1]{a_1 + \sqrt[n_2]{a_2 + \ldots + \sqrt[n_r]{a_r}}}$ has the above property if and only if the extension $\mathbb{Q}(\alpha)/\mathbb{Q}$ is not Cogalois. Examples of such numbers are $\sqrt{1+\sqrt{2}}$ and $\sqrt{2+\sqrt{2+\sqrt{2+\ldots+\sqrt{2}}}}$. Also, for any square-free integer $d \in \mathbb{N}$, $d \geqslant 2$, and any $n \in \mathbb{Z}^*$ such that $\sqrt{n^2-d} \notin \mathbb{Q}(\sqrt{d})$, the extension $\mathbb{Q}(\sqrt{n+\sqrt{d}})/\mathbb{Q}$ is not Cogalois (see Albu [2]), so $\sqrt{n+\sqrt{d}}$ is a number we are looking for.

6.8. When is a rational combination of powers from a given set of radicals of positive rational numbers itself a radical of a positive rational number?

Answer: *Let* $r, n_1, \dots, n_r \in \mathbb{N}^*$ *and* $a_1, \dots, a_r \in \mathbb{Q}_+^*$ *be given numbers, and let* $\alpha \in \mathbb{R}_+^*$ *be a finite sum of monomials of form* $c \cdot \sqrt[n_1]{a_1}^{\,j_1} \cdot \ldots \cdot \sqrt[n_r]{a_r}^{\,j_r}$*, with* $j_1, \dots, j_r \in \mathbb{N}$ *and* $c \in \mathbb{Q}^*$*. Then* $\alpha^m \in \mathbb{Q}$ *for some* $m \in \mathbb{N}^*$ *if and only if* α *is itself such a monomial.*

Proof. Set $E := \mathbb{Q}(\sqrt[n_1]{a_1}, \dots, \sqrt[n_r]{a_r})$ and $G := \mathbb{Q}^*\langle \sqrt[n_1]{a_1}, \dots, \sqrt[n_r]{a_r} \rangle$. Then the statement above on α can be reformulated as follows:

When is an element $\alpha \in E$ *such that* $\alpha \in T(E/\mathbb{Q})$*?*

By Corollary 4.6, we have $\mathrm{Cog}\,(E/\mathbb{Q}) = G/\mathbb{Q}^*$, i.e., $T(E/\mathbb{Q}) = G$; so $\alpha^m \in \mathbb{Q}$ if and only α is a monomial as described above. □

6.9. Radical extensions of prime exponent:

The finite G-radical extensions E/F with $\exp(G/F^*)$ a prime number $p > 0$ are extensions of the following type: $E = F(\sqrt[p]{a_1}, \dots, \sqrt[p]{a_r})$ where $r \in \mathbb{N}^*$, $a_1, \dots, a_r \in F^*$, and $\sqrt[p]{a_1}, \dots, \sqrt[p]{a_r} \in \Omega$ denote certain pth roots.

Such extensions are nicely controlled when some additional conditions are imposed, namely the characteristic of F is not p, and

$$[\, F(\sqrt[p]{a_1}, \dots, \sqrt[p]{a_r}) : F\,] = p^r.$$

With these assumptions, the extension E/F is $F^*\langle \sqrt[p]{a_1}, \dots, \sqrt[p]{a_r} \rangle$-Cogalois, and so $\sqrt[p]{a_1} + \cdots + \sqrt[p]{a_r}$ is a primitive element of it (see Albu [5], [6]).

The results of Kaplansky [23], Baker and Stark [14], and Albu [1] concerning very particular such radical extensions of exponent p, that were established by them in a more complicated way using the standard methods and tools of field theory, are now easy consequences of our Cogalois approach.

6.10. Simple radical separable extensions having the USP:

Following Vélez [30], a finite extension E/F is said to have the *unique subfield property*, abbreviated USP, if for every divisor m of $[\,E : F\,]$ there exists a unique intermediate field K of E/F such that $[\,K : F\,] = m$. The finite G-Cogalois extensions which have the USP are precisely those having cyclic Kneser groups (see Albu [4]). For simple radical separable extensions we have the following characterization of the USP.

Proposition (Albu [4]). *Let* F *be any field, and let* $u \in \Omega$ *be a root of an irreducible binomial* $X^n - a \in F[X]$*, with* $\gcd(n, e(F)) = 1$*. Then, the extension* $F(u)/F$ *has the* USP *if and only if it is* $F^*\langle u \rangle$*-Cogalois.*

Corollary (Albu [4]). *Let* F *be an arbitrary field, and let* $n \in \mathbb{N}^*$ *be such that* $\zeta_n \in F$ *and* $\gcd(n, e(F)) = 1$*. Let* $X^n - a$*,* $X^n - b$ *be irreducible polynomials in* $F[X]$ *with roots* $u, v \in \Omega$*, respectively. Then* $F(u) = F(v)$ *if and only if there exists* $c \in F$ *and* $j \in \mathbb{N}$ *with* $\gcd(j, n) = 1$ *and* $a = b^j c^n$*.*

7. Other applications

7.1. Binomial ideals and Gröbner bases

Let F be any field, $n \in \mathbb{N}^*$, and $F[\underline{X}] := F[X_1, \dots, X_n]$ be the polynomial ring in n indeterminates with coefficients in F. By a *monomial* in $F[\underline{X}]$ we mean any $c\prod_{1\leqslant i\leqslant n} X_i^{r_i}$ with $c \in F$ and $r_i \in \mathbb{N}$, $1 \leqslant i \leqslant n$, and a sum of two monomials, both of which may be zero is called *binomial*. An ideal $\mathfrak{a}$ of $F[\underline{X}]$ is said to be a *binomial ideal* if it can be generated by a set of binomials. An algorithm to detect whether a given ideal $\mathfrak{a}$ of $F[\underline{X}]$ is binomial involves the *Gröbner bases* (see Eisenbud and Sturmfels [18]). The most interesting binomial ideals are those associated with Kneser extensions of F (see Becker, Grobe, and Niermann [15]).

7.2. Hecke's systems of ideal numbers

The Kneser criterion is not only a powerful as well as indispensable tool in investigating radical field extensions, but, it has nice applications in proving some classical results of algebraic number theory. We present here one of them.

A classical construction from 1920 in algebraic number theory, originating with Hecke [21], is the following one: to every algebraic number field K one can associate a so-called *system of ideal numbers* S, which is a certain subgroup of the multiplicative group $\mathbb{C}^*$ of complex numbers such that $K^* \leqslant S$ and the quotient group S/K^* is canonically isomorphic to the ideal class group $\mathcal{C}\ell_K$ of K. The equality $[\,K(S) : K\,] = |\,\mathcal{C}\ell_K\,|$ was claimed by Hecke on page 122 of his monograph [22] published in 1948, but never proved by him. To the best of our knowledge, no proof of this assertion, excepting the very short one due to Albu and Nicolae [11], based on the Kneser criterion, is available in the literature.

References

[1] T. Albu, *Kummer extensions with few roots of unity*, J. Number Theory **41** (1992), 322–358.

[2] T. Albu, *Some examples in Cogalois Theory with applications to elementary Field Arithmetic*, J. Algebra Appl. **1** (2002), 1–29.

[3] T. Albu, *Infinite field extensions with Cogalois correspondence*, Comm. Algebra **30** (2002), 2335–2353.

[4] T. Albu, *Field extensions with the unique subfield property, and G-Cogalois extensions*, Turkish J. Math. **26** (2002), 433–445.

[5] T. Albu, *On radical field extensions of prime exponent*, J. Algebra Appl. **1** (2002), 365–373.

[6] T. Albu, *Corrigendum and Addendum to my paper concerning Kummer extensions with few roots of unity*, J. Number Theory **99** (2003), 222–224.

[7] T. Albu, "*Cogalois Theory*", A Series of Monographs and Textbooks, Vol. 252, Marcel Dekker, Inc., New York and Basel, 2003.

[8] T. Albu and Ş.A. Basarab, *An Abstract Cogalois Theory for profinite groups*, J. Pure Appl. Algebra **200** (2005), 227–250.

[9] T. Albu and F. Nicolae, *Kneser field extensions with Cogalois correspondence*, J. Number Theory **52** (1995), 299–318.

[10] T. Albu and F. Nicolae, *G-Cogalois field extensions and primitive elements*, in "Symposia Gaussiana", Conference A: Mathematics and Theoretical Physics, Eds. M. Behara, R. Fritsch, and R.G. Lintz, Walter de Gruyter & Co., Berlin New York, 1995, pp. 233–240.

[11] T. Albu and F. Nicolae, *Heckesche Systeme idealer Zahlen und Knesersche Körpererweiterungen*, Acta Arith. **73** (1995), 43–50.

[12] T. Albu, F. Nicolae, and M. Ţena, *Some remarks on G-Cogalois field extensions*, Rev. Roumaine Math. Pures Appl. **41** (1996), 145–153.

[13] T. Albu and M. Ţena, *Infinite Cogalois Theory*, Mathematical Reports **3 (53)** (2001), 105–132.

[14] A. Baker and H.M. Stark, *On a fundamental inequality in number theory*, Ann. of Math. **94** (1971), 190–199.

[15] E. Becker, R. Grobe, and M. Niermann, *Radicals of binomial ideals*, J. Pure Appl. Algebra **117 & 118** (1997), 41–79.

[16] A. Besicovitch, *On the linear independence of fractional powers of integers*, J. London Math. Soc. **15** (1940), 3–6.

[17] N. Bourbaki, "*Algèbre*", Chapitres 4 à 7, Masson, Paris, 1981.

[18] D. Eisenbud and B. Sturmfel, *Binomial ideals*, Duke Math. J. **84** (1996), 1–45.

[19] D. Gay and W.Y. Vélez, *On the degree of the splitting field of an irreducible binomial*, Pacific J. Math. **78** (1978), 117–120.

[20] C. Greither and D.K. Harrison, *A Galois correspondence for radical extensions of fields*, J. Pure Appl. Algebra **43** (1986), 257–270.

[21] E. Hecke, *Eine neue Art von Zetafunktionen und ihre Beziehungen zur Verteilung der Primzahlen* (*Zweite Mitteilung*), Math. Z. **4** (1920), 11–51.

[22] E. Hecke, "*Vorlesungen über die Theorie der algebraischen Zahlen*", Chelsea Publishing Company, New York, 1948.

[23] I. Kaplansky, "*Fields and Rings*", University of Chicago Press, Chicago, 1972.

[24] G. Karpilovsky, "*Topics in Field Theory*", North-Holland, Amsterdam, New York, Oxford, and Tokyo, 1989.

[25] M. Kneser, *Lineare Abhängigkeit von Wurzeln*, Acta Arith. **26** (1975), 307–308.

[26] S. Lang, "*Algebra*", Addison-Wesley Publishing Company, Reading, Massachusetts, 1965.

[27] L.J. Mordell, *On the linear independence of algebraic numbers*, Pacific J. Math. **3** (1953), 625–630.

[28] A. Schinzel, *On linear dependence of roots*, Acta Arith. **28** (1975), 161–175.

[29] C.L. Siegel, *Algebraische Abhängigkeit von Wurzeln*, Acta Arith. **21** (1972), 59–64.

[30] W.Y. Vélez, *On normal binomials*, Acta Arith. **36** (1980), 113–124.

Toma Albu
"Simion Stoilow" Institute of Mathematics of the Romanian Academy
P.O. Box 1-764
RO-010145 Bucharest 1, Romania
e-mail: `Toma.Albu@imar.ro`

Advances in Ring Theory
Trends in Mathematics, 19–36

On Big Lattices of Classes of R-modules Defined by Closure Properties

Alejandro Alvarado García, Hugo Alberto Rincón Mejía and José Ríos Montes

Abstract. In this paper we introduce the big lattices R-sext and R-qext consisting the former of classes of left R-modules closed under isomorphisms, submodules and extensions and the later of classes closed under homomorphic images and extensions, respectively. We work with these two big lattices and study the consequences of assuming that they are the same proper class. We also consider big lattices of R-modules defined by other closure properties.

Mathematics Subject Classification (2000). 16D90 (06C).

Keywords. Classes of modules, natural classes, torsion theories, big lattices.

1. Introduction

Following Stenström [18, p. 89] we call big lattice a proper class $\mathcal{C}$ with a partial order $\leq$, such that $\mathcal{C}$ with this order is a lattice except the fact that it is not a set. In recent works, big lattices have been considered. For example in [9], the big lattice of open classes is studied, where it is remarked that this in fact is not a set, but in all other respect it is a distributive complete lattice.

In [16] the big lattice of Serre classes is considered.

In [6] the big lattice of non-hereditary torsion theories is studied.

In [10]–[13] a detailed study is made about the big lattice of preradicals defined in R-mod.

In [15] the authors considered the big lattice of preradicals defined in the category $\sigma[M]$.

The main purpose of this work is to introduce and study some new big lattices of module classes, namely R-sext and R-qext. We also obtain information about other well-known lattices.

R will denote an associative ring with unitary element, and R-mod will denote the category of unitary left R-modules. R-simp will denote a family of representatives of isomorphism classes of left simple modules and $L(M)$ will denote the lattice of left R-submodules of a left R-module ${}_RM$.

We consider some closure properties of a class of modules, like being closed under submodules, quotients, extensions, direct sums, injective hulls, products or projective covers, we will use the symbols $\leq, \twoheadrightarrow, \text{ext}, \oplus, E\left(\right), \prod, P\left(\right)$ respectively, to abbreviate. If A denotes a set of these closure properties, we denote L_A the proper class of classes of modules closed under each closure property in A. So $L_{\{\leq\}}$ denotes the proper class of hereditary classes in R-mod, $L_{\{\leq,\twoheadrightarrow,\text{ext}\}}$ denotes the proper class of Serre classes, and so on.

We should notice that L_A becomes a complete big lattice with inclusion of classes as the order and with infima given by intersections.

If C is a class of modules, we will denote by $L_A(C)$ the least element in L_A which contains C as a subclass (notice that R-mod is the largest element in L_A.)

We say that D is a pseudocomplement for C in L_A if D is maximal such that $C \cap D = \{0\}$. We say that D is a strong pseudocomplement of C if D is the largest element of L_A such that $C \cap D = \{0\}$. We abbreviate saying that D is an S-pseudocomplement of C.

If $S \in L_A$, we denote by $S^{\perp_A}$ a pseudocomplement of S in L_A, when it exists.

With Skel(L_A) we denote the class of pseudocomplements in L_A.

Remark 1.1. We recall that N is a subquotient of M if there exists a diagram

$$\begin{array}{ccc} & & M \\ & & \downarrow^{\alpha} \\ N & \overset{\beta}{\rightarrowtail} & C \end{array},$$

where α is an epimorphism and β is a monomorphism. As is clear taking pullbacks (resp. taking pushouts) this is equivalent to ask for a diagram

$$\begin{array}{ccc} K & \overset{\mu}{\rightarrowtail} & M \\ \downarrow^{\lambda} & & \\ N & & \end{array},$$

where λ is epic and μ is monic.

In some of these big lattices it is easy to describe pseudocomplements. Recall the following examples.

Example. In $L_{\{\leq\}}$, the big lattice of hereditary classes of left R-modules, we have that

$$C^{\perp_{\{\leq\}}} = \{M \mid N \leq M, N \in C \Longrightarrow N = 0\}$$

In this case we notice that pseudocomplements are unique because they are in fact S-pseudocomplements.

In previous works [1], [2], we have denoted the big lattice $L_{\{\leq\}}$ with R-her, and the hereditary class generated by C, $L_{\{\leq\}}(C)$ with her(C).

Example. If $\mathfrak{C} \in L_{\{\leq,\twoheadrightarrow\}}$, then

$$\mathfrak{C}^{\perp_{\{\leq,\twoheadrightarrow\}}} = \{M \mid M \text{ has no non zero subquotients in } \mathfrak{C}\}.$$

Proof. Let us denote the described class by $\mathfrak{D}$. As submodules and quotients of M are subquotients of M it is clear that $\mathfrak{D}$ is closed under taking submodules and quotients. Now it follows directly that $\mathfrak{D}$ is an S-pseudocomplement for $\mathfrak{C}$ in $L_{\{\leq,\twoheadrightarrow\}}$. □

Definition 1.2. We say that a big lattice L is strongly-pseudocomplemented (S-pseudocomplemented, for short) if each $C \in L$ has an S-pseudocomplement $C^{\perp} \in L$.

Recall that some lattices are S-pseudocomplemented: R-tors, the frame of hereditary torsion theories, R-pr the big lattice of preradicals, R-nat the lattice of natural classes. On the opposite side, the lattice $L(M)$ of R-submodules of M is pseudocomplemented but in general it is not S-pseudocomplemented. See [7], [10], [14], [19], [18].

Remark 1.3. When L is S-pseudocomplemented, then $C \subseteq \left(C^{\perp}\right)^{\perp}$for each $C \in L$.

Theorem 1.4. *Suppose that both L_P and L_Q are S-pseudocomplemented, P, Q being sets of closure properties. If* $\operatorname{Skel}(L_P) \subseteq L_Q \subseteq L_P$ *then* $\operatorname{Skel}(L_Q) = \operatorname{Skel}(L_P)$.

Proof. Take $C^{\perp_Q}$, $C \in L_Q$. As $C^{\perp_Q} \in L_Q \subseteq L_P$, and $C^{\perp_Q} \wedge C = \{0\}$, we have that $C^{\perp_Q} \leq C^{\perp_P} \in \operatorname{Skel}(L_P) \subseteq L_Q$. As $C \wedge C^{\perp_P} = \{0\}$ and $C^{\perp_P} \in L_Q$, we have that $C^{\perp_P} \leq C^{\perp_Q}$. Then $C^{\perp_Q} = C^{\perp_P} \in \operatorname{Skel}(L_P)$, thus $\operatorname{Skel}(L_Q) \subseteq \operatorname{Skel}(L_P)$.

Now let us take $C^{\perp_P}$; we claim that this is an element of $\mathit{Skel}(L_Q)$. By Remark 1.3 we have that $C^{\perp_P} \leq \left(C^{\perp_P}\right)^{\perp_P\perp_P}$, also we have that $C \leq C^{\perp_P\perp_P}$ implies that $\left(C^{\perp_P\perp_P}\right)^{\perp_P} \leq C^{\perp_P}$ thus we have that $C^{\perp_P} = C^{\perp_P\perp_P\perp_P}$. Thus it suffices to show that $\left(C^{\perp_P\perp_P}\right)^{\perp_P} = \left(C^{\perp_P\perp_P}\right)^{\perp_Q}$. Let us take $D \in L_Q$ such that $D \wedge \left(C^{\perp_P\perp_P}\right) = \{0\}$; as $L_Q \subseteq L_P$ then $D \leq \left(C^{\perp_P\perp_P}\right)^{\perp_P} = C^{\perp_P}$. So $\left(C^{\perp_P\perp_P}\right)^{\perp_Q} \leq C^{\perp_P}$.

On the other hand, $C^{\perp_P} \in L_Q$, by the hypothesis. As $C^{\perp_P} \wedge C^{\perp_P\perp_P} = \{0\}$, then $C^{\perp_P} \leq \left(C^{\perp_P\perp_P}\right)^{\perp_Q}$. □

Corollary 1.5. *With the hypothesis of Theorem* 1.4, *for $C \in L_Q$ we have that $C^{\perp_Q} = C^{\perp_P}$.*

Proof. It is immediate. □

Theorem 1.6. *If $\mathit{Skel}(L_P) = L_Q$, with L_P and L_Q S-pseudocomplemented and P, Q being sets of closure properties, then for each $C \in L_P$ we have that $\left(C^{\perp_P}\right)^{\perp_P} = L_Q(C)$.*

Proof. By hypothesis $C^{\perp_P} \in L_Q$, thus by Corollary 1.5, $\left(C^{\perp_P}\right)^{\perp_P} = \left(C^{\perp_P}\right)^{\perp_Q} \in L_Q$. As $C \leq \left(C^{\perp_P}\right)^{\perp_P}$, we have that $L_Q(C) \leq \left(C^{\perp_P}\right)^{\perp_P}$.

Now, if $C \leq D \in L_{Q,}$ then $D = E^{\perp_P}$, for some $E \in L_P$, by hypothesis. Then $C \leq C^{\perp_P \perp_P} \leq D^{\perp_P \perp_P} = E^{\perp_P \perp_P \perp_P} = E^{\perp_P} = D$, this shows that $C^{\perp_P \perp_P} = L_Q(C)$. □

1.1. The skeletons of R-tors, R-Serre and R-op

As an application of Theorem 1.4 we notice that the skeletons of R-tors, (which is $L_{\{\leq, \twoheadrightarrow, \text{ext}, \oplus\}}$), of R-Serre (which is $L_{\{\leq, \twoheadrightarrow, \text{ext}\}}$) and of R-op (which is $L_{\{\leq, \twoheadrightarrow\}}$) are all the same.

In order to apply Theorem 1.4, we will show that a class $\mathfrak{C} \in \text{Skel}(R\text{-op})$ is also closed under extensions and direct sums.

The following lemma is proved in [9, Theorem 3]; we include a proof for reader's convenience.

Lemma 1.7. *Each* $\mathfrak{D} \in \text{Skel}(R\text{-op})$ *is closed under extensions and direct sums.*

Proof. Suppose $\mathfrak{D} = \mathfrak{C}^{\perp_{\{\leq, \twoheadrightarrow\}}}$.

Extensions. Let $0 \to L \overset{f}{\to} M \to N \overset{g}{\to} 0$ be an exact sequence with $L, N \in \mathfrak{C}^{\perp_{\{\leq, \twoheadrightarrow\}}}$. To show a contradiction, suppose that $0 \neq K \in \mathfrak{C}$ is a subquotient of M, as in the diagram

$$\begin{array}{ccc} & & M \\ & & \downarrow^{\alpha} \\ K & \overset{\beta}{\rightarrowtail} & C \end{array},$$

where α is epic and β is monic. As $\beta(K) \cap \alpha f(L)$ is a subquotient of both L and K, then $\beta(K) \cap \alpha f(L) = 0$. So we get a commutative diagram

$$\begin{array}{ccccccccc} 0 & \longrightarrow & L & \overset{f}{\to} & M & & \longrightarrow & N & \longrightarrow & 0 \\ & & & & \downarrow^{\alpha} & & & & \\ & & K & \overset{\beta}{\rightarrowtail} & C & & \swarrow^{\gamma} & & \\ & & & & \downarrow^{\pi} & & & & \\ & & & & C/\alpha f(L) & & & & \end{array}$$

with γ being an epimorphism, and π the natural epimorphism. Now consider the non-zero quotient $C/\alpha f(L)$, thus $C/\alpha f(L) \in \mathfrak{C}^{\perp_{\{\leq, \twoheadrightarrow\}}}$. Notice now that $\pi\beta$ is a monomorphism, so $\pi\beta(K) \in \mathfrak{C}^{\perp_{\{\leq, \twoheadrightarrow\}}}$, thus $0 \neq K \in \mathfrak{C} \cap \mathfrak{C}^{\perp_{\{\leq, \twoheadrightarrow\}}}$, a contradiction.

Direct sums. Let $\{M_i\}_I$ be a family in $\mathfrak{C}^{\perp_{\{\leq, \twoheadrightarrow\}}}$; we want to see that $\oplus\{M_i\}_I$ cannot have a non-zero subquotient in $\mathfrak{C}$. To show a contradiction, if $0 \neq N$ were a subquotient of $\oplus\{M_i\}_I$, with $N \in \mathfrak{C}$, there would be a diagram

$$\begin{array}{ccc} & & \oplus\{M_i\}_I \\ & & \downarrow^{\beta} \\ N & \overset{\alpha}{\rightarrowtail} & C \end{array}$$

with β epic and α monic. We can choose N as a cyclic module, changing N for a submodule if necessary. In fact, we can choose N as a simple module by using Remark 1.1. Let us take a simple module N, $N \in \mathfrak{C}$, then it is a subquotient of a finite direct sum $\oplus\{M_i\}_J$, $J \subseteq I$. But $\mathfrak{C}^{\perp_{\{\leq, \twoheadrightarrow\}}}$ is closed under finite direct sums because it is closed under extensions. Thus $0 \neq N \in \mathfrak{C} \cap \mathfrak{C}^{\perp_{\{\leq, \twoheadrightarrow\}}}$, a contradiction. □

Now we apply Theorem 1.4 to

$$\operatorname{Skel}\left(L_{\{\leq,\twoheadrightarrow\}}\right) \subseteq L_{\{\leq,\twoheadrightarrow,\oplus,\text{ext}\}} \subseteq L_{\{\leq,\twoheadrightarrow,\text{ext}\}} \subseteq L_{\{\leq,\twoheadrightarrow\}}$$

to conclude $\operatorname{Skel}\left(L_{\{\leq,\twoheadrightarrow\}}\right) = \operatorname{Skel}\left(L_{\{\leq,\twoheadrightarrow,\oplus,\text{ext}\}}\right) = \operatorname{Skel}\left(L_{\{\leq,\twoheadrightarrow,\text{ext}\}}\right)$.

Thus pseudocomplements of Serre classes and of open classes are always hereditary torsion classes belonging to the skeleton of R-tors.

As a consequence we also obtain a new description for the pseudocomplement of an hereditary torsion theory.

Corollary 1.8. *$\tau^{\perp}$ is the torsion theory whose torsion class is given by*

$$\mathbb{T}_{\tau^{\perp}} = \{M \mid M \textit{ has no nonzero } \tau\textit{-torsion subquotients}\}.$$

2. The big lattice R-sext

We shall say that a class of left R-modules $\mathfrak{C}$ is a class with zero if $\mathfrak{C}$ is closed under isomorphisms and contains the zero module.

R-her and $L_{\{\leq\}}$ both denote the same big lattice, in particular we denote by her(C) the hereditary module class generated by the class C (see [1]).

Notation. Let $\mathfrak{C}$, $\mathfrak{D}$ be two classes with zero. We denote

$$E(\mathfrak{C},\mathfrak{D}) = \left\{ M \in R\text{-}mod \mid \begin{array}{c}\text{there exists an exact sequence}\\ 0 \to C \to M \to D \to 0\\ \text{with } C \in \mathfrak{C} \text{ and } D \in \mathfrak{D}\end{array} \right\}.$$

Definition 2.1. We shall denote by R-sext the proper class of all classes of left R-modules closed under isomorphisms, submodules and extensions.

Thus R-sext means the same as $L_{\{\leq,\text{ext}\}}$.

In the following propositions we prove some facts that we will need later.

Proposition 2.2. *Let $\mathfrak{C},\mathfrak{D}$ and $\mathfrak{E}$ be three classes with zero, then*

$$E(E(\mathfrak{C},\mathfrak{D}),\mathfrak{E}) = E(\mathfrak{C},E(\mathfrak{D},\mathfrak{E})).$$

Proof. Take $M \in E(E(\mathfrak{C},\mathfrak{D}),\mathfrak{E})$, then we can assume that there exists an exact sequence

$$0 \to N \hookrightarrow M \twoheadrightarrow \frac{M}{N} \to 0$$

with $N \in E(\mathfrak{C},\mathfrak{D})$ and $\dfrac{M}{N} \in \mathfrak{E}$.

We have the following diagram for some $C \in \mathfrak{C}$ and $\dfrac{N}{C} \in \mathfrak{D}$:

$$\begin{array}{ccccccccc} & & 0 & & 0 & & & & \\ & & \downarrow & & \downarrow & & & & \\ & & C & = & C & & & & \\ & & {\scriptstyle \mathrm{inc}}\downarrow & & {\scriptstyle \mathrm{inc}}\downarrow & & & & \\ 0 & \to & N & \hookrightarrow & M & \twoheadrightarrow & \dfrac{M}{N} & \to & 0 \\ & & \downarrow & & \downarrow & & & & \\ 0 & \to & \dfrac{N}{C} & \hookrightarrow & \dfrac{M}{C} & \twoheadrightarrow & \dfrac{M}{N} & & \\ & & \downarrow & & \downarrow & & & & \\ & & 0 & & 0 & & & & \end{array}$$

Since $\frac{N}{C} \in \mathfrak{D}$ and $\frac{M}{N} \in \mathfrak{C}$, then $\frac{M}{C} \in \mathfrak{D} : \mathfrak{C}$. It follows that $M \in E(\mathfrak{C}, E(\mathfrak{D}, \mathfrak{C}))$. Conversely, take $M \in E(\mathfrak{C}, E(\mathfrak{D}, \mathfrak{C}))$, then there exists an exact sequence

$$0 \to L \hookrightarrow M \twoheadrightarrow \frac{M}{L} \to 0$$

with $L \in \mathfrak{C}$ and $\dfrac{M}{L} \in E(\mathfrak{D}, \mathfrak{C})$. So we have the following diagram with $\dfrac{K}{L} \in \mathfrak{D}$ and $\dfrac{M}{K} \in \mathfrak{C}$:

$$\begin{array}{ccccccccc} & & 0 & & 0 & & 0 & & \\ & & \downarrow & & \downarrow & & \downarrow & & \\ 0 & \to & L & \hookrightarrow & K & \twoheadrightarrow & \dfrac{K}{L} & \to & 0 \\ & & & & {\scriptstyle \mathrm{inc}}\downarrow & & {\scriptstyle \mathrm{inc}}\downarrow & & \\ 0 & \to & L & \hookrightarrow & M & \twoheadrightarrow & \dfrac{M}{L} & \to & 0 \\ & & & & \downarrow & & \downarrow & & \\ & & & & \dfrac{M}{K} & & \dfrac{M}{K} & & \\ & & & & \downarrow & & \downarrow & & \\ & & & & 0 & & 0 & & \end{array}$$

Since $L \in \mathfrak{C}$ and $\dfrac{K}{L} \in \mathfrak{D}$, then $K \in E(\mathfrak{C}, \mathfrak{D})$. Hence $M \in E(E(\mathfrak{C}, \mathfrak{D}), \mathfrak{C})$. □

Notice that for two classes with zero $\mathfrak{C}, \mathfrak{D}$, we have that $\mathfrak{C} \cup \mathfrak{D} \subseteq E(\mathfrak{C}, \mathfrak{D})$.

Definition 2.3. For a class with zero $\mathfrak{C}$, define $E(\mathfrak{C}, \mathfrak{C})^0 = \{0\}$ and $E(\mathfrak{C}, \mathfrak{C})^{n+1} = E(\mathfrak{C}, E(\mathfrak{C}, \mathfrak{C})^n)$, $n \in \mathbb{N}$.

Theorem 2.4. *If $\mathfrak{C}$ is a hereditary class, then $\bigcup_{n \in \mathbb{N}} E(\mathfrak{C}, \mathfrak{C})^n \in R$-sext.*

Proof. First we prove that $E(\mathfrak{C}, \mathfrak{C})^n$ is a hereditary class for each $n \in \mathbb{N}$.

The assertion is clear for $n = 0$. Let us take $n > 0$.

Suppose that $M \in E(\mathfrak{C},\mathfrak{C})^k$ for some $0 < k \in \mathbb{N}$, and let N be a submodule of M. Thus there exists an exact sequence

$$0 \to L \hookrightarrow M \twoheadrightarrow \frac{M}{L} \to 0$$

with $L \in \mathfrak{C}$, and $\frac{M}{L} \in E(\mathfrak{C},\mathfrak{C})^{k-1}$. So we have the following commutative diagram:

$$\begin{array}{ccccccccc} 0 & \to & L & \hookrightarrow & M & \twoheadrightarrow & \frac{M}{L} & \to & 0 \\ & \text{inc} & \uparrow & \text{inc} & \uparrow & \text{inc} & \uparrow & & \\ 0 & \to & L \cap N & \hookrightarrow & N & \twoheadrightarrow & \frac{N+L}{L} & \to & 0. \end{array}$$

Since $\mathfrak{C}$ and $E(\mathfrak{C},\mathfrak{C})^{k-1}$ are hereditary classes, then $N \in E(\mathfrak{C},\mathfrak{C})^k$.

Thus for each $n \in \mathbb{N}$, $E(\mathfrak{C},\mathfrak{C})^n$ is a hereditary class and it is immediate that $\bigcup_{n\in\mathbb{N}} E(\mathfrak{C},\mathfrak{C})^n$ is also hereditary.

Now we claim that $\bigcup_{n\in\mathbb{N}} E(\mathfrak{C},\mathfrak{C})^n$ is closed under extensions.

Consider the exact sequence

$$0 \to K \to M \to L \to 0$$

with $K \in E(\mathfrak{C},\mathfrak{C})^l$ and $L \in E(\mathfrak{C},\mathfrak{C})^m$. We will prove that $M \in E(\mathfrak{C},\mathfrak{C})^{l+m}$, by induction on l. If $l = 0$, there is nothing to prove. Let us suppose $l > 0$. We can take a diagram with exact rows and columns:

$$\begin{array}{ccccccccc} & & 0 & & 0 & & & & \\ & & \downarrow & & \downarrow & & & & \\ & & K_1 & = & K_1 & & & & \\ & & \downarrow & & \downarrow & & & & \\ 0 & \to & K & \to & M & \to & L & \to & 0 \\ & & \downarrow & & \downarrow & & & & \\ 0 & \to & \frac{K}{K_1} & \to & \frac{M}{K_1} & \to & L & \to & 0 \\ & & \downarrow & & \downarrow & & & & \\ & & 0 & & 0 & & & & \end{array}$$

where $\frac{K}{K_1} \in E(\mathfrak{C},\mathfrak{C})^{l-1}$ and $K_1 \in \mathfrak{C}$.

Since $L \in E(\mathfrak{C},\mathfrak{C})^m$ and $\frac{K}{K_1} \in E(\mathfrak{C},\mathfrak{C})^{l-1}$, we have that

$$\frac{M}{K_1} \in E(\mathfrak{C},\mathfrak{C})^{(l-1)+m}.$$

So we have that $M \in E(\mathfrak{C},\mathfrak{C})^{l+m}$ as desired. □

For each hereditary class $\mathfrak{H} \subseteq \mathfrak{C}$, with $\mathfrak{C} \in R$-sext we have that $E(\mathfrak{H},\mathfrak{H}) \subseteq \mathfrak{C}$. Thus, by induction, we get $\bigcup_{n\in\mathbb{N}} E(\mathfrak{H},\mathfrak{H})^n \subseteq \mathfrak{C}$. So we obtain the following result.

Corollary 2.5. *If $\mathfrak{A}$ is a class of modules, then $\bigcup_{n\in\mathbb{N}} E(\mathrm{her}(\mathfrak{A}), \mathrm{her}(\mathfrak{A}))^n$ is the class in R-sext generated by $\mathfrak{A}$.*

For each class of modules $\mathfrak{A}$ we will denote $\bigcup_{n\in\mathbb{N}} E(\operatorname{her}(\mathfrak{A}), \operatorname{her}(\mathfrak{A}))^n$ by $\operatorname{sext}(\mathfrak{A})$.

From the above we have that R-sext is a complete big lattice where for each set X and any family $\{\mathfrak{C}_\alpha\}_{\alpha\in X}$ of elements in R-sext we have that

$$\begin{aligned} \bigwedge \{\mathfrak{C}_\alpha\}_{\alpha\in X} &= \bigcap \{\mathfrak{C}_\alpha\}_{\alpha\in X} \\ \bigvee \{\mathfrak{C}_\alpha\}_{\alpha\in X} &= \operatorname{sext}\left(\bigcup \{\mathfrak{C}_\alpha\}_{\alpha\in X}\right). \end{aligned}$$

Another important property that R-sext has is given in the following:

Theorem 2.6. *R-sext is S-pseudocomplemented.*

Proof. Let $\mathfrak{C} \in R$-sext. We will prove that

$$\mathfrak{C}^{\perp_{\text{sext}}} = \{M \in R\text{-}\ \operatorname{mod} \mid \operatorname{her}(M) \cap \mathfrak{C} = \{0\}\}$$

is the S-pseudocomplement of $\mathfrak{C}$ in R-sext.

Let us define $\mathfrak{H} = \{M \in R\text{-}\ \operatorname{mod} \mid \operatorname{her}(M) \cap \mathfrak{C} = \{0\}\}$.

It is clear that $\mathfrak{C} \cap \mathfrak{H} = \{0\}$. Now take $M \in \mathfrak{H}$ and $N \leq M$. Then $\operatorname{her}(N) \subseteq \operatorname{her}(M)$, so we have that $\operatorname{her}(N) \cap \mathfrak{C} \subseteq \operatorname{her}(M) \cap \mathfrak{C} = \{0\}$, thus $N \in \mathfrak{H}$, hence $\mathfrak{H}$ is a hereditary class.

Now, consider the exact sequence $0 \to K \to M \xrightarrow{p} L \to 0$ with K and L in $\mathfrak{H}$ and suppose $M \notin \mathfrak{H}$. Then there exists $0 \neq N \leq M$ such that $N \in \mathfrak{C}$, thus $N \cap K \in \mathfrak{H} \cap \mathfrak{C} = \{0\}$ which implies that $p_{|N} : N \to L$ is a monomorphism. As $L \in \mathfrak{H}$ we obtain $N \in \mathfrak{H} \cap \mathfrak{C} = \{0\}$, a contradiction. Thus $\mathfrak{H}$ is closed under extensions.

Finally we claim that $\mathfrak{H}$ contains each $\mathfrak{D}$ such that $\mathfrak{C} \cap \mathfrak{D} = \{0\}$. If not, take $\mathfrak{D} \in R$-sext such that $\mathfrak{D} \cap \mathfrak{C} = \{0\}$ and $\mathfrak{D} \nsubseteq \mathfrak{H}$, then there exists $0 \neq M \in \mathfrak{D} \setminus \mathfrak{H}$, thus also there exists $0 \neq N \leq M$ with $N \in \mathfrak{D} \cap \mathfrak{C}$, a contradiction.

We conclude that $\mathfrak{H}$ is the S-pseudocomplement for $\mathfrak{C}$. □

Theorem 2.7. $\operatorname{sext}(R) = \operatorname{sext}(R\text{-simp})$ *if and only if R is left artinian and R contains a copy of each simple module.*

Proof. Let us assume that $\operatorname{sext}(R) = \operatorname{sext}(R\text{-simp})$.

$\operatorname{sext}(R\text{-simp})$ consists of finitely generated semiartinian modules in view of Corollary 2.5. As $\operatorname{sext}(R)$ is the class closed under extensions generated by the left ideals the hypothesis implies that each left ideal is semiartinian and finitely generated. Thus R is left noetherian and left semiartinian. Hence R is left artinian. If S is a simple module, then $S \in \operatorname{sext}(R)$, thus $S \in E(L(R), L(R))^n$ for some minimal $n \in \mathbb{N}$. So there exists an exact sequence $0 \to I \longrightarrow S \longrightarrow K \longrightarrow 0$ where $I \in L(R)$ and $K \in E(L(R), L(R))^{n-1}$. As S is simple this implies that $S \cong I$.

Conversely, assume that R is left artinian thus it is noetherian and semiartinian. As usual, let us define $\operatorname{soc}_1(R) = \operatorname{soc}(R)$ and $\operatorname{soc}_{n+1}(R) / \operatorname{soc}_n(R) = \operatorname{soc}(R/\operatorname{soc}_n(R))$. It follows that $\operatorname{soc}_n(R) \in \operatorname{sext}(R\text{-simp})$, for each n. Also there exists an m such that $\operatorname{soc}_{m+1}(R) = \operatorname{soc}_m(R)$ because R is left noetherian. Thus $\operatorname{soc}(R/\operatorname{soc}_m(R)) = 0$ which implies that $R/\operatorname{soc}_m(R) = 0$ because R is semiartinian. Hence we see that $R \in \operatorname{sext}(R\text{-simp})$. The same argument can be used to prove that each left ideal belongs to sext(R-simp). Thus we get that sext$(R) \subseteq$ sext(R-simp). The converse inclusion follows directly from the hypothesis. □

2.1. R-sext and R-nat

We recall that a natural class of R-modules is a class of modules closed under submodules, direct sums and injective hulls. The class R-nat of natural classes is in fact a boolean lattice (in particular, R-nat is a set). See [19].

In [1] we proved that R-nat = Skel(R-her).

Theorem 2.8. *The skeleton of* R-sext *is* R-nat.

Proof. As R-nat = Skel(R-her) $\subseteq$ R-sext $\subseteq$ R-her, applying Theorem 1.4, we have that Skel(R-her) = Skel(R-sext). □

Corollary 2.9. *If* $\mathfrak{N} \in R$-nat, *then* $\mathfrak{N}^{\perp_{\text{sext}}\perp_{\text{sext}}} = \mathfrak{N}$.

Proof. Follows directly from Theorem 1.6. □

From Theorem 1.6 we can make the following remark.

Remark 2.10. If $\mathfrak{C} \in R$-sext, then $\mathfrak{C}^{\perp_{\text{sext}}\perp_{\text{sext}}} = \text{nat}(\mathfrak{C})$, the natural class generated by $\mathfrak{C}$.

Theorem 2.11. *If* $\mathfrak{C}$ *and* $\mathfrak{D}$ *are in* R-sext, *then:*

$$(\mathfrak{C} \vee \mathfrak{D})^{\perp_{\text{sext}}} = \mathfrak{C}^{\perp_{\text{sext}}} \wedge \mathfrak{D}^{\perp_{\text{sext}}}$$

and

$$(\mathfrak{C} \wedge \mathfrak{D})^{\perp_{\text{sext}}} = \mathfrak{C}^{\perp_{\text{sext}}} \vee \mathfrak{D}^{\perp_{\text{sext}}}$$

Proof. For the first statement, we always have $\mathfrak{C} \leq \mathfrak{C} \vee \mathfrak{D}$ and then $(\mathfrak{C} \vee \mathfrak{D})^{\perp_{\text{sext}}} \leq \mathfrak{C}^{\perp_{\text{sext}}}$. Analogously $(\mathfrak{C} \vee \mathfrak{D})^{\perp_{\text{sext}}} \leq \mathfrak{D}^{\perp_{\text{sext}}}$, and then $(\mathfrak{C} \vee \mathfrak{D})^{\perp_{\text{sext}}} \leq \mathfrak{C}^{\perp_{\text{sext}}} \wedge \mathfrak{D}^{\perp_{\text{sext}}}$ always happens.

On the other side, suppose there exists

$$0 \neq M \in \left(\mathfrak{C}^{\perp_{\text{sext}}} \wedge \mathfrak{D}^{\perp_{\text{sext}}}\right) \setminus (\mathfrak{C} \vee \mathfrak{D})^{\perp_{\text{sext}}}.$$

Then $\text{her}(M) \cap \mathfrak{C} = \{0\}$, $\text{her}(M) \cap \mathfrak{D} = \{0\}$ and $\text{her}(M) \cap (\mathfrak{C} \vee \mathfrak{D}) \neq \{0\}$. Then, there exists $0 \neq N \leq M$ such that $N \in (\mathfrak{C} \vee \mathfrak{D})$, and hence there exists an exact sequence

$$0 \to C \rightarrowtail N \twoheadrightarrow L \to 0$$

with $0 \neq C \in \mathfrak{C} \cup \mathfrak{D}$. Since $C \in \text{her}(M)$ too, we have $\text{her}(M) \cap \mathfrak{C} \neq \{0\}$ or $\text{her}(M) \cap \mathfrak{D} \neq \{0\}$, a contradiction, so $(\mathfrak{C} \vee \mathfrak{D})^{\perp_{\text{sext}}} = \mathfrak{C}^{\perp_{\text{sext}}} \wedge \mathfrak{D}^{\perp_{\text{sext}}}$.

For the second statement, it always happens $\mathfrak{C} \wedge \mathfrak{D} \leq \mathfrak{C}$, then $\mathfrak{C}^{\perp_{\text{sext}}} \leq (\mathfrak{C} \wedge \mathfrak{D})^{\perp_{\text{sext}}}$.

Analogously $\mathfrak{D}^{\perp_{\text{sext}}} \leq (\mathfrak{C} \wedge \mathfrak{D})^{\perp_{\text{sext}}}$, hence $\mathfrak{C}^{\perp_{\text{sext}}} \vee \mathfrak{D}^{\perp_{\text{sext}}} \leq (\mathfrak{C} \wedge \mathfrak{D})^{\perp_{\text{sext}}}$.

Now, take $M \in (\mathfrak{C} \wedge \mathfrak{D})^{\perp_{\text{sext}}}$, then $\text{her}(M) \cap (\mathfrak{C} \wedge \mathfrak{D}) = \{0\}$. Since $(\mathfrak{C} \wedge \mathfrak{D})^{\perp_{\text{sext}}} \in R$-nat we can suppose that M is injective. Let C be a maximal submodule of M such that $C \in \mathfrak{C}^{\perp_{\text{sext}}}$, then C is an essentially closed submodule of M (see [20, Section 1.]), thus $M = C \oplus D$ for some $0 \neq D \leq M$. If

$D \notin \mathfrak{D}^{\perp_{\text{sext}}}$, then there exists $0 \neq E \leq D$ such that $E \in \mathfrak{D}$ and $E \notin \mathfrak{C}^{\perp_{\text{sext}}}$, then there exists $0 \neq F \leq E$ with $F \in \mathfrak{C}$. Then $F \in \text{her}(M) \cap (\mathfrak{C} \wedge \mathfrak{D}) = \{0\}$, a contradiction. So $M = C \oplus D$ with $C \in \mathfrak{C}^{\perp_{\text{sext}}}$ and $D \in \mathfrak{D}^{\perp_{\text{sext}}}$ which implies $M \in \mathfrak{C}^{\perp_{\text{sext}}} \vee \mathfrak{D}^{\perp_{\text{sext}}}$. □

3. The big lattice R-qext

We shall denote by R-qext the proper class of all classes of left R-modules closed under isomorphisms, quotients and extensions.

Analogously to Theorem 2.4, Corollary 2.5 and to Theorem 2.6, we have the following results: (Notice that $R\text{-quot} = L_{\{\twoheadrightarrow\}}$).

Theorem 3.1. *If $\mathfrak{Q}$ is a cohereditary class, then $\bigcup_{n\in\mathbb{N}} E(\mathfrak{Q},\mathfrak{Q})^n \in R$-quot.*

Corollary 3.2. *If $\mathfrak{U}$ is a class of modules, then $\bigcup_{n\in\mathbb{N}} E(\text{quot}(\mathfrak{U}), \text{quot}(\mathfrak{U}))^n$ is the class in R-qext generated by $\mathfrak{U}$.*

Denoting $\text{qext}(\mathfrak{U}) = \bigcup_{n\in\mathbb{N}} E(\text{quot}(\mathfrak{U}), \text{quot}(\mathfrak{U}))^n$ we have that R-qext is a complete big lattice where for each family $\{\mathcal{C}_\alpha\}_{\alpha\in X}$ in R-qext:

$$\begin{aligned} \bigwedge \{\mathfrak{C}_\alpha\}_{\alpha\in X} &= \bigcap \{\mathfrak{C}_\alpha\}_{\alpha\in X} \\ \bigvee \{\mathfrak{C}_\alpha\}_{\alpha\in X} &= \text{qext}\left(\bigcup \{\mathfrak{C}_\alpha\}_{\alpha\in X}\right). \end{aligned}$$

Theorem 3.3. *If $\mathfrak{Q} \in R$-qext, then $\mathfrak{Q}$ has a unique pseudocomplement in R-qext given by*

$$\mathfrak{Q}^{\perp_{\text{qext}}} = \{M \in R\text{-}\ \ \text{mod} \mid \text{quot}(M) \cap \mathfrak{Q} = \{0\}\}.$$

Example. $\text{qext}(R) = \{{}_R M \mid M \text{ is finitely generated}\}$.

Proof. $\text{qext}(R)$ is the class closed under extensions generated by the cyclic modules, thus it is clear that $\text{qext}(R)$ contains just finitely generated modules. On the other hand, each finitely generated free module R^n belongs to $\text{qext}(R)$, thus each finitely generated module also belongs to $\text{qext}(R)$. □

Lemma 3.4. $\text{qext}(R) = \text{qext}(R\text{-simp})$ *implies that R is left semiartinian.*

Proof. It is clear that qext(R-simp) consists of semiartinian finitely generated modules. Thus the hypothesis implies that each finitely generated module is semiartinian, thus R is semiartinian. □

Recall that for two preradicals ρ, σ in R-mod, $[\rho : \sigma]$ is the preradical defined by $\dfrac{[\rho:\sigma](M)}{\rho(M)} = \sigma\left(\dfrac{M}{\rho(M)}\right)$. See [18].

Theorem 3.5. *R is left artinian if and only if R is left noetherian and* $\text{qext}(R) = \text{qext}(R\text{-simp})$.

Proof. Let us suppose R is left artinian, then R is noetherian and semiartinian. If we consider the sequence of preradicals $\mathrm{soc}, \mathrm{soc}_2 = [\mathrm{soc} : \mathrm{soc}], \ldots, \mathrm{soc}_{n+1} = [\mathrm{soc} : \mathrm{soc}_n]$, then the corresponding sequence of ideals $\mathrm{soc}(R) \leq \cdots \leq \mathrm{soc}_n(R)$ stabilizes, say at $\mathrm{soc}_n(R)$.

This means that $\mathrm{soc}(R/\mathrm{soc}_n(R)) = 0$ and thus $R/\mathrm{soc}_n(R) = 0$, since R is semiartinian. Then $R = \mathrm{soc}_n(R) \in \mathrm{qext}(R\text{-simp})$. So we have that $\mathrm{qext}(R) \subseteq \mathrm{qext}(R\text{-simp})$. The converse inclusion holds because each simple module is a quotient of R.

Assuming that $\mathrm{qext}(R) = \mathrm{qext}(R\text{-simp})$ and R noetherian, using Lemma 3.4, we have that R is left semiartinian. Thus R is left artinian. □

3.1. R-qext and R-conat

In [1] we proved that the skeleton of R-her is R-nat and we defined R-conat as the skeleton of R-quot. An element of R-conat is called a conatural class. In [2] we proved that R-conat is also a boolean lattice. Also we showed that a class $Q \in R$-quot is a conatural class if and only if it satisfies the following CN-condition: $Q = Q^{\perp_{\{\twoheadrightarrow\}}\perp_{\{\twoheadrightarrow\}}}$, where $\perp_{\{\twoheadrightarrow\}}$ denotes pseudocomplements in the big lattice $L_{\{\twoheadrightarrow\}}$ consisting of the module classes closed under quotients. (This big lattice is denoted R-quot in [1].)

In [1], we described the pseudocomplement in R-quot of a class Q, as $Q^{\perp_{\{\twoheadrightarrow\}}} = \{M \mid M \text{ has no non zero quotients in } Q\}$. It is easy to see from this description that pseudocomplements in R-quot are in fact S-pseudocomplements. Also, we have already seen in Lemma 3.3 that R-qext is S-pseudocomplemented and it is easy to see that pseudocomplements in R-qext are the same as the pseudocomplements in R-quot. To see this, just recall that $\mathrm{Skel}(R\text{-quot}) \subseteq R\text{-qext}$.

Now we obtain the following consequence from Theorem 1.4.

Theorem 3.6. R-conat $= \mathrm{Skel}(R\text{-qext})$.

Also we obtain the following result related with Corollary 2.9.

Corollary 3.7. *For a module class* $Q \in R$-qext, $Q^{\perp_{\{\twoheadrightarrow,\mathrm{ext}\}}\perp_{\{\twoheadrightarrow,\mathrm{ext}\}}} = \mathrm{conat}(Q)$, *where* $\mathrm{conat}(Q)$ *denotes the conatural class generated by* Q.

Remark 3.8. Notice that $\mathrm{conat}(Q) = Q^{\perp_{\{\twoheadrightarrow\}}\perp_{\{\twoheadrightarrow\}}}$ *if* $Q \in R$-quot. Thus we can describe the conatural class generated by an arbitrary family of modules $\mathcal{A}$ as

$$\mathrm{conat}(\mathcal{A}) = \left\{ M \mid \begin{array}{l} \forall \text{ epic } M \overset{f\neq 0}{\twoheadrightarrow} N, \exists\ N \overset{g\neq 0}{\twoheadrightarrow} C \text{ epic}, \\ \text{with } C \text{ a quotient of an element of } \mathcal{A} \end{array} \right\}.$$

4. R-nat and R-conat

In this section we study the consequences of assuming that R-nat $= R$-conat. We begin with the following.

Theorem 4.1. *For a ring R are equivalent:*

(1) R-nat $= R$-conat.
(2) $\text{nat}(M) = \text{conat}(M)$ *for each* $M \in R\text{-}\ \text{mod}$.

Proof. (1) $\Longrightarrow$ (2) Let M be an R-module, as $\text{nat}(M) \in R\text{-nat} \subseteq R\text{-conat}$, then $\text{nat}(M)$ is a conatural class containing M, thus $\text{conat}(M) \subseteq \text{nat}(M)$. Symmetrically, $\text{nat}(M) \subseteq \text{conat}(M)$.

(2) $\Longrightarrow$ (1) Suppose (2) and let us take a natural class $\mathfrak{C}$. If $M \in \mathfrak{C}$, as $\text{conat}(M) = \text{nat}(M) \subseteq \mathfrak{C}$, it follows that $\mathfrak{C}$ is closed under quotients. To show that $\mathfrak{C}$ is a conatural class, it suffices to prove that a module M such that all of its nonzero quotients have a non zero quotient in $\mathfrak{C}$, must belong to $\mathfrak{C}$ (for this is equivalent to the CN condition mentioned at the beginning of the preceding section). Let us take a module M such that for each nonzero epimorphism $f : M \twoheadrightarrow N$, there exists a nonzero epimorphism $g : N \twoheadrightarrow C$, $C \in \mathfrak{C}$. We want to prove that $M \in \mathfrak{C}$.

There exists maximal submodules of M belonging to $\mathfrak{C}$, see [4]. Let U be one of them. If U was a proper submodule of M, it could not be an essential submodule, because a natural class is closed under essential extensions. Hence we can assume that U is essentially closed. If V is a pseudocomplement of U in M, then U is also a pseudocomplement of V. Thus ${}_R0 \neq V$ embeds in $\frac{M}{U}$ as an essential submodule. The choice of U and V implies that $V \in \mathfrak{C}^{\perp_{\text{nat}}}$, where $\mathfrak{C}^{\perp_{\text{nat}}}$ denotes the complement of $\mathfrak{C}$ in R-nat. Thus $0 \neq \frac{M}{U} \in \mathfrak{C}^{\perp_{\text{nat}}}$, and by hypothesis, $\frac{M}{U}$ has a nonzero quotient in $\mathfrak{C}$, which also belongs to $\mathfrak{C}^{\perp_{\text{nat}}}$, because we have noted that a natural class is closed under quotients. We have obtained a contradiction. Hence $M \in \mathfrak{C}$.

Thus R-nat $\subseteq R$-conat.

For the converse inclusion, let us take a conatural class $\mathfrak{C}$. For $M \in \mathfrak{C}$, we have that $\text{nat}(M) = \text{conat}(M) \subseteq \mathfrak{C}$, then $\mathfrak{C}$ is closed under submodules.

Thus each conatural class is a class closed under submodules and quotients (i.e., it is an open class). The pseudocomplement of $\mathfrak{C}$ in R-quot, which is described as

$$\{N \mid N \text{ has no nonzero quotients in } \mathfrak{C}\}$$

is also a conatural class. Therefore

$$\{N \mid N \text{ has no nonzero quotients in } \mathfrak{C}\}$$

is closed under submodules and for this reason it coincides with

$$\{N \mid N \text{ has no nonzero subquotients in } \mathfrak{C}\},$$

the pseudocomplement of the open class $\mathfrak{C}$ in the big lattice of open classes. But this is a hereditary torsion class by Lemma 1.7. In particular it is closed under taking direct sums.

In order to see that a conatural class $\mathfrak{C}$ is natural, it suffices to prove that a module M such that each one of its nonzero submodules has a nonzero submodule in $\mathfrak{C}$, must belong to $\mathfrak{C}$.

Let us assume that M has the mentioned property. Then it must contain a maximal independent family of submodules in $\mathfrak{C}$, whose direct sum U, say, must be an essential submodule of M. Thus $U \in \mathfrak{C}$, as we remarked above. As U is essential in M, then $M \in \operatorname{nat}(U) = \operatorname{conat}(U) \subseteq \mathfrak{C}$. □

Theorem 4.2. *The following conditions are equivalent for a ring R:*

(1) R-nat $=$ R-conat.
(2) *R is isomorphic to a finite direct product of right perfect, left local rings.*

Proof. (1) $\Longrightarrow$ (2) Each hereditary torsion free class is a natural class and then each class is closed under quotients. So by [17], R is a finite direct product of right perfect left local rings.

(2) $\Longrightarrow$ (1) By [3, Theorem I.9.I] we can assume that R is a right perfect left local ring. Now, since R is left semiartinian, each module is an essential extension of its socle. So each natural class is generated by a family of left simple modules. As by hypothesis $|R\text{-simp}| = 1$, then R-nat $= \{\{0\}, R\text{-}\operatorname{mod}\}$.

On the other hand, each conatural class is also generated by a family of left simple modules (see [1]), thus R-conat $= \{\{0\}, R\text{-}\operatorname{mod}\}$. □

In [2] we proved that the conditions: (1) R-her $=$ R-quot and (2) R is a finite product of artinian principal ideal rings are equivalent. In the following example we give a ring R where R-nat $=$ R-conat but R-her $\neq$ R-quot.

Example. By [5, Chapter 24, Exercise 4] we have that the ring $R = \dfrac{A}{(x^2, y^2)}$, where $A = k[x, y]$ is the polynomial ring in two indeterminates x and y over a field k, is a QF local (commutative) algebra over k such that $\dfrac{R}{\operatorname{soc}(R)}$ is not QF. By [5, Chapter 24, Exercise 3] R is not an artinian principal ideal ring. By [2, Theorem 38] R-her $\neq$ R-quot (Indeed, $R\bar{x} + R\bar{y}$ is an ideal which is not principal, thus it can not be a quotient of R, thus $\operatorname{her}(R) \neq \operatorname{quot}(R)$). But we do have that R-nat $=$ R-conat by the previous theorem.

5. R-sext and R-qext

In this section we study what can we say about the ring when we suppose that R-sext $=$ R-qext, that is, when every class in R-sext is a class in R-qext and viceversa.

We begin with the following lemma:

Lemma 5.1. *If every class in* R-sext (R-qext) *belongs to* R-qext (R-sext), *then* $\operatorname{qext}(M) \subseteq \operatorname{sext}(M)$ ($\operatorname{sext}(M) \subseteq \operatorname{qext}(M)$) *for each* $M \in R\text{-}\operatorname{mod}$. *Thus for a ring R the following conditions are equivalent:*

(1) R-sext $=$ R-qext
(2) $\forall M \in R\text{-}\operatorname{mod}$, $\operatorname{sext}(M) = \operatorname{qext}(M)$.

Proof. It is straightforward from the properties of generating classes in R-sext and R-qext. □

Lemma 5.2. *If every class in* R-sext *belongs to* R-qext*, then* R *is isomorphic to a finite direct product of right perfect left local rings.*

Proof. If R-sext $\subseteq$ R-qext, as every hereditary torsion free class $\mathbb{F}_\tau$ belongs to R-sext, we have that all of these are also closed under quotients, we conclude by [17]. □

Lemma 5.3. *If every class in* R-sext *belongs to* R-qext*, then every simple left* R*-module embeds in* R.

Proof. If R-sext $\subseteq$ R-qext, then for the R-module ${}_RR$ we have that $\operatorname{qext}(R) \subseteq \operatorname{sext}(R)$ and then each simple left R-module S is in $\operatorname{sext}(R)$, so there exist a minimal $n \in \mathbb{N}$ and modules $A \in L(R)$ and $B \in E(L(R), L(R))^{n-1}$ such that the sequence

$$0 \to A \to S \to B \to 0$$

is exact. Hence $S \cong A$, and A is contained in R. □

Lemma 5.4. *If every class in* R-qext *belongs to* R-sext*, then* R *is a left noetherian ring.*

Proof. Suppose that R-qext $\subseteq$ R-sext, then

$$\operatorname{sext}(R) \subseteq \operatorname{qext}(R) = \{M \in R\text{-}\ \ \operatorname{mod} \mid M \text{ is finitely generated}\}.$$

So that every left ideal of R is finitely generated. □

Proposition 5.5. *If* R-sext $=$ R-qext*, then* R *is isomorphic to a finite direct product of left artinian, left local rings.*

Proof. By Lemma 5.2, R is a product of finitely many right perfect left local rings. By Lemma 5.4 R is also a left noetherian ring. Now, a right perfect ring is left semiartinian. Thus the right perfect factors of R are left semiartinian and left noetherian, thus they are left artinian. □

It should be noticed that in the following lemmas about local left artinian rings we could put "left local" instead of "local" in view of [3, Theorem V.2.3] and by straightforward uses of Morita equivalence theory.

Lemma 5.6. *If* R *is a left artinian, local ring such that* $E\left(R/\operatorname{Rad}(R)\right)$ *is finitely generated, then* $\operatorname{sext}(R) = \operatorname{qext}(R)$.

Proof. We have that $\operatorname{sext}(R)$ is the class of all modules closed under extensions generated by the left ideals of R. Since R is left noetherian, then each left ideal is finitely generated, thus every left ideal of R belongs to $\operatorname{qext}(R)$. As $\operatorname{qext}(R)$ is the class of all finitely generated left R-modules, then $\operatorname{sext}(R) \subseteq \operatorname{qext}(R)$. The other inclusion follows from the fact that, for each finitely generated module ${}_RM$, $\operatorname{qext}(M) = \operatorname{qext}(S)$, where S is the unique simple module in R-simp. Indeed $M = \operatorname{soc}_n(M) \in \operatorname{sext}(S)$ for some $n \in \mathbb{N}$. (See the proof of Theorem 2.7.) □

Lemma 5.7. *If $N \in \text{sext}(M)$ and X is a set, then $N^{(X)} \in \text{sext}\left(M^{(X)}\right)$.*

Proof. Suppose that $0 \neq N$ to be $\text{sext}(M)$, then $N \in E(L(M), L(M))^n$ for some $n \in \mathbb{N}$, which we can assume minimal. We shall proceed by induction on n.

If $n = 1$, then $N \leq M$ and it is clear that $N^{(X)} \leq M^{(X)}$. Now let us suppose that $n > 1$, thus there exists a short exact sequence $0 \longrightarrow K \longrightarrow N \longrightarrow T \longrightarrow 0$, with $K \leq M$ and $T \in E(L(M), L(M))^{n-1}$, so we obtain a short exact sequence $0 \longrightarrow K^{(X)} \longrightarrow N^{(X)} \longrightarrow T^{(X)} \longrightarrow 0$. By induction hypothesis, $T^{(X)} \in \text{sext}\left(M^{(X)}\right)$, thus we have $N^{(X)} \in \text{sext}\left(M^{(X)}\right)$. □

With an analogous argument we obtain the following lemma.

Lemma 5.8. *If $N \in \text{qext}(M)$ and X is a set, then $N^{(X)} \in \text{qext}\left(M^{(X)}\right)$.*

Corollary 5.9. *If X is a set, then:*

(1) $\text{sext}(N) = \text{sext}(M) \Longrightarrow \text{sext}\left(N^{(X)}\right) = \text{sext}\left(M^{(X)}\right)$.
(2) $\text{qext}(N) = \text{sext}(M) \Longrightarrow \text{qext}\left(N^{(X)}\right) = \text{qext}\left(M^{(X)}\right)$.

Lemma 5.10. *If R is a local left artinian ring such that $E(R/\text{Rad}(R))$ is finitely generated, then $\text{sext}(M) = \text{qext}(M)$ for every non finitely generated left R-module M.*

Proof. Notice that, under the current hypothesis, every left R-module has a projective cover and every projective left R-module is free. Now, let us suppose that $0 \longrightarrow K \longrightarrow R^{(X)} \twoheadrightarrow M \longrightarrow 0$ is a projective cover of a non zero left R-module M, then $M/\text{Rad}(M)$ is a semisimple module of the form $S^{(Z)}$ for some set Z. Since $0 \longrightarrow \text{Rad}(R) \longrightarrow R \longrightarrow S \longrightarrow 0$ is a projective cover, then it induces a projective cover $0 \longrightarrow K' \longrightarrow R^{(Z)} \twoheadrightarrow S^{(Z)}$ of $M/\text{Rad}(M)$. From the fact that the rows of the diagram

$$\begin{array}{ccccccccc} 0 & \longrightarrow & K & \longrightarrow & R^{(X)} & \twoheadrightarrow & M & \longrightarrow & 0 \\ & & & & & & \downarrow & & \\ 0 & \longrightarrow & K' & \longrightarrow & R^{(Z)} & \twoheadrightarrow & M/\text{Rad}(M) & \longrightarrow & 0 \end{array}$$

are projective covers, we obtain that $R^{(X)} \cong R^{(Z)}$, and consequently we have that $|X| = |Z|$. Thus $M \in \text{qext}\left(R^{(X)}\right)$. Now, since $\text{qext}(R) = \text{qext}(S)$, we have that $\text{qext}\left(R^{(X)}\right) = \text{qext}\left(S^{(X)}\right)$ from Corollary 5.9.

Now, it is clear that

$$\text{qext}(M) \leq \text{qext}\left(R^{(X)}\right) = \text{qext}\left(S^{(X)}\right) = \text{qext}(M/\text{Rad}(M)) \leq \text{qext}(M),$$

so that $\text{qext}(M) = \text{qext}(M/\text{Rad}(M)) = \text{qext}\left(S^{(X)}\right)$.

From the above, we conclude that the projective cover of a left module M determines $\text{qext}(M)$ since its projective cover has as many direct summands as $M/\text{Rad}(M)$. □

Lemma 5.11. *If R is a local left artinian ring such that $E(R/\text{Rad}(R))$ is finitely generated, then $\text{sext}(M) = \text{sext}(\text{soc}(M))$, for each left module M.*

Proof. Assume that $\operatorname{soc}(M) \cong S^{(Z)}$ for some set Z. Since R is semiartinian, then $\operatorname{soc}(M) \leq_{es} M$, and consequently M and $\operatorname{soc}(M)$ have isomorphic injective hulls; thus we have that $E(M) \cong E\left(S^{(Z)}\right) \cong (E(S))^{(Z)}$. As $E(S)$ is finitely generated, we conclude that $E(S) \in \operatorname{qext}(R) = \operatorname{qext}(S)$. But $\operatorname{sext}(S) = \operatorname{qext}(S)$, so that $E(S) \in \operatorname{sext}(S)$. Notice that, from Lemma 5.7, $E(S) \in \operatorname{sext}(S)$ implies that $(E(S))^{(Z)} \in \operatorname{sext}\left(S^{(Z)}\right)$, thus we have

$$\operatorname{sext}(\operatorname{soc}(M)) = \operatorname{sext}\left(S^{(Z)}\right) = \operatorname{sext}\left((E(S))^{(Z)}\right) = \operatorname{sext}(E(M))$$

which implies that

$$\operatorname{sext}(\operatorname{soc}(M)) = \operatorname{sext}(M) = \operatorname{sext}(E(M)) = \operatorname{sext}\left(S^{(Z)}\right).$$

□

Corollary 5.12. *If R is a local left artinian ring such that $E(R/\operatorname{Rad}(R))$ is finitely generated, then:*
(1) $\operatorname{sext}\left(S^{(X)}\right) = \left\{M \mid \operatorname{soc}(M) \cong S^{(X)}\right\}$
(2) $\operatorname{qext}\left(S^{(X)}\right) = \left\{M \mid M/\operatorname{Rad}(M) \cong S^{(X)}\right\}$.

Proof. (1) This is a direct consequence of Lemma 5.11.

(2) In the proof of Lemma 5.10, we noticed that $\operatorname{qext}(M) = \operatorname{qext}(M/\operatorname{Rad}(M))$. □

Lemma 5.13. *If R is a local left artinian ring such that $E(R/\operatorname{Rad}(R))$ is finitely generated, then*

$$\operatorname{soc}(M) \cong M/\operatorname{Rad}(M).$$

for each non finitely generated left module M.

Proof. From Lemma 5.10, we have that $\operatorname{sext}(M) = \operatorname{qext}(M)$. Now, if $\operatorname{soc}(M) \cong S^{(X)}$ and $M/\operatorname{Rad}(M) \cong S^{(Y)}$, then

$$\operatorname{sext}\left(S^{(X)}\right) = \operatorname{sext}(M) = \operatorname{qext}(M/\operatorname{Rad}(M)) = \operatorname{qext}\left(S^{(Y)}\right) = \operatorname{sext}\left(S^{(Y)}\right).$$

From this we conclude that $|X| = |Y|$ and then $\operatorname{soc}(M) \cong M/\operatorname{Rad}(M)$. □

Theorem 5.14. *Suppose that a ring R is such that the injective hull of every simple left module if finitely generated. Then the following conditions are equivalent:*

1. $\operatorname{sext}(M) = \operatorname{qext}(M)$, $\forall M \in R$- mod .
2. R-sext $=$ R-qext.
3. *R is isomorphic to a finite direct product of left local, left artinian rings.*
4. *R is isomorphic to a finite direct product of left local and both left and right perfect rings with the property that $\operatorname{soc}(M) \cong M/\operatorname{Rad}(M)$ for each non-finitely generated left module M.*

Proof. The equivalence of (1) and (2) is given by Lemma 5.1 and (1) $\Longrightarrow$ (3) is Proposition 5.5.

(3) $\Longrightarrow$ (4) A left artinian ring is left and right perfect and also it is left noetherian. By [3, Theorem V.2.3], we can assume R is local. Then it follows from Lemma 5.13 that $\operatorname{soc}(M) \cong M/\operatorname{Rad}(M)$.

(4) $\Longrightarrow$ (1) Assume that R is a left local and left and right perfect ring, we want to show that $\operatorname{sext}(M) = \operatorname{qext}(M)$ for each left module M.

Indeed, by Lemma 5.11, we have that $\operatorname{sext}(M) = \operatorname{sext}(\operatorname{soc}(M))$ and as in the proof of Lemma 5.10 we can show that $\operatorname{qext}(M) = \operatorname{qext}(M/\operatorname{Rad}(M))$. Thus we have the conclusion for every non finitely generated left module M since, in this case,

$$\begin{aligned} \operatorname{sext}(M) &= \operatorname{sext}(soc(M)) = \operatorname{sext}(M/\operatorname{Rad}(M)) \\ &= \operatorname{qext}(M/\operatorname{Rad}(M)) = \operatorname{qext}(M). \end{aligned}$$

Now, if M is finitely generated, then $\operatorname{soc}(M)$ and $M/\operatorname{Rad}(M)$ are semisimple finitely generated modules, then

$$\operatorname{sext}(M) = \operatorname{sext}(S) = \operatorname{qext}(S) = \operatorname{qext}(M/\operatorname{Rad}(M)) = \operatorname{qext}(M),$$

where S denotes the unique element of R-simp. □

Example. A ring R such that R-sext $=$ R-qext but R-her $\neq$ R-quot: By [8, Chapter 13, Exercise 5] a commutative artinian ring R has the property that its finitely generated modules are closed under injective hulls. Thus the ring of Example 4, satisfies the required properties, as any other commutative artinian local ring with a non principal ideal.

Example. A commutative perfect non artinian ring satisfies that R-nat $=$ R-conat but R-sext $\neq$ R-qext. Take, for example, the trivial extension of a field F by an infinite-dimensional vector space ${}_FV$. A particular case is the trivial extension $\mathbb{Q} \ltimes \mathbb{R}$.

References

[1] A. Alvarado, H. Rincón and J. Ríos, *On the lattices of natural and conatural classes in R-mod*, Comm. Algebra. 29 (2) (2001) 541–556.

[2] A. Alvarado, H. Rincón and J. Ríos, *On Some Lattices of Module Classes,* Journal of Algebra and its Applications. 2006, 105–117.

[3] L. Bican, T. Kepka, P. Němec, *Rings, modules, and preradicals.* Lecture Notes in Pure and Applied Mathematics, 75. Marcel Dekker, Inc., New York, 1982.

[4] J. Dauns, Y. Zhou, *Classes of Modules.* Pure and Applied Mathematics (Boca Raton), 281. Chapman & Hall/CRC, Boca Raton, FL., 2006.

[5] C. Faith, *Algebra II, Ring Theory,* Grundlehren der Mathematischen Wissenschaften 191 (Springer-Verlag, 1976).

[6] Fernández-Alonso Rogelio, Raggi, Francisco, *The lattice structure of nonhereditary torsion theories.* Comm. Algebra 26 (1998), no. 6, 1851–1861.

[7] Golan, Jonathan, *Torsion theories.* Pitman Monographs and Surveys in Pure and Applied Mathematics, 29. Longman Scientific & Technical, Harlow; John Wiley & Sons, Inc., New York, 1986.

[8] F. Kasch, *Modules and Rings*, Academic Press Inc. (London) LTD. 1982.

[9] Raggi, Francisco, Rincón, Hugo, Signoret, Carlos, *On some classes of R-modules and congruences in R-tors.* Comm. Algebra 27 (1999), no. 2, 889–901.

[10] Raggi, Francisco, Ríos, José, Rincón, Hugo, Fernández-Alonso, Rogelio, Signoret, Carlos, *The lattice structure of preradicals.* Comm. Algebra 30 (2002), no. 3, 1533–1544.

[11] Raggi, Francisco, Ríos, José, Rincón, Hugo, Fernández-Alonso, Rogelio, Signoret, Carlos, *The lattice structure of preradicals II. Partitions.* J. Algebra Appl. 1 (2002), no. 2, 201–214.

[12] Raggi, Francisco, Ríos, José, Rincón, Hugo, Fernández-Alonso, Rogelio, Signoret, Carlos, *The lattice structure of preradicals III. Operators.* J. Pure Appl. Algebra 190 (2004), no. 1-3, 251–265.

[13] Raggi, Francisco, Ríos, José, Rincón, Hugo, Fernández-Alonso, Rogelio, Signoret, Carlos, *Prime and irreducible preradicals.* J. Algebra Appl. 4 (2005), no. 4, 451–466.

[14] Raggi, Francisco, Ríos, José, Wisbauer, Robert, *The lattice structure of hereditary pretorsion classes.* Comm. Algebra 29(2001), no. 1, 131–140.

[15] Raggi, Francisco, Ríos Montes, José, Wisbauer, Robert, *Coprime preradicals and modules.* J. Pure Appl. Algebra 200 (2005), no. 1-2, 51–69.

[16] Raggi, Francisco F., Signoret P., Carlos J.E., *Serre subcategories of* R-mod. Comm. Algebra 24 (1996), no. 9, 2877–2886.

[17] R. Bronowitz and M. Teply, *Torsion theories of simple type,* J. Pure Appl. Algebra 3 (1973), 329–336.

[18] B. Stenström, *Rings of Quotients*, Springer-Verlag, New York, 1975.

[19] Y. Zhou, *The Lattice of natural classes of modules,* Comm. Algebra 24 (5) (1996) 1637–1648.

[20] Y. Zhou, *Decomposing modules into direct sums of submodules with types.* J. Pure Appl. Algebra 138 (1999), no. 1, 83–97.

A. Alvarado García and H.A. Rincón Mejía
Departamento de Matemáticas
Facultad de Ciencias
Universidad Nacional Autónoma de México
Circuito Exterior, Ciudad Universitaria
04510 México D.F., México
e-mail: `alejandroalvaradogarcia@gmail.com`
`hurincon@gmail.com`

José Ríos Montes Instituto de Matemáticas
Universidad Nacional Autónoma de México
Área de la Investigación Científica
Circuito Exterior, Ciudad Universitaria
04510 México D.F., México
e-mail: `jrios@matem.unam.mx`

Advances in Ring Theory
Trends in Mathematics, 37–46

Reversible and Duo Group Rings

Howard E. Bell and Yuanlin Li

Abstract. We summarize recent results on reversible group rings, duo group rings, and graded reversible group rings; and we mention several open problems.

Mathematics Subject Classification (2000). Primary 16S34; Secondary 16U80.
Keywords. Group rings; reversible rings; duo rings: graded reversible rings.

1. Introduction

Let R be an associative ring with identity. R is called reversible if $\alpha\beta = 0$ implies $\beta\alpha = 0$, and it is called symmetric if $\alpha\beta\gamma = 0$ implies $\alpha\gamma\beta = 0$ for all $\alpha, \beta, \gamma \in R$. The reversibility property, a natural generalization of commutativity, has been exploited by various authors over the years; but apparently the name was introduced by Cohn [3], who noted that the Köthe conjecture holds for the class of reversible rings.

Marks [10] has discussed the relationship between symmetric and reversible rings. Symmetric rings are clearly reversible, but the converse is not true. In fact, Marks showed that the group algebra $\mathbb{Z}_2 Q_8$ of the quaternion group of order 8 over the two-element field is reversible but not symmetric. In [5], Gutan and Kisielewicz characterized reversible group algebras KG of torsion groups G over fields K. In particular, they described all finite reversible group algebras which are not symmetric.

In this expository paper, we present some of results in [5], together with extensions to group rings RG over commutative rings R with 1. We deal briefly with the question of when reversible group rings are not symmetric, noting that $\mathbb{Z}_2 Q_8$ is the minimal reversible group ring which is not symmetric. We investigate when a group ring RG is a duo ring, where R is either a field or an integral domain. Finally, we present some results on the more general notion of graded reversibility.

This research was supported in part by Discovery Grants from the Natural Sciences and Engineering Research Council of Canada.
Corresponding author: Yuanlin Li.

For a discussion of reversibility of semigroups and rings not assumed to have an identity, we refer to a recent paper [6], in which it was proved that symmetric rings, not necessarily with identity, satisfy the Köthe conjecture.

2. Reversibility in group rings

In this section we discuss recent developments regarding reversibility in group rings. We deal first with group algebras KG over fields K, and then with group rings RG over commutative rings R. Finally, we discuss minimal reversible group rings which are not symmetric.

Consider a group ring RG of a torsion group G over an associative ring R with identity. If RG is reversible, then the structure of G is very restricted; in fact, G is either an abelian group or a Hamiltonian group. To see this, we need only to verify that every cyclic subgroup $\langle g\rangle$ of G is normal. Let $h \in G$ and $\bar{g} = 1 + g + g^2 + \cdots + g^{o(g)-1}$. Since $h(1-g)\bar{g} = 0$ and RG is reversible, we have $\bar{g}h(1-g) = 0$ and thus $\bar{g}h = \bar{g}hg$, so $h = g^i hg$ for some i. Hence, $hgh^{-1} = g^{-i} \in \langle g\rangle$, implying that $\langle g\rangle$ is normal.

2.1. Reversibility in group algebras KG

Marks [10] showed that the group algebra $\mathbb{Z}_2Q_8$ of the quaternion group of order 8 over the two-element field is reversible, but not symmetric. In [5], Gutan and Kisielewicz characterized all reversible group algebras KG of torsion groups G over fields K. In particular, they described all finite reversible group algebras which are not symmetric, extending a result of Marks. If G is abelian, clearly KG is commutative, hence symmetric, so the interesting case is when G is a Hamiltonian group, i.e., $G = Q_8 \times E_2 \times E_2'$, where Q_8 is the quaternion group of order 8, E_2 is an elementary abelian 2-group, and E_2' is an abelian group all of whose elements are of odd order. Gutan and Kisielewicz first considered when KQ_8 is reversible and obtained the following two results.

Theorem 2.1. *Let K be a field of characteristic $\neq 2$. Then KQ_8 is reversible if and only if the equation $1 + x^2 + y^2 = 0$ has no solutions in K.*

Note that if K is a field of characteristic $p > 2$, then $1 + x^2 + y^2 = 0$ has a solution in K. Consequently, if K is a field of characteristic $p > 2$, then KQ_8 is not reversible.

Theorem 2.2. *Let K be a field of characteristic $= 2$. Then KQ_8 is reversible if and only if the equation $1 + x + x^2 = 0$ has no solutions in K.*

By using a result of Perlis and Walker ([12, Prop. II. 2.6]) together with the above two theorems, Gutan and Kisielewicz obtained the following general characterization theorem.

Theorem 2.3. *Let K be a field and let G be a torsion group. Then KG is a reversible ring if and only if one of the following conditions holds.*

(1) *G is abelian.*
(2) $G = Q_8 \times E_2 \times E_2'$ *is Hamiltonian, the characteristic of* K *is* 0*, and the equation* $1 + x^2 + y^2 = 0$ *has no solutions in any cyclotomic field* $K(\xi_d)$ *for any odd* d *which is an order of an element of* E_2'.
(3) $G = Q_8 \times E_2'$*, the characteristic of* K *is* 2*, and the equation* $1+x+x^2 = 0$ *has no solutions in any cyclotomic field* $K(\xi_d)$ *for any odd* d *which is an order of an element of* E_2'.

As a consequence, they characterized all finite reversible group algebras.

Theorem 2.4. *A finite group ring* KG *of a non-abelian group* G *over a field* K *is reversible if and only if* $K = GF(2^n)$ *with an odd* $n \geq 1$ *and* $G = Q_8 \times E_2'$*, where the order of* E_2' *divides* $2^m - 1$ *for some odd* $m > 1$.

The following result from [5] addresses the question of which reversible group algebras KG are not symmetric.

Corollary 2.5. *A reversible group algebra* KG *of a non-abelian torsion group* G *over a field* K *is not symmetric if and only if* $\mathrm{char}(K) = 2$*, and* $G = Q_8 \times E_2'$.

Note that $(R_1 \times R_2)G$ is reversible if and only if both R_1G and R_2G are reversible. It now follows from Theorem 2.1 and Corollary 2.5 that $(\mathbb{Q} \times \mathbb{Z}_2)Q_8$ is reversible but not symmetric.

2.2. Reversible group rings over commutative rings

We now present some recent results from [8] regarding reversibility of group rings RG over commutative rings R, which extend the above mentioned results of Gutan and Kisielewicz. The following two preliminary results are useful.

Lemma 2.6. *Let* R *be a ring with identity. If* R *contains a nonzero nilpotent element* r *such that* $2r = 0$*, then* RQ_8 *is not reversible.*

Theorem 2.7. $\mathbb{Z}_nQ_8$ *is reversible if and only if* $n = 2$.

Note that if RQ_8 is reversible and $\mathrm{char}(R) = n > 0$, then the subring $\mathbb{Z}_nQ_8$ of RQ_8 is also reversible, and thus $n = 2$. This tells us that if RQ_8 is reversible, then the characteristic of R is either 0 or 2.

For R a commutative ring with characteristic 2, a necessary and sufficient condition for RQ_8 to be reversible was given by Parmenter and the second author in [8].

Theorem 2.8. *Let* R *be a commutative ring of characteristic* 2*. Then* RQ_8 *is reversible if and only if the equation* $x^2 + xy + y^2 = 0$ *has no nonzero solutions in* R.

Note that all of the above reversible group rings RQ_8 are not symmetric because RQ_8 has a non-symmetric subring $\mathbb{Z}_2Q_8$. Note also that in a field K of characteristic 2, the equation $x^2+xy+y^2 = 0$ has no nonzero solutions if and only if $1 + x + x^2 = 0$ has no solutions. Thus the above theorem extends Theorem 2.2.

As a consequence of Theorem 2.8, we obtain the following necessary and sufficient condition for group ring RQ_8 over a commutative Artinian R of characteristic 2 to be reversible.

Corollary 2.9. *If R is a commutative Artinian ring of characteristic 2, then RQ_8 is reversible if and only if $R = \prod K_i$, where each K_i is a field of characteristic 2, in which the equation $1 + x + x^2 = 0$ has no solutions.*

Next we discuss the case where R is a commutative ring of characteristic 0. It is interesting to note that while the most complex argument of Gutan and Kisielewicz's proof of reversibility of group algebras KQ_8 occurs when K has characteristic 2, the most complicated situation with regard to general group rings RG appears to be the case when R has characteristic 0.

Let $R_2 = \{ x \in R \mid 2^l x = 0 \text{ for some } l > 0\}$ denote the 2-torsion of R and $\text{ann}\{2\} = \{ x \in R \mid 2x = 0\}$ be the annihilator of 2 in R. Clearly $\text{ann}\{2\} \subseteq R_2$. If RQ_8 is reversible, then by Lemma 2.6, R_2 has no nonzero nilpotent elements, so $R_2 = \text{ann}\{2\}$.

In [8], Parmenter and the second author were able to prove the following two results, which extend Theorem 2.1. The first shows that when investigating the reversibility of RG over a commutative ring R of characteristic 0, one may always assume that $R_2 = 0$ (i.e., R has no 2-torsion).

Proposition 2.10. *Let R be a commutative ring of characteristic 0. Then the following statements are equivalent:*

(1) *RQ_8 is reversible.*
(2) *R_2 has no nonzero nilpotent elements, and both R_2Q_8 and $(R/R_2)Q_8$ are reversible.*
(3) *The equation $x^2+xy+y^2 = 0$ has no nonzero solutions in R_2, and $(R/R_2)Q_8$ is reversible.*

The next theorem characterizes all reversible group rings RQ_8 when R has no nonzero nilpotent elements.

Theorem 2.11. *Let R be a commutative ring of characteristic 0. Assume that $R_2 = 0$ and R has no nonzero nilpotent elements. Then the following statements are equivalent.*

(1) *RQ_8 is reversible.*
(2) *The equation $x^2 + y^2 + z^2 = 0$ has no nonzero solutions in R.*
(3) *RQ_8 has no nonzero nilpotent elements.*

The following example given in [8] shows that even when R has nonzero nilpotent elements, it is still possible that RQ_8 is reversible.

Example 2.12. *Let $R = \left\{ \begin{bmatrix} x & y \\ 0 & x \end{bmatrix} \middle| x, y \in \mathbb{Q} \right\}$, where $\mathbb{Q}$ is the field of rational numbers. Then R is a commutative ring of characteristic 0 with nonzero nilpotent elements, and RQ_8 is reversible.*

We close this subsection by proposing a few research problems.

Problem 2.13. *Let R be a commutative ring of characteristic* 0 *(not necessarily without nonzero nilpotent elements) and $R_2 = 0$. Find a necessary and sufficient condition such that RQ_8 is reversible.*

After solving the above problem, one may attack the following.

Problem 2.14. *Let R be a commutative ring of characteristic* 0 *(not necessarily without nonzero nilpotent elements) and $R_2 = 0$. Find a necessary and sufficient condition such that $R(Q_8 \times E_2 \times E_2')$ is reversible.*

When R is non-commutative, very little is known about reversibility of RG. Perhaps, one may first study the following question.

Question 2.15. *Let R be a non-commutative division ring of characteristic* 0 *and $G = Q_8$ or C_n. When is RG reversible?*

2.3. Minimal reversible group rings

In [10], Marks asks whether $\mathbb{Z}_2Q_8$ is the smallest ring which is reversible but not symmetric. In [5] Gutan and Kisielewicz asserted that it is the minimal group ring over a field with this property. Using an argument on the orders of group rings, one can prove that this is the case when group rings over commutative rings are considered. The following theorem proved in [7] confirms that $\mathbb{Z}_2Q_8$ is indeed the smallest group ring with this property, thereby providing a partial answer to the question raised by Marks.

Theorem 2.16. *$\mathbb{Z}_2Q_8$ is the smallest reversible group ring which is not symmetric.*

To see this, we only need to show that every reversible group ring RG having $|RG| \leq |\mathbb{Z}_2Q_8| = 256$ is symmetric except for $RG = \mathbb{Z}_2Q_8$. If RG is reversible, then R is reversible and G is either abelian or Hamiltonian; and since Q_8 is the smallest Hamiltonian group, $\mathbb{Z}_2Q_8$ is the minimal reversible non-symmetric group ring with G Hamiltonian. Thus, we may suppose that G is abelian, R is reversible but not commutative, and $|RG| \leq 256$. If $|G| \geq 3$, then $|R| < 7$; therefore, R is commutative, hence RG is both reversible and symmetric. Thus we may assume that $G = C_2$. If $|RC_2| < 256$, then $|R| \leq 15$ and thus R is either commutative or non-reversible. Hence, we need only to consider the case RC_2 with R non-commutative and reversible, and $|R| = 16$. It was proved in [7] that there is a unique non-commutative reversible ring R_0 with 1 of order 16. Moreover, R_0C_2 is not reversible, so Theorem 2.16 follows.

3. Duo group rings

An associative ring R is called left (right) duo if every left (right) ideal is an ideal, and R is said to be duo if it is both left and right duo. Say that R has the "SI" property if $\alpha\beta = 0$ implies $\alpha R\beta = \{0\}$ for all $\alpha, \beta \in R$.

Let R be a commutative ring with identity and G be any group. Using the standard involution $*$ on the group ring RG, defined by $(\sum a_i g_i)^* = \sum a_i g_i^{-1}$ for all $a_i \in R$ and $g_i \in G$, we see that RG is left duo if and only if it is right duo.

Marks [10] has clarified the relationships among duo, reversible and symmetric rings. Moreover, he proved the following result.

Proposition 3.1. *Let R be a commutative ring with identity, and let G be a finite group. Then the group ring RG is reversible if and only if RG has the "SI" property.*

It was pointed out in [2] that this result remains valid for an arbitrary group G. Since the "SI" property is simply the statement that left annihilators and right annihilators are ideals, it is obvious that duo rings have the "SI" property. It now follows from Proposition 3.1 that if RG is a duo ring, then it is reversible. However, the converse is not true, as the following example shows.

Example 3.2. *Let $Q_8 = \langle a, b | a^4 = 1, a^2 = b^2, a^b = a^{-1} \rangle$ be the quaternion group of order 8. The integral group ring $\mathbb{Z}Q_8$ is a reversible ring, but not a duo ring.*

It follows from Theorem 2.1 that the rational group algebra $\mathbb{Q}Q_8$ is reversible. As a subring of the rational algebra, clearly, the integral group ring $\mathbb{Z}Q_8$ is reversible.

To show that $\mathbb{Z}Q_8$ is not a duo ring, one needs only to verify that the left ideal $R(a + 2b)$ generated by $a + 2b$ is not a right ideal.

Remark 3.3. *$\mathbb{Z}Q_8$ is, in fact, symmetric; hence this example shows that "symmetric" does not imply "duo".*

3.1. Duo group algebras

As mentioned earlier, if a group ring RG over a commutative ring is duo, then it is reversible. All reversible group rings of torsion groups over fields were characterized by Gutan and Kisielewicz (Theorem 2.8). A natural question which arises is whether a reversible group algebra KG is also duo. An affirmative answer was given by Bell and the Li in [2]. The following result proved in [2] characterizes when a group algebra KQ_8 is duo.

Theorem 3.4. *The following statements are equivalent:*

(1) *KQ_8 is duo.*
(2) *The equation $1 + x^2 + y^2 = 0$ has no solutions in K when $\text{char}(K) \neq 2$, or the equation $1 + x + x^2 = 0$ has no solutions in K when $\text{char}(K) = 2$.*
(3) *KQ_8 is reversible.*

The above theorem together with a result of Perlis and Walker ([12, Prop. II.2.6]) gives a characterization of when a group algebra is duo.

Theorem 3.5. *Let K be a field and let G be a torsion group. Then KG is a duo ring if and only if one of the following conditions holds.*

(1) *G is abelian.*

(2) $G = Q_8 \times E_2 \times E_2'$ *is Hamiltonian, the characteristic of* K *is* 0 *and the equation* $1 + x^2 + y^2 = 0$ *has no solutions in any cyclotomic field* $K(\xi_d)$ *for any odd* d *which is an order of an element of* E_2'.

(3) $G = Q_8 \times E_2'$, *the characteristic of* K *is* 2 *and the equation* $1 + x + x^2 = 0$ *has no solutions in any cyclotomic field* $K(\xi_d)$ *for any odd* d *which is an order of an element of* E_2'.

As a consequence, we have the following:

Corollary 3.6. *Let* K *a field and let* G *be a torsion group. Then* KG *is duo if and only if* KG *is reversible.*

Remark 3.7. *It was brought to our attention recently that some theorems equivalent to Theorem* 2.3 *and Corollary* 3.6 *were proved by Menal* [11] *with different methods and different terminology.*

3.2. Duo group rings over integral domains

We now deal with the question of when a group ring RG is duo, where R is an integral domain and G is a non-abelian torsion group. Note that if RG is duo, then RG is reversible and thus $G = Q_8 \times E_2 \times E_2'$ is a Hamiltonian group. Therefore, as a homomorphic image of a duo ring $RG = (RQ_8)(E_2 \times E_2')$, RQ_8 is duo. Thus determining when RG is duo essentially reduces to determining when RQ_8 is duo.

As mentioned earlier, the integral group ring $\mathbb{Z}Q_8$ is a reversible ring but not a duo ring, while $\mathbb{Q}Q_8$ is a duo ring. A natural question which arises is as follows:

Question 3.8. *Is there any ring* R *with identity between* $\mathbb{Z}$ *and* $\mathbb{Q}$ *(excluding* $\mathbb{Q}$*), such that* RQ_8 *is duo?*

We also propose the following general question.

Question 3.9. *Let* R *be a integral domain and* G *be a non-abelian torsion group. When is* RG *duo?*

Note that if RQ_8 is duo, then it is reversible, so either $\mathrm{char}(R) = 2$ or $\mathrm{char}(R) = 0$. In the latter case, it follows from Theorem 2.11 that for all $x, y \in R$, $1 + x^2 + y^2 \neq 0$. Moreover, the following result due to Gao and Li [4] shows that $1 + x^2 + y^2$ is, in fact, invertible in R, giving a necessary condition for RQ_8 to be duo.

Lemma 3.10. *Let* R *be an integral domain such that* RQ_8 *is duo. If* $1 + x^2 + y^2 \neq 0$ *for some* $x, y \in R$, *then* $1 + x^2 + y^2$ *is invertible in* R. *Moreover, either* $\mathrm{char}(R) = 2$ *or* $\mathrm{char}(R) = 0$. *In the latter case,* $1 + x^2 + y^2 \in \mathcal{U}(R)$ *for all* $x, y \in R$.

Using this lemma, Gao and Li were able to prove the following main result, providing a negative answer to Question 3.8.

Theorem 3.11. *Let* R *be an integral domain such that* RQ_8 *is duo. Then the following statements hold.*

(1) *If* $\text{char}(R) \neq 0$, *then* R *must be a field.*
(2) *If* S *is a ring of algebraic integers with quotient field* K_S *and* $S \subseteq R \subseteq K_S$, *then* $R = K_S$. *In particular, if* $\mathbb{Z} \subseteq R \subseteq \mathbb{Q}$, *then* $R = \mathbb{Q}$.

In view of this theorem, one might conjecture that if RQ_8 is duo, then R is a field. However, the following proposition shows that this is not the case.

Proposition 3.12. *Let* $S = \mathbb{Q}[x]$ *be the polynomial ring over the rational field, and* S_P *be the localization of* S *at the maximal ideal* $P = \langle x \rangle$. *Then* $R = S_P$ *is a local integral domain of characteristic* 0, *but not a field, such that* RQ_8 *is duo.*

Remark 3.13. *We note that the ring* R *in Proposition* 3.12 *is a principal local integral domain such that* RQ_8 *is duo. However, for any prime* p, *the localization* $\mathbb{Z}_{(p)}$ *of* $\mathbb{Z}$ *at the ideal generated by* p *is a principal local integral domain, but* $\mathbb{Z}_{(p)}Q_8$ *is not duo.*

Note that Theorem 3.11 together with Theorem 2.3 provides a complete answer to Question 3.9 when $\text{char}(R) = 2$ and a partial answer when $\text{char}(R) = 0$.

Theorem 3.14. *If* R *is an integral domain with* $\text{char}(R) \neq 0$ *and* G *is a non-abelian torsion group, then the following statements are equivalent:*

(1) RG *is duo.*
(2) R *is a field and* RG *is reversible.*
(3) $G = Q_8 \times E_2'$, $R = K$ *is a field of characteristic* 2 *and the equation* $1+x+x^2 = 0$ *has no solutions in any cyclotomic field* $K(\xi_d)$ *for any odd* d *which is an order of an element of* E_2'.

Theorem 3.15. *If* R *is an integral domain with* $\text{char}(R) = 0$ *such that* $S \subseteq R \subseteq K_S$, *where* S *is a ring of algebraic integers, and* G *is a non-abelian torsion group, then the following statements are equivalent:*

(1) RG *is duo.*
(2) R *is a field and* RG *is reversible.*
(3) $G = Q_8 \times E_2 \times E_2'$, $R = K$ *is a field of characteristic* 0 *and the equation* $1 + x^2 + y^2 = 0$ *has no solutions in any cyclotomic field* $K(\xi_d)$ *for any odd* d *which is an order of an element of* E_2'.

We note that if $\text{char}(R) = 0$, a necessary condition for RQ_8 to be duo is given in Lemma 3.10, i.e., $1 + x^2 + y^2 \in \mathcal{U}(R)$ for all $x, y \in R$. We are not aware of any example of an integral domain R with $\text{char}(R) = 0$ satisfying this necessary condition for which RQ_8 is not duo. We close this subsection by proposing the following question.

Question 3.16. *Assume that* R *is an integral domain with* $\text{char}(R) = 0$ *such that* $1 + x^2 + y^2 \in \mathcal{U}(R)$ *for all* $x, y \in R$. *Is* RQ_8 *duo?*

4. Graded reversibility in integral group rings

This section deals with graded reversibility in integral group rings and presents the results obtained in [9]. Let R be an S-algebra graded by a group A. Call R graded reversible with respect to the grading if $ab = 0$ implies $ba = 0$, where a, b are homogeneous elements of R. In the following we will be interested in graded reversibility when $R = \mathbb{Z}G$ (viewed as a $\mathbb{Z}$-algebra) in the important special case where $A = C_2$, the cyclic group of order 2. To be very specific, this means $\mathbb{Z}G = R_0 \bigoplus R_1$, where R_0, R_1 are subgroups of $(\mathbb{Z}G, +)$ satisfying $R_0R_0 \subseteq R_0, R_0R_1 \subseteq R_1, R_1R_0 \subseteq R_1, R_1R_1 \subseteq R_0$ and the reversibility condition applies to elements a, b where $a \in R_i$ and $b \in R_j$ for some i, j.

If G has a subgroup H of index 2 and $g \in G - H$, then a C_2-grading of $\mathbb{Z}G$ can be given as follows:

$$\mathbb{Z}G = \mathbb{Z}H \oplus (\mathbb{Z}H)g.$$

Note that any automorphism α of $\mathbb{Z}G$ gives another C_2-grading since $\mathbb{Z}G = \alpha(\mathbb{Z}H) \oplus \alpha((\mathbb{Z}H)g)$. It is an open question as to whether this method gives all C_2-gradings of $\mathbb{Z}G$ (see [1] for further information about this problem). We will focus exclusively on gradings of the type $\mathbb{Z}G = \mathbb{Z}H \oplus (\mathbb{Z}H)g$ and try to determine when $\mathbb{Z}G$ is graded reversible. While the reversibility of integral group rings $\mathbb{Z}G$ is completely determined ([9, Theorem 1.1]), very little is known about graded reversibility. The following result due to Li and Parmenter gives a necessary and sufficient condition for $\mathbb{Z}G$ to be graded reversible.

Proposition 4.1. *Assume that $\mathbb{Z}G = \mathbb{Z}H \oplus (\mathbb{Z}H)g$ is a C_2-grading. Then $\mathbb{Z}G$ is graded reversible if and only if both of the following hold.*

(i) *$\mathbb{Z}H$ is reversible.*
(ii) *Whenever $\alpha_1, \alpha_2 \in \mathbb{Z}H$ satisfy $\alpha_1\alpha_2 = 0$, then $\alpha_2\alpha_1^g = 0$ (where $\alpha_1^g = g\alpha_1 g^{-1}$).*

Corollary 4.2. *Assume $\mathbb{Z}G = \mathbb{Z}H \oplus (\mathbb{Z}H)g$ is graded reversible. Then every finite subgroup of H is normal in G.*

It seems to be an open question as to whether the conclusion of Corollary 4.2 can actually replace condition (ii) in Proposition 4.1. In the special case where H is abelian this would say that $\mathbb{Z}G$ is graded reversible if and only if every finite subgroup of H is normal in G. The latter condition is automatically satisfied when H is cyclic, and Li and Parmenter showed that $\mathbb{Z}G$ is indeed graded reversible in that case. We remark that cyclotomic polynomials play a very crucial role in proving the following main result (see [9, Lemma 2.3] for details).

Theorem 4.3. *Let $\langle a \rangle$ be a finite cyclic group of order n and let s be a positive integer such that $(s, n) = 1$. For $\alpha \in \mathbb{Z}\langle a \rangle$, let α^f denote the image of α under the automorphism of $\mathbb{Z}\langle a \rangle$ which maps a to a^s. If $\alpha_1\alpha_2 = 0$ in $\mathbb{Z}\langle a \rangle$, then $\alpha_1^f\alpha_2 = 0$.*

It follows immediately from the above theorem that condition (ii) of Proposition 4.1 is satisfied when H is finite cyclic. Since $\mathbb{Z}H$ has no zero divisors when H

is infinite cyclic, condition (ii) is also satisfied. Therefore, $\mathbb{Z}G$ is graded reversible whenever H is cyclic. Note that if $\mathbb{Z}G$ is reversible, clearly it is graded reversible; however the converse is not true. The above observation provides many examples of groups G where $\mathbb{Z}G$ is not reversible but can be made graded reversible over C_2 (e.g., all meta-cyclic groups G having a cyclic normal subgroup H of index 2, including dihedral groups).

References

[1] Y. Bathurin and M.M. Parmenter, *Group gradings on integral group rings*, Groups, Rings and Group Rings. Lecture Notes in Pure and Applied Mathematics, **248** (2006), 25–32,. Chapman & Hall Boca Raton.

[2] H.E. Bell and Y. Li, *On duo group rings*, J. Pure Appl. Algebra **209** (2007), 833–838.

[3] M.P. Cohn, *Reversible rings*, Bull. London Math. Soc. **31** (1999), 641–648.

[4] W. Gao and Y. Li, *On duo group rings*, Algebra Colloq. In press, 2008.

[5] M. Gutan and A. Kisielewicz, *Reversible group rings*, J. Algebra **279** (2004), 280–291.

[6] M. Gutan and A. Kisielewicz, *Rings and semigroups with permutable zero products*, J. Pure Appl. Algebra **206** (2006), 355–369.

[7] Y. Li, H.E. Bell and C. Phipps, *On reversible group rings*, Bull. Austral. Math. Soc. **74** (2006), 139–142.

[8] Y. Li and M.M. Parmenter, *Reversible group rings over commutative rings*, Comm. Algebra **35** (2007), 4096–4104.

[9] Y. Li and M.M. Parmenter, *Graded reversibility in integral group rings*, Acta Appl. Math. **108** (2009), 129–133.

[10] G. Marks, *Reversible and symmetric rings*, J. Pure Appl. Algebra **174** (2002), 311–318.

[11] P. Menal, *Group rings in which every left ideal is a right ideal*, Proc. Amer. Math. Soc. **76** (1979) 204–208.

[12] S.K. Sehgal, *Topics in Group Rings*, Marcel Dekker, New York, 1978.

Howard E. Bell and Yuanlin Li
Department of Mathematics
Brock University
St. Catharines, Ontario, Canada L2S 3A1
e-mail: hbell@brocku.ca
yli@brocku.ca

Advances in Ring Theory
Trends in Mathematics, 47–61

Principally Quasi-Baer Ring Hulls

Gary F. Birkenmeier, Jae Keol Park and S. Tariq Rizvi

Dedicated to Professor S.K. Jain on his seventieth birthday

Abstract. We show the existence of principally (and finitely generated) right FI-extending right ring hulls for semiprime rings. From this result, we prove that right principally quasi-Baer (i.e., right p.q.-Baer) right ring hulls always exist for semiprime rings. This existence of right p.q.-Baer right ring hull for a semiprime ring unifies the result by Burgess and Raphael on the existence of a closely related unique smallest overring for a von Neumann regular ring with bounded index and the result of Dobbs and Picavet showing the existence of a weak Baer envelope for a commutative semiprime ring. As applications, we illustrate the transference of certain properties between a semiprime ring and its right p.q.-Baer right ring hull, and we explicitly describe a structure theorem for the right p.q.-Baer right ring hull of a semiprime ring with only finitely many minimal prime ideals. The existence of PP right ring hulls for reduced rings is also obtained. Further application to ring extensions such as monoid rings, matrix, and triangular matrix rings are investigated. Moreover, examples and counterexamples are provided.

Mathematics Subject Classification (2000). Primary 16N60; Secondary 16S20, 16P70.

Keywords. FI-extending, right ring hulls, right rings of quotients, p.q.-Baer rings, quasi-Baer rings.

Throughout all rings are associative rings with unity. Ideals without the adjectives "right" or "left" mean two-sided ideals.

In this paper, we prove the existence of principally (and finitely generated) right FI-extending right ring hulls for semiprime rings by using the concepts of distinguished extending classes (or $\mathfrak{D}$-$\mathfrak{E}$ classes), pseudo right ring hulls, and techniques studied in [12]. From this result, we obtain the existence of right p.q.-Baer right ring hulls for semiprime rings. Thereby, the existence of right p.q.-Baer right ring hulls for semiprime rings unifies the results on the existence of a closely related unique smallest overring for a von Neumann regular ring with bounded index by

Burgess and Raphael [16], and that of the weak Baer envelope for a commutative semiprime ring by Dobbs and Picavet [18]. As applications, (i) we investigate the transference of properties between a semiprime ring and its right p.q.-Baer right ring hull; (ii) a structure theorem for the right p.q.-Baer right ring hull of a semiprime ring with only finitely many minimal prime ideals is described; (iii) we establish the existence of PP right ring hulls for reduced rings; and (iii) the existence of right p.q.-Baer right ring hulls of ring extensions such as monoid rings, matrix, and triangular matrix rings are studied. Furthermore, examples and counterexamples are provided.

Recall from [9] that a ring R is called *right p.q.-Baer* (i.e., right principally quasi-Baer) if the right annihilator of a principal ideal of R is generated by an idempotent as a right ideal. Equivalently, R is right p.q.-Baer if R modulo the right annihilator of each principal right ideal is projective. We let $\mathfrak{pq}\mathfrak{B}$ denote the class of right p.q.-Baer rings. Similarly, left p.q.-Baer rings can be defined. If a ring R is both right and left p.q.-Baer, then we say that R is *p.q.-Baer.* A ring R is called *right PP* if the right annihilator of every singleton subset of R is generated by an idempotent as a right ideal. Note that the definition of a right PP ring is equivalent to every principal right ideal of R being projective (these rings are also called right *Rickart* rings). A ring R is called PP if R is both right and left PP.

Recall from [4] that a ring R is called *quasi-Baer* if the right annihilator of every right ideal is generated by an idempotent (see [4], [5], [6], and [8] for more details on quasi-Baer rings). The class of p.q.-Baer rings includes biregular rings, quasi-Baer rings and abelian (i.e., every idempotent is central) PP rings. Also recall that a ring R is called *right (FI)-extending* if every right ideal (ideal) is essential as a right R-module in an idempotent generated right ideal of R. We let $\mathfrak{E}$ and $\mathfrak{F}\mathfrak{I}$ denote the class of right extending rings and that of right FI-extending rings, respectively.

We say that a ring R is *principally right FI-extending* (resp., *finitely generated right FI-extending*) if every principal ideal (resp., finitely generated ideal) of R is essential as a right R-module in a right ideal of R generated by an idempotent. We use $\mathfrak{p}\mathfrak{F}\mathfrak{I}$ (resp., $\mathfrak{fg}\mathfrak{F}\mathfrak{I}$) to denote the class of principally (resp., finitely generated) right FI-extending rings.

An overring S of a ring R is said to be a *right ring of quotients* (resp., *right essential overring*) of R if R_R is dense (resp., essential) in S_R. Thus every right ring of quotients of R is a right essential overring of R.

For a right R-module M_R, we use $N_R \leq M_R$, $N_R \trianglelefteq M_R$, $N_R \leq^{\text{ess}} M_R$, and $N_R \leq^{\text{den}} M_R$ to denote that N_R is a submodule of M_R, N_R is a fully invariant submodule of M_R, N_R is an essential submodule of M_R, and N_R is a dense (or rational) submodule of M_R, respectively. We use $\mathbf{I}(R)$, $\mathbf{B}(R)$, $\text{Cen}(R)$, $\text{Mat}_n(R)$, and $T_n(R)$ to denote the set of all idempotents of R, the set of all central idempotents of R, the center of R, the n-by-n matrix ring over R, and the n-by-n upper triangular matrix ring over R, respectively. For a nonempty subset Y of a ring R,

$\langle Y\rangle_R$, $\ell_R(Y)$, and $r_R(Y)$ denote the subring of R generated by Y, the left annihilator of Y in R, and the right annihilator of Y in R, respectively. The notion $I \trianglelefteq R$ means that I is an ideal of a ring R.

We let $Q(R)$, $E(R_R)$, and $\mathcal{E}_R$ denote the maximal right ring of quotients of R, the injective hull of R_R, and the endomorphism ring $\text{End}(E(R_R)_R)$, respectively. Let $\mathcal{Q}_R = \text{End}({}_{\mathcal{E}_R}E(R_R))$. Note that $Q(R) = 1 \cdot \mathcal{Q}_R$ (i.e., the canonical image of $\mathcal{Q}_R$ in $E(R_R)$) and that $\mathbf{B}(\mathcal{Q}_R) = \mathbf{B}(\mathcal{E}_R)$ [21, pp. 94–96]. Also, $\mathbf{B}(Q(R)) = \{b(1) \mid b \in \mathbf{B}(\mathcal{Q}_R)\}$ [20, p.366]. Thus $R\mathbf{B}(\mathcal{E}_R) = R\mathbf{B}(Q(R))$, the subring of $Q(R)$ generated by R and $\mathbf{B}(Q(R))$. If R is semiprime, then $\text{Cen}(Q(R)) = \text{Cen}(Q^m(R))$ [20, pp. 389–390], where $Q^m(R)$ is the Martindale right ring of quotients of R.

Proposition 1.

(i) ([5, Proposition 1.8] *and* [9, Proposition 1.12]) *The center of a quasi-Baer (resp., right p.q.-Baer) ring is Baer (resp., PP).*

(ii) ([9, Proposition 3.11]) *Assume that a ring R is semiprime. Then R is quasi-Baer if and only if R is p.q.-Baer and the center of R is Baer.*

(iii) ([26, pp. 78–79] *and* [5, Theorem 3.5]) *Let a ring R be von Neumann regular (resp., biregular). Then R is Baer (resp., quasi-Baer) if and only if the lattice of principal right ideals (resp., principal ideals) is complete.*

(iv) *A ring R is biregular if and only if R is right (or left) p.q.-Baer ring and $r_R(\ell_R(RaR)) = RaR$, for all $a \in R$.*

Proof. The proof of part (iv) is straightforward. □

Let R be a ring and $e = e^2 \in R$. Recall from [3] that e is called *left* (resp., *right*) *semicentral* if $exe = xe$ (resp., $exe = ex$) for every $x \in R$. Note that $e = e^2 \in R$ is left (resp., right) semicentral if and only if eR (resp., Re) is an ideal of R. We use $\mathbf{S}_\ell(R)$ (resp., $\mathbf{S}_r(R)$) to denote the set of all left (resp., right) semicentral idempotents of R. See [7, Propositions 1.1 and 1.3] for more details on left (or right) semicentral idempotents.

Proposition 2.

(i) *Let R be a ring, K_i an ideal of R, and $e_i \in \mathbf{S}_\ell(R)$ such that $K_{iR} \leq^{ess} e_iR_R$ for $i = 1, 2, \ldots, n$. Then there exists $g \in \mathbf{S}_\ell(R)$ such that $(\sum_{i=1}^{n} K_i)_R \leq^{ess} gR_R$.*

(ii) *Let R be a right nonsingular ring. Then R is principally right FI-extending if and only if R is finitely generated right FI-extending.*

Proof. (i) We will first prove the result for $n = 2$. Let $A = K_1$, $B = K_2$, $e = e_1$, and $f = e_2$. Then $A_R \leq^{\text{ess}} eR_R$, $B_R \leq^{\text{ess}} fR_R$, and $e, f \in \mathbf{S}_\ell(R)$. Since $A+B$ is an ideal of R, we have that $A+B = [(A+B)\cap eR] \oplus [(A+B)\cap(1-e)R]$. Note that $(A+B)\cap(1-e)R = B\cap(1-e)R$. Thus $A+B = [(A+B)\cap eR]\oplus[B\cap(1-e)R]$. Now $[(A+B)\cap eR]_R \leq^{\text{ess}} eR_R$. Also $[B\cap(1-e)R]_R \leq^{\text{ess}} fR_R\cap(1-e)R_R = (1-e)fR_R$ because $B_R \leq^{\text{ess}} fR_R$ and $fR\cap(1-e)R = (1-e)fR$. So $(A+B)_R \leq^{\text{ess}} (eR + (1-e)fR)_R = (e+f-ef)R_R$. In this case, we see that $e+f-ef \in \mathbf{S}_\ell(R)$. Now an induction argument can be used to complete the proof.

(ii) This part follows from part (i) and [10, Proposition 1.10]. □

We include the following result from [9], for the convenience of the reader, which shows the connections between the right p.q.-Baer condition and some "finitely generated" right FI-extending conditions for semiprime rings.

Lemma 3. ([9, Corollary 1.11]) *Let R be a semiprime ring. Then the following conditions are equivalent.*

(i) *R is right p.q.-Baer.*
(ii) *R is principally right FI-extending.*
(iii) *R is finitely generated right FI-extending.*

Definition 4. (cf. [12, Definition 2.1]) Let $\mathfrak{K}$ denote a class of rings. For a ring R, $\widehat{Q}_{\mathfrak{K}}(R)$ denotes the smallest right ring of quotients of R which is in $\mathfrak{K}$. Further, let $Q_{\mathfrak{K}}(R)$ be the smallest right essential overring of R which is in $\mathfrak{K}$. We say that $Q_{\mathfrak{K}}(R)$ is the *absolute $\mathfrak{K}$ right ring hull* of R. Note that if $Q(R) = E(R_R)$, then $\widehat{Q}_{\mathfrak{K}}(R) = Q_{\mathfrak{K}}(R)$. In this paper, we call $\widehat{Q}_{\mathfrak{K}}(R)$ the *$\mathfrak{K}$ right ring hull* of R.

Since our interest is primarily in classes of rings which are defined by properties on the set of right ideals of the rings in the classes, we recall the following definition.

Definition 5. ([12, Definition 1.6]) Let $\mathfrak{R}$ be a class of rings, $\mathfrak{K}$ a subclass of $\mathfrak{R}$, and $\mathfrak{X}$ a class containing all subsets of every ring. We say that $\mathfrak{K}$ is a class *determined by a property on right ideals* if there exist an assignment $\mathfrak{D}_{\mathfrak{K}} : \mathfrak{R} \to \mathfrak{X}$ such that $\mathfrak{D}_{\mathfrak{K}}(R) \subseteq \{\text{right ideals of } R\}$ and a property P such that each element of $\mathfrak{D}_{\mathfrak{K}}(R)$ has P if and only if $R \in \mathfrak{K}$.

If $\mathfrak{K}$ is a class determined by the particular property P such that a right ideal is essential in an idempotent generated right ideal, then we say that $\mathfrak{K}$ is a *$\mathfrak{D}$-$\mathfrak{E}$ class* and use $\mathfrak{C}$ to designate a $\mathfrak{D}$-$\mathfrak{E}$ class. Note that every $\mathfrak{D}$-$\mathfrak{E}$ class contains the class $\mathfrak{E}$ of right extending (hence right self-injective) rings. Recall from [10] that a ring R is right FI-extending if every ideal is essential in an idempotent generated right ideal. Thus the class $\mathfrak{FI}$ of right FI-extending rings is a $\mathfrak{D}$-$\mathfrak{E}$ class. Furthermore, from their definitions, we see that $\mathfrak{pFI}$ and $\mathfrak{fgFI}$ are $\mathfrak{D}$-$\mathfrak{E}$ classes.

Some examples illustrating Definition 5 are (see [12]):

(1) $\mathfrak{K}$ is the class of right Noetherian rings, $\mathfrak{D}_{\mathfrak{K}}(R) = \{\text{right ideals of } R\}$, and P is the property that a right ideal is finitely generated.
(2) $\mathfrak{K}$ is the class of von Neumann regular rings, $\mathfrak{D}_{\mathfrak{K}}(R) = \{\text{principal right ideals of } R\}$, and P is the property that a right ideal is generated by an idempotent.
(3) $\mathfrak{K} = \mathfrak{pqB}, \mathfrak{D}_{\mathfrak{pqB}}(R) = \{r_R(xR) \mid x \in R\}$, and P is the property that a right ideal is generated by an idempotent.
(4) $\mathfrak{C} = \mathfrak{E}$ (resp., $\mathfrak{C} = \mathfrak{FI}$), $\mathfrak{D}_{\mathfrak{E}}(R) = \{I \mid I_R \leq R_R\}$ (resp., $\mathfrak{D}_{\mathfrak{FI}}(R) = \{I \mid I \trianglelefteq R\}$). (Recall that $\mathfrak{E}$ is the class of right extending rings and $\mathfrak{FI}$ is the class of right FI-extending rings.)
(5) $\mathfrak{C} = \mathfrak{pFI}$, $\mathfrak{D}_{\mathfrak{pFI}}(R) = \{\text{principal ideal of } R\}$.
(6) $\mathfrak{C} = \mathfrak{fgFI}$, $\mathfrak{D}_{\mathfrak{fgFI}}(R) = \{\text{finitely generated ideal of } R\}$.

Next, we consider generating a right essential overring in a class $\mathfrak{K}$ from a base ring R and some subset of $\mathcal{E}_R$. By using equivalence relations, in [12] we reduce the size of the subsets of $\mathcal{E}_R$ needed to generate a right essential overring of R in a $\mathfrak{D}$-$\mathfrak{E}$ class of rings $\mathfrak{C}$. Also in [12], to develop the theory of pseudo right ring hulls for $\mathfrak{D}$-$\mathfrak{E}$ classes $\mathfrak{C}$, we fix $\mathfrak{D}_{\mathfrak{C}}(R)$ for each ring R and define

$$\delta_{\mathfrak{C}}(R) = \{e \in \mathbf{I}(\mathcal{E}_R) \mid V_R \leq^{\text{ess}} eE(R_R) \text{ for some } V \in \mathfrak{D}_{\mathfrak{C}}(R)\}.$$

We set $\delta_{\mathfrak{C}}(R)(1) = \{e(1) \mid e \in \delta_{\mathfrak{C}}(R)\}$.

Definition 6. (cf. [12, Definition 2.2]) Let S be a right essential overring of R. If $\delta_{\mathfrak{C}}(R)(1) \subseteq S$ and $\langle R \cup \delta_{\mathfrak{C}}(R)(1)\rangle_S \in \mathfrak{C}$, then we call $\langle R \cup \delta_{\mathfrak{C}}(R)(1)\rangle_S$ the *pseudo right ring hull of R with respect to S* and denote it by $R(\mathfrak{C}, S)$. If $S = R(\mathfrak{C}, S)$, then we say that S is a $\mathfrak{C}$ *pseudo right ring hull of R*.

To find a right essential overring S of R such that $S \in \mathfrak{C}$, one might naturally look for a right essential overring T of R with $\delta_{\mathfrak{C}}(R)(1) \subseteq T$ and take $S = \langle R \cup \delta_{\mathfrak{C}}(R)(1)\rangle_T$. Indeed, under some mild conditions, this choice of S can be in $\mathfrak{C}$. However, in order to obtain a right essential overring with some hull-like behavior, we need to determine subsets Λ of $\delta_{\mathfrak{C}}(R)(1)$ for which $\langle R \cup \Lambda\rangle_T \in \mathfrak{C}$ in some minimal sense. Moreover, to facilitate the transfer of information between R and $\langle R \cup \Lambda\rangle_T$, one would want to include in Λ enough of $\delta_{\mathfrak{C}}(R)(1)$ so that for all (or almost all) $V \in \mathfrak{D}_{\mathfrak{C}}(R)$ there is $e \in \delta_{\mathfrak{C}}(R)$ with $V_R \leq^{\text{ess}} e(1) \cdot (\langle R \cup \Lambda\rangle_T)_R$ and $e(1) \in \Lambda$.

Lemma 7. *Let $\{e_1, \ldots, e_n\} \subseteq \mathbf{B}(T)$, where T is an overring of a ring R. Then there exists a set of orthogonal idempotents $\{f_1, \ldots, f_m\} \subseteq \mathbf{B}(T)$ such that $\sum_{i=1}^n e_iR \subseteq \sum_{i=1}^m f_iR$.*

Proof. The proof is similar to that of [23, Lemma 3.2]. □

For a semiprime ring R, the concepts of (right) FI-extending and quasi-Baer coincide by [10, Theorem 4.7]. Recall that the existence of the quasi-Baer right ring hull and that of right FI-extending right ring hull of a semiprime ring were shown in [14, Theorem 3.3]. It was also proved in [14, Theorem 3.3] that the quasi-Baer right ring hull is precisely the same as its right FI-extending right ring hull for a semiprime ring. In view of this result, it is natural to ask: *Do the right principally quasi-Baer right ring hull and the principally right FI-extending right ring hull exist for a semiprime ring and if they do, are they equal?* In our next result, we provide affirmative answers to these two questions.

Burgess and Raphael [16] study ring extensions of von Neumann regular rings with bounded index. In particular for a von Neumann regular ring R with bounded index, they obtain a closely related unique smallest overring, $R^{\#}$, which is "almost biregular" (see [16, p. 76 and Theorem 1.7]). The next result shows that their ring $R^{\#}$ is precisely our principally right FI-extending pseudo right ring hull of a von Neumann regular ring R with bounded index (see also [14, Theorem 3.8]). When R is a commutative semiprime ring, the "weak Baer envelope" defined in [18] is exactly the right p.q.-Baer right ring hull $\widehat{Q}_{\mathfrak{pq}\mathfrak{B}}(R)$.

Theorem 8. *Let R be a semiprime ring. Then we have the following.*

(i) $\langle R \cup \delta_{\mathfrak{p}\mathfrak{F}\mathfrak{I}}(R)(1)\rangle_{Q(R)} = \widehat{Q}_{\mathfrak{p}\mathfrak{F}\mathfrak{I}}(R) = R(\mathfrak{p}\mathfrak{F}\mathfrak{I},\, Q(R))$.

(ii) $\langle R \cup \delta_{\mathfrak{p}\mathfrak{F}\mathfrak{I}}(R)(1)\rangle_{Q(R)} = \widehat{Q}_{\mathfrak{p}\mathfrak{q}\mathfrak{B}}(R)$.

(iii) $\langle R \cup \delta_{\mathfrak{p}\mathfrak{F}\mathfrak{I}}(R)(1)\rangle_{Q(R)} = \widehat{Q}_{\mathfrak{f}\mathfrak{g}\mathfrak{F}\mathfrak{I}}(R) = R(\mathfrak{f}\mathfrak{g}\mathfrak{F}\mathfrak{I},\, Q(R))$.

Proof. (i) Let $\mathbf{B}_p(Q(R)) = \{c \in \mathbf{B}(Q(R)) \mid$ there exists $x \in R$ with $RxR_R \leq^{\text{ess}} cR_R\}$. We first *claim* that

$$\mathbf{B}_p(Q(R)) = \delta_{\mathfrak{p}\mathfrak{F}\mathfrak{I}}(R)(1).$$

For this claim, note that by [1, Theorem 7], $\delta_{\mathfrak{p}\mathfrak{F}\mathfrak{I}}(R) \subseteq \mathbf{B}(\mathcal{E}_R)$. Thus $\delta_{\mathfrak{p}\mathfrak{F}\mathfrak{I}}(R)(1) \subseteq \mathbf{B}(Q(R))$. To prove the claim, let $e(1) \in \delta_{\mathfrak{p}\mathfrak{F}\mathfrak{I}}(R)(1)$ with $e \in \delta_{\mathfrak{p}\mathfrak{F}\mathfrak{I}}(R)$. Then there exists $x \in R$ such that $RxR_R \leq^{\text{ess}} eE(R_R)$. Thus $RxR = eRxR = e(1)RxR \subseteq e(1)R = eR$. So $RxR_R \leq^{\text{ess}} eR_R = e(1)R_R$. Hence $e(1) \in \mathbf{B}_p(Q(R))$ because $e(1) \in \delta_{\mathfrak{p}\mathfrak{F}\mathfrak{I}}(R)(1) \subseteq \mathbf{B}(Q(R))$. Conversely, let $c \in \mathbf{B}_p(Q(R))$. Then there exists $b \in \mathbf{B}(\mathcal{E}_R)$ such that $c = b(1)$. Also there is $x \in R$ such that $RxR_R \leq^{\text{ess}} cR_R = b(1)R_R = bR_R$. Thus $RxR_R \leq^{\text{ess}} bE(R_R)$. So $b \in \delta_{\mathfrak{p}\mathfrak{F}\mathfrak{I}}(R)$. Hence $c = b(1) \in \delta_{\mathfrak{p}\mathfrak{F}\mathfrak{I}}(R)(1)$. Therefore $\mathbf{B}_p(Q(R)) = \delta_{\mathfrak{p}\mathfrak{F}\mathfrak{I}}(R)(1)$.

Let $S = \langle R \cup \delta_{\mathfrak{p}\mathfrak{F}\mathfrak{I}}(R)(1)\rangle_{Q(R)}$. Take $0 \neq s \in S$. From Lemma 7, $s = \sum r_i b_i$, where each $r_i \in R$ and the b_i are mutually orthogonal idempotents in $\mathbf{B}(S)$. There exists $c_i \in \delta_{\mathfrak{p}\mathfrak{F}\mathfrak{I}}(R)(1)$ such that $Rr_iR_R \leq^{\text{ess}} c_iR_R$ for each i. Hence $s = \sum r_i e_i$, where $e_i = b_i c_i$ for each i. Observe that the e_i are mutually orthogonal idempotents in $\mathbf{B}(S)$ since $c_i \in \delta_{\mathfrak{p}\mathfrak{F}\mathfrak{I}}(R)(1) = \mathbf{B}_p(Q(R))$ and $SsS \subseteq D = \bigoplus e_iS$. Now we *claim* that $SsS_S \leq^{\text{ess}} D_S$. Let $0 \neq y \in D$. There exist $y_i \in S$ such that $y = \sum e_i y_i$. In this case, there is $e_j y_j \neq 0$ for some j and $v \in R$ with $0 \neq e_j y_j v \in R$. Since $ye_jv = e_jy_jv = b_jc_jy_jv \in c_jR$ and $Rr_jR_R \leq^{\text{ess}} c_jR_R$, there exists $w \in R$ such that $0 \neq ye_jvw \in Rr_jR$. Hence $0 \neq e_jy_jvw \in Rr_je_jR = Rse_jR \subseteq SsS$ because $se_j = r_je_j$ and $e_j = b_jc_j \in S$. Since $e = \sum e_i \in \mathbf{B}(S)$ and $SsS_S \leq^{\text{ess}} D_S = \bigoplus e_iS_S = eS_S$, it follows that $S \in \mathfrak{p}\mathfrak{F}\mathfrak{I}$. Hence $S = R(\mathfrak{p}\mathfrak{F}\mathfrak{I},\, Q(R))$.

Next we assume that T is a right ring of quotients of R and $T \in \mathfrak{p}\mathfrak{F}\mathfrak{I}$. Take $e \in \delta_{\mathfrak{p}\mathfrak{F}\mathfrak{I}}(R)$. Then by the above claim, $e(1) \in \mathbf{B}_p(Q(R))$. So there is $x \in R$ such that $RxR_R \leq^{\text{ess}} e(1)R_R$. Hence $RxR_R \leq^{\text{ess}} e(1)Q(R)_R$. Note that $TxT = T(RxR)T \subseteq T(e(1)Q(R))T = e(1)Q(R)$. Thus $TxT_R \leq^{\text{ess}} e(1)Q(R)_R$, so $TxT_R \leq^{\text{ess}} e(1)Q(R)_R$. Hence $TxT_T \leq e(1)Q(R)_T$ from [12, Lemma 1.4(i)] because $R_R \leq^{\text{den}} T_R$. Therefore $TxT_T \leq^{\text{ess}} e(1)T_T$. On the other hand, since $T \in \mathfrak{p}\mathfrak{F}\mathfrak{I}$, there exists $c = c^2 \in T$ such that $TxT_T \leq^{\text{ess}} cT_T$. Thus $e(1) = c$ because $e(1) \in \mathbf{B}(Q(R))$. Hence $e(1) \in T$ for each $e(1) \in \delta_{\mathfrak{p}\mathfrak{F}\mathfrak{I}}(R)(1)$. So S is a subring of T. Therefore $S = \widehat{Q}_{\mathfrak{p}\mathfrak{F}\mathfrak{I}}(R)$.

(ii) It is a direct consequence of part (i) and Lemma 3.

(iii) As in the proof of part (i), we can verify that $\delta_{\mathfrak{f}\mathfrak{g}\mathfrak{F}\mathfrak{I}}(R)(1) = \{e \in \mathbf{B}(Q(R)) \mid$ there is a finitely generated ideal I of R with $I_R \leq^{\text{ess}} eR_R\}$. A proof similar to that used in part (i) yields that

$$\langle R \cup \delta_{\mathfrak{f}\mathfrak{g}\mathfrak{F}\mathfrak{I}}(R)(1)\rangle_{Q(R)} = R(\mathfrak{f}\mathfrak{g}\mathfrak{F}\mathfrak{I}, Q(R)) = \widehat{Q}_{\mathfrak{f}\mathfrak{g}\mathfrak{F}\mathfrak{I}}(R).$$

Since $\delta_{\mathfrak{p}\mathfrak{F}\mathfrak{I}}(R)(1) \subseteq \delta_{\mathfrak{f}\mathfrak{g}\mathfrak{F}\mathfrak{I}}(R)(1)$, $\widehat{Q}_{\mathfrak{p}\mathfrak{F}\mathfrak{I}}(R) \subseteq \widehat{Q}_{\mathfrak{f}\mathfrak{g}\mathfrak{F}\mathfrak{I}}(R)$. By Lemma 3, $\widehat{Q}_{\mathfrak{p}\mathfrak{F}\mathfrak{I}}(R) \in \mathfrak{f}\mathfrak{g}\mathfrak{F}\mathfrak{I}$, so $\widehat{Q}_{\mathfrak{f}\mathfrak{g}\mathfrak{F}\mathfrak{I}}(R) \subseteq \widehat{Q}_{\mathfrak{p}\mathfrak{F}\mathfrak{I}}(R)$. Thus $\widehat{Q}_{\mathfrak{f}\mathfrak{g}\mathfrak{F}\mathfrak{I}}(R) = \widehat{Q}_{\mathfrak{p}\mathfrak{F}\mathfrak{I}}(R)$. □

Recall that a ring R is *left π-regular* if for each $a \in R$ there exist $b \in R$ and a positive integer n such that $a^n = ba^{n+1}$. Note from [17] that the class of special radicals includes most well-known radicals (e.g., the prime radical, the Jacobson radical, the Brown-McCoy radical, the nil radical, the generalized nil radical, etc.). For a ring R, the *classical Krull dimension* $\mathrm{kdim}(R)$ is the supremum of all lengths of chains of prime ideals of R.

By Theorem 8, if R is a semiprime ring, then $\widehat{Q}_{\mathfrak{p}\mathfrak{q}\mathfrak{B}}(R) = R\mathbf{B}_p(Q(R))$, the subring of $Q(R)$ generated by R and $\mathbf{B}_p(Q(R))$. Thus we have the following corollaries which show the transference of certain properties between R and $\widehat{Q}_{\mathfrak{p}\mathfrak{q}\mathfrak{B}}(R)$. We use LO, GU, and INC for "lying over", "going up", and "incomparability", respectively (see [25, p. 292]).

Corollary 9. *Let R be a semiprime ring.*

(i) *If K is a prime ideal of $\widehat{Q}_{\mathfrak{p}\mathfrak{q}\mathfrak{B}}(R)$, then $\widehat{Q}_{\mathfrak{p}\mathfrak{q}\mathfrak{B}}(R)/K \cong R/(K \cap R)$.*
(ii) *LO, GU, and INC hold between R and $\widehat{Q}_{\mathfrak{p}\mathfrak{q}\mathfrak{B}}(R)$.*

Proof. The proof follows from Theorem 8 and [14, Lemma 2.1]. □

Corollary 10. *Assume that R is a semiprime ring. Then:*

(i) *$\varrho(R) = \varrho(\widehat{Q}_{\mathfrak{p}\mathfrak{q}\mathfrak{B}}(R)) \cap R$, where $\varrho(-)$ is a special radical of a ring.*
(ii) *R is left π-regular if and only if $\widehat{Q}_{\mathfrak{p}\mathfrak{q}\mathfrak{B}}(R)$ is left π-regular.*
(iii) *$kdim(R) = kdim(\widehat{Q}_{\mathfrak{p}\mathfrak{q}\mathfrak{B}}(R))$.*

Proof. Theorem 8 and [14, Theorem 2.2] yield this result. □

Corollary 11. *Let R be a semiprime ring. Then:*

(i) *R is von Neumann regular if and only if $Q_{\mathfrak{p}\mathfrak{q}\mathfrak{B}}(R)$ is von Neumann regular.*
(ii) *R is strongly regular if and only if $Q_{\mathfrak{p}\mathfrak{q}\mathfrak{B}}(R)$ is strongly regular.*
(iii) *R has bounded index at most n if and only if $Q_{\mathfrak{p}\mathfrak{q}\mathfrak{B}}(R)$ has bounded index at most n.*

Proof. This can be verified by Theorem 8 and similar arguments as used in the proof of [14, Corollary 3.6 and Theorem 3.8]. □

Let $\mathfrak{q}\mathfrak{B}$ be the class of quasi-Baer rings. In [14, Theorem 3.3], it is shown that there exist $\widehat{Q}_{\mathfrak{q}\mathfrak{B}}(R)$ and $\widehat{Q}_{\mathfrak{F}\mathfrak{I}}(R)$ for each semiprime ring R.

Theorem 12. (cf. [14, Theorem 3.3]) *Let R be a semiprime ring. Then $\widehat{Q}_{\mathfrak{F}\mathfrak{I}}(R) = R\mathbf{B}(Q(R)) = R(\mathfrak{F}\mathfrak{I}, Q(R))$.*

From Theorem 12 and [5, Theorem 3.5], one can see that for a semiprime ring R, $\widehat{Q}_{\mathfrak{q}\mathfrak{B}}(R)$ is the smallest right ring of quotients of R which is right p.q.-Baer and has a complete lattice of annihilator ideals. However, in general, $\widehat{Q}_{\mathfrak{p}\mathfrak{q}\mathfrak{B}}(R)$ is a proper subring of $\widehat{Q}_{\mathfrak{q}\mathfrak{B}}(R)$ as in the next example.

Example 13.

(i) Let F be a field and let $F_n = F$ for all positive integer n. Put

$$R = \left\{ (a_n)_{n=1}^{\infty} \in \prod_{n=1}^{\infty} F_n \mid a_n \text{ is eventually constant} \right\},$$

which is a subring of $\prod_{n=1}^{\infty} F_n$. Then $\widehat{Q}_{\mathfrak{pq}\mathfrak{B}}(R) = R$, but $\widehat{Q}_{\mathfrak{q}\mathfrak{B}}(R) = \prod_{n=1}^{\infty} F_n$.

(ii) Let R be a biregular ring (i.e., every principal ideal of R is generated by a central idempotent). Then $R = \widehat{Q}_{\mathfrak{pq}\mathfrak{B}}(R)$ and if its lattice of principal ideals is not complete then $R \neq \widehat{Q}_{\mathfrak{q}\mathfrak{B}}(R)$ (see [5, Theorem 3.5]). In fact, let $R = \{(d_n) \in \prod_{n=1}^{\infty} D_n \mid d_n \text{ is eventually constant}\}$, a subring of $\prod_{n=1}^{\infty} D_n$ where $D_n = D$ is a division ring for all n. Then R is biregular, so $R = \widehat{Q}_{\mathfrak{pq}\mathfrak{B}}(R)$, but $R \neq \widehat{Q}_{\mathfrak{q}\mathfrak{B}}(R)$ by Theorem 8 because $\mathbf{B}(Q(R)) \not\subseteq R$ or by [5, Theorem 3.5].

Despite Example 13, we have the following result in which $\widehat{Q}_{\mathfrak{pq}\mathfrak{B}}(R)$ does coincide with $\widehat{Q}_{\mathfrak{q}\mathfrak{B}}(R)$. Recall that the *extended centroid* of R is $\mathrm{Cen}(Q(R))$.

Theorem 14. *Assume that R is a semiprime ring with only finitely many minimal prime ideals, say $P_1, \ldots, P_n$. Then $\widehat{Q}_{\mathfrak{pq}\mathfrak{B}}(R) = \widehat{Q}_{\mathfrak{q}\mathfrak{B}}(R)$ and $\widehat{Q}_{\mathfrak{pq}\mathfrak{B}}(R) \cong R/P_1 \oplus \cdots \oplus R/P_n$.*

Proof. Since R has exactly n minimal prime ideals, the extended centroid $\mathrm{Cen}(Q(R))$ of R has a complete set of primitive idempotents with n elements by [1, Theorem 11]. Note that the extended centroid of R is equal to that of $\widehat{Q}_{\mathfrak{pq}\mathfrak{B}}(R)$. Thus $\widehat{Q}_{\mathfrak{pq}\mathfrak{B}}(R)$ also has exactly n minimal prime ideals by [1, Theorem 11]. By [11, Theorem 3.4] and [9, Theorem 3.7], $\widehat{Q}_{\mathfrak{pq}\mathfrak{B}}(R)$ is quasi-Baer and so $\widehat{Q}_{\mathfrak{pq}\mathfrak{B}}(R) = \widehat{Q}_{\mathfrak{q}\mathfrak{B}}(R)$. The rest of the proof follows from [13, Theorem 3.15]. □

Theorem 15. *Let R be a reduced ring. Then $Q_{\mathfrak{pq}\mathfrak{B}}(R)$ exists and is the PP absolute right ring hull of R.*

Proof. Note that since R is reduced, then $Q(R) = E(R_R)$; and so $\widehat{Q}_{\mathfrak{K}}(R) = Q_{\mathfrak{K}}(R)$ for any class $\mathfrak{K}$ of rings. By Theorem 8, $Q_{\mathfrak{p}\mathfrak{F}\mathfrak{I}}(R) = Q_{\mathfrak{pq}\mathfrak{B}}(R)$. Let $S = Q_{\mathfrak{p}\mathfrak{F}\mathfrak{I}}(R) = Q_{\mathfrak{pq}\mathfrak{B}}(R)$. From [9, Corollary 1.15], S is right (and left) PP.

Suppose A is a right ring of quotients of R which is right PP. Let $e \in \delta_{\mathfrak{p}\mathfrak{F}\mathfrak{I}}(R)(1)$. (Note that $\delta_{\mathfrak{p}\mathfrak{F}\mathfrak{I}}(R)(1) = \mathbf{B}_p(Q(R)) \subseteq \mathbf{B}(Q(R))$ as in the proof of Theorem 8.) Then there exists $x \in R$ such that $RxR_R \leq^{\text{ess}} eR_R$. So we have that $SxS_S \leq^{\text{ess}} eS_S$. Since S is semiprime and e is a central idempotent in S, it follows that $\ell_{eS}(SxS) = r_{eS}(SxS) = 0$ by noting that the ring S is semiprime. Therefore $r_S(SxS) = (1-e)S$. Moreover, since $Q_{\mathfrak{q}\mathfrak{B}}(R)$ is reduced by [14, Theorem 3.8], so is S ($\subseteq Q_{\mathfrak{q}\mathfrak{B}}(R)$). Thus $r_S(x) = r_S(SxS) = (1-e)S$. Since A is right PP, there exists $f \in \mathbf{I}(A)$ such that $r_A(x) = fA$. Then $r_R(x) = (1-e)S \cap R$

and $r_R(x) = r_A(x) \cap R$. Hence $r_R(x)_R \leq^{\text{ess}} (1-e)S_R \leq^{\text{ess}} (1-e)Q(R)_R$ and $r_R(x)_R \leq^{\text{ess}} fA_R \leq^{\text{ess}} fQ(R)_R$. Therefore

$$r_R(x)_R \leq^{\text{ess}} ((1-e)Q(R) \cap fQ(R))_R = f(1-e)Q(R)_R$$

because $1-e$ is central. Thus $(1-e)Q(R) = f(1-e)Q(R) = fQ(R)$, so $1-e = f$. Therefore $e = 1-f \in A$, hence $Q_{\mathfrak{pq}\mathfrak{B}}(R) = S \subseteq A$ by Theorem 8. □

Note that Theorem 15 shows that when R is a commutative semiprime ring, $Q_{\mathfrak{pq}\mathfrak{B}}(R)$ is related to the *Baer extension* considered in [19]. Also note that the generalized nil radical, $\mathbf{N}_g$ [17], is the radical whose semisimple class is the class of reduced rings. Hence for every ring R such that $R \neq \mathbf{N}_g(R)$, R has a nontrivial homomorphic image, $R/\mathbf{N}_g(R)$, which has a Baer absolute right ring hull and a right PP absolute right ring hull.

A monoid G is called a *u.p.-monoid* (unique product monoid) if for any two nonempty finite subsets A, $B \subseteq G$ there exists an element $x \in G$ uniquely presented in the form ab, where $a \in A$ and $b \in B$. The class of u.p.-monoids is quite large and important (see [24] and [22]). For example, this class includes the right or left ordered monoids, submonoids of a free group, and torsion-free nilpotent groups. Every u.p.-monoid is cancellative, and every u.p.-group is torsion-free.

Theorem 16. *Let $R[G]$ be a semiprime monoid ring of a monoid G over a ring R. Then:*

(i) $\widehat{Q}_{\mathfrak{pq}\mathfrak{B}}(R)[G] \subseteq \widehat{Q}_{\mathfrak{pq}\mathfrak{B}}(R[G])$.

(ii) *If G is a u.p.-monoid, then $\widehat{Q}_{\mathfrak{pq}\mathfrak{B}}(R[G]) = \widehat{Q}_{\mathfrak{pq}\mathfrak{B}}(R)[G]$.*

Proof. (i) To show that $\widehat{Q}_{\mathfrak{pq}\mathfrak{B}}(R)[G] \subseteq \widehat{Q}_{\mathfrak{pq}\mathfrak{B}}(R[G])$, we *claim* that $\mathbf{B}_p(Q(R)) \subseteq \mathbf{B}_p(Q(R[G]))$. To prove the claim, let $e \in \mathbf{B}_p(Q(R))$. Then there exists $a \in R$ such that $RaR_R \leq^{\text{ess}} eR_R$. Since $R[G]$ is a free right R-module, a routine argument shows that $(RaR)[G]_R \leq^{\text{ess}} eR[G]_R$. Thus $(RaR)[G]_{R[G]} \leq^{\text{ess}} eR[G]_{R[G]}$. Since $\mathbf{B}_p(Q(R)) \subseteq \mathbf{B}(Q(R[G]))$ from the proof of part (i), $e \in \mathbf{B}(Q(R[G]))$. So $e \in \mathbf{B}_p(Q(R[G]))$ because $(RaR)[G] = R[G]aR[G]$. Hence $\mathbf{B}_p(Q(R)) \subseteq \mathbf{B}_p(Q(R[G]))$. Theorem 8 shows that $\widehat{Q}_{\mathfrak{pq}\mathfrak{B}}(R)[G] \subseteq \widehat{Q}_{\mathfrak{pq}\mathfrak{B}}(R[G])$.

(ii) This is a consequence of part (i) and [11, Theorem 1.2]. □

Corollary 17. *Let R be a semiprime ring. Then $\widehat{Q}_{\mathfrak{pq}\mathfrak{B}}(R[x,x^{-1}]) = \widehat{Q}_{\mathfrak{pq}\mathfrak{B}}(R)[x,x^{-1}]$ and $\widehat{Q}_{\mathfrak{pq}\mathfrak{B}}(R[X]) = \widehat{Q}_{\mathfrak{pq}\mathfrak{B}}(R)[X]$, where X a nonempty set of not necessarily commuting indeterminates.*

Proof. Note that $R[x,x^{-1}] \cong R[C_\infty]$, which is semiprime, where C_∞ is the infinite cyclic group. Since R is semiprime, so is $R[X]$. Thus $\widehat{Q}_{\mathfrak{pq}\mathfrak{B}}(R[x,x^{-1}]) = \widehat{Q}_{\mathfrak{pq}\mathfrak{B}}(R)[x,x^{-1}]$ and $\widehat{Q}_{\mathfrak{pq}\mathfrak{B}}(R[X]) = \widehat{Q}_{\mathfrak{pq}\mathfrak{B}}(R)[X]$ follow from Theorem 16. □

Example 18. There is a semiprime ring R such that $\widehat{Q}_{\mathfrak{pq}\mathfrak{B}}(R[[x]]) \neq \widehat{Q}_{\mathfrak{pq}\mathfrak{B}}(R)[[x]]$. In [6, Example 2.3], there is a commutative von Neumann regular ring R (hence right p.q.-Baer), but the ring $R[[x]]$ is not right p.q.-Baer. Thus $\widehat{Q}_{\mathfrak{pq}\mathfrak{B}}(R) = R$

and so $\widehat{Q}_{\mathfrak{pq}\mathfrak{B}}(R)[[x]] = R[[x]]$. Since $R[[x]]$ is not right p.q.-Baer, $\widehat{Q}_{\mathfrak{pq}\mathfrak{B}}(R[[x]]) \neq \widehat{Q}_{\mathfrak{pq}\mathfrak{B}}(R)[[x]]$.

Let R be a ring. Then the subring $R\mathbf{B}(Q(R))$ of $Q(R)$ generated by R and $\mathbf{B}(Q(R))$ is called the *idempotent closure* of R (see [2]). From the following lemma, one can see that the idempotent closure of $\mathrm{Mat}_n(R)$ is the matrix ring of n-by-n matrices over the idempotent closure of R and similarly for $T_n(R)$. Let 1_n denote the unity of $\mathrm{Mat}_n(R)$.

Lemma 19. *Let $\delta \subseteq \mathbf{B}(Q(R))$ and $\Delta = \{1_n c \mid c \in \delta\}$. Then:*

(i) $Mat_n(\langle R \cup \delta\rangle_{Q(R)}) = \langle Mat_n(R) \cup \Delta\rangle_{Q(Mat_n(R))}$.
(ii) $Q(T_n(R)) = Q(Mat_n(R)) = Mat_n(Q(R))$.
(iii) $T_n(\langle R \cup \delta\rangle_{Q(R)}) = \langle T_n(R) \cup \Delta\rangle_{Q(Mat_n(R))}$.

Proof. (i) This part follows from straightforward calculation.

(ii) Let $T = T_n(R)$. By routine calculations, T_T is dense in $\mathrm{Mat}_n(R)_T$. So we have that $Q(T_n(R)) = Q(\mathrm{Mat}_n(R))$. From [27, 2.3], $Q(\mathrm{Mat}_n(R)) = \mathrm{Mat}_n(Q(R))$. Thus it follows that $Q(T_n(R)) = Q(\mathrm{Mat}_n(R)) = \mathrm{Mat}_n(Q(R))$.

(iii) This follows from part (ii) and a routine calculation. □

Theorem 20. *Let R be a semiprime ring. Then $\widehat{Q}_{\mathfrak{K}}(Mat_n(R)) = Mat_n(\widehat{Q}_{\mathfrak{K}}(R))$, where $\mathfrak{K} = \mathfrak{pq}\mathfrak{B}$, $\mathfrak{p}\mathfrak{F}\mathfrak{I}$, or $\mathfrak{fg}\mathfrak{F}\mathfrak{I}$.*

Proof. Assume that $\mathfrak{K} = \mathfrak{pq}\mathfrak{B}$, $\mathfrak{p}\mathfrak{F}\mathfrak{I}$, or $\mathfrak{fg}\mathfrak{F}\mathfrak{I}$. By Theorem 8, it follows that $\widehat{Q}_{\mathfrak{K}}(\mathrm{Mat}_n(R)) = \langle \mathrm{Mat}_n(R) \cup \delta_{\mathfrak{fg}\mathfrak{F}\mathfrak{I}}(\mathrm{Mat}_n(R))(1_n)\rangle_{Q(\mathrm{Mat}_n(R))}$. Observe that if J is a finitely generated ideal of $\mathrm{Mat}_n(R)$, then there is a finitely generated ideal I of R such that $J = \mathrm{Mat}_n(I)$. Thus $\delta_{\mathfrak{fg}\mathfrak{F}\mathfrak{I}}(\mathrm{Mat}_n(R))(1_n) = \{1_n c \mid c \in \delta_{\mathfrak{fg}\mathfrak{F}\mathfrak{I}}(R)(1_n)\}$. So Lemma 19 and Theorem 8 yield that $\widehat{Q}_{\mathfrak{K}}(\mathrm{Mat}_n(R)) = \mathrm{Mat}_n(\widehat{Q}_{\mathfrak{K}}(R))$. □

Theorem 21. *Let R be a semiprime ring. Then $\widehat{Q}_{\mathfrak{pq}\mathfrak{B}}(T_n(R)) = T_n(\widehat{Q}_{\mathfrak{pq}\mathfrak{B}}(R))$.*

Proof. Let $T = T_n(R)$ and S be a right ring of quotients of T. From [9, Proposition 2.6], $T_n(\widehat{Q}_{\mathfrak{pq}\mathfrak{B}}(R))$ is a right p.q.-Baer ring. Assume that S is a right p.q.-Baer ring. Take $e \in \mathbf{B}_p(Q(R))$. Then there exists $x \in R$ such that $RxR_R \leq^{\text{ess}} eR_R$, hence $RxR_R \leq^{\text{ess}} eQ(R)_R$. Therefore $Q(R)xQ(R)_{Q(R)} \leq^{\text{ess}} eQ(R)_{Q(R)}$. Thus $eQ(R)xQ(R)e_{eQ(R)e} \leq^{\text{ess}} eQ(R)e_{eQ(R)e}$ because $e \in \mathbf{B}_p(Q(R)) \subseteq \mathbf{B}(Q(R))$. Since $eQ(R)e$ is a semiprime ring, $0 = r_{eQ(R)e}(eQ(R)xQ(R)e) = r_{Q(R)}(eQ(R)xq(R)e) \cap eQ(R)e = r_{Q(R)}(Q(R)xQ(R)) \cap eQ(R)$.

So we have that $r_{Q(R)}(Q(R)xQ(R))eQ(R) = r_{Q(R)}(Q(R)xQ(R))Q(R)e = 0$. Hence $r_{Q(R)}(Q(R)xQ(R)) \subseteq (1-e)Q(R)$.

Obviously, $(1-e)Q(R) \subseteq r_{Q(R)}(Q(R)xQ(R))$. Thus $r_{Q(R)}(Q(R)xQ(R)) = (1-e)Q(R)$.

Next we show that $r_{Q(R)}(RxR) = (1-e)Q(R)$. For this, first note that $(1-e)Q(R) = r_{Q(R)}(Q(R)xQ(R)) \subseteq r_{Q(R)}(RxR)$. Thus by the modular law, $r_{Q(R)}(RxR) = (1-e)Q(R) \oplus [r_{Q(R)}(RxR) \cap eQ(R)]$. Assume to the contrary that $r_{Q(R)}(RxR) \cap eQ(R) \neq 0$. Take $0 \neq eq \in r_{Q(R)}(RxR) \cap eQ(R)$ with $q \in Q(R)$.

Since $RxR_R \leq^{\text{ess}} eQ(R)_R$, there exists $r \in R$ such that $0 \neq eqr \in RxR$. Thus $eqr \in r_{Q(R)}(RxR) \cap R = r_R(RxR)$. So $eqr \in RxR \cap r_R(RxR) = 0$ because R is semiprime. This is absurd. So $r_{Q(R)}(RxR) \cap eQ(R) = 0$. Therefore $r_{Q(R)}(RxR) = (1-e)Q(R)$.

Let $\theta \in T = T_n(R)$ be the n-by-n matrix with x in the (1,1)-position and 0 elsewhere. Thus $T\theta T$ is the n-by-n matrix with RxR throughout the top row and 0 elsewhere. Moreover, $Q(T)\theta Q(T) = \text{Mat}_n(Q(R)xQ(R))$. Since $T\theta T \subseteq S\theta S \subseteq Q(T)\theta Q(T)$ and $r_{Q(R)}(RxR) = (1-e)Q(R)$, we have that

$$(1-f)Q(T) = r_{Q(T)}(Q(T)\theta Q(T)) \subseteq r_{Q(T)}(S\theta S) \subseteq r_{Q(T)}(T\theta T) = (1-f)Q(T),$$

where f is the diagonal matrix with e on the diagonal. Since S is right p.q.-Baer, there exists $c = c^2 \in S$ such that $cS = r_S(S\theta S) = S \cap r_{Q(R)}(S\theta S) = S \cap (1-f)Q(T)$. Thus $cQ(T) \subseteq (1-f)Q(T)$. Let $0 \neq (1-f)q \in (1-f)Q(T)$ with $q \in Q(T)$. Then $0 \neq (1-f)qQ(T) \cap S \subseteq (1-f)Q(T) \cap S = cS \subseteq cQ(T)$. Hence $0 \neq (1-f)q\alpha \in cQ(T)$ with $\alpha \in Q(T)$. So $cQ(T)_{Q(T)} \leq^{\text{ess}} (1-f)Q(T)_{Q(T)}$ and hence $c = 1-f$. Thus $f = 1-c \in S$. Therefore $T_n(\widehat{Q}_{\mathfrak{pq}\mathfrak{B}}(R)) \subseteq S$ by Theorem 8. So $\widehat{Q}_{\mathfrak{pq}\mathfrak{B}}(T)$ also exists and $\widehat{Q}_{\mathfrak{pq}\mathfrak{B}}(T) = T_n(\widehat{Q}_{\mathfrak{pq}\mathfrak{B}}(R))$. □

For a ring R and a nonempty set Γ, $\text{CFM}_\Gamma(R), \text{RFM}_\Gamma(R)$, and $\text{CRFM}_\Gamma(R)$ denote the column finite, the row finite, and the column and row finite matrix rings over R indexed by Γ, respectively.

Theorem 22. ([15, Theorem 19])

(i) *$R \in \mathfrak{q}\mathfrak{B}$ if and only if $CFM_\Gamma(R)$ (resp., $RFM_\Gamma(R)$ and $CRFM_\Gamma(R)) \in \mathfrak{q}\mathfrak{B}$.*
(ii) *If $R \in \mathfrak{F}\mathfrak{I}$, then $CFM_\Gamma(R)$ (resp., $CRFM_\Gamma(R)) \in \mathfrak{F}\mathfrak{I}$.*
(iii) *If R is semiprime, then we have that $\widehat{Q}_{\mathfrak{q}\mathfrak{B}}(CFM_\Gamma(R)) \subseteq CFM_\Gamma(\widehat{Q}_{\mathfrak{q}\mathfrak{B}}(R))$, $\widehat{Q}_{\mathfrak{q}\mathfrak{B}}(RFM_\Gamma(R)) \subseteq RFM_\Gamma(\widehat{Q}_{\mathfrak{q}\mathfrak{B}}(R))$, and $\widehat{Q}_{\mathfrak{q}\mathfrak{B}}(CRFM_\Gamma(R)) \subseteq CRFM_\Gamma(\widehat{Q}_{\mathfrak{q}\mathfrak{B}}(R))$.*

Theorems 15 and 21, and the fact that the right p.q.-Baer condition is a Morita invariant property [9, Theorem 2.2] motivate the following questions:

(1) *Is the right p.q.-Baer property preserved under the various infinite matrix ring extensions?*
(2) *Does $\widehat{Q}_{\mathfrak{pq}\mathfrak{B}}(R)$ of a ring R have behavior similar to that of $\widehat{Q}_{\mathfrak{q}\mathfrak{B}}(R)$ for the various infinite matrix ring extensions?*

Our next example provides negative answers to these questions.

Example 23. Let F be a field and $F_n = F$ for $n = 1, 2 \ldots$. Put

$$R = \left\{ (q_n)_{n=1}^{\infty} \in \prod_{n=1}^{\infty} F_n \mid q_n \text{ is eventually constant} \right\},$$

which is a subring of $\prod_{n=1}^{\infty} F_n$. Then R is a commutative von Neumann regular ring. Hence R is a right p.q.-Baer ring. Let $S = \text{CFM}_\Gamma(R)$, where $\Gamma = \{1, 2, \ldots\}$.

Take

$$a_1 = (0,1,0,0,\dots),\ a_2 = (0,1,0,1,0,0,\dots),\ a_3 = (0,1,0,1,0,1,0,0,\dots),$$

and so on, in R. Let x be the element in S with a_n in the (n,n)-position for $n = 1,2,\dots$ and 0 elsewhere, and let

$$e = (q_n)_{n=1}^{\infty} \in Q(R) = \prod_{n=1}^{\infty} F_n$$

such that $q_{2n} = 1$ and $q_{2n-1} = 0$ for $n = 1,2,\dots$. Then $e = e^2 \in \mathbf{B}(Q(R))$, hence

$$eI \in \mathrm{CFM}_\Gamma(\widehat{Q}_{\mathfrak{q}\mathfrak{B}}(R)) \subseteq Q(S)$$

because $\widehat{Q}_{\mathfrak{q}\mathfrak{B}}(R) = R\mathbf{B}(Q(R))$ from Theorem 12, where I is the unity matrix in S. Therefore $eI \in \mathbf{B}(Q(S))$. Also note that $(\sum a_iR)_R \leq^{\text{ess}} eR_R$. We *claim* that

$$SxS_S \leq^{\text{ess}} (eI)S_S.$$

For convenience, let E_{ij} be the matrix in S with 1 in the (i,j)-position and 0 elsewhere. Take $0 \neq (eI)s \in (eI)S$ with $s = (r_{ij}) \in S$. Then there is a nonzero column, say the m-th column, of $(eI)s$. In this case the m-th column of $(eI)s$ is the same as the first column of $(eI)sE_{m1}$. Thus the first column of $(eI)E_{m1}$ is nonzero and all other columns except the first column of $(eI)E_{m1}$ are zero. So without loss of generality, we may assume that the first column of the matrix $(eI)s$ is nonzero and all the other columns except the first column are zero. In the first column of $(eI)s$, there are only finitely many nonzero entries, say

$$er_{k_1 1}, er_{k_2 1}, \dots, er_{k_n 1}$$

with

$$k_1 < k_2 < \cdots < k_n.$$

To show that $SxS_S \leq^{\text{ess}} (eI)S_S$, we proceed by induction. Suppose $n = 1$. Since $(\sum a_iR)_R \leq^{\text{ess}} eR_R$, there exist $b_1, \lambda_1, \dots, \lambda_m \in R$ such that $0 \neq er_{k_1 1}b_1 = a_1\lambda_1 + \cdots + a_m\lambda_m$. Thus $0 \neq (eI)s(b_1E_{11}) = (\lambda_1E_{k_1 1} + \cdots + \lambda_mE_{k_1 m})\cdot x \cdot (E_{11} + \cdots + E_{m1}) \in SxS$.

Next consider the case for $n > 1$. Since $(\sum a_iR)_R \leq^{\text{ess}} eR_R$, there is $b_1 \in R$ such that $0 \neq er_{k_1 1}b_1 \in \sum a_iR$. If $er_{k_i 1}b_1 = 0$ for some i with $1 < i \leq n$, then we are done by induction. So $er_{k_i 1}b_1 \neq 0$ for all $i = 1,2,\dots,n$. Assume that $er_{k_2 1}b_1 \notin \sum a_iR$. There exists $b_2 \in R$ with $0 \neq er_{k_2 1}b_1b_2 \in \sum a_iR$. In this case, note that $er_{k_1 1}b_1b_2 \in \sum a_iR$. Suppose $er_{k_i 1}b_1b_2 = 0$ for some $i \neq 2$. Again we are done by induction. Next if $er_{k_3 1}b_1b_2 \notin R$, then there is $b_3 \in R$ such that $0 \neq er_{k_3 1}b_1b_2b_3 \in \sum a_iR$ and $er_{k_i 1}b_1b_2b_3 \neq 0$ for all i. Also note that $er_{k_1 1}b_1b_2b_3$, $er_{k_2 1}b_1b_2b_3$, $er_{k_3 1}b_1b_2b_3 \in \sum a_iR$. Continue this process, it follows that there are $b_1, b_1, \dots, b_n \in R$ with $er_{k_i 1}b_1b_2\cdots b_n \neq 0$ and $er_{k_i 1}b_1b_2\cdots b_n \in \sum a_iR$ for all i. Let $b = b_1b_2\cdots b_n$. Then there is a positive integer ℓ and $\lambda_{ij} \in R$ such that

$$er_{k_1 1}b = a_1\lambda_{11} + a_2\lambda_{12} + \cdots + a_\ell\lambda_{1\ell},\ er_{k_2 1}b = a_1\lambda_{21} + a_2\lambda_{22} + \cdots + a_\ell\lambda_{2\ell}, \dots,$$

and

$$er_{k_n1}b = a_1\lambda_{n1} + a_2\lambda_{n2} + \cdots + a_\ell\lambda_{n\ell}.$$

Thus

$$0 \neq (eI)s(bE_{11}) = (\lambda_{11}E_{k_11} + \cdots + \lambda_{1\ell}E_{k_1\ell} + \lambda_{21}E_{k_21} + \cdots + \lambda_{2\ell}E_{k_2\ell} + \cdots$$

$$\cdots + \lambda_{n1}E_{k_n1} + \cdots + \lambda_{n\ell}E_{k_n\ell}) \cdot x \cdot (E_{11} + \cdots E_{\ell 1}) \in SxS.$$

Therefore $SxS_S \leq^{\text{ess}} (eI)S_S$, hence $eI \in \mathbf{B}_p(Q(S))$. But note that $eI \notin S$. Observe that S is a semiprime ring because R is semiprime. Thus the ring S is not right p.q.-Baer by Theorem 8(ii). Furthermore, since R is right p.q.-Baer, $\widehat{Q}_{\mathfrak{pq}\mathfrak{B}}(R) = R$. Thus we have that $\widehat{Q}_{\mathfrak{pq}\mathfrak{B}}(\mathrm{CFM}_\Gamma(R)) \not\subseteq \mathrm{CFM}_\Gamma(\widehat{Q}_{\mathfrak{pq}\mathfrak{B}}(R))$. Also $\mathrm{CFM}_\Gamma(\widehat{Q}_{\mathfrak{pq}\mathfrak{B}}(R))$ is not right p.q.-Baer.

For $\widehat{Q}_{\mathfrak{pq}\mathfrak{B}}(\mathrm{CRFM}_\Gamma(R)) \not\subseteq \mathrm{CRFM}_\Gamma(\widehat{Q}_{\mathfrak{pq}\mathfrak{B}}(R))$, let x and e be as in the case of the column finite matrix ring. Then, by the same method, we can show that $eI \in \mathbf{B}_p(Q(\mathrm{CRFM}_\Gamma(R)))$; but $eI \notin \mathrm{CRFM}_\Gamma(R)$. So $\mathrm{CRFM}_\Gamma(R)$ $(= \mathrm{CRFM}_\Gamma(\widehat{Q}_{\mathfrak{pq}\mathfrak{B}}(R))$ is not right p.q.-Baer by Theorem 8(ii). Also we have that

$$\widehat{Q}_{\mathfrak{pq}\mathfrak{B}}(\mathrm{CRFM}_\Gamma(R)) \not\subseteq \mathrm{CRFM}_\Gamma(\widehat{Q}_{\mathfrak{pq}\mathfrak{B}}(R)).$$

Finally for $\widehat{Q}_{\mathfrak{pq}\mathfrak{B}}(\mathrm{RFM}_\Gamma(R)) \not\subseteq \mathrm{RFM}_\Gamma(\widehat{Q}_{\mathfrak{pq}\mathfrak{B}}(R))$, let $U = \mathrm{RFM}_\Gamma(R)$ and x, e be as before. Then by modifying the method used for the case of column finite matrix rings, it can be shown that

$$_UUxU \leq^{\text{ess}} {}_U(eI)U,$$

where I is the unity matrix in U. Note eI is a central idempotent. So we have that $_{(eI)U(eI)}UxU \leq^{\text{ess}} {}_{(eI)U(eI)}(eI)U(eI)$. Since UxU is an ideal of the semiprime ring $(eI)U(eI)$, $r_{(eI)U(eI)}(UxU) = \ell_{(eI)U(eI)}(UxU) = 0$. So $UxU_{(eI)U(eI)} \leq^{\text{ess}} (eI)U(eI)_{(eI)U(eI)}$. Thus $UxU_U \leq^{\text{ess}} (eI)U_U$. Moreover, since $e \in \mathbf{B}(Q(R)) = \mathbf{B}(Q^m(R))$, there exists $J \trianglelefteq R$ such that $\ell_R(J) = 0$ and $eJ \subseteq R$. Thus

$$\mathrm{RFM}_\Gamma(J) \trianglelefteq \mathrm{RFM}_\Gamma(R),\ \ell_{\mathrm{RFM}_\Gamma(R)}(\mathrm{RFM}_\Gamma(J)) = 0,$$

and

$$(eI)\mathrm{RFM}_\Gamma(J) \subseteq \mathrm{RFM}_\Gamma(R),$$

where I is the unity matrix in $\mathrm{RFM}_\Gamma(R)$. So $eI \in Q^m(\mathrm{RFM}_\Gamma(R))$. Hence we have that $eI \in \mathbf{B}(Q^m(\mathrm{RFM}_\Gamma(R)))$. So $eI \in \mathbf{B}(Q(U))$, hence $eI \in \mathbf{B}_p(Q(U))$. But $eI \notin U$. Therefore $U = \mathrm{RFM}_\Gamma(R)$ $(= \mathrm{RFM}_\Gamma(\widehat{Q}_{\mathfrak{pq}\mathfrak{B}}(R))$ is not right p.q.-Baer by Theorem 8. Thus $\widehat{Q}_{\mathfrak{pq}\mathfrak{B}}(\mathrm{RFM}_\Gamma(R)) \not\subseteq \mathrm{RFM}_\Gamma(\widehat{Q}_{\mathfrak{pq}\mathfrak{B}}(R))$.

Acknowledgment

The authors are grateful for the support they received from the Mathematics Research Institute, Columbus, Ohio and for the kind hospitality and support of Busan National University, South Korea, the Ohio State University at Lima, and the University of Louisiana at Lafayette.

References

[1] S.A. Amitsur, *On rings of quotients.* Symp. Math. **8** Academic Press, London (1972), 149–164.

[2] K. Beidar and R. Wisbauer, *Strongly and properly semiprime modules and rings.* Ring Theory, Proc. Ohio State-Denison Conf. (S.K. Jain and S.T. Rizvi (eds.)), World Scientific, Singapore (1993), 58–94.

[3] G.F. Birkenmeier,*Idempotents and completely semiprime ideals.* Comm. Algebra **11** (1983), 567–580.

[4] G.F. Birkenmeier, H.E. Heatherly, J.Y. Kim and J.K. Park, *Triangular matrix representations.* J. Algebra **230** (2000), 558–595.

[5] G.F. Birkenmeier, J.Y. Kim and J.K. Park, *Quasi-Baer ring extensions and biregular rings.* Bull. Austral. Math. Soc. **61** (2000), 39–52.

[6] G.F. Birkenmeier, J.Y. Kim and J.K. Park, *On quasi-Baer rings.* Algebras and Its Applications (D.V. Huynh, S.K. Jain, and S.R. López-Permouth (eds.)), Contemp. Math. **259**, Amer. Math. Soc., Providence, 2000, 67–92.

[7] G.F. Birkenmeier, J.Y. Kim and J.K. Park, *Semicentral reduced algebras.* International Symposium on Ring Theory (G.F. Birkenmeier, J.K. Park, and Y.S. Park (eds.)), Trends in Math. Birkhäuser, Boston, 2001, 67–84.

[8] G.F. Birkenmeier, J.Y. Kim and J.K. Park, *Polynomial extensions of Baer and quasi-Baer rings.* J. Pure Appl. Algebra **259** (2001), 25–42.

[9] G.F. Birkenmeier, J.Y. Kim and J.K. Park, *Principally quasi-Baer rings.* Comm. Algebra **29** (2001), 639–660.

[10] G.F. Birkenmeier, B.J. Müller and S.T. Rizvi, *Modules in which every fully invariant submodule is essential in a direct summand.* Comm. Algebra **30** (2002), 1395–1415.

[11] G.F. Birkenmeier and J.K. Park, *Triangular matrix representations of normalizing extensions.* J. Algebra **265** (2003), 457–477.

[12] G.F. Birkenmeier, J.K. Park and S.T. Rizvi, *Ring hulls and applications.* J. Algebra **304** (2006), 633–665.

[13] G.F. Birkenmeier, J.K. Park and S.T. Rizvi, *The structure of rings of quotients.* J. Algebra **321** (2009), 2545–2566.

[14] G.F. Birkenmeier, J.K. Park and S.T. Rizvi, *Hulls of semiprime rings with applications to C^*-algebras.* J. Algebra **322** (2009), 327–352.

[15] G.F. Birkenmeier, J.K. Park and S.T. Rizvi, *Hulls of ring extensions.* Canad. Math. Bull., to appear.

[16] W.D. Burgess and R.M. Raphael, *On extensions of regular rings of finite index by central elements.* Advances in Ring Theory (S.K. Jain and S.T. Rizvi (eds.)), Trends in Math., Birkhäuser, Boston (1997), 73–86.

[17] N. Divinsky, *Rings and Radicals.* Univ. Toronto Press, 1965.

[18] D.E. Dobbs and G. Picavet, *Weak Baer going-down rings.* Houston J. Math. **29** (2003), 559–581.

[19] J. Kist, *Minimal prime ideals in commutative semigroups.* Proc. London Math. Soc. **13** (1963), 31–50.

[20] T.Y. Lam, *Lectures on Modules and Rings.* Springer-Verlag, Berlin-Heidelberg-New York, 1999.

[21] J. Lambek, *Lectures on Rings and Modules.* Chelsea, New York, 1986.

[22] J. Okniński, *Semigroup Algebras.* Marcel Dekker, New York, 1991.

[23] K. Oshiro, *On torsion free modules over regular rings.* Math. J. Okayama Univ. **16** (1973), 107–114.

[24] D.S. Passman, *The Algebraic Structure of Group Rings.* Wiley, New York, 1977.

[25] L.H. Rowen, *Ring Theory I.* Academic Press, San Diego, 1988.

[26] B. Stenström, *Rings of Quotients.* Springer-Verlag, Berlin-Heidelberg-New York, 1975.

[27] Y. Utumi, *On quotient rings.* Osaka Math. J. **8** (1956), 1–18.

Gary F. Birkenmeier
Department of Mathematics
University of Louisiana at Lafayette
Lafayette, Louisiana 70504-1010, USA
e-mail: `gfb1127@louisiana.edu`

Jae Keol Park
Department of Mathematics
Busan National University
Busan 609-735, South Korea
e-mail: `jkpark@pusan.ac.kr`

S. Tariq Rizvi
Department of Mathematics
Ohio State University
Lima, Ohio 45804-3576, USA
e-mail: `rizvi.1@osu.edu`

Advances in Ring Theory
Trends in Mathematics, 63–68

Strongly Prime Ideals of Near-rings of Continuous Functions

G.L. Booth

Abstract. In this paper we investigate strongly prime ideals in the near-ring $\mathbb{N}_0(\mathbb{R}^n)$ of continuous, zero-preserving self-maps of $\mathbb{R}^n$. The strongly prime and uniformly strongly prime radicals of these near-rings are characterized. The Peano space-filling curves play a crucial rôle in this investigation. We also consider strongly prime ideals in $N_0(\mathbb{R}^\omega)$, where ω denotes the first transfinite cardinal.

Mathematics Subject Classification (2000). 16Y30, 22A05.

Keywords. near-ring of continuous functions, strongly prime, uniformly strongly prime.

1. Preliminaries

In this note, all near-rings are right distributive. For all relevant definitions, we refer to Pilz [9]. Strongly prime rings were introduced by Handelman and Lawrence [4], and the concept was extended to near-rings by Groenewald [3]. A near-ring N is called *strongly prime* if for all $0 \neq a \in N$ there exists a finite subset F of N such that $aFx = 0$ implies $x = 0$, for all $x \in N$. F is called an *insulator* of a. If F is independent of the choice of a, then N is said to be *uniformly strongly prime*. An ideal A of N is called (uniformly) strongly prime if the factor near-ring N/A is (uniformly) strongly prime. The *strongly prime radical* $\mathcal{P}_s(N)$ (resp. *uniformly strongly prime radical* $\mathcal{P}_u(N)$) is the intersection of the strongly prime (resp. uniformly strongly prime) ideals of N. N is said to be (*uniformly*) *strongly prime of bound* 1 if is (uniformly) strongly prime and the insulator F always contains exactly one element. (Uniformly) strongly prime ideals of N of bound 1 are defined in the obvious manner.

Let G be an additive (but not necessarily abelian) topological group. The set $N_0(G)$ of continuous self maps f of G such that $f(0) = 0$ is easily seen to be a zero-symmetric near-ring with pointwise addition and composition of mappings.

For surveys of results on near-rings of continuous functions, we refer to [7] and [8]. Let

$$P_G = \{f \in N_0(G) \mid \exists U \text{ open in } G \text{ such that } 0 \in U, f(U) = 0\}.$$

Then P_G is an ideal of $N_0(G)$, which may be nontrivial [5]. Hence $N_0(G)$ need not be simple, in contrast to the situation for the near-ring $M_0(G)$ of all zero-preserving self-maps of G. Investigations of strongly prime ideals in near-rings of continuous functions commenced in [2], and continued in [1].

Proposition 1.1. [1, Proposition 2.8] *$P_{\mathbb{R}}$ is a strongly prime ideal of $N_0(\mathbb{R})$ which is contained in every strongly prime ideal of $N_0(\mathbb{R})$, where $\mathbb{R}$ denotes the real numbers with the usual topology.*

Remark 1.2. In the proof of the above proposition, the required insulator for any $a \in N_0(\mathbb{R}) \backslash P_{\mathbb{R}}$ was $F = \{f, g\}$, where $f(x) = x^2$, and $g(x) = -x^2$ for all $x \in \mathbb{R}$. An examination of the proof shows that F could be substituted by $G = \{h\}$ where

$$h(x) := \begin{cases} x \sin(1/x) & \text{if } x \neq 0 \\ 0 & \text{if } x = 0 \end{cases}.$$

Hence Proposition 1.1 can be sharpened to

Proposition 1.3. *$P_{\mathbb{R}}$ is a uniformly strongly prime ideal of bound 1 in $N_0(\mathbb{R})$, which is contained in every strongly prime ideal of $N_0(\mathbb{R})$.*

Corollary 1.4. $\mathcal{P}_s(N_0(\mathbb{R})) = \mathcal{P}_u(N_0(\mathbb{R})) = P_{\mathbb{R}}$.

In the sequel, we will consider strongly prime ideals of $N_0(\mathbb{R}^n)$, where $n \in \mathbb{N}$ and $\mathbb{R}^n$ has the usual topology. We will also investigate $N_0(\mathbb{R}^\omega)$, where ω denotes the first transfinite cardinal, and $\mathbb{R}^\omega$ has the usual product topology. The Peano *space-filling curves* will play a fundamental rôle in the investigation of $N_0(\mathbb{R}^n)$. A space-filling curve is a continuous, surjective mapping f from $I = [0,1]$ onto I^n. The existence of such curves for all $n \in \mathbb{N}$ is well known. From such curves, it is easy to construct continuous, surjective mappings σ from I onto $\overline{B}(0,1) = \{x \in R^n : |x| \leq 1\}$ such that $\sigma(0) = \sigma(1) = 0$, and we will make frequent use of such mappings.

If $a, b, c, d \in \mathbb{R}$, $a < b, c < d$, the *standard homeomorphism* of $[a, b]$ onto $[c, d]$ is the mapping τ defined by $\tau(x) = \frac{d-c}{b-a}(x - b) + d$. For all undefined topological concepts, we refer to any of the standard texts, for example [6].

2. Strongly prime ideals in $N_0(\mathbb{R}^n)$

In this section, we investigate strongly prime ideals in $N_0(\mathbb{R}^n)$, and characterise $\mathcal{P}_s(N_0(\mathbb{R}^n))$ and $\mathcal{P}_u(N_0(\mathbb{R}^n))$.

Lemma 2.1. *Let G be a topological group with a countable, monotone decreasing open base at 0 consisting of arcwise connected sets. If $f \in N_0G)$, then $f \notin P_G$ if and only if there exists an arc α such that $\alpha(0) = 0, \alpha(1) \neq 0$ and for all $\varepsilon > 0$ there exists $0 < t < \varepsilon$ such that $f(\alpha(t)) \neq 0$.*

Proof. Suppose that there exists an arc α which satisfies the conditions of the lemma. Let U be an open set of G which contains 0. Since α is continuous, there exists $\varepsilon > 0$ such that $\alpha([0,\varepsilon)) \subseteq U$. Then there exists $0 < t < \varepsilon$ such that $f(\alpha(t)) \neq 0$. Let $g = \alpha(t)$. Then $f(g) \neq 0$, so $f \notin P_G$.

Conversely, suppose that $f \notin P_G$. Let $\mathcal{B} = \{B_n : n \in \mathbb{N}\}$ be a countable, monotone decreasing open base at 0 which consists of arcwise connected sets. Since $f \notin P_G$, for each $n \in \mathbb{N}$ there exists $g_n \in B_n$ such that $f(g_n) \neq 0$. Let α_n be an arc in B_n such that $\alpha_n(0) = g_n$ and $\alpha_n(1) = g_{n+1}$ for each $n \in \mathbb{N}$. Define the arc α by

$$\alpha(t) = \begin{cases} \alpha_n \tau_n(t) & \text{if } t \in \left(\frac{1}{n+1}, \frac{1}{n}\right] \\ 0 & \text{if } t = 0 \end{cases},$$

where τ_n denotes the standard homeomorphism of $\left(\frac{1}{n+1}, \frac{1}{n}\right]$ onto $(0,1]$. It follows from the continuity of α_n and τ_n that α is continuous on $(0,1]$. Let U be an open set in G which contains 0. Then there exists $n \in \mathbb{N}$ such that $B_n \subseteq U$. It follows from the definition of α that $\alpha(t) \in B_n \subseteq U$ for $0 < t < \frac{1}{n}$. Moreover, $\alpha(0) = 0 \in U$. Hence, $\alpha([0,\frac{1}{n})) \subseteq U$, so α is continuous at 0. Finally, $\alpha(0) = 0$ and $\alpha(1) = g_1 \neq 0$, so α is the required arc. □

Proposition 2.2. *For each $n \in \mathbb{N}$, $P_{\mathbb{R}^n}$ is a uniformly strongly prime ideal of $\mathrm{N}_0(\mathbb{R}^n)$ of bound 1.*

Proof. Let α be an arc in $\mathbb{R}^n$ whose range is the unit closed ball $\overline{B}(0,1)$, such that $\alpha(0) = 0 = \alpha(1)$, and let $\beta : [0,1] \to \mathbb{R}^n$ be defined by

$$\beta(t) = \begin{cases} \frac{1}{n}\alpha\tau_n(t) & \text{if } t \in \left(\frac{1}{n+1}, \frac{1}{n}\right] \\ 0 & \text{if } t = 0 \end{cases},$$

where τ_n is the standard homeomorphism of $\left(\frac{1}{n+1}, \frac{1}{n}\right]$ onto $(0,1]$. It is easily verified that β is continuous on $[0,1]$. Moreover $\beta([0,\frac{1}{n}])$ contains the open ball $B(0,\frac{1}{n})$ for each $n \in \mathbb{N}$. Let $\gamma(x) = \frac{2}{\pi}\arctan|x|$ for all $x \in \mathbb{R}^n$. Then γ maps $\mathbb{R}^n$ continuously into $[0,1]$ and $\gamma(0) = 0$ if and only if $x = 0$. Let $f = \beta\gamma$. Let $a, b \in N_0(\mathbb{R}^n)\backslash P_{\mathbb{R}^n}$. If $\varepsilon > 0$, $a(B(0,\varepsilon)) \neq 0$, so $\gamma a(B(0,\varepsilon)) \neq 0$. It follows from the continuity of the functions and the connectedness of $B(0,\varepsilon)$ that $\gamma a(B(0,\varepsilon))$ contains an interval $[0,\delta)$ for some $\delta > 0$. Let $n \in \mathbb{N}$ be such that $\frac{1}{n} < \delta$. Then $B(0,\frac{1}{n}) \subseteq \beta([0,\delta))$ and so $B(0,\frac{1}{n}) \subseteq \beta\gamma a(B(0,\varepsilon)) = fa(B(0,\varepsilon))$. Since $b \in N_0(\mathbb{R}^n)\backslash P_{\mathbb{R}^n}$, $b(B(0,\frac{1}{n})) \neq 0$ and so $bfa(B(0,\varepsilon)) \neq 0$. Hence $bfa \in N_0(\mathbb{R}^n)\backslash P_{\mathbb{R}^n}$ Thus $\{f\}$ is the required insulator. □

Let c be any cardinal (finite or infinite), and . If $a \in N_0(\mathbb{R})$, define $\overline{a} : \mathbb{R}^c \to \mathbb{R}^c$ by $\pi_i\overline{a}(x) = a(\pi_i(x))$ for all i. Clearly, $\overline{a} \in N_0(\mathbb{R}^c)$.

Lemma 2.3. *Let I be an ideal of $N_0(\mathbb{R}^c)$. If I contains all $\overline{a}$ such that a is bounded on $\mathbb{R}$, then $I = N_0(\mathbb{R}^c)$.*

Proof. Let $m(x) := \begin{cases} x^2 & x \geq 0 \\ -x^2 & x < 0 \end{cases}$, $n(x) := x, a(x) :=$ arctan x. Then a is bounded on $\mathbb{R}$ and so $\overline{a} \in I$. Let $b := m(a+n) - mn \in I$. It may be shown that $b(x) = \begin{cases} 2x \arctan x + (\arctan x)^2 & \text{if } x \geq 0 \\ -2x \arctan x - (\arctan x)^2 & \text{if } x < 0 \end{cases}$. Moreover, b is unbounded and strictly monotone increasing and hence is a bijection of $\mathbb{R}$ onto itself. Hence b has a continuous inverse b^{-1}. Clearly, $\overline{b} = \overline{m}(\overline{a} + \overline{n}) - \overline{mn} \in I$ and $(\overline{b})^{-1} = (\overline{b^{-1}})$. From this we deduce that $I = N_0(\mathbb{R}^c)$. □

Proposition 2.4. *Let I be a strongly prime ideal of $\mathrm{N}_0(\mathbb{R}^n)$. Then $P_{\mathbb{R}^n} \subseteq I$.*

Proof. Suppose to the contrary that $P_{\mathbb{R}^n} \nsubseteq I$. Let $a \in P_{\mathbb{R}^n} \backslash I$. Then there exists an open set U of $\mathbb{R}^n$ such that $a(U) = 0$. Let $f_i, \ldots, f_m \in N_0(\mathbb{R}^n)$. Let $V := \bigcap_{i=1}^{m} f_i^{-1}(U)$. Then V is open in $\mathbb{R}^n$ and $0 \in V$. Hence there exists $\delta > 0$ such that $B(0, \delta) \subseteq V$. By Lemma 2.3 there exists a bounded function $b \in \mathrm{N}_0(\mathbb{R})$ such that $\overline{b} \in \mathrm{N}_0(\mathbb{R}^n) \backslash I$. For a suitable choice of $\varepsilon > 0$, we have that $\left|\varepsilon \overline{b}(x)\right| < \delta$ for all $x \in \mathbb{R}^n$. Moreover, $\varepsilon \overline{b} \notin I$, otherwise $\overline{b} = \varepsilon^{-1}(\varepsilon \overline{b}) \in I$, contradicting our choice of b. Then $\varepsilon \overline{b}(x) \in V$ for all $x \in \mathbb{R}^n$. Hence for $1 \leq i \leq n$, it holds that $f_i(\varepsilon \overline{b})(x) \in U$ and so $a f_i(\varepsilon \overline{b})(x) = 0$. Thus $a f_i(\varepsilon \overline{b}) = 0 \in I$, so $\{f_1, \ldots, f_m\}$ is not an insulator for I. Hence I is not a strongly prime ideal of $\mathrm{N}_0(\mathbb{R}^n)$. This concludes the proof. □

Theorem 2.5. $\mathcal{P}_u(\mathrm{N}_0(\mathbb{R}^n)) = \mathcal{P}_s(\mathrm{N}_0(\mathbb{R}^n)) = P_{\mathbb{R}^n}$ *for all* $n \in \mathbb{N}$.

Proof. Follows immediately from Propositions 2.2 and 2.4 and the fact that every uniformly strongly prime ideal is strongly prime. □

3. Strongly prime ideals in $N_0(\mathbb{R}^\omega)$

In this section we investigate strongly prime ideals in $N_0(\mathbb{R}^\omega)$, where ω is the first transfinite cardinal, and $\mathbb{R}^\omega$ has the usual (Tychonoff) product topology. Recall that $\mathbb{R}^\omega$ metrizable, with metric d defined by $d(x,y) := \sum_{i=1}^{\infty} \frac{|x_i - y_i|}{2^i(1+|x_i-y_i|)}$, where $x := (x_i)_{i \in \mathbb{N}}$ and $y := (y_i)_{i \in \mathbb{N}}$.

Proposition 3.1. *$P_{\mathbb{R}^\omega}$ is a strongly prime ideal of $N_0(\mathbb{R}^\omega)$, and is contained in every strongly prime ideal of $N_0(\mathbb{R}^\omega)$.*

Proof. Let $a \in N_0(\mathbb{R}^\omega) \backslash P_{\mathbb{R}^\omega}$. Then $B := \{B(0, \frac{1}{n}) \mid n \in \mathbb{N}\}$ is a countable, monotone decreasing local basis at 0 which consists of arcwise connected sets. Hence by Lemma 2.1, here exists an arc α such that $\alpha(0) = 0$, $\alpha(1) \neq 0$, and for all $\varepsilon > 0$ there exists $0 < t < \varepsilon$ such that $a(\alpha(t)) \neq 0$. Let $\beta(X) := d(x, 0)$ for all $x \in \mathbb{R}^\omega$. Let $f := \alpha\beta$. If $b \in N_0(\mathbb{R}^\omega) \backslash P_{\mathbb{R}^\omega}$, and $\varepsilon > 0$, then $b(B(0,\varepsilon)) \neq 0$, so $\beta b(B(0,\varepsilon)) \neq 0$. It follows from the continuity of β and b that $\beta b(B(0,\varepsilon))$ contains an interval $[0, \delta)$ for some $\delta > 0$. Then there exists $t \in [0, \delta)$ such that $a(\alpha(t)) \neq 0$. Let $t = \beta b(x)$, where $x \in B(0, \varepsilon)$. Then $a(\alpha\beta)b(x) \neq 0$, i.e., $afb(x) \neq 0$. Hence $afb \notin P_{\mathbb{R}^\omega}$, and so $\{f\}$ is the required insulator for $P_{\mathbb{R}^\omega}$. Hence $P_{\mathbb{R}^\omega}$ is strongly prime.

Now let I be an ideal of such that $P_{\mathbb{R}^\omega} \nsubseteq I$. We show that I is not strongly prime. Let $a \in P_{\mathbb{R}^n} \backslash I$. Then there exists an open set U of $\mathbb{R}^n$ such that $a(U) = 0$. Let $f_1, \ldots, f_m \in N_0(\mathbb{R}^\omega)$. Let $V := \bigcap_{i=1}^{m} f_i^{-1}(U)$. Then V is open in $\mathbb{R}^n$ and $0 \in V$. By definition of the product topology, there exist $\delta > 0$ and $n \in \mathbb{N}$ such that $0 \in W \subseteq V$, where $V = \prod_{i=1}^{\infty} W_i$ with $W_i = (-\delta, \delta)$ for $i \leq n$ and $W_i = \mathbb{R}$ for $i > n$. By Lemma 2.3 there exists a bounded function $b \in \mathrm{N}_0(\mathbb{R})$ such that $\overline{b} \in \mathrm{N}_0(\mathbb{R}^\omega)\backslash I$. For a suitable choice of $\varepsilon > 0$, we have that $|\varepsilon b(x)| < \delta$ for all $x \in \mathbb{R}$. Moreover, $\varepsilon\overline{b} \notin I$, otherwise $\overline{b} = \varepsilon^{-1}(\varepsilon\overline{b}) \in I$, contradicting our choice of b. Then $\varepsilon\overline{b}(y) \in V$ for all $y \in \mathbb{R}^\omega$. It is clear that $af_i(\varepsilon\overline{b})(y) = 0$ for all $y \in \mathbb{R}^\omega$ and $1 \leq i \leq m$. Hence $aF(\varepsilon\overline{b}) = 0 \in I$, so I is not strongly prime, as required. □

As an immediate consequence of Proposition 3.1 we have:

Theorem 3.2. $\mathcal{P}_s(N_0(\mathbb{R}^\omega)) = P_{\mathbb{R}^\omega}$.

Proposition 3.3. *Let G be a T_0 topological group which is first countable, contains an arc and is not locally compact. Suppose that there exists a continuous mapping $\beta : G \to \mathbb{R}$ such that $\beta(0) = 0$ and $\beta(U) \neq 0$ for every open set U which contains 0. Then P_G is not a uniformly strongly prime ideal of $N_0(G)$.*

Proof. Let $\alpha : [0,1] \to G$ be an arc in G. We may assume without loss of generality that $\alpha(0) = 0$ and $\alpha(1) \neq 0$. We may also assume that $\alpha(t) \neq 0$ for $0 < t \leq 1$. For otherwise, replace α with γ, where γ is defined as follows. Let $s := \sup\{t \in [0,1] \mid \alpha(t) = 0\}$. Since $[0,1]$ is closed, $s \in [0,1]$. By continuity of α, $\alpha(s) = 0$. Let $\gamma = \alpha\tau$, where τ is the standard homeomorphism of $[0,1]$ onto $[s,1]$. Let $f_1, \ldots, f_n \in N_0(G)$. Let $S_i := f_i\alpha([0,1])$ for $1 \leq i \leq n$. Since $[0,1]$ is compact, so is S_i. Let $S := \bigcup_{i=1}^{n} S_i$. Then S is a compact subset of G. Since G is T_0 (and hence Hausdorff), S is closed.

Since G is first countable, there exists a monotone decreasing open basis $\mathcal{B} := \{B_n \mid n \in \mathbb{N}\}$ at 0. Since G is not locally compact, B_n is not contained in S for all $n \in \mathbb{N}$. Let $g_n \in B_n \backslash S$ for each $n \in \mathbb{N}$. Since G is T_0, and hence completely regular, there exist a continuous mapping $\delta_i : G \to [0,1]$ such that $\delta_i(S) = 0$ and $\delta_i(g_i) = 1$. Let $\delta := \sum_{i=1}^{\infty} \frac{\delta_i}{2^i}$. It follows easily from the Weierstrass M-test that $\sum_{i=1}^{\infty} \frac{\delta_i}{2^i}$ converges uniformly, and hence that δ is continuous. Moreover, $\delta(S) = 0$ and $\delta(g_i) > 0$ for all $i \in \mathbb{N}$. Let $a := \alpha\delta$. Then $a(g_i) \neq 0$ and hence $a(B_i) \neq 0$ for all $i \in \mathbb{N}$. Hence $a \notin P_G$.

Let $b := \alpha\beta$. It is easily verified that $b \notin P_G$. Moreover $af_ib = \alpha\delta f_i\alpha\beta = 0$ for $1 \leq i \leq n$. Hence P_G is not a uniformly strongly prime ideal of $N_0(G)$. □

Corollary 3.4. *Let G be a metrizable group which contains an arc and is not locally compact. Then P_G is not a uniformly strongly prime ideal of $N_0(G)$. In particular, $P_{\mathbb{R}^\omega}$ is not a uniformly strongly prime ideal of $N_0(\mathbb{R}^\omega)$.*

Proof. Clearly, G is T_0 and first countable. Let d be a metric which induces the topology on G, and let $\beta(g) = d(0, g)$ for all $g \in G$. Then β is continuous, $\beta(0) = 0$ and $\beta(U) \neq 0$ for all open sets U which contain 0. Hence P_G is not uniformly strongly prime by of Proposition 3.3. Finally, we note that $\mathbb{R}^\omega$ metrizable, not locally compact, and contains an arc: Let g and h be distinct elements of $\mathbb{R}^\omega$, and let $\alpha(t) := tg + (1-t)h$ for $0 \leq t \leq 1$. Hence $P_{\mathbb{R}^\omega}$ is not a uniformly strongly prime ideal of $N_0(\mathbb{R}^\omega)$ by the first part of the corollary. □

Remark 3.5. We have not been able to establish whether $N_0(\mathbb{R}^\omega)$ contains any proper uniformly strongly prime ideals, nor to characterize $\mathcal{P}_u(N_0(\mathbb{R}^\omega))$. This is a matter of further investigation.

References

[1] G.L. Booth: Primeness and radicals of near-rings of continuous functions, *Nearrings and Nearfields, Proceedings of the Conference on Nearrings and Nearfields, Hamburg, July* 27–*August* 3, 2003, Springer, Dordrecht, 2005, 171–176.

[2] G.L. Booth and P.R. Hall: Primeness in near-rings of continuous functions, *Beiträge Alg. Geom.* 45 (2004), No. 1, 21–27.

[3] N.J. Groenewald: Strongly prime near-rings, *Proc. Edinburgh Math.* Soc. 31 (1988), No. 3, 337–343.

[4] D. Handelman and J. Lawrence: Strongly prime rings, *Trans. Amer. Math. Soc.* 211 (1975), 209–223.

[5] R.D. Hofer: Near-rings of continuous functions on disconnected groups, *J. Austral. Math. Soc. Ser. A* 28 (1979), 433–451.

[6] J.L. Kelley: *General topology*, Graduate Texts in Mathematics, No. 27, Springer Verlag, Berlin, 1975.

[7] K.D. Magill: Near-rings of continuous self-maps: a brief survey and some open problems, *Proc. Conf. San Bernadetto del Tronto, 1981*, 25–47, 1982.

[8] K.D. Magill: A survey of topological nearrings and nearrings of continuous functions, *Proc. Tenn. Top. Conf.*, World Scientific Pub. Co., Singapore, 1997, 121–140.

[9] G. Pilz: *Near-rings*, 2nd ed., North-Holland, Amsterdam, 1983.

G.L. Booth
Nelson Mandela Metropolitan University
Port Elizabeth, South Africa

Advances in Ring Theory
Trends in Mathematics, 69–81

Elements of Minimal Prime Ideals in General Rings

W.D. Burgess, A. Lashgari and A. Mojiri

Dedicated to S.K. Jain on his seventieth birthday

Abstract. Let R be any ring; $a \in R$ is called a *weak zero-divisor* if there are $r, s \in R$ with $ras = 0$ and $rs \neq 0$. It is shown that, in any ring R, the elements of a minimal prime ideal are weak zero-divisors. Examples show that a minimal prime ideal may have elements which are neither left nor right zero-divisors. However, every R has a minimal prime ideal consisting of left zero-divisors and one of right zero-divisors. The union of the minimal prime ideals is studied in 2-primal rings and the union of the minimal strongly prime ideals (in the sense of Rowen) in NI-rings.

Mathematics Subject Classification (2000). Primary: 16D25; Secondary: 16N40, 16U99.

Keywords. Minimal prime ideal, zero-divisors, 2-primal ring, NI-ring.

Introduction

E. Armendariz asked, during a conference lecture, if, in any ring, the elements of a minimal prime ideal were zero-divisors of some sort. In what follows this question will be answered in the positive with an appropriate interpretation of "zero-divisor".

Two very basic statements about minimal prime ideals hold in a commutative ring R: (I) If P is a minimal prime ideal, then the elements of P are zero-divisors, and (II) the union of the minimal prime ideals is $M = \{a \in R \mid \exists\, r \in R \text{ with } ar \in \mathbf{N}_*(R) \text{ but } r \notin \mathbf{N}_*(R)\}$, where $\mathbf{N}_*(R)$ is the prime radical. We will see that (I), suitably interpreted, is true for all rings. The statement (II) is false in general non-commutative rings but a version of it does hold in rings where the set of nilpotent elements forms an ideal.

In a commutative ring R we always have that $R/\mathbf{N}_*(R)$ is reduced (i.e., has no non-zero nilpotent elements); this fails in the non-commutative case. Hence we can expect "commutative-like" behaviour when, for a non-commutative ring R, $R/\mathbf{N}_*(R)$ is reduced; these rings are called *2-primal* and have been extensively studied. Statement (II), above, holds for these rings. A larger class of rings is where the set of nilpotent elements, $\mathbf{N}(R)$, forms an ideal (called *NI-rings*). Once again statements (I) (Corollary 2.9) and (II) (Corollary 2.11) hold when "minimal prime ideals" are replaced by "minimal **r**-strongly prime ideals" whose definition is recalled below. (The two types of prime ideal coincide in commutative rings.)

Various weakened forms of commutativity yield results which show that minimal prime ideals consist of (left or right) zero-divisors. A thorough study of this is in [2, e.g., Corollary 2.7]. Our purpose here is to look at minimal prime ideals in general where elements need not be zero-divisors but always are what we call *weak zero-divisors* (Theorem 2.2); an element a in a ring R is a *weak zero-divisor* if there are $r, s \in R$ with $ras = 0$ and $rs \neq 0$. It will also be seen that, in special cases, other sorts of prime ideals consist of weak zero-divisors. Examples will show that "weak zero-divisor" cannot be replaced by "left (or right) zero-divisor" (Examples 3.2, 3.3 and the semiprime Example 3.4), however, in any ring R there is a minimal prime ideal consisting of left zero-divisors and one consisting of right zero-divisors (Proposition 2.7).

Terminology. For a ring R (always unital) the prime radical is denoted $\mathbf{N}_*(R)$, the upper nil radical $\mathbf{N}^*(R)$ and the set of nilpotent elements $\mathbf{N}(R)$. As usual, R is called *semiprime* if $\mathbf{N}_*(R) = \mathbf{0}$, while R is called an *NI-ring* if $\mathbf{N}^*(R) = \mathbf{N}(R)$. Recall that an ideal P in a ring R is called *completely prime* if R/P is a domain.

There are several uses of the term "strongly prime". In the sequel we will use the definition chosen by Rowen (see [13] and [6]). In order to avoid confusion we will say that a prime ideal P in a ring R is an **r**-*strongly prime ideal* if R/P has no non-zero nil ideals. (Since every maximal ideal of R is an **r**-strongly prime ideal, there are **r**-strongly prime ideals which are not completely prime.) A ring in which every minimal prime ideal is completely prime is called *2-primal*. The 2-primal rings are special cases of NI-rings.

The (two-sided) ideal of a ring R generated by a subset X is written $\langle X\rangle$ or by an element $a \in R$ written $\langle a\rangle$.

Section 1 is devoted to a brief look at **r**-strongly prime ideals. Section 2 contains the main results and Section 3 is devoted to examples, counterexamples and special cases.

1. On r-strongly prime ideals

The main topic will be deferred to the next section. Since **r**-strongly prime ideals will show up in several places we first briefly study these ideals. We get a description of **r**-strongly prime ideals in terms of special sorts of m-systems. Recall that an m-system S in a ring R is a subset of $R \setminus \{0\}$ such that $1 \in S$, and for $r, s \in S$ there is $t \in R$ such that $rts \in S$. The complement of a prime ideal is an m-system and an ideal maximal with respect to not meeting an m-system S is a prime ideal (e.g., [11, §10]). A subset S of $R \setminus \{0\}$ containing 1 and which is closed under multiplication is an example of an m-system.

A ring R, viewed as an algebra over $\mathbb{Z}$, has an *enveloping algebra* $R^e = R \otimes_{\mathbb{Z}} R^{op}$. The bimodule ${}_R R_R$ can be thought of as a left R^e-module. The ring $\mathbf{M}(R) = R^e / \operatorname{ann}_{R^e} R$ is called the *multiplication ring* of R. Then, R is a faithful $\mathbf{M}(R)$-module. For $\lambda \in \mathbf{M}(R)$ we can lift λ to some $\sum_{i=1}^n r_i \otimes s_i \in R^e$ and, for $a \in R$, think of λa as $\sum_{i=1}^n r_i a s_i$. We now formalize the definitions (cf. [13, Definition 2.6.5]). (In [9], the multiplication algebra was used in the definition of a different sort of "strongly prime" ideal.)

Definition 1.1. Let R be a ring.

(1) A prime ideal P of R is called an **r**-*strongly prime ideal* if R/P has no non-zero nil ideals.
(2) A subset S of $R \setminus \{0\}$ is called an *nm-system* if
  (i) S is an m-system and
  (ii) for $t \in S$ there is $\lambda \in \mathbf{M}(R)$, depending on t, such that $(\lambda t)^n \in S$ for all $n \geq 1$.

It is readily seen that any **r**-strongly prime ideal contains an **r**-strongly prime ideal which is minimal among **r**-strongly prime ideals. The intersection of the (minimal) **r**-strongly prime ideals of a ring R is $\mathbf{N}^*(R)$ (see [13, Proposition 2.6.7]). The connection between **r**-strongly prime ideals and nm-systems is clear. The basic information is contained in the following.

Proposition 1.2. *Let R be a ring. Then*

(i) *If $S \subseteq R \setminus \{0\}$ with $1 \in S$ is multiplicatively closed, then S is an nm-system.*
(ii) *If P is an* **r**-*strongly prime ideal, then $R \setminus P$ is an nm-system.*
(iii) *If S is an nm-system and I is an ideal maximal with respect to not meeting S, then I is an* **r**-*strongly prime ideal.*
(iv) *Every* **r**-*strongly prime ideal in R contains a minimal* **r**-*strongly prime ideal (i.e., minimal among the* **r**-*strongly prime ideals).*

Proof. (i) This is clear since for $t \in S$ we can use $\lambda = 1 \in \mathbf{M}(R)$ and then $(\lambda t)^n = t^n \in S$ for all $n \geq 1$.

(ii) If P is an **r**-strongly prime ideal and $S = R \setminus P$, S is an m-system and because R/P has no non-zero nil ideals, for $t \in S$ there is $\lambda \in \mathbf{M}(R)$ such that λt is not nil modulo P, which is exactly the defining feature of an mn-system.

(iii) If S is an nm-system and I an ideal maximal with respect to not meeting S, then I is prime since S is an m-system. Suppose that $x \notin I$ generates an ideal which is nil modulo I. Consider the ideal $I + \langle x \rangle$. Using maximality we pick $t \in (I + \langle x \rangle) \cap S$ and write $t = a + y$, $a \in I$, $y \in \langle x \rangle$. There is $\lambda \in \mathbf{M}(R)$ such that $(\lambda t)^n \in S$ for all $n \geq 1$. Now $(\lambda t)^n = (\lambda a + \lambda y)^n = b + (\lambda y)^n$, where $b \in I$. However, for some $m \geq 1$, $(\lambda y)^m \in I$, which is impossible. Hence, I is an **r**-strongly prime ideal.

(iv) Clear. □

The result [6, Lemma 2.2], using a multiplicatively closed set for S, is a special case of Proposition 1.2(iii).

In a commutative ring R a multiplicatively closed set $S \subseteq R \setminus \{0\}$, $1 \in S$, has a "saturation" $T = \{t \in R \mid \langle t \rangle \cap S \neq \emptyset\}$ which is a multiplicatively closed set and is the complement of the union of the prime ideals maximal with respect to not meeting S. There is a similar result, [9, Proposition 3.6], in connection with the "strongly prime ideals" of that paper. However, there is no "saturation" for nm-systems, in general. A given nm-system can in some cases be enlarged but Example 3.1 will show that there is not always a "saturation".

Remark 1.3. Let R be a ring and $S \subseteq R \setminus \{0\}$ an nm-system. Define $T = \{t \in R \mid \exists\, r, s \in R \text{ with } rts \in S\}$. Then, T is an nm-system whose complement contains the same ideals as the complement of S.

Proof. We first show that T is an m-system. If $t, u \in T$, there are $r, s, r's' \in R$ with $rts, r'us' \in S$. Since S is, in particular, an m-system there is $x \in R$ with $rtsxr'us' \in S$. It follows that $tsxr'u \in T$, showing that T is an m-system. Moreover, if $rts \in S$, there is $\lambda \in M(R)$ with $(\lambda(rts))^n \in S$, for all $n \in \mathbb{N}$. However, $rts \in T$. □

Theorem 2.10, below, gives examples of multiplicatively closed sets which are saturated. As a final remark in this section we have the following companion to a result of Shin, [14, Proposition 1.11]: R is 2-primal if and only if each minimal prime ideal is completely prime.

Proposition 1.4. *A ring R is an NI-ring if and only if each minimal* $\mathbf{r}$*-strongly prime ideal is completely prime.*

Proof. If R is an NI-ring, then each minimal $\mathbf{r}$-strongly prime ideal is completely prime by [6, Theorem 2.3(1)]. In the other direction, if each minimal $\mathbf{r}$-strongly prime ideal is completely prime, then $R/\mathbf{N}^*(R)$ is reduced. This means that $\mathbf{N}^*(R) = \mathbf{N}(R)$. □

2. Weak zero-divisors

The following definition contains some terminology to be used throughout.

Definition 2.1. Let R be a ring.

(i) An element $a \in R$ is called a *left zero-divisor* if there is $0 \neq r \in R$ with $ar = 0$. The set of elements which are not left zero-divisors is denoted S_{nl}. (Similarly for *right zero-divisors* and S_{nr}.)

(ii) An element $a \in R$ is called a *weak zero-divisor* if there are $r, s \in R$ with $ras = 0$ and $rs \neq 0$. The set of elements of R which are not weak zero-divisors is denoted by S_{nw}.

The notion of a weak zero-divisor is what is needed to answer the question about elements of minimal primes.

Theorem 2.2. *Let R be a ring and P a minimal prime ideal of R. Then, for each $a \in P$, a is a weak zero-divisor.*

Proof. Let P be a minimal prime ideal and put $S = R \setminus P$. Suppose, on the contrary, that $a \in P$ is not a weak zero-divisor. Consider the set

$$T = \{r_1 a^{i_1} r_2 \cdots r_k a^{i_k} r_{k+1} \mid k \in \mathbb{N}, i_j \geq 0, r_1 \cdots r_{k+1} \in S\}.$$

It is clear that $T \supseteq S$. The claim is that T is an m-system. It first must be shown that $0 \notin T$. If $0 = r_1 a^{i_1} r_2 \cdots r_k a^{i_k} r_{k+1} \in T$, then the product remains 0 if any factors a are removed since $a \in S_{\text{nw}}$; once all the factors are removed from the expression we get $r_1 \cdots r_{k+1} = 0$, which is not possible since that product is in S. It is next shown that T is an m-system: given two elements of T, $r_1 a^{i_1} r_2 \cdots r_k a^{i_k} r_{k+1}$ and $s_1 a^{j_1} s_2 \cdots s_l a^{j_l} s_{l+1}$, we know that there is $t \in R$ such that $r_1 \cdots r_{k+1} t s_1 \cdots s_{l+1} \in S$. From that, $r_1 a^{i_1} r_2 \cdots r_k a^{i_k} r_{k+1} t s_1 a^{j_1} s_2 \cdots s_l a^{j_l} s_{l+1} \in T$, as required. □

Examples 3.2 and 3.3, below, show that left or right zero-divisors cannot replace weak zero-divisors in Theorem 2.2. However, in a reduced ring weak zero-divisors are both left and right zero-divisors.

Corollary 2.3. *In a ring R, if a is an element of a minimal prime ideal, then there are $r, s \in R$ such that $ras \in \mathbf{N}_*(R)$ and $rs \notin \mathbf{N}_*(R)$. If $R/\mathbf{N}_*(R)$ is reduced (i.e., $\mathbf{N}_*(R) = \mathbf{N}(R)$), then there is $r \notin \mathbf{N}_*(R)$ such that $ra \in \mathbf{N}_*(R)$ and $ar \in \mathbf{N}_*(R)$.*

Proof. The first part is Theorem 2.2 applied to $R/\mathbf{N}_*(R)$. The second follows since in a reduced ring S, $abc = 0$ implies $acb = bac = 0$. □

Corollary 2.3 can, of course, be restated for any ideal I of R in place of $\mathbf{N}_*(R)$ and using the prime ideals minimal over I.

The following simple lemma will be used here and again later.

Lemma 2.4. *Let R be any ring and X a subset of R. Set $M(X) = \{a \in R \mid \exists r, s \in R$ with $ras \in X$ but $rs \notin X\}$ and $M_r(X) = \{a \in R \mid \exists r \in R$ with $ar \in X$ but $r \notin X\}$. Then, $R \setminus M(X)$ and $R \setminus M_r(X)$ are multiplicatively closed and both contain 1.*

Proof. We write M for $M(X)$ and M_r for $M_r(X)$. Suppose $a, b \in R \setminus M$ and $ab \in M$. Then, there are $r, s \in R$ with $rabs \in X$ while $rs \notin X$. Since $a \notin M$, $rbs \in X$ and then $b \in M$. This contradiction shows $ab \notin M$. The statements about M_r are proved similarly. □

In Lemma 2.4 there is an analogous statement for $M_l = M_l(X) = \{a \in R \mid \exists r \in R$ with $ra \in X$ but $r \notin X\}$. Results about M_r for various sets X can be restated for M_l.

Corollary 2.5. *Let R be a ring. Then, S_{nw} and S_{nl} are closed under multiplication and contain 1; in particular, S_{nw} and S_{nl} are nm-systems with $S_{\text{nw}} \subseteq S_{\text{nl}}$.*

Proof. In Lemma 2.4 we take $X = \{0\}$. Moreover, if a is a left zero-divisor then it is a weak zero-divisor. □

Remark 2.6. Let R be a ring. If the set of weak zero-divisors in R forms an ideal W then W is a completely prime ideal. Moreover, if a minimal prime ideal P contains all the weak zero-divisors, then P is completely prime and $P = \mathbf{N}_*(R) = \mathbf{N}(R)$.

Proof. By Corollary 2.5, $R \setminus W$ is a multiplicatively closed set. Hence, W is prime and if $rs \in W$ then $r \in W$ or $s \in W$.

For the remaining part, the minimal prime ideal P is the only minimal prime and is, hence, $\mathbf{N}_*(R)$. □

Remark 2.6 can be illustrated by a trivial extension of a domain. If R is the ring of column finite $\aleph_0 \times \aleph_0$ upper triangular matrices with constant diagonal over a domain D, then R is an example of the situation of Remark 2.6 with $P = \mathbf{N}_*(R)$ nil but not nilpotent. Rings of the type in Remark 2.6 are the subject of [7].

According to Proposition 1.2(iii) or [6, Lemma 2.2], if S is a multiplicatively closed set in a ring R with $0 \notin S$ and $1 \in S$, then an ideal maximal with respect to not meeting S is an $\mathbf{r}$-strongly prime ideal.

Proposition 2.7. *Let R be any ring.*

(i) *There is an $\mathbf{r}$-strongly prime ideal consisting of weak zero-divisors.*
(ii) *There is an $\mathbf{r}$-strongly prime ideal consisting of left zero-divisors. There is a minimal prime ideal consisting of left zero-divisors. Similarly for right zero-divisors.*

Proof. By Proposition 1.2(i) and (iii), an ideal maximal with respect to not meeting the multiplicatively closed set S_{nw} is an $\mathbf{r}$-strongly prime ideal. Similarly for S_{nl}. Moreover, among the prime ideals not meeting S_{nl} there are minimal prime ideals. □

The ring of Example 3.2 has two minimal prime ideals, one consists of elements which are both left and right zero-divisors while the other has weak zero-divisors which are not left or right zero-divisors. See also Example 3.6.

Proposition 2.8. *Let P be a completely prime ideal in a ring R which is minimal among $\mathbf{r}$-strongly prime ideals. Then, the elements of P are weak zero-divisors.*

Proof. We use Proposition 1.2(iii) and put $S = R \setminus P$. The argument of Theorem 2.2 is modified. If $a \in P$ is not a weak zero-divisor then put

$$T = \{r_1 a^{i_1} r_2 \cdots r_k a^{i_k} r_{k+1} \mid i_j \geq 0, r_j \in S, j = 1, \ldots, k\}.$$

It follows that T is a multiplicatively closed set strictly containing S and with $0 \notin T$. An ideal maximal with respect to not meeting T is an $\mathbf{r}$-strongly prime and is contained in P. This is not possible. □

There is an example, [6, Proposition 1.3], based on [6, Example 1.2], of a prime NI-ring R in which $\mathbf{N}^*(R) \neq \mathbf{0}$. Hence, there are $\mathbf{r}$-strongly prime ideals minimal among $\mathbf{r}$-strongly prime ideals but which are not minimal prime ideals ($\mathbf{0}$ is the only minimal prime ideal). Moreover, by [6, Theorem 2.3(1)], these minimal $\mathbf{r}$-strongly prime ideals are completely prime; then Proposition 2.8 applies and the elements of such ideals are weak zero-divisors. We will use the construction of [6, Example 1.2] and our Example 3.2 to show that in Proposition 2.8 weak zero-divisors are required (see Example 3.3, below). We collect some of the remarks above as follows.

Corollary 2.9. *Let R be an NI-ring and P a minimal* $\mathbf{r}$*-strongly prime ideal. Then P consists of weak zero-divisors.*

Proof. As already mentioned, [6, Theorem 2.3(1)] says that Proposition 2.8 applies. □

It also follows from [6, Example 1.2] that, unlike the commutative semiprime case, the union of the minimal primes is not the set of weak zero-divisors. However, in an NI-ring there is an analogous result. Recall (e.g., [12, §2.1, Exercise 11]), that, in a commutative ring R we always have $\mathbf{N}_*(R) = \mathbf{N}(R)$ and, also, the union of the minimal prime ideals is $\{r \in R \mid \exists s \notin \mathbf{N}_*(R) \text{ such that } rs \in \mathbf{N}_*(R)\}$.

Recall that an ideal I of R is called a *completely semiprime ideal* if R/I is a reduced ring; if I is a completely semiprime ideal then the prime ideals minimal over I are completely prime. The next result mimics the commutative case.

Theorem 2.10. *Let R be a ring and I a completely semiprime ideal. Then, $M_r(I) = \{a \in R \mid \exists r \in R \text{ with } ar \in I \text{ but } r \notin I\}$ is the union of the completely prime ideals minimal with respect to containing I. In addition, $R \setminus M_r(I)$ is multiplicatively closed and contains* 1*. The sets $M_r(I)$ and $M_l(I)$ coincide.*

Proof. Let $\mathcal{P}$ be the set of completely prime ideals minimal over I. Suppose $a \in P$ for some $P \in \mathcal{P}$ and we can suppose $a \notin I$. In the reduced ring R/I, $a+I$ is a left zero-divisor. I.e., there is $r \notin I$ such that $ar \in I$, showing that $a \in M_r(I)$.

In the other direction, if we have $a \in M_r(I)$ with $r \notin I$ and $ar \in I$ but $a \notin P$ for each $P \in \mathcal{P}$, then, since these primes are completely prime, r would be in I, which is impossible. Hence, $M_r(I) = \bigcup_{P \in \mathcal{P}} P$.

The next part is an application of the second part of Lemma 2.4 applied to $X = I$. The last observation follows since left and right zero-divisors coincide in a reduced ring. □

The set $R \setminus M_r(I)$ in Theorem 2.10 is a saturated nm-system as discussed at the end of Section 1.

When R is an NI-ring Theorem 2.10 yields a result analogous with the commutative case.

Corollary 2.11. *Let R be an NI-ring and $M_r = M_r(\mathbf{N}(R)) = \{a \in R \mid \exists r \in R \text{ with } ar \in \mathbf{N}(R) \text{ but } r \notin \mathbf{N}(R)\}$. Then, M_r is the union of the minimal* $\mathbf{r}$*-strongly prime ideals of R. Moreover, $R \setminus M_r$ is closed under multiplication and contains* 1.

Proof. We need only invoke Theorem 2.10 with $I = \mathbf{N}(R)$ and the fact that in an NI-ring $\mathbf{N}(R)$ is completely semiprime (e.g., [6, Lemma 2.1]). □

In the special case of a 2-primal ring, the minimal **r**-strongly prime ideals of Corollary 2.11 are, in fact, the minimal prime ideals; in a 2-primal ring Corollary 2.11 is exactly as for commutative rings.

It is remarked in [5, page 4869] that if R is a PI-ring or a ring of bounded index, then R is a NI-ring if and only if R is 2-primal.

The conclusion of Corollary 2.11 need not hold when the ring is not an NI-ring: see Example 3.5, below.

3. Examples and special rings

Our first example is to illustrate how an nm-system can fail to have a saturation.

Example 3.1. There is a ring R such that $T = \{t \in R \mid \langle t\rangle = R\} = \{t \in R \mid \langle t\rangle \cap S \neq \emptyset\}$ is not an m-system and is not a saturation for $S = \{1\}$.

Proof. Let K be a field and $F = K\langle Y, X_1, X_2\rangle$ a free algebra in 3 variables. Set I to be the ideal of F generated by $\rho = X_1YX_2 - 1$ and $R = F/I$. We write the images $Y + I = y, X_1 + I = x_1$ and $X_2 + I = x_2$. By construction, $y \in T$ (as are x_1 and x_2). However, in order to have $v, u_j, w_j \in F$, $j = 1, \ldots, m$ with $\sum_j u_jYvYw_j - 1 \in I$ we would need an equation of the form $\sum_j u_jYvYw_j - 1 = \sum_i r_i\rho s_i$ for some $r_i, s_i \in F$, $i = 1, \ldots, n$. The equation shows that for some k, $1 \leq k \leq n$, r_ks_k has a non-zero constant term. The corresponding $r_kX_1YX_2s_k$, when split into monomial terms, has a monomial term with only one copy of Y. No such term can exist in the other expression. Hence, no element of yRy is in T. This shows that T is not an m-system. □

When a semiprime ring R has only finitely many minimal prime ideals (see [10, Theorem 11.43] for characterizations of such rings) then each element of a minimal prime is a left and a right zero-divisor. The following example shows that even when there are only finitely many minimal prime ideals weak zero-divisors may be required when the ring is not semiprime.

Example 3.2. Let K be a field and $R = K\langle X, Y\rangle/I$ where I is generated by the monomials $XY^iX, i \geq 1$. Write $X + I = x$ and $Y + I = y$. Then, $\langle y\rangle$ is a minimal prime of R, R has only two minimal primes and $\mathbf{N}_*(R) \neq \mathbf{0}$. Moreover, y is neither a left nor a right zero-divisor but $xyx = 0$ while $x^2 \neq 0$.

Proof. Since $R/\langle y\rangle \cong K[X]$, $\langle y\rangle$ is a prime ideal, and, similarly, $\langle x\rangle$ is a prime ideal. Put $L = \langle x\rangle \cap \langle y\rangle$. Then, $L^3 = 0$. Moreover, R/L is reduced since if $r^2 \in L$ and r is written as a polynomial with no terms containing a factor $xy^ix, i \geq 1$, then $r \notin L$ would mean that r has a term purely in x or in y. Then, r^2 would also have such a term. It follows that any prime ideal Q of R contains L and, hence, $\langle x\rangle\langle y\rangle \subseteq Q$. Hence, the minimal primes are $\langle x\rangle$ and $\langle y\rangle$. However, y is not a left or a right zero-divisor while $xyx = 0$ and $x^2 \neq 0$. On the other hand the elements of $\langle x\rangle$ are all left and right zero-divisors. □

The ring R in Example 3.2 is an NI-ring (even 2-primal) because $\langle x\rangle \cap \langle y\rangle = \mathbf{N}(R) = \mathbf{N}_*(R)$. The set M_r from Corollary 2.11 is $\langle x\rangle \cup \langle y\rangle$ and $R/\mathbf{N}(R)$ is the reduced ring $K\langle X,Y\rangle/K$, where K is generated by $\{XY, YX\}$. Moreover (cf., Corollary 2.5), $S_{\mathrm{nw}} = R\setminus(\langle x\rangle \cup \langle y\rangle)$ and $S_{\mathrm{nl}} = R\setminus\langle x\rangle = S_{\mathrm{nr}}$.

Example 3.3. There is an example of an NI-ring R such that $\mathbf{N}_*(R) \neq \mathbf{N}(R)$ in which there is a prime ideal minimal over $\mathbf{N}(R)$ whose elements are neither left nor right zero-divisors (they are weak zero-divisors).

Proof. We rename the ring from Example 3.2 as S and use it as the seed ring in the construction of [6, Example 1.2]. To recall the construction: for each $n \in \mathbb{N}$ let S_n be the ring of $2^n \times 2^n$ upper triangular matrices over S, and S_n is embedded in S_{n+1} by sending $A \in S_n$ to $\left(\begin{smallmatrix} A & 0 \\ 0 & A\end{smallmatrix}\right)$. Then, R is the direct limit of this system of rings. According to [6, Example 1.2], R is an NI-ring but $\mathbf{N}_*(R) \neq \mathbf{N}(R)$.

Now let P be the set of elements r from R which from some $n \in \mathbb{N}$, the matrices representing r have an element of $\langle y\rangle$ in the $(1,1)$ position. The claim is that P is a prime ideal minimal over $\mathbf{N}(R)$. It is clear that it is an ideal. Moreover, $R/P \cong S/\langle y\rangle \cong K[x]$, a prime ring. Just as in Example 3.2, a prime ideal contained in P and containing $\mathbf{N}(R)$ would have to contain the elements with $(1,1)$ entry equal to y. Again as in Example 3.2, if we take for $a \in P$ an element represented by $\left(\begin{smallmatrix} y & 0 \\ 0 & y\end{smallmatrix}\right) \in S_1$, then the equation $ar = 0$ in R with $r \neq 0$ would imply that there is a representative of r in, say, S_n. The element corresponding to a in S_n is yI_{2^n} where I_{2^n} is the identity matrix. Then the product $ar = 0$ in S_n multiplies each row of r by y. Since y is not a left zero-divisor, we have a contradiction. Similarly, a is not a right zero-divisor. $\square$

In the ring R of Example 3.3, $r \in \mathbf{N}(R)$ if and only if r has a representative whose diagonal elements are in $\mathbf{N}(S)$. Examples 3.2 and 3.3 are not semiprime; the next example is of a semiprime ring which has a minimal prime whose elements are neither left nor right zero-divisors.

Example 3.4. There is a semiprime ring R and a minimal prime ideal P along with $a \in P$ such that a is neither a left nor a right zero-divisor.

Proof. We again use the ring of Example 3.2 as a starting point. We will here call that ring R_0. Let R_1 be the ring $K\langle X,Y,Z\rangle/I$ where I is generated by $\{XY^iX \mid i \geq 1\}$, the same defining relations as for R_0. There is a natural embedding of R_0 into R_1. However, R_1 is a prime ring.

The ring R is defined as follows: R is the ring of all sequences $r = (r_n)$ from R_1 such that for some $k \in \mathbb{N}$, depending on r, $r_j \in R_0$ is constant for all $j \geq k$. The ring R is semiprime. To see this, if $r = (r_n) \in R$ and, for some $k \in \mathbb{N}$, $r_k \neq 0$ then $r_k R_1 r_k \neq \mathbf{0}$, showing that $rRr \neq \mathbf{0}$.

We define $P = \{r = (r_n) \in R \mid r_n \text{ is eventually constant and in } \langle y\rangle\}$. Since $R/P \cong R_0/\langle y\rangle$, P is a prime ideal. It now needs to be shown that P is a minimal prime ideal. Suppose that $Q \subseteq P$ is a prime ideal. For any idempotent $e \in R$ (all the idempotents in R are central), $eR(1-e) = \mathbf{0}$ means that $e \in Q$ or $1 - e \in Q$.

However, if e is eventually 1, $e \notin P$ and, hence, $e \notin Q$. Thus $\bigoplus_{i\in\mathbb{N}} R_1 \subseteq Q$. For $u \in R_0$, let $\hat{u}$ denote the element of R which is constantly u. We will see that $\hat{x}R\hat{y}\hat{x} \subseteq Q$. Indeed, for $v \in R$, we may assume that $v \notin \bigoplus_{i\in\mathbb{N}} R_1$ and, hence, that v has the form $v = (0, \ldots, 0, w, w, \ldots)$, where $w \in R_0$. Then, as in the proof of Example 3.2, $\hat{x}v\hat{y}\hat{x} \in Q$. The rest of the proof follows as in the proof of Example 3.2, showing that $\hat{y} \in Q$ and that $Q = P$.

Finally, $\hat{y}$ is neither a left nor a ring zero-divisor but is, of course, a weak zero-divisor. □

It can also be seen that the ring of Example 3.4 is left and right non-singular. See also Proposition 3.9 for more about constructions related to that in Example 3.4.

Example 3.5. There is a ring R with $\mathbf{N}^*(R) = \mathbf{0}$ where $M_r = M_r(\mathbf{N}(R)) = \{a \in R \mid \exists r \notin \mathbf{N}(R) \text{ with } ar \in \mathbf{N}(R)\}$ is not the union of the minimal (**r**-strongly) prime ideals.

Proof. Consider a division ring D and the ring R of sequences of 2×2 matrices over D which are eventually a constant diagonal matrix (e.g., [14, Example 5.6]). Then the von Neumann regular ring R has no nonzero nil ideals and the minimal **r**-strongly prime ideals are also the minimal prime ideals; they are the maximal ideals (i) I_n of sequences zero in the nth component, and (ii) the ideals P_i, $i = 1, 2$, of sequences, eventually a constant diagonal matrix which is zero in the ii position. Consider $a \in R$ where, for $i = 1, \ldots, n$, $n \geq 1$, the ith component of a, a_i, is nonzero but there is $0 \neq r_i$, which is not nilpotent, with $a_i r_i = 0$, while the constant part of a can be the identity matrix. Put $r \in R$ to be r_i for $i = 1, \ldots, n$ and 0 beyond. Then, $ar = 0$ but a is not in the union of the minimal (**r**-strongly)prime ideals. For example, $a = (\left(\begin{smallmatrix} 1 & 0 \\ 0 & 0 \end{smallmatrix}\right), \left(\begin{smallmatrix} 1 & 0 \\ 0 & 1 \end{smallmatrix}\right), \left(\begin{smallmatrix} 1 & 0 \\ 0 & 1 \end{smallmatrix}\right), \ldots)$ and $r = (\left(\begin{smallmatrix} 0 & 0 \\ 0 & 1 \end{smallmatrix}\right), \left(\begin{smallmatrix} 0 & 0 \\ 0 & 0 \end{smallmatrix}\right), \left(\begin{smallmatrix} 0 & 0 \\ 0 & 0 \end{smallmatrix}\right), \ldots)$. Hence, $a \in M_r$ but is not in the union of the prime ideals. Similarly, a is in the set $M_l = M_l(\mathbf{N}(R))$.

On the other hand, the union of the prime ideals is contained in $M_r \cap M_l$. □

In the ring R of Example 3.5, the minimal prime ideals consist of left (and right) zero-divisors. The set of elements of R with constant part 0 is a completely semiprime ideal, call it K. The minimal prime ideals containing K are P_1 and P_2 whose union is, according to Theorem 2.10, $M_r(K) = M_l(K)$. As in any von Neumann regular ring, the set of left zero-divisors is $\{a \in R \mid Ra \neq R\} = M_r(\mathbf{0})$ and that of right zero-divisors in $\{a \in R \mid aR \neq R\} = M_l(\mathbf{0})$; these coincide if, as in our example, the ring is directly finite. However, elements of a proper ideal in a von Neumann regular ring are all left and right zero-divisors. (See also Proposition 3.8, below, for information about a related class of rings to that of the von Neumann regular ones.)

The next example illustrates the left and right versions of Proposition 2.7.

Example 3.6. Let A be a domain which is neither left nor right Ore and $R = \left(\begin{smallmatrix} A & A \\ \mathbf{0} & A \end{smallmatrix}\right)$. Then, R has two minimal prime ideals, one consists of left zero-divisors and the other of right zero-divisors; neither consists of both.

Proof. The two minimal prime ideals are $I = \left(\begin{smallmatrix} A & A \\ \mathbf{0} & \mathbf{0} \end{smallmatrix}\right)$ and $J = \left(\begin{smallmatrix} \mathbf{0} & A \\ \mathbf{0} & A \end{smallmatrix}\right)$. □

See also [2, Example 2.6] and its references for information on minimal prime ideals and zero-divisors of rings of the form of that of Example 3.6.

The ring $K\langle X, Y\rangle/I$, where $I = \langle XY\rangle$, of [2, Example 2.8] shows the same phenomenon as that of Example 3.6.

As in the commutative case, zero-divisors of all sorts do not behave well with respect to homomorphic images. Some information can be gleaned.

Proposition 3.7. *Suppose R is an NI-ring.*

(i) *If $a + \mathbf{N}(R) \in R/\mathbf{N}(R)$ is a weak zero-divisor, then a is a weak zero-divisor in R.*

(ii) *If every element of a proper ideal of $R/\mathbf{N}(R)$ is a weak zero-divisor, then every element of a proper ideal of R is a weak zero-divisor.*

Proof. (i) Suppose that $a \in R$ is such that $a + \mathbf{N}(R)$ is a weak zero-divisor. Then, there are $r, s \in R$ such that $ras \in \mathbf{N}(R)$ and $rs \notin \mathbf{N}(R)$. For some minimal $m \in \mathbb{N}$, $(ras)^m = 0$. If some of the factors a in $(ras)^m$ can be removed to get a non-zero element, the proof is complete. Removing all the factors a, if necessary, leaves $(rs)^m \neq 0$, which gives the result. (ii) This follows directly from (i). □

The converse of Proposition 3.7 is false even in the commutative case. Consider a field K and the ring $R = K[X, Y]/I$, where $I = \langle\{X^n, XY\}\rangle$, for some $n \geq 2$. Then $Y + I$ is a zero-divisor in R but not modulo $\mathbf{N}_*(R) = \langle X + I\rangle$.

The argument in Proposition 3.7(i) does not work for left zero-divisors and, in fact, the conclusion is false for left (or right) zero-divisors. In Example 3.3, the element a shown to be a weak zero-divisor but neither a left nor a right zero-divisor, is both a left and right zero-divisor modulo $\mathbf{N}(R)$.

There are various weak forms of von Neumann regularity which guarantee that elements of proper ideals are in fact zero-divisors. Recall that a ring R is *right weakly π-regular* if for every $a \in R$ there is $m \in \mathbb{N}$ such that $a^m \in a^m\langle a^m\rangle$.

Proposition 3.8. *Let R be a right weakly π-regular ring. Then, every element of a proper ideal is a left zero-divisor.*

Proof. Let $a \in R$ be in a proper ideal and we may assume that a is not nilpotent. We can write, for some $m \in \mathbb{N}$, $a^m = a^m \sum_{i=1}^n r_i a^m s^i$ and $a^m(1 - \sum r_i a^m s_i) = 0$. We know that $\sum r_i a^m s_i \neq 1$ and, thus, there is a minimal $k \geq 1$ such that $a^k(1 - \sum r_i a^m s^i) = 0$. Then, $a^{k-1}(1 - \sum r_i a^m s_i) \in \operatorname{rann} a$. □

Proposition 3.7 applies to rings not covered by Proposition 3.8. Using [1, Theorem 2.6], one needs to find NI-rings R which do not satisfy the idempotent condition WCI ([1, Definition 2.1]), and, hence, is not right weakly π-regular, but for which $R/\mathbf{N}(R)$ is right weakly π-regular. One such is [1, Example 1.7].

For a von Neumann regular ring satisfying general comparability ([4, Definition, page 83]), the minimal prime ideals are generated by central idempotents ([4, Theorem 8.26]) and, hence, an element of a minimal prime ideal is annihilated by a non-zero central idempotent. More generally the observation applies to any ring in which the minimal primes are generated by central idempotents. We will not go into details here but the condition that each minimal prime of a ring R is generated by central idempotents is equivalent to saying that the Pierce sheaf of R has prime stalks (see [8, V 2] or [3]). Biregular rings have this property as do full products of prime rings.

More generally we have the following which will help in the construction of examples. The key property of a Pierce sheaf of a ring R which we will use is that if for some $x \in \operatorname{Spec} \mathbf{B}(R)$ and $r, s \in R$ we have $r_x = s_x$, then there is $e \in \mathbf{B}(R) \setminus x$ such that $re = se$.

Proposition 3.9. *Let R be a ring whose Pierce sheaf has stalks R_x which have the property that each minimal prime ideal of R_x consists of left or of right zero-divisors. Then each minimal prime ideal of R consists of left or of right zero divisors.*

Sketch of proof. Let R_x be a stalk of R (x refers to a maximal ideal of the boolean algebra $\mathbf{B}(R)$ of central idempotents of R and $R_x = R/Rx$).

Since for any prime ideal P of R, $P \cap \mathbf{B}(R) = x$, for some $x \in \operatorname{Spec} \mathbf{B}(R)$ and $R \to R_x = R/Rx$ is surjective, a minimal prime ideal P of R has the following form. For $x = P \cap \mathbf{B}(R)$ and $Q = P_x = P/Rx$, $P = \{r \in R \mid r_x \in Q\}$. Moreover, each such pair (x, Q) yields a minimal prime ideal of R.

Then, if Q, a minimal prime ideal of R_x, consists, say, of left zero-divisors, for $u \in P$, as constructed above, there is $r \in R$ with $r_x \neq 0_x$ and $u_x r_x = 0$. For some $e \in \mathbf{B}(R) \setminus x$, $ure = 0$. Since $re \neq 0$, u is a left zero-divisor. □

The converse is true in a ring like that in Example 3.4 but a small change in that example shows that it is false in general.

Example 3.10. There is a ring R which has a Pierce stalk R_x so that R_x has a minimal prime ideal with an element which is neither a left nor a right zero-divisor but the corresponding minimal prime ideal of R consists of zero-divisors.

Proof. Let R_0 be the ring of Example 3.2 and $S = K\langle x, y, z\rangle/I$, where I is generated by $\{xy^i x \mid i \geq 1\}$ and $\{zxy, zx^2\}$. Then, let R be the ring of sequences from S which are eventually constant and in R_0. The Pierce stalks of R are $R_n = S$, $n \in \mathbb{N}$, and $R_\infty = R_0$. Let $P = \{r \in R \mid r_\infty \in \langle y\rangle\}$; P is a minimal prime ideal of R. It can be seen that the elements of P are all right zero-divisors even though P_∞ has an element which is not a right zero-divisor. Indeed, any monomial in $\langle y\rangle$ is annihilated on the left by $zx \neq 0$. Now let $r \in P$ be such that $r_\infty = u \in R_0$, $u \neq 0$. Then, for some $n \in \mathbb{N}$, $r_n = u$. Let $e \in R$ be such that $e_m = 0$ if $m \neq n$ and $e_n = 1$. Then, $zxer = 0$, while $zxe \neq 0$. □

References

[1] G.K. Birkenmeier, J.Y. Kim and J.K. Park, *Regularity conditions and the simplicity of prime factor rings*, J. Pure Appl. Algebra **115** (1997), 213–230.

[2] G.K. Birkenmeier, J.Y. Kim and J.K. Park, *A characterization of minimal prime ideals*, Glasgow Math. J. **40** (1998), 223–236.

[3] W.D. Burgess and W. Stephenson, *Pierce sheaves of non-commutative rings*, Comm. Algebra **4** (1976), 51–75.

[4] K.R. Goodearl, *Von Neumann Regular Rings.* Krieger, 1991.

[5] C.Y. Hong and T.K. Kwak, *On minimal strongly prime ideals*, Comm. Algebra **28** (2000), 4867–4878.

[6] S.U. Hwang, Y.C. Jeon and Y. Lee, *Structure and topological conditions of NI rings*, J. Algebra **302** (2006), 186–199.

[7] K.-H. Kang, B.-O. Kim, S.-J. Nam and S.-H. Sohn, *Rings whose prime radicals are completely prime*, Commun. Korean Math. Soc. **20** (2005), 457–466.

[8] P.T. Johnstone, *Stone Spaces.* Cambridge Studies in Advanced Mathematics 3, Cambridge University Press, 1982.

[9] A. Kaučikas and R. Wisbauer, *On strongly prime rings and ideals*, Comm. Algebra **28** (2000), 5461–5473.

[10] T.Y. Lam, *Lectures on Modules and Rings.* Springer-Verlag, 1998.

[11] T.Y. Lam, *A First Course in Ring Theory.* Second Edition, Springer-Verlag, 2001.

[12] J. Lambek, *Lectures on Rings and Modules.* Chelsea Publication Co., 1986.

[13] L.H. Rowen, *Ring Theory.* Student Edition, Academic Press, 1991.

[14] G. Shin, *Prime ideals and sheaf representations of a pseudo symmetric ring*, Trans. Amer. Math. Soc. **184** (1973), 43–60.

W.D. Burgess
Department of Mathematics and Statistics
University of Ottawa
Ottawa, ON, K1N 6N5
Canada
e-mail: `wburgess@uottawa.ca`

A. Lashgari
Department of Mathematics
California State University, Fullerton
Fullerton, CA 92834
USA
e-mail: `allashgari@yahoo.com`

A. Mojiri
Department of Mathematics
Saint Xavier University
Chicago, IL 60655
USA
e-mail: `mojiri@gmail.com`

Advances in Ring Theory
Trends in Mathematics, 83–84

On a Theorem of Camps and Dicks

Victor Camillo and Pace P. Nielsen

Abstract. We provide a short, intuitive proof of a theorem of Camps and Dicks [1].

Mathematics Subject Classification (2000). Primary 16P20, Secondary 16D60.
Keywords. Semi-local ring, Artinian module.

1. The theorem

Below, rings are associative with 1, but possibly noncommutative. Modules are unital. We also make use of the well-known fact that a ring R is semi-simple if and only if every maximal left ideal is a summand.

Theorem 1. *Let R and S be rings and let ${}_RM_S$ be an R-S-bimodule. If M_S has finite uniform dimension and for $r \in R$ the equality $\mathrm{ann}_M(r) = (0)$ implies $r \in U(R)$ then R is semi-local.*

Proof. Let $\overline{R} = R/J(R)$, and let ${}_{\overline{R}}A$ be a maximal submodule of ${}_{\overline{R}}\overline{R}$. We wish to show that A is a direct summand. Since M_S has finite uniform dimension there exists an element $b \in R$, $\overline{b} \notin A$, such that $\mathrm{ann}_M(b) \subseteq M_S$ has maximal uniform dimension (with respect to the restriction $\overline{b} \notin A$).

Let $x \in R$ be such that $\overline{xb} \in A$. Notice the containment $\mathrm{ann}_M(b - bxb) \supseteq \mathrm{ann}_M(b) \oplus \mathrm{ann}_M(1 - xb)$ (in fact, equality holds, although we do not need that information). But $\overline{b - bxb} \notin A$, so by the maximality condition on b we conclude $\mathrm{ann}_M(1-xb) = (0)$. Therefore $1-xb \in U(R)$. Repeating the argument, we see that $1-yxb \in U(R)$ for all $y \in R$, so $xb \in J(R)$. We have thus shown that $A \cap \overline{Rb} = (\overline{0})$. By maximality of A we have $A \oplus \overline{Rb} = \overline{R}$, finishing the proof. □

Corollary 2. *If S is a ring and M_S is an Artinian right S-module then $R = \mathrm{End}(M_S)$ is a semi-local ring.*

Notice that in the proof of Theorem 1, we could weaken the condition "$r \in U(R)$" to "r is left invertible." We also remark that in the original proof given by Camps and Dicks in [1], they showed that R is semi-local if and only if there exists an integer $n \geq 0$ and a function $d : R \to \{1, 2, \ldots, n\}$ satisfying $d(b - bxb) = d(b) + d(1 - xb)$, and if $d(a) = 0$ then $a \in U(R)$. One can recover this fact by letting $d(a)$ denote the composition length of the right annihilator of $\overline{a} \in \overline{R}$ and following the ideas in the proof of Theorem 1.

Acknowledgement

This paper was written while the second author was partially supported by the University of Iowa Department of Mathematics NSF VIGRE grant DMS-0602242.

References

[1] Rosa Camps and Warren Dicks, On semilocal rings, *Israel J. Math.* **81** (1993), no. 1–2, 203–211.

Victor Camillo
Department of Mathematics
University of Iowa
Iowa City, IA 52242,USA
e-mail: camillo@math.uiowa.edu

Pace P. Nielsen
Department of Mathematics
Brigham Young University
Provo, UT 84602, USA
e-mail: pace@math.byu.edu

Advances in Ring Theory
Trends in Mathematics, 85–111

Applications of the Stone Duality in the Theory of Precompact Boolean Rings

Mitrofan M. Choban and Mihail I.Ursul

Abstract. In this paper we study connections between topological and algebraical properties of a Boolean ring. We discuss some properties of the lattice of precompact topologies on a Boolean ring. A characterization of precompact topologies on a Boolean ring is given. The classes of extremally disconnected spaces and F-spaces are characterized in terms of injectivity.

The study of precompact topologies is related to the C_p-theory. Many results of the C_p-theory have natural analogous assertions in the theory of precompact Boolean rings.

Mathematics Subject Classification (2000). 06E15, 54H13, 06B20, 13A15, 06E20, 54C40.

Keywords. Boolean ring, Stone space, Bohr topology, totally bounded topology, self-injectivity, extremally disconnectedness, F-space, compactness.

1. Preliminaries

Boolean rings form a remarkable subclass of the class of associative rings. Ring topologies on Boolean rings have many unexpected and subtle properties. For instance, if $(R, \mathcal{T})$ is a Boolean topological ring, then:

(i) components coincide with quasicomponents;
(ii) the product of two neighborhoods of zero is a neighborhood of zero;
(iii) $(R, \mathcal{T})$ is locally connected provided if it is a bounded connected ring.

Recall that a *Boolean ring* is an associative ring R with identity satisfying the identity $x^2 = x$. We consider $x \leq y$ if and only if $xy = x$. A Boolean ring R is called *trivial* if it consists of only one element. Any Boolean ring consisting of two elements is isomorphic to $\mathbb{F}_2 = \mathbb{Z}/2\mathbb{Z} = \{0, 1\}$.

Methods and ideas of Boolean rings are used in the mathematical logics, measure theory, Stone duality, Banach algebras and other domains. An outstanding role in the theory of Boolean rings belongs to the theory of Boolean algebras, established by M.H. Stone [4, 21, 22, 23, 24]. A Boolean ring R endowed with the operations $x \vee y = x + y + xy, x \wedge y = xy, x' = 1 + x$ is a *Boolean algebra.* The element x' is called the *complement* of x. Always $x' \wedge x = 0, x' \vee x = 1$ and $(x')' = x$.

The paper uses the terminology from [3, 11, 21, 25]. As usual, ω stands for the set of all natural numbers $\{0, 1, 2, \ldots\}$ or for the first infinite ordinal. As a rule, we will simply say "a space" instead of "a topological space". By $\mathbb{R}$ we denote the topological ring of reals. Symbol $|A|$ denotes the cardinality of a set A and $w(X)$ denotes the weight of a space X. The closure of a subset A of a space X is denoted by $cl_X(A)$ or, briefly, by clA. A *clopen* subset is a closed-and-open subset of the space.

A zero-dimensional non-empty compact space is called a *Stone space.*

Denote by $C(X)$ the ring of all continuous functions of X into $\mathbb{F}_2$ for every Stone space X. Put $M_Y = \{f | f \in C(X), f(Y) = 0\}$ for every subset $Y \subseteq X$. Obviously, M_Y is an ideal of $C(X)$ and $M_Y = M_{clY}$ for every subset $Y \subseteq X$. The ideal $M_x = M_{\{x\}}$ of $C(X)$ is maximal and $M_Y = \cap\{M_y | y \in Y\}$. Every maximal ideal of the ring $C(X)$ has the form M_x.

If A is a subset of a space X, then $1_A : A \to \mathbb{F}_2$ denotes the characteristic function of A for which $1_A(x) = 1$ if and only if $x \in A$.

Proposition 1.1. *If A, B are closed subsets of a Stone space X, then:*

(i) $M_A \subseteq M_B$ *if and only if* $B \subseteq A$.
(ii) $M_A = M_B$ *if and only if* $A = B$.

Proof. (i) Let $B \subseteq A$. Then $M_A = \cap\{M_x | x \in A\} \subseteq \cap\{M_x | x \in B\} = M_B$. Suppose that $x_0 \in A \setminus B$. There exists a clopen subset U of X such that $x_0 \in U \subseteq X \setminus B$. Then $1_U \in M_B \setminus M_A$. Thus $B \subseteq A$ provided $M_B \subseteq M_A$. The assertion (i) is proved. The assertion (ii) follows immediately from (i). □

Let R be a Boolean ring. Denote by $X(R)$ the set of all maximal ideals of R. We put $V(L) = \{I | I \in X(R), L \subseteq I\}$ for every $L \subseteq R$. The family $\{V(L) | L \subseteq R\}$ is a closed base for the *Stone topology* on $X(R)$. The space $X(R)$ becomes a Stone space, rings R and $C(X(R))$ are isomorphic.

If $\mathfrak{S}$ is the category of all Stone spaces and $\mathfrak{B}$ is the category of all Boolean rings, then $X : \mathfrak{B} \to \mathfrak{S}$ and $C : \mathfrak{S} \to \mathfrak{B}$ are contravariant functors such that $C(X(R)) = R, X(C(X)) = X$ for all $R \in \mathfrak{B}$ and $X \in \mathfrak{S}$. The Stone Duality Theorem is composed from above given facts (see [4, 6, 20, 21, 23, 24]).

For every property P (S, respectively) of the category $\mathfrak{S}$ of all Stone spaces (of the category $\mathfrak{B}$ of all Boolean rings, respectively) there exists a unique property $C(P)$ ($X(S)$, respectively) of Boolean rings (of Stone spaces, respectively) such that:

(i) The properties P and $X(C(P))$ are equivalent.
(ii) The properties S and $C(X(S))$ are equivalent.
(iii) A Stone space X has the property P if and only if the Boolean ring $R = C(X)$ has the property $S = C(P)$.

One principal aim is to find characterizations of the properties $C(P)$ and $X(S)$ for the given properties P and S. We should mention that some properties of Boolean rings may be described in the terms:

(a) of the properties of the lattice of precompact topologies on a ring;
(b) of the properties of the upper semilattice of totally bounded topologies on a ring;
(c) of the properties of the Bohr topology on a ring.

An *atom* of a Boolean ring R is an element $a \in R$, $a \neq 0$, such that if $0 \neq b \leq a$, then $b = a$, equivalently, the principal ideal of R generated by a is equal to $\{0, a\}$. A Boolean ring R is called *atomless* provided it has no atoms. An element $y \in R$ *contains an atom* provided $x \leq y$ for some atom $x \in R$. A Boolean ring R is called *atomic* provided each of its non-zero element contains an atom. Thus $1_b = 1_{\{b\}}$ is an atom in $C(X)$ if and only if b is an isolated point of X.

Remark 1.1. We will consider that $\emptyset$ is a compact zero-dimensional space for which $C(\emptyset) = 0$.

2. Topologies on a Boolean ring

Let R be a Boolean ring. We will consider that $R = C(X)$ for some non-empty Stone space X.

Denote by $\mathcal{T}(R)$ the set of all ring topologies on R and by $\mathcal{T}_\rho(R)$ – the set of all Hausdorff ring topologies on R.

A topology $\mathcal{T} \in \mathcal{T}(R)$ is called *precompact* provided for every non-empty set $U \in \mathcal{T}$ there exists a finite subset F such that $R = F + U$. A Hausdorff precompact topology is called *totally bounded*.

Set $\mathcal{T}^p(R) = \{\mathcal{T} | \mathcal{T} \in \mathcal{T}(R), \mathcal{T} \text{ is precompact}\}$ and $\mathcal{T}^p_\rho(R) = \mathcal{T}^p(R) \cap \mathcal{T}_\rho(R)$. We associate to every subset $Y \subseteq X$ a precompact ring topology $\mathcal{T}_Y$ on $R = C(X)$ having the family $\{M_x | x \in Y\}$ as a subbase at 0.

Proposition 2.1. *Let Y, Z be subsets of a Stone space X and $R = C(X)$. Then:*

(i) $\mathcal{T}_Y \in \mathcal{T}^p(R)$.
(ii) $\mathcal{T}_Y$ *is Hausdorff if and only if Y is dense in X.*
(iii) $\mathcal{T}_Y \subseteq \mathcal{T}_Z$ *if and only if $Y \subseteq Z$.*
(iv) $\mathcal{T}_Y = \mathcal{T}_Z$ *if and only if $Y = Z$.*
(v) $\mathcal{T}^p(R) = \{\mathcal{T}_H | H \subseteq X\}$.

Proof. (i) For every $x \in X$ there exists a unique homomorphism $h_x : R \to \mathbb{F}_2$ such that $h_x^{-1}(0) = M_x$. The homomorphism

$$h_X : R \to \mathbb{F}_2^X, \quad \text{where } r \mapsto (h_x(r))_{x \in X},$$

is an isomorphism of R onto $h(R) \subseteq \mathbb{F}_2^X$.

Let $Y \subseteq X$ and $\pi_Y : \mathbb{F}_2^X \to \mathbb{F}_2^Y$ be the natural projection. Obviously, $h_Y = \pi_Y \circ h_X$. Consider the topology of Tychonoff product on the ring $\mathbb{F}_2^Y$. The ring

$\mathbb{F}_2^Y$ endowed with this topology is a compact Boolean ring. By construction, $\mathcal{T}_Y = \{h_Y^{-1}(U)|U \text{ is open in } \mathbb{F}_2^Y\}$. This implies $\mathcal{T}_Y \in \mathcal{T}^p(R)$.

(ii) The homomorphism $h_Y : R \to \mathbb{F}_2^Y$ is injective if and only if the set Y is dense in X.

(iii) If $Y \subseteq Z$, then $\mathcal{T}_Y \subseteq \mathcal{T}_Z$. Suppose that $b \in Y \setminus Z$ and $\mathcal{T}_Y \subseteq \mathcal{T}_Z$. For every $y \in M_b$ there exists a finite subset $F(y) \subseteq Z$ such that $y \in \cap\{M_z|z \in F(y)\} \subseteq M_b$. There exists a function $f : X \to \mathbb{F}_2$ such that $f(b) = 1$ and $F(y) \subseteq f^{-1}(0)$. Then $f \notin M_b$ and $f \in \cap\{M_z|z \in F(y)\} \subseteq M_b$, contradiction.

(iv) Follows from (iii).

(v) Suppose that $\mathcal{T} \in \mathcal{T}^p(R)$. If $\mathcal{T} = \{\emptyset, R\}$, then $\mathcal{T} = \mathcal{T}_\emptyset$. Suppose that $\mathcal{T} \neq \mathcal{T}_\emptyset$. There exist a Hausdorff compact Boolean ring S and a continuous homomorphism $g : R \to S$ such that $\mathcal{T} = \{g^{-1}(U)|U \text{ is an open subset of } S\}$. It is obvious that $0 = g(0) \neq g(1) = 1$. There exists an open subbase $\{H_\alpha|\alpha \in \Omega\}$ at 0_S such that H_α is a maximal ideal of S for each $\alpha \in \Omega$. For every $\alpha \in \Omega$ there exists a unique $x_\alpha \in X$ such that $M_{x_\alpha} = g^{-1}(H_\alpha)$. If $H = \{x_\alpha|\alpha \in \Omega\}$, then $\mathcal{T} = \mathcal{T}_H$. □

Remark 2.1. We call the homomorphism $h_Y : R \to F_2^Y$ canonical.

Corollary 2.2. *If $|X| = \tau$ and $R = C(X)$, then $|\mathcal{T}^p(R)| = 2^\tau$.*

Corollary 2.3. *If $R = C(X)$, then the set $\mathcal{T}^p(R)$ is a complete lattice, $\mathcal{T}_X$ is the maximal element and $\mathcal{T}_\emptyset$ is the minimal element of $\mathcal{T}^p(R)$.*

Corollary 2.4. *For every topology $\mathcal{T} \in \mathcal{T}^p(R)$ there exists a unique topology $\mathcal{T}' \in \mathcal{T}^p(R)$ such that $\mathcal{T} \vee \mathcal{T}' = \mathcal{T}_X$ and $\mathcal{T} \wedge \mathcal{T}' = \mathcal{T}_\emptyset$.*

Remark 2.2. The lattice $\mathcal{T}^p(R)$ considered as a Boolean algebra is isomorphic to the Boolean algebra $\mathcal{P}(X)$ of all subsets of X. Denote by $\mathcal{T}^{mp}$ the maximal element of $\mathcal{T}^p(R)$.

Remark 2.3. For two spaces X and Y denote by $C_p(X, Y)$ the space of all continuous functions of X into Y with the topology of pointwise convergence. Clearly, $C_p(X, Y)$ is a subspace of the space Y^X. If Z is a closed subspace of Y, then $C_p(X, Z)$ is a closed subspace of the space $C_p(X, Y)$ (see [3], Section 0.3; [11], Section 2.6). We consider $\mathbb{F}_2$ as a subspace (but not a subring) of the space $\mathbb{R}$. Under this convention $\mathcal{T}^{mp}$ is the topology of the space $C_p(X) = C_p(X, \mathbb{F}_2)$ and $C_p(X)$ is a closed subspace of the space $C_p(X, \mathbb{R})$. We identify $C_p(X)$ with $(C(X), \mathcal{T}^{mp})$. Hence, if $\mathcal{P}$ is a topological property hereditary related to closed subspaces and $C_p(X, \mathbb{R}) \in \mathcal{P}$, then $C_p(X) \in \mathcal{P}$ too.

Remark 2.4. Let Y be a subspace of a space X and $C(Y|X) = \{f|Y|f \in C(X)\}$. Then $(C(Y|X), \mathcal{T}_Y)$ is a subring and a subspace of the topological ring $C_p(Y)$.

3. Countably linearly compact Boolean rings

The concept of a linearly compact ring (module) is a natural generalization of compactness in the class of topological rings (modules, respectively)(see, for instance, the historical notes in ([5], p. 675), [7], [28], [17], [18] and [15]).

The following concept was introduced in ([26], p. 39). A topological ring R is called *countably linearly compact* if the intersection of every countable filter base consisting of the sets of the form $x + I$, $x \in R$ and I is a closed left ideal of R, is non-empty. This concept is a generalization of countably compactness to the class of topological rings.

Lemma 3.1. *Let $\{R_\alpha | \alpha \in \Omega\}$ be a family of Boolean topological rings and $R = \prod_{\alpha \in \Omega} R_\alpha$. Then every closed ideal I has the form $I = \prod_{\alpha \in \Omega} I_\alpha$, where each I_α is a closed ideal of R_α.*

Proof. Set $I_\alpha = pr_\alpha I$, where pr_α is the projection of R on $R_\alpha, \alpha \in \Omega$. Obviously, $I \subseteq \prod_{\alpha \in \Omega} I_\alpha$. Conversely, let $x = (x_\alpha) \in \prod_{\alpha \in \Omega} I_\alpha$. Fix $\alpha \in \Omega$. There exists $y \in I$ such that $pr_\alpha(y) = x_\alpha$. Then $x_\alpha \times \prod_{\beta \neq \alpha} 0_\beta = y(x_\alpha \times \prod_{\beta \neq \alpha} 0_\beta) \in I$. This implies that $x_K = \prod_{\alpha \in K} x_\alpha \times \prod_{\beta \notin K} 0_\beta \in I$ for every finite subset $K \subseteq \Omega$. Then $x \in cl\{x_K | K \text{ is a finite subset of } \Omega\} \subseteq I$. □

The problem of the countably compactness of the product of two countably compact groups is not completely solved: under some set-theoretical assumptions there were constructed two countably compact abelian groups A and B whose product $A \times B$ is not countably compact [10]. In this context the theorem below may be of some interest.

Theorem 3.2. *Let $\{R_\alpha | \alpha \in \Omega\}$ be a family of Boolean topological countably linearly compact rings. Then $R = \prod_{\alpha \in \Omega} R_\alpha$ is a countably linearly compact ring.*

Proof. Let $r_0 + J_0 \supseteq r_1 + J_1 \supseteq \cdots \supseteq r_n + J_n \supseteq \cdots$ be a non-increasing family of closed subsets, where J_i is a closed ideal of $R, i \in \omega$. Let $r_i = (x^i_\alpha)$ for every $i \in \omega$. According to Lemma 3.1, there exists the closed ideals $I_{\alpha,i}$ of R_α, $\alpha \in \Omega$, such that $J_i = \prod_{\alpha \in \Omega} I_{\alpha,i}$. Obviously, $x^0_\alpha + I_{\alpha,0} \supseteq x^1_\alpha + I_{\alpha,1} \supseteq \cdots$ for every $\alpha \in \Omega$. If $x_\alpha \in \cap_{i \in \omega}(x^i_\alpha + I_{\alpha,i}), \alpha \in \Omega$, then $(x_\alpha) \in \cap_{i \in \omega}(r_i + J_i)$, i.e., R is a countably linearly compact ring. □

Remark 3.1.

(i) Every countably linearly compact Boolean ring is precompact.

(ii) The class of countably linearly compact Boolean rings is closed relative to the following operations: products, closed ideals and continuous homomorphic images.

(iii) The underlying topological space of a countably linearly compact ring is a Baire space ([26], Lemma I.4.22, p. 40).

4. On minimal topologies

A ring topology $\mathcal{T}$ on a ring R is called *minimal* provided it is a Hausdorff topology and there is no strictly coarser Hausdorff ring topology on R [19, 2, 8, 10].

Lemma 4.1. *Let $(R, \mathcal{T})$ be a Hausdorff (not necessarily associative) topological ring and $b \in R$. Then the set $N_b = \{x | x \in R, bx = b\}$ is closed in the topology $\mathcal{T}$.*

Proof. Follows immediately from continuity of the operations of the topological ring. □

Proposition 4.2. *Let X be a Stone space, $R = C(X)$ and $b \in X$. The following assertions are equivalent:*

(i) *b is an isolated point of X;*
(ii) *$M_b \in \mathcal{T}$ for each $\mathcal{T} \in \mathcal{T}_\rho(R)$;*
(iii) *$M_b \in \mathcal{T}$ for each $\mathcal{T} \in \mathcal{T}^p_\rho(R)$;*
(iv) *$R \setminus M_b \in \mathcal{T}$ for each $\mathcal{T} \in \mathcal{T}_\rho(R)$;*
(v) *$R \setminus M_b \in \mathcal{T}$ for each $\mathcal{T} \in \mathcal{T}^p_\rho(R)$.*

Proof. The implications (ii)⇒(iii) and (iv)⇒(v) are obvious. The implications (ii)⇒(iv), (iii)⇒(v) follow from the fact that any open subgroup of a topological group is closed.

Let I be a maximal ideal of R and 1_X be the identity of R. Then $R \setminus I = 1_X + I$. Hence the ideal I is closed in the space $(R, \mathcal{T})$ if and only if $I \in \mathcal{T}$. Therefore we have proved the equivalences (iv)⇔(ii) and (v)⇔(iii).

(iii)⇒(i). Let b be a non-isolated point of X. Then $Y = X \setminus \{b\}$ is a dense subset of X, $\mathcal{T}_Y \in \mathcal{T}^p_\rho(R)$ and $M_b \notin \mathcal{T}_Y$.

(i)⇒(ii). Let b be an isolated point of X. Consider $r = 1_b \in R$. If $\mathcal{T} \in \mathcal{T}_\rho(R)$, then $M_b = R \setminus N_r \in \mathcal{T}$. □

Proposition 4.3. *The sets $\mathcal{T}_\rho(R)$ and $\mathcal{T}^p_\rho(R)$ are complete upper semilattices for every Boolean ring R.*

Proof. Obviously. □

Theorem 4.4. *Let X be a Stone space, $R = C(X)$ and $\overset{0}{X}$ be the set of all isolated points of X. The following statements are equivalent:*

(i) *$\mathcal{T}_\rho(R)$ is a complete lattice, i.e., there exists a topology $\mathcal{T}_0 \in \mathcal{T}_\rho(R)$ such that $\mathcal{T}_0 \leq \mathcal{T}$ for every $\mathcal{T} \in \mathcal{T}_\rho(R)$;*
(ii) *$\mathcal{T}^p_\rho(R)$ is a complete lattice, i.e., there exists a topology $\mathcal{T}_0 \in \mathcal{T}^p_\rho(R)$ such that $\mathcal{T}_0 \leq \mathcal{T}$ for every $\mathcal{T} \in \mathcal{T}^p_\rho(R)$;*
(iii) *R has a minimal totally bounded topology;*
(iv) *$\overset{0}{X}$ is dense in X.*

Proof. The implication (iv)⇒(i) follows from Proposition 4.2 (see [27], Theorem 9). Let $\mathcal{T}_0 \in \mathcal{T}_\rho(R)$ and $\mathcal{T}_0 \le \mathcal{T}$ for every $\mathcal{T} \in \mathcal{T}_\rho^p(R)$. Since $\mathcal{T}_\rho^p(R) \neq \emptyset$, we obtain $\mathcal{T}_0 \in \mathcal{T}_\rho^p(R)$.

The implication (ii)⇒(iii) is obvious.

(iii)⇒(iv). Let $\mathcal{T}_0$ be a minimal totally bounded topology. According to Proposition 2.1, there exists a dense subset Y of X such that $\mathcal{T}_0 = \mathcal{T}_Y$. Obviously, $\overset{0}{X} \subseteq Y$. If $y \in Y \setminus \overset{0}{X}$, then the set $Z = Y \setminus \{y\}$ is dense in $X, \mathcal{T}_Z \in \mathcal{T}_\rho^p(R), \mathcal{T}_Z \le \mathcal{T}_0$ and $\mathcal{T}_Z \neq \mathcal{T}_0$, a contradiction. □

Corollary 4.5. *There exists no more than one totally bounded minimal topology on a Boolean ring.*

Corollary 4.6. *Let X be a Stone space, $\overset{0}{X}$ dense in X, $R = C(X)$ and $Z = X \setminus \overset{0}{X}$. Then:*

(i) $\mathcal{T}_\rho^p(R) = \{\mathcal{T}_Y | \overset{0}{X} \subseteq Y \subseteq X\}$.
(ii) $|\mathcal{T}_\rho^p(R)| = 2^\tau$, *where* $\tau = |Z|$.
(iii) $\mathcal{T}_\rho^p(R)$ *is a Boolean lattice.*
(iv) *The lattice $\mathcal{T}_\rho^p(R)$ is finite if and only if the set of non-isolated points Z of X is finite.*

Corollary 4.7. *Let X be a Stone space, $R = C(X)$ and $\mathcal{T}_{mp} = \mathcal{T}_{\overset{0}{X}}$. Then $\mathcal{T}_{mp} = \cap\mathcal{T}_\rho^p(R) = \cap\mathcal{T}_\rho(R)$.*

5. Intersection of totally bounded topologies

Let $\overset{0}{X}$ be the set of all isolated points of a Stone space X and $R = C(X)$. Then $\mathcal{T}^p(R)$ is a lattice with zero and identity. Namely, the identity is $\mathcal{T}^{mp} = \mathcal{T}_X$ and the zero is $\mathcal{T}_m = \mathcal{T}_\emptyset = \{\emptyset, R\}$.

If R is an atomic ring, then $\mathcal{T}_{mp}$ is the least element of $\mathcal{T}_\rho^p(R)$.

Definition 5.1. A topology $\mathcal{T} \in \mathcal{T}(R)$ is called atomic provided M_x is open in $\mathcal{T}$ for every isolated point x of X.

Let $\mathcal{T}_a(R)$ be the set of all atomic topologies on R and $\mathcal{T}_a^p(R) = \mathcal{T}_a(R) \cap \mathcal{T}^p(R)$.

Theorem 5.2. *Let R be a Boolean ring and $\mathcal{T} \in \mathcal{T}^p(R)$. The following assertions are equivalent:*

(i) $\mathcal{T} \in \mathcal{T}_a(R)$.
(ii) *There exist two totally bounded topologies $\mathcal{T}', \mathcal{T}'' \in \mathcal{T}_\rho^p(R)$ such that $\mathcal{T} = \mathcal{T}' \cap \mathcal{T}''$ and $\mathcal{T}' \cup \mathcal{T}'' = \mathcal{T}^{mp}$.*

Proof. Assume that $R = C(X)$, where X is a Stone space. The implication (ii)⇒(i) follows from Corollary 4.7.

We fix $\mathcal{T} \in \mathcal{T}_a^p(R)$. According to Proposition 2.1, there exists a subset $H \subseteq X$ such that $\mathcal{T} = \mathcal{T}_H$. According to Definition 5.1 and Corollary 4.6, we have $\overset{0}{X} \subseteq H$.

Case 1. H is dense in X. In this case $\mathcal{T} \in \mathcal{T}_\rho^p(R)$, $\mathcal{T}' = \mathcal{T}$ and $\mathcal{T}'' = \mathcal{T}_X$.

Case 2. H is not dense in X. Then $X_1 = X \setminus cl_X H$ is a locally compact subspace without isolated points and there exist two subsets Y_1 and Z_1 of X_1 such that $Y_1 \cup Z_1 = X_1, X_1 \subseteq cl_X Y_1 = cl_X Z_1$ and $Y_1 \cap Z_1 = \emptyset$. Let $Y = H \cup Y_1$ and $Z = H \cup (X \setminus Y_1)$. Then Y and Z are dense subsets of X, $X = Y \cup Z$, $\overset{0}{X} \subseteq Y \cap Z = H$ and $\mathcal{T}' = \mathcal{T}_Y$, $\mathcal{T}'' = \mathcal{T}_Z$ are the searched totally bounded topologies. □

Corollary 5.3. *Let R be an atomless Boolean ring. Then for every $\mathcal{T} \in \mathcal{T}^p(R)$ there exist $\mathcal{T}', \mathcal{T}'' \in \mathcal{T}_\rho^p(R)$ such that $\mathcal{T} = \mathcal{T}' \cap \mathcal{T}''$ and $\mathcal{T}' \cup \mathcal{T}'' = \mathcal{T}^{mp}$.*

Remark 5.1. Let R be a Boolean ring, X be a Stone space and $R = C(X)$.

1. If $\overset{0}{X}$ is dense in X, then $\mathcal{T}_a(R) = \mathcal{T}_\rho(R)$.
2. If $\overset{0}{X} = \emptyset$, then $\mathcal{T}_a(R) = \mathcal{T}(R)$.
3. A topology $\mathcal{T} = \mathcal{T}_Y$ is atomic if and only if $\overset{0}{X} \subseteq Y$.
4. The set $\mathcal{T}_a^p(R)$ is a complete lattice with the minimal topology $\mathcal{T}_{mp}$ and the maximal topology $\mathcal{T}^{mp}$.
5. The lattice $\mathcal{T}_a^p(R)$ considered as a Boolean algebra is isomorphic to the Boolean algebra $\mathcal{P}(Z)$ of all subsets of $Z = X \setminus \overset{0}{X}$.

6. The Bohr topology on a Boolean ring

The maximal totally bounded topology $\mathcal{T}^{mp}$ on a Boolean ring R is the *Bohr topology* on R.

It is well known that the set of all maximal ideals of R is a subbase and the set of all cofinite ideals of R, respectively, is a base for the space $(R, \mathcal{T}^{mp})$ at zero.

A ring R is called a *minimally almost periodic ring* provided its Bohr topology is the coarsest possible Hausdorff ring topology on R (see [14]).

Theorem 6.1. *Every minimally almost periodic Boolean ring R is finite.*

Proof. Let $R = C(X)$ and X be a Stone space. Since the topology $\mathcal{T}^{mp}$ is the minimal element of the lattice $\mathcal{T}_\rho^p(R)$, the ring R is atomic and the set $\overset{0}{X}$ is dense in X. Moreover, in this case $\mathcal{T}^{mp} = \mathcal{T}_{mp}$ and $|\mathcal{T}_\rho^p(R)| = 1$. Thus $X \setminus \overset{0}{X} = \emptyset$ and the set $\overset{0}{X} = X$ is finite. Therefore the ring $R = \mathbb{F}_2^X$ is finite too. □

Theorem 6.2. *For a Boolean ring R the following assertions are equivalent:*

(i) *R is finite;*
(ii) *R is minimally almost periodic;*
(iii) *The Bohr topology on R is compact;*
(iv) *The Bohr topology on R is countably compact;*
(v) *The Bohr topology on R is pseudocompact.*

Proof. Implications (i)⇒(iii)⇒(ii) and (iii)⇒(iv)⇒(v) are obvious. Implications (ii)⇒(iii) and (ii)⇒(i) follow from Theorem 6.1. Implication (v)⇒(iii) follows from M.O. Asanov-N.V. Velichko's generalization of Grothendieck's Theorem ([3], Theorem 3.4.1). □

Recall that that the tightness $t(X)$ of a space X is countable provided $cl_X L = \cup\{cl_X H | H \subseteq L, |H| \leq \omega\}$ for every subset $L \subseteq X$.

Theorem 6.3. *If R is a Boolean ring endowed with the Bohr topology, then the tightness $t(R)$ of R is countable.*

Proof. The proof is similar to the Arhangel'skii's proof of Theorem 2.1.1 from [3]. Let $A \subseteq C(X) = R$, where X is the Stone space corresponding to the ring R. Denote by A' the closure of A in $(R, \mathcal{T}^{mp})$.

We fix $f \in A'$. If $n \geq 1$, then for every $\xi = (x_1, \ldots, x_n) \in X^n$ there exists $g_\xi \in A$ such that $g_\xi(x_i) = f(x_i)$ for every $i \leq n$. We put $V_{x_i} = g_\xi^{-1}(g_\xi(x_i)) \cap f^{-1}(f(x_i))$ and $V_\xi = \prod\{V_{x_i} | i \leq n\}$. Then V_ξ is an open set and $\xi \in V_\xi$. Thus there exists a finite set $B_n \subseteq X^n$ such that $X^n = \cup\{V_\xi | \xi \in B_n\}$. Let $L = \{g_\xi | \xi \in \cup\{B_n | n \in \omega\}\}$. Then $f \in clL$ and $L \subseteq A$. □

Theorem 6.4. *Let X be the Stone space of a Boolean ring R with the Bohr topology $\mathcal{T}^{mp}$. The following conditions are equivalent:*

(i) *X is a scattered space;*
(ii) *R is a superatomic ring;*
(iii) *R is a k-space;*
(iv) *R is a sequential space;*
(v) *R is a Fréchet-Urysohn space.*

Proof. Implications (v)⇒(iv)⇒(iii) are obvious. Implications (i)⇒(ii)⇒(i) are well known.

We can consider that the space $(R, \mathcal{T}^{mp})$ is a closed subspace of the space $C_p(X, \mathbb{R})$ of all real-valued continuous functions with the topology of pointwise convergence (see Remark 2.3). Thus the implication (i)⇒(v) follows from ([3], Theorem 3.1.2). Denote by Y the Cantor subset of reals. If X is not scattered, then there exists a continuous mapping of X onto Y and the ring $C(Y)$ is a subring of $R = C(X)$. The ring $C(Y)$ with the Bohr topology is a closed subspace of $(R, \mathcal{T}^{mp})$. Thus the ring $C(Y)$ with the Bohr topology is a Fréchet-Urysohn space. We claim that $C(Y)$ is not a Fréchet-Urysohn space. There exists on Y a σ-additive Borel measure μ such that $\mu(Y) = 1$ and $\mu(U) > 0$ for every non-empty subset $U \subseteq Y$.

For every $n \in \omega$ there exists a finite cover $\xi_n = \{V_i|i \in A_n\}$ such that V_i^n is clopen, $\mathrm{diam}(V_i^n) < 2^{-n}$ and $\mu(V_i^n) < \frac{1}{(n+1)^2}$. Let $\xi'_n = \{\cup\{V_i^n|i \in B\}|B \subseteq A_n, |B| \leq n\}$ and $\xi = \cup\{\xi'_n|n \in \omega\} = \{W_m|m \in \omega\}$, where:

$P1$. $\mu(W_{m+1}) \leq \mu(W_m)$;

$P2$. If $\varepsilon > 0, F$ is a finite subset of Y, U is open in Y and $F \subseteq U$, then there exists $m \in \omega$ such that $F \subseteq W_m \subseteq U$ and $\mu(W_m) < \varepsilon$.

Thus ξ is a clopen base for Y and $\lim \mu(W_m) = 0$.

We fix a point $r_n \in W_n$. Then $\{r_n|n \in \omega\}$ is a dense subset of Y and for each $m \in \omega$ there exists $f_m \in C(Y)$ such that $\{r_i|i \leq m\} \cup W_m \subseteq f_m^{-1}(0)$ and $\mu(f_m^{-1}(0)) < 2\mu(W_m)$.

Let $g(x) = 0$ for every $x \in X$. From $P2$ it follows that g is an accumulation point of $Z = \{f_m|m \in \omega\}$ in R. For every sequence from Z there exists a subsequence $\{f_{m_k}|k \in \omega\}$ such that $\mu(f_{m_k}^{-1}(0)) \leq 2^{-k}$ for every $k \in \omega$. There exists, in this case, a F_σ-subset $H \subseteq Y$ such that $\mu(H) = 1$ and $\lim f_{m_k}^{-1}(x) = 1$ for any $x \in H$. Thus a subsequence $\{f_{m_n}|n \in \omega\}$ of Z for which $g = \lim f_{m_n}$ does not exist. Hence $C(Y)$ and $C(X)$ are not Fréchet-Urysohn spaces. □

7. Compact topologies on Boolean rings

A space X is called *extremally disconnected* provided the closure of each its open subset is open (see [11, 13, 20]).

Theorem 7.1. *Let X be a Stone space and $R = C(X)$. The following conditions are equivalent:*

(i) *There exists a compact ring topology in $\mathcal{T}_\rho(R)$;*

(ii) *There exists a compact ring topology in $\mathcal{T}_\rho^p(R)$;*

(iii) *R is atomic and X is extremally disconnected;*

(iv) *There exists a discrete space D_τ such that $X = \beta D_\tau$.*

Proof. The implications (i)⇒(ii)⇒(i) and (iii)⇒(iv) are obvious.

(iv)⇒(i) Let Y be a discrete space and $X = \beta Y$. Then $\overset{0}{X} = Y$ and the space X is extremally disconnected. Every mapping $f : Y \to \mathbb{F}_2$ has a continuous extension $\beta f : \beta Y \to \mathbb{F}_2$. Thus $h_Y : R \to \mathbb{F}_2^Y$ is an isomorphism of R onto $\mathbb{F}_2^Y$ and $\mathcal{T}_Y$ is a compact ring topology on R.

(ii)⇒(iii) Let $\mathcal{T}$ be a compact ring topology on R. Then $\mathcal{T} = \mathcal{T}_{mp} = \mathcal{T}_{\overset{0}{X}} = \mathcal{T}_Y$.

Thus X is a compactification of the discrete space $Y = \overset{0}{X}$ and the canonical mapping $h_Y : F \to \mathbb{F}_2^Y$ is an isomorphism onto $\mathbb{F}_2^Y$. In particular, for every function $f : Y \to \mathbb{F}_2$ there exists a unique continuous mapping $bf : X \to \mathbb{F}_2$ such that $f = bf|Y$. Then $cl_X f^{-1}(0) \cap cl_X f^{-1}(1) = \emptyset$, i.e., $X = \beta Y$. □

Remark 7.1. The Stone-Čech compactification of a discrete space is called a free compact space (see [20], p. 246).

8. Dense ideals of a ring

The following concept is central in the theory of complete rings of quotients (see, for instance, [16]). Let R be a commutative ring with identity. An ideal I of R is called *dense* provided $rI = 0$ implies $r = 0$.

Proposition 8.1. *Let Y be a closed subset of a Stone space X and $R = C(X)$. The following conditions are equivalent:*

(i) *The ideal M_Y is dense in R;*
(ii) *The set $X \setminus Y$ is dense in X.*

Proof. (i)⇒(ii) By definition, $M_Y = \cap\{M_y | y \in Y\} = \{f | f \in R, Y \subseteq f^{-1}(0)\}$.

Assume that the set $Z = X \setminus Y$ is not dense in X. There exists $r \in R$ such that $r^{-1}(1) \subseteq Y$ and $r^{-1}(1) \neq \emptyset$. Then $r \neq 0$ and $r \cdot f = 0$ for every $f \in M_Y$.

(ii)⇒(i) Assume that the set $Z = X \setminus Y$ is dense and $f \cdot r = 0$ for every $f \in M_Y$. There exists a clopen subset U of X such that $U = r^{-1}(1)$. If $r \neq 0$, then $U \neq \emptyset$ and there exists an element $t \in U \cap Z$. Since Y is closed in X we can assume that $U \cap Y = \emptyset$ and $r \in M_Y$. Then $r \cdot r = r \neq 0$, a contradiction. □

9. Self-injective Boolean rings

Let X be a Stone space and $R = C(X)$. An open set $U \subseteq X$ is called *regular* provided $U = \mathrm{Int}\, cl_X U = X \setminus cl_X(X \setminus cl_X U)$.

A mapping $f : X \to \mathbb{F}_2$ is called *semicontinuous* provided $f^{-1}(1)$ is a regular open subset of X. The set $E(f) = f^{-1}(1) \cup \mathrm{Int}\, f^{-1}(0)$ is open and dense in X.

Denote by $B(X)$ the set of all semicontinuous mappings $f : X \to \mathbb{F}_2$.

For each mappings $f, g \in B(X)$ there exist an open dense subset Y of X and two uniquely determined mappings $\varphi, \psi \in B(X)$ such that $\varphi(x) = f(x) + g(x)$ and $\psi(x) = f(x) \cdot g(x)$ for every $x \in U$. We put $\varphi = f + g$ and $\psi = f \cdot g$. Then $B(X)$ will be a Boolean ring and $C(X)$ a subring of $B(X)$. The space X is extremally disconnected if and only if $C(X) = B(X)$.

It is well known that $B(X)$ is a complete Boolean ring and the complete ring of quotients of $C(X)$ (see [6, 12, 20, 24]). Recall that a continuous mapping $f : X \to Y$ of a space X on a space Y is called *irreducible* provided $f(A) \neq Y$ for every proper closed subset A of X. A pair (aX, π_X) is called the *projective resolution* or the *projective envelope,* or the *absolute* of the space X if aX is an extremally disconnected compact space and $\pi_X : aX \to X$ is a continuous irreducible mapping onto X (see [11, 20]).

If $i_X(f)(t) = f(\pi_X(t))$ for any $f \in B(X)$ and $t \in \pi_X^{-1}(E(f))$, then $i_X : B(X) \to C(aX)$ is an isomorphism.

Every ideal I of a ring R is considered as an R-module. We put $S(f) = f^{-1}(1)$ for any $f \in B(X)$.

We fix for an ideal I of R an R-module homomorphism $\varphi : I \to R$. We fix $V(f, \varphi) = S(\varphi(f))$ and $W(f, \varphi) = S(f) \setminus V(f, \varphi)$ for each $f \in I$. We put also $V(\varphi) = \cup\{V(f, \varphi) | f \in I\}$ and $W(\varphi) = \cup\{W(f, \varphi) | f \in I\}$.

We note that $S(f)$, $V(f,\varphi)$ and $W(f,\varphi)$, $f \in I$, are clopen subsets of the space X. This implies that $V(\varphi)$ and $W(\varphi)$ are open subsets of X.

We mention also that $S(f \cdot g) = S(f) \cap S(g)$ for all $f, g \in R$.

Lemma 9.1. *$V(f,\varphi) \subseteq S(f)$ for every $f \in I$.*

Proof. Since $\varphi(f) = \varphi(f \cdot f) = f\varphi(f)$ we have $V(f,\varphi) = S(\varphi(f)) = S(f\varphi(f)) = S(f) \cap S(\varphi(f)) \subseteq S(f)$. □

Lemma 9.2. *If $f, g \in B(X)$, then $f \leq g$ if and only if $S(f) \subseteq S(g)$.*

Proof. Obviously. □

Lemma 9.3. *If $f, g \in I$ and $f \leq g$, then $V(f,\varphi) \subseteq V(g,\varphi)$.*

Proof. Indeed, $V(f,\varphi) = S(\varphi(f)) = S(\varphi(fg)) = S(f\varphi(g)) = S(f) \cap S(\varphi(g)) \subseteq S(\varphi(g)) = V(g,\varphi)$. □

Lemma 9.4. *$\varphi(\varphi(f)) = \varphi(f)$ for every $f \in I$.*

Proof. Since $\varphi(f) = \varphi(f \cdot f) = f\varphi(f) \in I$, we have $\varphi(\varphi(f)) = \varphi(\varphi(f \cdot f)) = \varphi(f\varphi(f)) = \varphi(f)\varphi(f) = \varphi(f)$. □

Lemma 9.5. *If $f, g \in I$, then $S(f) \cap S(\varphi(g)) = S(\varphi(f)) \cap S(\varphi(g))$.*

Proof. According to Lemma 9.4 we have $S(f) \cap S(\varphi(g)) = S(f \cdot \varphi(g)) = S[\varphi(f \cdot \varphi(g))] = S[\varphi(\varphi(g) \cdot f)] = S(\varphi(g) \cdot \varphi(f)) = S(\varphi(f)) \cap S(\varphi(g))$. □

Theorem 9.6. *$V(f,\varphi) = S(f) \cap V(\varphi)$ for every $f \in I$.*

Proof. Obviously, $V(f,\varphi) \subseteq S(f) \cap V(\varphi)$.

Conversely, let $t \in S(f) \cap V(\varphi)$. There exists $g \in I$ such that $t \in V(g,\varphi)$. According to Lemma 9.5, we have $t \in S(f) \cap S(\varphi(g)) = S(\varphi(f)) \cap S(\varphi(g)) \subseteq V(f,\varphi)$. □

Corollary 9.7. *Let $D(I) = \cap\{f^{-1}(0) | f \in I\}$. Then $V(\varphi) \cap W(\varphi) = \emptyset$ and $V(\varphi) \cup W(\varphi) = X \setminus D(I)$.*

Proof. From Theorem 9.6 it follows that $V(\varphi) \cap W(\varphi) = \emptyset$. If $f \in I$, then $D(I) \cap V(f,\varphi) \subseteq f^{-1}(0) \cap f^{-1}(1) = \emptyset$ and $V(f,\varphi) \cup W(f,\varphi) = S(f) = X \setminus f^{-1}(0)$. Thus $V(\varphi) \cup W(\varphi) \subseteq X \setminus D(I)$. For every $t \in X \setminus D(I)$ there exists a function $g \in I$ such that $g(t) = 1$. Hence, $t \in S(f)$. □

Theorem 9.8. *If $\varphi \in \mathrm{Hom}_R(I,R)$, then the following statements are equivalent:*

(i) *φ can be extended to an R-module homomorphism $\psi : R \to R$;*
(ii) *There exists a clopen subset $H \subseteq X$ such that $V(\varphi) \subseteq H \subseteq X \setminus W(\varphi)$.*

Proof. (i)⇒(ii) Let $1_X(t) = 1$ for every $t \in X$. If $\psi : R \to R$ is an R-module homomorphism and $\varphi = \psi|I$, then we put $H = S(\psi(1_X))$. The set H is clopen in X and $v(f, \varphi) = S(\varphi(f)) = S(\varphi(f \cdot 1_X)) = S(\psi(f \cdot 1_X))$, $S(f \cdot \psi(1_X)) = S(f) \cap S(\psi(1_X)) \subseteq H$ for every $f \in I$. This implies that $V(\varphi) \subseteq H$. Furthermore, $H \cap W(f, \varphi) = \emptyset$ for every $f \in I$. Assume on the contrary that there exists $t \in H \cap W(f, \varphi)$. Then $\psi(1_X)(t) = 1, f(t) = 1$ and $\varphi(f)(t) = 0$. Therefore $0 = \varphi(f)(t) = \psi(f)(t) = \psi(1_X)(t) \cdot f(t) = 1$, a contradiction.

(ii)⇒(i) Let H be a clopen subset of X and $V(\varphi) \subseteq H \subseteq X \setminus W(\varphi)$. Consider the mapping $\psi : R \to R, r \mapsto r \cdot 1_H$. If $A \subseteq X \setminus D(I)$, then $A \cap V(\varphi) = A \cap H$. Since $V(f, \varphi) = S(f) \cap V = S(f) \cap H$, we obtain $\varphi(f) = \psi(f)$ for every $f \in I$. □

Remark 9.1. The clopen subset $H = V(\psi)$ from Theorem 9.8 is called the kernel of the extension ψ of the homomorphism φ. Every extension ψ of the homomorphism φ is determined in this way by some kernel H. Thus, the set of extensions of the homomorphism φ to R can be enumerated by kernels of type H. If $X \setminus D(I)$ is dense in X, then φ has at most one extension.

Remark 9.2. Let $\varphi_1, \varphi_2 : R \to R$ be the extensions of $\varphi : I \to R$. We put $\varphi_1 \leq \varphi_2$ if $\varphi_1(g) \leq \varphi_2(g)$ for any $g \in R$, i.e., $V(\varphi_1) \subseteq V(\varphi_2)$. Denote $H_{\min} = cl_X V(\varphi)$ and $H_{\max} = X \setminus cl_X W(\varphi)$. If ψ is an extension of the homomorphism φ, then $H_{\min} \subseteq V(\psi) \subseteq H_{\max}$. Suppose now that the space X is extremally disconnected. There exist two R-module homomorphisms $\varphi_{\max}, \varphi_{\min} : R \to R$ for which $V(\varphi_{\min}) = H_{\min}$ and $V(\varphi_{\max}) = H_{\max}$. In this case the mappings $\varphi_{\max}, \varphi_{\min}$ are the maximal and the minimal extensions of the homomorphism φ.

Theorem 9.9. *Let (V, W) be a pair of open subsets of X such that $V \cap W = \emptyset$ and $V \cup W = X \setminus D(I)$. Then:*

(i) *The mapping $\varphi : I \to R$, where $V(f, \varphi) = S(f) \cap V$ for every $f \in I$, is an element of* $\mathrm{Hom}_R(I, R)$ *and $V(\varphi) = V$, $W(\varphi) = W$.*
(ii) *If $\varphi_1 \in \mathrm{Hom}_R(I, R), V(\varphi_1) = V$ and $W(\varphi_1) = W$, then $\varphi_1 = \varphi$.*

Proof. (i) *Let $f \in I$. Then $S(f) \cap V$ is a clopen subset. It is obvious that $S(f) \cap D(I) = \emptyset$ and $S(f) \subseteq V \cup W$.*

By construction, $\varphi(f) = f \cdot 1_V$ for any $f \in I$. Since $S(\varphi(f)) = S(f) \cap V$ is a clopen subset, $\varphi(f) \in R$. Thus φ is a mapping of I in R. If $f \in I$ and $g \in R$, then $S(f \cdot g) = f^{-1}(1) \cap g^{-1}(1) \cap V = f^{-1}(1) \cap (g^{-1}(1) \cap V)$. Thus $\varphi(g \cdot f) = g \cdot \varphi(f)$. Let $f, g \in I$ and $h = f + g$. Then $V(h, \varphi) = h^{-1}(1) \cap V$ $W(\varphi) = W$.

(ii) Follows from Theorem 9.6. □

Theorem 9.10. *The following assertions are equivalent for a Boolean ring R:*

(i) *R is self-injective.*
(ii) *The Stone space X of R is extremally disconnected.*
(iii) *The ring R considered as a Boolean algebra is complete.*

Proof. (i)⇒(ii) We will use the well-known Baer's Test for the self-injectivity of R: if I is an ideal of R and $\varphi : I \to R$ is an R-module homomorphism, then there exists an R-module homomorphism $\psi : R \to R$ such that $\varphi = \psi|I$ (see [16], §4.2, Lemma 1, p. 88).

Let R be a self-injective ring, V be an open subset of X, $W = X \setminus cl_X V$, $Y = X \setminus (V \cup U)$ and $I = M_Y$. Then $D(I) = Y$. We consider the homomorphism $\varphi \in \mathrm{Hom}_R(I, R)$ such that $V(\varphi) = V$ and $W(\varphi) = W$. There exists an R-module homomorphism $\psi : R \to R$ such that $\varphi = \psi|I$. The set $H = S(\psi(1_X))$ is clopen and $V \subseteq H \subseteq X \setminus W$. By construction, $H = cl_X V$. Thus the closure of an open set is open.

(ii)⇒(i) Let X be an extremally disconnected space and $\varphi : I \to R$ be an R-module homomorphism. The set $cl_X V(\varphi) = H$ is clopen in X and $V(\varphi) \subseteq H \subseteq X \setminus W(\varphi)$. Theorem 9.8 finishes the proof.

The equivalence (ii)⇔(iii) is well known ([21], p. 140). □

Corollary 9.11. *The ring $B(X)$ is self-injective for every compact space X.*

Corollary 9.12. *There are no Boolean self-injective countable rings. In particular, every infinite Boolean ring contains a non self-injective subring.*

10. Zero-dimensional F-spaces

A compact space Z is called an *F-space* provided for each pair of disjoint open F_σ-sets V and W their closures are disjoint.

Every extremally disconnected compact space is an F-space and each closed subspace of an F-space is an F-space (see [20], Proposition 24.2.5 and Notes 24.2.12).

Definition 10.1. A commutative ring R with identity is called ω-self-injective provided for every countably generated ideal I of R every $\varphi \in \mathrm{Hom}_R(I, R)$ can be extended to an endomorphism $\psi \in \mathrm{Hom}_R(R, R)$.

Remark 10.1. If X is a compact zero-dimensional space and Y is a closed G_δ-subspace of X, then the ideal M_Y is countably generated.

Indeed, we may assume without loss in generality that $Y = \cap\{U_i | i \in \omega\}$, where U_i are clopen subsets of X and $U_0 \supseteq U_1 \supseteq \cdots$. Let $f \in C(X)$ and $f_n^{-1}(0) = U_n$ for any $n \in \omega$. Then the ideal I is generated by the set $\{f_n | n \in \omega\}$ and $D(I) = Y$.

Theorem 10.2. *Let X be a Stone space and $R = C(X)$. The following assertions are equivalent:*

(i) *X is an F-space;*
(ii) *R is ω-self-injective.*

Proof. (i)⇒(ii) An ideal I of R is countably generated if and only if $D(I)$ is a G_δ-set.

Let X be an F-space and I be a countably generated ideal of R. There exists a sequence $(f_n)_{n\in\omega}$ such that $D(I) = \cap\{f_i^{-1}(0) | i \in \omega\}$ and $f_{i+1}^{-1}(0) \subseteq f_i^{-1}(0)$ for any $i \in \omega$. Then $X \setminus D(I) = \cup\{S(f_i)|i \in \omega\}$. We fix an R-module homomorphism $\varphi : I \to R$. The sets $V_i(\varphi) = V(\varphi) \cap S(f_i)$ and $W_i(\varphi) = W(\varphi) \cap S(f_i)$ are clopen in X for every $i \in \omega$. Since $V(\varphi) = \cup\{V_i(\varphi)|i \in \omega\}$ and $W(\varphi) = \cup\{W_i(\varphi)|i \in \omega\}$, the sets $V(\varphi)$ and $W(\varphi)$ are open F_σ-subsets of X. Thus $cl_X V(\varphi) \cap cl_X W(\varphi) = \emptyset$ and there exists a clopen subset H of X such that $V(\varphi) \subseteq X \subseteq X \setminus W(\varphi)$. From Theorem 9.8 it follows that φ can be extended to an R-module homomorphism $\psi : R \to R$.

(ii)⇒(i) Assume that R is an ω-self-injective ring, V and W are open F_σ-sets of X and $V \cap W = \emptyset$. The set $Y \setminus (V \cup W)$ is a closed G_δ-set of X. According to Remark 10.1, the ideal $I = M_Y$ is countably generated. According to Theorem 9.9, there exists $\varphi \in \mathrm{Hom}_R(I, R)$ such that $V(\varphi) = V$ and $W(\varphi) = W$. By condition, φ can be extended to a homomorphism $\psi \in \mathrm{Hom}_R(R, R)$. According to Theorem 9.8, there exists a clopen subset H of X such that $V \subseteq H \subseteq X \setminus W = X \setminus W(\varphi)$. Therefore the closures of the sets V and W are disjoint. We proved that X is an F-space. □

Let $\mathfrak{m}$ be an infinite cardinal. The union of $\mathfrak{m}$ closed subsets is called *an $F_\mathfrak{m}$-set.*

Definition 10.3. A space X is called an $F(\mathfrak{m})$-space provided the closures of each two disjoint open $F_\mathfrak{m}$-sets are disjoint.

Definition 10.4. A commutative ring R with identity is called $\mathfrak{m}$-self-injective provided for every ideal I of R generated by a subset of cardinality $\leq \mathfrak{m}$ every $\varphi \in \mathrm{Hom}_R(I, R)$ can be extended to a homomorphism $\psi \in \mathrm{Hom}_R(R, R)$.

Theorem 10.5. *Let X be a Stone space and $R = C(X)$. The following assertions are equivalent:*

(i) *The Stone space X is an $F(\mathfrak{m})$-space;*
(ii) *R is $\mathfrak{m}$-self-injective.*

Proof. The proof is similar to the proof of Theorem 10.2. □

11. Necessary conditions for countably compactness

Lemma 11.1. *Let Y be a dense subspace of a Stone space $X, \mathcal{T}_Y$ be a countably compact topology on $R = C(X)$, $\{U_n|n \in \omega\}$ be a sequence of clopen subsets of the space X and the family $\{U_n \cap Y|n \in \omega\}$ be discrete in the space Y. Then there exists a clopen subset U of X such that $U \cap Y = \cup\{U_n \cap Y|n \in \omega\}$. Moreover, the set $cl_X(\cup\{U_n|n \in \omega\})$ is open in X.*

Proof. The assertions are true if the set $\{n|n \in \omega, U_n \neq \emptyset\}$ is finite. We may suppose that $U_n \neq \emptyset$ for any $n \in \omega$. For every $n \in \omega$ there exists a function $r_n \in C(X)$ such that $r_n^{-1}(1) = \cup\{U_i|i \leq n\}$. Let r be an accumulation point of the set $B = \{r_n|n \in \omega\}$. By construction, $\cup\{Y \cap U_n|n \in \omega\} = Y \cap r^{-1}(1)$. Thus for $U = r^{-1}(1)$ we have $U \cap Y = \bigcup\{Y \cap U_n|n \in \omega\}$ and $U = cl_X \cup \{U_n|n \in \omega\}$. □

The cardinal number $c(X) = \sup\{|\gamma| | \gamma$ is a family of pairwise disjoint non-empty open subsets of $X\}$ is called the *Souslin number* or the *cellularity* of the space X ([11], p. 86).

Theorem 11.2. *Let X be an infinite Stone space and $R = C(X)$. If there exists a countably compact Hausdorff topology on R, then there exists a closed G_δ-subspace Z on X with the properties:*

(i) *Z is an F-space;*
(ii) *the Souslin number $c(Z) \geq 2^\omega$.*

Proof. Since a countably compact topology is precompact there exists a dense subspace Y of X such that the topology $\mathcal{T}_Y$ is countably compact. Let $C_p(Y|X)$ be the set $\{f|Y|f \in C(X)\}$ in the topology of pointwise convergence and $\mathbb{R}$ be the space of reals. Then $C_p(Y|X)$ may be considered as a subspace of the space $C_p(Y, \mathbb{R})$ of all real valued functions in the topology of pointwise convergence (Remarks 2.4 and 2.3). The space $(R, \mathcal{T}_Y)$ is homeomorphic to the space $C_p(Y|X)$.
Case 1. The space $C_p(Y|X)$ is compact. It follows from Theorem 7.1 that Y is a discrete subspace and $X = \beta Y$. Thus X is extremally disconnected. If Y_1 is a countable subspace of Y, then $Z = clY_1 \setminus Y_1$ is the searched space.
Case 2. The space $C_p(Y|X)$ is not compact. In this case the subspace Y is not pseudocompact (see [3], Theorem 3.4.23). There exist a real-valued function f on Y and a sequence $(y_n)_{n\in\omega}$ in Y such that $f(y_1) = 1$ and $f(y_{n+1}) \geq f(y_n) + 3$ for any $n \in \omega$. We fix for every $n \in \omega$ a clopen subset U_n of X such that $y_n \in U_n \cap Y \subseteq f^{-1}(f(y_n) - 1, f(y_n) + 1)$. By construction, $\{U_n \cap Y | n \in \omega\}$ is a discrete family of non-empty subsets of Y. Obviously, $Y_1 = \cup\{U_n | n \in \omega\}$ is an open σ-compact subspace of the space X. By Lemma 11.1, the space $X_1 = clY_1$ is open and closed in X. Let $Z = X_1 \setminus Y_1$. We claim that $X_1 = \beta Y_1$. Let Φ_1 and Φ_2 be two disjoint closed subsets of the space Y_1. For every $n \in \omega$ there exists a clopen subset V_n of X such that $\Phi_1 \cap U_n \subseteq V_n \subseteq U_n$ and $V_n \cap \Phi_2 = \emptyset$. Thus $V' = \cup\{V_n | n \in \omega\}$ is an open subset of $X_1, \Phi_1 \subseteq V'$ and $clV' \cap \Phi_2 = \emptyset$. According to Lemma 11.1, the set $V = cl_X V'$ is clopen in X. Therefore $cl\Phi_1 \cap cl\Phi_1 = \emptyset$ and $X_1 = \beta Y_1$. The space $Z = \beta Y_1 \setminus Y_1 = X_1 \setminus Y_1$ is an F-space (see [13], Theorem 14.27, p. 210). Clearly, Z is a G_δ-subset of X. There exists a family $\{N_\beta | \beta \in B\}$ of infinite subsets of ω such that the intersection $N_\alpha \cap N_\beta$ is finite for every pair α, β of distinct numbers of the set $B = [0, 1]$ (see [11], Example 3.6.18, p. 229). Then $\{W_\beta | W_\beta = Z \cap cl_X(\cup\{U_n | n \in N_\beta\}), \beta \in B\}$ is a disjoint family of non-empty clopen subsets of Z of cardinality 2^ω. □

Corollary 11.3. *Let $\mathfrak{m}$ be an infinite regular number, $\mathfrak{m} < 2^\omega, \{X_\beta | \beta \in B\}$ be a family of non-empty Stone spaces, the density $d(X_\beta) \leq \mathfrak{m}$ for any $\beta \in B, |X_\beta| \geq 2$ for any $\beta \in B$ and an infinite Stone space X is a continuous homomorphic image of the product $\prod\{X_\beta | \beta \in B\}$. Then every topology $\mathcal{T} \in \mathcal{T}_\rho(R)$ on $R = C(X)$ is not countably compact.*

Proof. Use Theorem 11.2 and Theorem 2.3.17 from ([11], p. 112). □

Corollary 11.4. *Let X be an infinite zero-dimensional dyadic space. Then no Hausdorff topology $\mathcal{T} \in \mathcal{T}_\rho(R)$ on $R = C(X)$ is countably compact.*

Corollary 11.5. *Let $R = C(X)$ be an infinite free Boolean ring. Then no Hausdorff topology $\mathcal{T} \in \mathcal{T}_\rho(R)$ on $R = C(X)$ is countably compact and X is a dyadic space.*

A Boolean ring P is called a *projective* Boolean ring if for any two Boolean rings A, B and any homomorphisms $g : P \to B$ and $f : A \to B$, for which $f(A) = B$, there exists a homomorphism $h : P \to A$ such that $g = fh$.

Corollary 11.6. *Let $R = C(X)$ be an infinite projective Boolean ring. Then no Hausdorff topology $\mathcal{T} \in \mathcal{T}_\rho(R)$ on $R = C(X)$ is countably compact and X is a dyadic space.*

A space X is called *a perfectly-κ-normal* if for each open subset U of X there exists a continuous function $f \in C(X, \mathbb{R})$ such that $f^{-1}(0) = cl_X U$ (see [3], Section 0.3).

Remark 11.1. Let R be a Boolean ring. Since $(R, \mathcal{T}^{mp})$ is a dense subspace of the space $\mathbb{F}_2^\tau$ for some cardinal number τ, we have:

(i) For each $\mathcal{T} \in \mathcal{T}^p(R)$ the Souslin number $c(R, \mathcal{T})$ is countable.
(ii) The space $(R, \mathcal{T}^{mp})$ is perfectly-κ-normal.

12. Basically disconnected spaces

Let $\mathfrak{m}$ be an infinite cardinal. A space X is called $\mathfrak{m}$-*basically disconnected* if the closure of every open $F_\mathfrak{m}$-set is open. If $\mathfrak{m} = \omega$, then an $\mathfrak{m}$-basically disconnected space is called *basically disconnected* or ω-*extremally disconnected* (see [11, 13, 20]).

A space is extremally disconnected if and only if it is τ-basically disconnected for every cardinal τ.

Every $\mathfrak{m}$-basically disconnected space is an $F(\mathfrak{m})$-space.

A lattice E is called $\mathfrak{m}$-complete if every non-empty subset $H \subseteq E$ of the cardinality $|H| \leq \mathfrak{m}$ has the supremum $\vee H$ and infimum $\wedge H$.

Let X be a Stone space and $R = C(X)$. The ring R is $\mathfrak{m}$-complete if and only if X is $\mathfrak{m}$-basically disconnected (see [21]).

Let $\mathfrak{m}$ be an infinite cardinal. A space X is called:

- $\mathfrak{m}$-*compact* if every open cover of X of cardinality $\leq \mathfrak{m}$ contains a finite subcover;
- $\omega(\mathfrak{m})$-*bounded* if for every subset $H \subseteq X$ of cardinality $\leq \mathfrak{m}$ the closure $cl_X H$ is compact;
- $\mathfrak{m}$-*pseudocompact* if X is completely regular and every completely regular continuous image of X of weight $\leq \mathfrak{m}$ is compact.

Every $\omega(\mathfrak{m})$-bounded space is $\mathfrak{m}$-compact and every $\mathfrak{m}$-compact space is $\mathfrak{m}$-pseudocompact. If $\mathfrak{m} = \omega$, then an $\omega(\mathfrak{m})$-bounded space is ω-bounded, an $\mathfrak{m}$-compact space is countably compact and an $\mathfrak{m}$-pseudocompact space is pseudocompact.

Theorem 12.1. *Let X be a Stone space, $R = C(X)$ an atomic $\mathfrak{m}$-complete Boolean ring, $Y = \{x | x \in X, x$ is an isolated point of $X\}, \tau = |Y|$ and $\mathfrak{m}$ an infinite cardinal.*

(i) *There exists a dense subset S of the space $(R, \mathcal{T}_Y)$ such that:*
   (i_1) *The set S is dense in the space $(R, \mathcal{T}^{\mathfrak{m}p})$ and S contains all atoms of R;*
   (i_2) *S is an $\mathfrak{m}$-complete atomic subring of R;*
   (i_3) *S is $\omega(\mathfrak{m})$-bounded as a subspace of $(R, \mathcal{T}_Y)$.*

(ii) *The space $(R, \mathcal{T}_Y)$ is $\mathfrak{m}$-pseudocompact.*

(iii) *If $\tau^{\mathfrak{m}} = \tau$, then there exists a non-empty subset $Z \subseteq X \setminus Y$ such that for every finite subset $\Phi \subseteq Z$ the topology $\mathcal{T}_{Y \cup \Phi}$ is $\mathfrak{m}$-pseudocompact.*

Proof. (i) Let $\tau \leq \mathfrak{m}$. In this case $X = \beta Y$ is an extremally disconnected space, the space $(R, \mathcal{T}_Y)$ is compact and $S = R$ is the searched subring. Thus we may consider that $\mathfrak{m} < \tau$. Denote by S_0 the set of all functions $r \in C(X)$ such that $|r^{-1}(0) \cap Y| \leq \mathfrak{m}$ and by S_1 the set of all functions $r \in C(X)$ such that $|r^{-1}(1) \cap Y| \leq \mathfrak{m}$ and set $S = S_0 \cup S_1$. Then S is an $\mathfrak{m}$-complete atomic subring of R. It is obvious that $S_1 = 1 + S_0$. The subspaces S_0, S_1, S are $\omega(\mathfrak{m})$-bounded respectively to the topology $\mathcal{T}_Y$.

If a space contains a dense $\mathfrak{m}$-pseudocompact subspace, then it is $\mathfrak{m}$-pseudocompact. Thus (ii) follows from (i).

(iii) Suppose that $\tau^{\mathfrak{m}} = \tau$. Then the set $Z = X \setminus \cup\{cl_X H | H \subseteq Y, |H| \leq \mathfrak{m}\}$ is non-empty. If $L \subseteq Z \cup Y$, the set $L \cap Z$ is finite and $|L| \leq \mathfrak{m}$, then the subspace L is C^*-embedded in X and $cl_X L$ is the Stone–Čech compactificaton of L. Thus, if $Y \subseteq L \subseteq Y \cup Z$ and $L \cap Z$ is finite, then $S_L = \{r | r \in C(X), |r^{-1} \cap L| \leq \mathfrak{m}\}$ is an $\omega(\mathfrak{m})$-bounded subspace of the space $(R, \mathcal{T}_L)$. □

Theorem 12.2. *Let Y be an infinite dense discrete subspace of a Stone space X, $R = C(X)$, $\mathfrak{m}$ be an infinite cardinal number, $\mathfrak{m} < |Y|$, and for every set $Z \subseteq Y$ the set $cl_X Z$ is open if and only if $\min\{|Z|, |Y \setminus Z|\} \leq \mathfrak{m}$.*

(i) *The space $(R, \mathcal{T}_Y)$ is $\omega(\mathfrak{m})$-bounded.*

(ii) *If τ is a cardinal, $\mathcal{T} \in \mathcal{T}_\rho(R)$ and the space $(R, \mathcal{T})$ is τ-compact, then $\tau \leq \mathfrak{m}$.*

Proof. The space X is τ-basically disconnected if and only if $\tau \leq \mathfrak{m}$. Moreover, the space X is not extremally disconnected.

(i) In this case $S = R$, where S is the set constructed in the proof of Theorem 12.1.

(ii) Suppose that the topology $\mathcal{T} \in \mathcal{T}_\rho(R)$ is τ-compact. Then the topology $\mathcal{T}_Y \subseteq \mathcal{T}$ is τ-compact too. If $\tau \geq |Y|$, then the topology $\mathcal{T}_Y$ is compact, contradiction (see Theorem 7.1). Thus $\tau < |Y|$. We fix a subset $A \subseteq Y$, where $|A| = \tau$ and $|Y \setminus A| = |Y|$. If $B \subseteq Y$ and $|B| \leq \mathfrak{m}$, then there exists a unique function $f_B \in C(X)$ such that $f_B^{-1}(1) = cl_X B$. Let $H = \{f_B | B \subseteq A, |B| < \omega\}$. Then $|H| = \tau$ and there exists a function $f \in C(X)$ such that if $U \in \mathcal{T}_Y$ and $f \in U$, then $|U \cap H| = \tau$. By construction, $f^{-1}(1) = cl_X A$ is an open subset of X. Thus $\tau = |A| \leq \mathfrak{m}$. □

A subset $L \subseteq X$ of a topological space X is called *bounded* provided every continuous function $f : X \to \mathbb{R}$ is bounded on L.

By $C_p(X, \mathbb{R})$ it is denoted the set of all continuous real valued functions furnished with the topology of pointwise convergence. Let $Y \subseteq X$ and $C_p(Y|X) = \{f|Y|f \in C(X, \mathbb{F}_2)\}$. We consider $\mathbb{F}_2 = \{0, 1\}$ as a discrete subspace of the reals $\mathbb{R}$ and $C_p(Y|X)$ as a subspace of the space $C(Y, \mathbb{R})$. By construction, $C_p(Y|X)$ is a subring of the Boolean ring $C(Y)$ and it is not a subring of the ring $C_p(Y, \mathbb{R})$. By construction, $C_p(Y|X)$ is a subring of the ring $\mathbb{F}_2^Y$.

Proposition 12.3. *Let Y be a subspace of the space X and* $\operatorname{ind} X = 0$. *Then:*

(i) *$C_p(Y|X)$ is a dense subspace of the space $\mathbb{F}_2^Y$.*

(ii) *If $C_p(Y|X)$ contains a non-empty compact subset Φ of countable character in $C_p(Y|X)$, then there exists a countable subset $H \subseteq Y$ such that the subspace $Y_0 = Y \setminus H$ is discrete and C^*-embedded in X.*

(iii) *If $C_p(Y|X)$ contains a dense Čech complete subspace, then Y is a discrete C^*-embedded subspace of the space X.*

Proof. The assertion (i) is obvious.

(ii) Let Φ be a non-empty compact subset of countable character in $C_p(Y|X)$. We fix $x_0 = (x_{0y}|y \in Y) \in \Phi \subseteq C_p(Y|X) \subseteq \mathbb{F}_2^Y$. There exists a sequence $\{U_n|n \in \omega\}$ of open subsets of $\mathbb{F}_2^Y$ such that $U_{n+1} \subseteq U_n$ for every $n \in \omega$, and for every open set $U \supseteq \Phi$ there exists $m \in \omega$ such that $\Phi \subseteq U_m \subseteq U$. There exists a countable subset $H \subseteq Y$ such that $\Phi_1 = \{x = (x_y|y \in Y)|x_y = x_{0y}$ for all $y \in H\} \subseteq \cap\{U_n|n \in \omega\}$.

Let $g : Y_0 \to \mathbb{F}_2^Y$ be a function. Then there exists $f \in C(X)$ and $x_1 = (x_{1y}|y \in Y) \in \Phi_1$ such that $f(y) = g(y) = x_{1y}$ for every $y \in Y_0$. Thus g is a continuous function and Y_0 is C^*-embedded in X. Therefore Y_0 is a discrete subspace of the space X.

Suppose that Z is a Čech complete dense subspace of the space $C_p(Y|X)$. Thus Z is a dense G_δ-subset of the compact space $\mathbb{F}_2^Y$. We claim that $C_p(Y|X) = \mathbb{F}_2^Y$. Suppose that $g \in \mathbb{F}_2^Y \setminus C_p(Y|X)$. Then $L = \{f + g|f \in C_p(Y|X)\}$ is a dense G_δ-subset of the compact space $\mathbb{F}_2^Y$ and $Z \cap L \subseteq L \cap C_p(Y|X) = \emptyset$, a contradiction, since in a compact space the intersection of two dense G_δ-subsets is dense. □

Theorem 12.4. *Let Y be a subspace of a space X. Then:*

(i) *If Y is a pseudocompact space and $C_p(Y|X)$ is countably compact, then $C_p(Y|X)$ is compact.*

(ii) *If Y is a countably compact space and $C_p(Y|X)$ is pseudocompact, then $C_p(Y|X)$ is compact.*

(iii) *If Y is a countably compact space and $C_p(Y|X)$ is a closed bounded subset of the space $C_p(Y) = C_p(Y|Y)$, then $C_p(Y|X)$ is compact.*

Proof. We consider $C_p(Y)$ as a closed subspace of the space $C_p(Y, \mathbb{R})$.

If $C_p(Y|X)$ is a bounded closed subset of the space $C_p(Y)$, then $C_p(Y|X)$ is a closed bounded subset of $C_p(Y, \mathbb{R})$. Thus (iii) follows from the Asanov-Velichko's generalization of Grothendieck's Theorem ([3], Theorem 3.4.1). The assertion (i) follows from ([3], Theorem 3.4.23). The assertion (ii) follows from (iii) and Theorem of Preiss-Simon ([3], Theorem 4.5.5). □

Remark 12.1. If Y is a σ-pseudocompact subspace of the space X and $C_p(Y|X)$ is compact, then $C_p(Y|X)$ is an Eberlein compact (see [3], Theorem 3.4.23).

Denote by A_τ the one-point compactification of the discrete space D_τ of cardinality τ.

The cardinal $p(Y) = \sup\{|\xi||\xi$ is a point-finite family of non-empty open subsets of $Y\}$ is the *Alexandroff number* of the space Y. It is obvious that $c(Y) \leq p(Y)$. If Y is a Baire space, then $c(Y) = p(Y)$.

Theorem 12.5. *Let Y be an infinite subspace of a space X and ind $X = 0$. Then $p(Y) = \sup\{\tau|A_\tau$ is embedded in $C_p(Y|X)\} = \sup\{w(Z)|Z$ is a compact subspace of $C_p(Y|X)\}$.*

Proof. Let $\xi = \{U_\alpha|\alpha \in D_\tau\}$ be a point-finite family of non-empty open subsets of Y. We fix for every $\alpha \in D_\tau$ a non-empty clopen subset V_α of X and $f_\alpha : X \to \mathbb{F}_2$ such that $\emptyset \neq Y \cap V_\alpha \subseteq U_\alpha$ and $f_\alpha^{-1}(1) = V_\alpha$. Consider that $f(x) = 0$ for all $x \in X$. Then the subspace $\{f|Y\} \cup \{f_\alpha|Y|\alpha \in D_\tau\}$ of $C_p(Y|X)$ is homeomorphic to A_τ. Thus $p(Y) \leq \sup\{\tau|A_\tau$ is embedded in $C_p(Y|X)\}$. It is well known that $p(Y) = \sup\{\tau|A_\tau$ is embedded in $C_p(Y, \mathbb{R})\}$ (see [3], Proposition 3.3.2 and Theorem 3.5.9). Thus $\sup\{\tau|A_\tau$ is embedded in $C_p(Y|X)\} \leq p(Y)$. □

We say that the spaces X and Y are S-equivalent if the topological spaces $C_p(X)$ and $C_p(Y)$ are homeomorphic.

Corollary 12.6. *Let X and Y be S-equivalent Stone spaces. Then:*

(i) $c(X) = c(Y)$.
(ii) *The space X is scattered if and only if the space Y is scattered.*

Construction 12.1 (D.B. Shakhmatov for $E = [0,1]$ and $\mathfrak{m} = \omega$, [3], Example 1.2.5). Let τ and $\mathfrak{m}$ be infinite cardinals, E be a compact space of the weight $\leq \mathfrak{m}$, $\tau^{\mathfrak{m}} = \tau$ and $|E| \geq 2$.

Denote by M the set of all ordinals of cardinality $< \tau$.

We put $E_\alpha = E$ for every $\alpha \in M$. If $B \subseteq M$, then $E^B = \prod\{E_\alpha|\alpha \in B\}$ and let $\pi_B : E^M \to E^B$ stands for the natural projection.

We fix $x_0, x_1 \in E, x_0 \neq x_1$. Let $G_{\mathfrak{m}} = \{x|x \in E^M, |\{\alpha|\alpha \in M, \pi_\alpha(x) \neq x_0\}| \leq \mathfrak{m}\}$. Then $G_{\mathfrak{m}}$ is a subspace of E^M and $|G_{\mathfrak{m}}| = \tau^{\mathfrak{m}} = \tau$. There exists an enumeration $\{g_\alpha|\alpha \in M\}$ of $G_{\mathfrak{m}}$ such that $|\{\alpha|\alpha \in M, g = g_\alpha\}| = \tau$ for every $g \in G_{\mathfrak{m}}$.

Let $\gamma = \{A \subseteq M||A| \leq \mathfrak{m}\}$. Consequently, $|\gamma| = \tau^m = \tau$. We fix an enumeration $\{A_\beta|\beta \in M\}$ of γ such that $|\{\beta|A = A_\beta\}| = \tau$ for every $A \in \gamma$.

We consider for every $\alpha \in M$ the point $x_\alpha \in E^M$, where

$$\pi_\mu(x_\alpha) = \begin{cases} \pi_\mu(g_\alpha), & \text{if } \mu \leq \alpha; \\ x_1, & \text{if } \mu > \alpha \text{ and } \alpha \in A_\mu; \\ x_0, & \text{if } \mu > \alpha \text{ and } \alpha \notin A_\mu. \end{cases}$$

Now we put $X_{\tau\mathfrak{m}} = \{x_\alpha | \alpha \in M\} \subseteq E^M$.

Property 1. *If $B \subseteq M$ and $|B| \leq \mathfrak{m}$, then $\pi_B(X_{\tau\mathfrak{m}}) = E^B$.*

There exists $\alpha > \sup\{\delta | \delta \in B\}$ such that $g = \pi_B(g_\alpha)$. Then $\pi_B(x_\alpha) = g$. Thus $\pi_B(X_\tau) = E^B$.

Property 2. *The space $X_{\tau\mathfrak{m}}$ is dense in E^M.*

This assertion follows from Property 1.

Property 3. *Let Y be a dense subspace of E^M. The space Y is $\mathfrak{m}$-pseudocompact if and only if $\pi_B(Y) = E^B$ provided $B \subseteq M$ and $|B| \leq \mathfrak{m}$.*

Let Y be an $\mathfrak{m}$-pseudocompact space, $B \subseteq M$ and $|B| \leq \mathfrak{m}$. Then $\pi_B(Y)$ is a dense compact subset of E^B. Therefore $\pi_B(Y) = E^B$.

Suppose that $\pi_B(Y) = E^B$, where $|B| \leq \mathfrak{m}$. Let $\varphi : Y \to Z$ be a continuous mapping and $w(Z) \leq \mathfrak{m}$. Since Y is dense in E^M and $w(Z) \leq \mathfrak{m}$, there exist a set $B \subseteq M$ and a continuous mapping $g : E^B \to Z$ such that $|B| \leq \mathfrak{m}$ and $\varphi = g \circ \pi_B$ (see [11], Problems 2.7.12 and 2.7.13 for $\mathfrak{m} = \omega$). Then $\varphi(Y) = g(E^B)$ is a compact space. Thus Y is $\mathfrak{m}$-pseudocompact.

Property 4. *$X_{\tau\mathfrak{m}}$ is an $\mathfrak{m}$-pseudocompact space and $\beta X_{\tau\mathfrak{m}} = E^B$.*

It follows from Properties 1–3.

Property 5. *Let H and L be the subsets of $X_{\tau\mathfrak{m}}$ and $|H \cup L| \leq \mathfrak{m}$. If $H \cap L = \emptyset$, then $cl_{E^M} H \cap cl_{E^M} L = \emptyset$.*

Suppose, that $H = \{x_\alpha | \alpha \in M_1\}$ and $L = \{x_\alpha | \alpha \in M_2\}$, where $M_1 \cup M_2 \subseteq M$ and $M_1 \cap M_2 = \emptyset$. We fix $\theta \in M$ for which $\theta > \sup(M_1 \cup M_2)$. Let $S_\alpha = S$ for $\alpha \in M$. Then $C_p(X_{\tau\mathfrak{m}}, S)$ is a subspace of the space $S^M = \prod\{S_\alpha | \alpha \in M\}$. We fix $B \subseteq M$, where $|B| \leq \mathfrak{m}$. We consider the natural projection $\pi_B : S^M \to S^B$. We put $L = \{x_\alpha | \alpha \in B\} \subseteq X_{\tau\mathfrak{m}}$. If $g \in S^B$, then g is a mapping of L into B. Since $|B| \leq \mathfrak{m}$, there exists a continuous function $f : X_{\tau\mathfrak{m}} \to S$ such that $g = f|L$. Thus $f \in C_p(X_{\tau\mathfrak{m}}, S)$ and $g = \pi_B(f)$. Since $C_p(X_{\tau\mathfrak{m}}, S)$ is dense in S^M, from Property 3 it follows that $C_p(X_{\tau\mathfrak{m}}, S)$ is an $\mathfrak{m}$-pseudocompact space. By construction, $\pi_\theta(x_\alpha) = x_1$, if $\alpha \in M_1$, and $\pi_\theta(x_\alpha) = x_0$, if $\alpha \in M_2$. Thus $cl_{E^M} H \subseteq \pi_\theta^{-1}(x_1), cl_{E^M} L \subseteq \pi_\theta^{-1}(x_0)$ and $cl_{E^M} H \cap cl_{E^M} L = \emptyset$.

Property 6. *Let $Z \subseteq X_{\tau\mathfrak{m}}$ and $|Z| \leq \mathfrak{m}$. Then Z is a discrete closed subspace of the space X_τ and the subspace $cl_{E^M} Z$ is homeomorphic to the Stone-Čech compactification βZ of Z.*

Property 6 follows from Property 5. The following property is obvious.

Property 7. *Let S be a closed subspace of the space $[0,1], \{0,1\} \subseteq S$ and $S = [0,1]$ if $indS \geq 1$. Denote by $C_p(Y,S)$ the space of all continuous mappings of Y in S with the topology of pointwise convergence. Then $C_p(X_{\tau\mathfrak{m}}, S)$ is an $\mathfrak{m}$-pseudocompact space.*

Example 12.1. Let τ and $\mathfrak{m}$ be infinite cardinals, $\tau^{\mathfrak{m}} = \tau, D = \{0,1\}$ be the two-point discrete space, $X = D^\tau$ and $R = C(X)$. Then :

(i) There exists no minimal totally bounded topology on R.
(ii) The Boolean ring R is atomless, free and has τ generators.
(iii) There exists some $\mathfrak{m}$-pseudocompact topology $\mathcal{T} \in \mathcal{T}^p_\rho(R)$ which is not countably compact.
(iv) There exists a dense subset Y of X such that the topology $\mathcal{T}_Y$ is $\mathfrak{m}$-pseudocompact and $\mathcal{T}_Z$ is a $\mathfrak{m}$-pseudocompact topology on R provided $Z \subseteq Y$ and $|Y \setminus Z| \leq \mathfrak{m}$.
(v) If $\mathcal{T} \in \mathcal{T}_\rho(X)$, then the topology $\mathcal{T}$ is not countably compact.

Construction. From Construction 12.1 it follows that there exists a dense subspace $Y = X_{\tau\mathfrak{m}} \subseteq D^\tau = X$ such that the space $C_p(Y,D)$ is $\mathfrak{m}$-pseudocompact and $\beta Y = X$. The space $(C(X), \mathcal{T}_Y)$ is homeomorphic to the space $C_p(Y,D)$. The assertion (iv) follows from Property 6. If $Z \subseteq Y$ and $|Y \setminus Z| \leq \mathfrak{m}$, then Z is dense in X and $\mathcal{T}_Z \subseteq \mathcal{T}_Y$. The assertion (iii) follows from the assertion (iv). The assertions (i) and (ii) are obvious, since X is without isolated points.

Example 12.2. Let τ and $\mathfrak{m}$ be infinite cardinals and $\mathfrak{m} < \tau$. Then there exists a Stone space X such that:

(i) X is $\mathfrak{m}$-basically disconnected.
(ii) $R = C(X)$ is an atomic $\mathfrak{m}$-complete Boolean ring.
(iii) The minimal topology $\mathcal{T}_{mp} \in \mathcal{T}^p_\rho(R)$ is $\omega(\mathfrak{m})$-bounded.

Construction. Let D_τ be a discrete space of cardinality τ. Denote by $U(H)$ the closure of H in βD_τ for every $H \subseteq D_\tau$. By definition, $U(H)$ is a clopen subset of βD_τ. Let $\Phi = \beta D_\tau \setminus \cup \{U(H) | H \subseteq D_\tau, |H| \leq \mathfrak{m}\}$. Obviously, Φ is a non-empty subset of βD_τ and $Y = \beta D_\tau \setminus \Phi$ is a locally compact space. Denote by $X = Y \cup \{b\}$ the one-point Alexandroff compactification of the space Y and let $p : \beta D_\tau \to X$ be the natural projection, where $p(y) = y, y \in Y$.

Property 1. *Y is an $\omega(\mathfrak{m})$-bounded space.*

Indeed, let $L \subseteq Y$ and $|L| \leq \mathfrak{m}$. For every $y \in L$ there exists $H_y \subseteq D_\tau$ such that $|H_y| \leq \mathfrak{m}$ and $y \in U(H_y)$. Let $H = \cup\{H_y | y \in L\}$. Then $|H| \leq \mathfrak{m}$ and $L \subseteq U(H)$. Thus $cl_Y L$ is a closed subset of the compact set $U(H)$ and hence it is compact.

Property 2. *The space X is $\mathfrak{m}$-basically disconnected.*

Let V be an open $F_{\mathfrak{m}}$-subset of the space X. There exists a family $\{P_\alpha | \alpha \in A\}$ of compact subsets of X such that $V = \cup\{P_\alpha | \alpha \in A\}$ and $|A| \leq \mathfrak{m}$. Suppose that $b \notin V$. Then for every $\alpha \in A$ there exists a subset $H_\alpha \subseteq D_\tau$ such that $P_\alpha \subseteq U(H_\alpha) \subseteq V$. We put $H = \cup\{H_\alpha | \alpha \in A\}$. Then $U(H)$ is a clopen subset of X

and $cl_X V = U(H)$. Suppose now that $b \in V$. There exists a subset H of D_τ such that $b \in X \backslash U(H) \subseteq V$ and $|H| \leq \mathfrak{m}$. The set $X \backslash U(H)$ is clopen. If $H' = H \cap V$, then $cl_X V = U(H') \cup (X \backslash U(H))$ is a clopen subset of X.

Property 3. *Let* $I_\mathfrak{m} = \{f | f \in C(X), |D_\tau \cap f^{-1}(0)| \leq \mathfrak{m}\}$ *and* $\Phi_\mathfrak{m} = \{f | f \in C(X), |D_\tau \cap f^{-1}(1)| \leq \mathfrak{m}\}$. *Then* $C(X) = I_\mathfrak{m} \cap \Phi_\mathfrak{m}$ *and* $I_\mathfrak{m} \cup \Phi_\mathfrak{m} = \emptyset$.

Obviously, $I_\mathfrak{m} \cap \Phi_\mathfrak{m} = \emptyset$. We fix $f \in C(X)$. If $f(b) = 1$, then there exists a subset $H \subseteq D_\tau$ such that $|H| \leq \mathfrak{m}$ and $b \in X \backslash U(H) \subseteq f^{-1}(1)$; thus $f \in \Phi_\mathfrak{m}$. If $f(1) = 0$, then $b \in X \backslash U(H) \subseteq f^{-1}(0)$ and $f \in I_\mathfrak{m}$.

Property 4. $I_\mathfrak{m}$ *is an ideal of* $C(X)$.

The proof is obvious.

Property 5. $R = C(X)$ *is atomic.*

The set D_τ of isolated points of X is dense in X. Thus the ring R is atomic.

Property 6. *If* $Z = D_\tau$, *then* $\mathcal{T}_{mp} = \mathcal{T}_Z$ *and the topology* $\mathcal{T}_Z$ *is* $\omega(\mathfrak{m})$*-bounded.*

We consider the projection $\pi : \mathbb{F}_2^X \to \mathbb{F}_2^L$. Then

$$\pi(I_\mathfrak{m}) = \{f : Z \to \mathbb{F}_2 \| f^{-1}(0)| \leq \mathfrak{m}\} \quad \text{and} \quad \pi(\Phi_\mathfrak{m}) = \{f : Z \to \mathbb{F}_2 \| f^{-1}(1)| \leq \mathfrak{m}\}.$$

Obviously, $\pi(I_\mathfrak{m})$ and $\pi(\Phi_\mathfrak{m})$ are $\omega(\mathfrak{m})$-bounded subspaces of $\mathbb{F}^Z$. Thus the subspace $S = \pi(\Phi_\mathfrak{m}) \cup \pi(I_\mathfrak{m})$ of $\mathbb{F}_2^Z$ is $\omega(\mathfrak{m})$-bounded. The space $(C(X), \mathcal{T}_Z)$ is homeomorphic to the space S.

Example 12.3. Let τ be an infinite cardinal and D_τ be a discrete space of cardinality τ. Denote by βD_τ the Stone-Čech compactification of the space D_τ. Then:

(i) βD_τ is a free compact space.
(ii) βD_τ is extremally disconnected.
(iii) The ring $C(\beta D_\tau)$ is self-injective and atomic.
(iv) The set $\mathcal{T}_\rho^p(C(\beta D_\tau))$ is a complete lattice and the topology $\mathcal{T}_{mp} \in \mathcal{T}_\rho^p(C(\beta D_\tau))$ is compact.
(v) The set $\mathcal{T}_\rho(C(\beta D_\tau))$ is a complete lattice with the minimal element $\mathcal{T}_{mp}$.
(vi) If R is a Boolean ring of cardinality $\leq \tau$, then we can consider that R is a subring of $C(\beta D_\tau)$.

We deduce that every Boolean ring is a subring of a self-injective atomic Boolean ring.

Example 12.4. Let D_τ be a discrete space of an infinite cardinality τ and $X = \beta D_\tau \backslash D_\tau$. Then X is an F-space which is not extremally disconnected. The ring $R = C(X)$ is an ω-self-injective but not self-injective. An ideal I of R and a non extendable homomorphism $\varphi : I \to R$ can be constructed as follows:

We fix a countable subset $N \subseteq D_\tau$ and a mapping $q : N \to [0,1]$ such that the set $q(N)$ is dense in $[0,1]$. We fix for every $t \in [0,1]$ an infinite sequence $(t(n) \in q(N) | n \in \omega)$ such that $|t - t(n+1)| < |t - t(n)| < 2^{-n}$ for every $n \in \omega$. Then $t = \lim t(n)$. We may consider that $t(n) < t(n+1) < t$ for $t > 0$. We put $A_t = \{a_t(n) | n \in \omega\}$. If $t, t' \in [0,1]$ and $t \neq t'$, then the set $A_t \cap A_{t'}$ is finite (see

[11]). There exists a maximal family $\{A_\beta|\beta \in B\}$ of infinite subsets of N with the properties:

- $[0,1] \subseteq B$;
- if $\alpha, \beta \in B$ and $\alpha \neq \beta$, then the set $A_\alpha \cap A_\beta$ is finite.

The subset $U_\alpha = X \cap cl_{\beta D_\tau} A_\alpha$ is clopen in X. The set $U = X \cap cl_{\beta D_\tau} N = cl_{\beta D_\tau} N \backslash D_\tau$ is clopen in X. If $\alpha \neq \beta$, then $U_\alpha \cap U_\beta = \emptyset$.

Therefore $\{U_\alpha|\alpha \in B\}$ is a family of disjoint clopen subsets of U. The set $\cup\{U_\alpha|\alpha \in B\}$ is dense in U. For any $\alpha \in B$ fix a non-empty clopen subset V_α of U_α such that $W_\alpha = U_\alpha \backslash V_\alpha \neq \emptyset$. We put $V = \cup\{U_\alpha|\alpha \in B\}$, $W = (X \setminus U) \cup \bigcup\{W_\alpha|\alpha \in B\}$ and $Y = X \setminus (V \cup W)$. There is no clopen subset H of X such that $V \subseteq H \subseteq X \setminus W$. There exist an ideal I of $R = C(X)$ and an R-module homomorphism $\varphi : I \to R$ such that $D(I) = Y, V(\varphi) = V$ and $W(\varphi) = W$. According to Theorem 9.8, the homomorphism φ is not extendable.

Example 12.5. Let X be an infinite perfectly normal zero-dimensional Stone space. Therefore every closed subset of X is a G_δ-set and every ideal of $R = C(X)$ is countably generated. The space X is not an F-space. Hence by Theorem 10.2 the ring R is not ω-self-injective. We fix a non-isolated point $b \in X$ and a closed subset Y of X such that $b \in cl_X(X\backslash Y)$ and $b \in Y$. There exists a sequence $\{b_n \in X\backslash Y|n \in \mathbb{N}\}$ such that $b = \lim b_n$ and $b_n \neq b_m$ for $n \neq m$. There exist two sequences $\{U_n|n \in \omega\}$ and $\{H_n|n \in \omega\}$ of clopen subsets of X such that $Y = \cap\{U_n|n \in \mathbb{N}\}, b_n \in H_n \subseteq U_n\backslash U_{n+1}$ for any $n \in \omega$. Let $V = \cup\{H_{2n}|n \in \omega\}$ and $W = X\backslash cl_X(V \cup Y)$. Then $V \cap W = \emptyset$ and $X\backslash Y = V \cup W$. There exists an R-module homomorphism $\varphi : I \to R$, where $I = M_Y, V(\varphi) = V$ and $W(\varphi) = W$. Since it does not exist a clopen set H for which $V \subseteq H \subseteq X\backslash W$, the homomorphism φ is not extendable on R. The ideal $I = M_b$ is maximal.

Example 12.6. Let X be an infinite compact scattered space. Then X is not an F-space. Denote by $\overset{0}{X}$ the set of all isolated points of X. We fix an isolated point b of the space $X_1 = X\backslash\overset{0}{X}$. There exists a sequence $\{U_n|n \in \mathbb{N}\}$ of clopen subsets of X such that $b \in Y = \cap\{U_n|n \in \mathbb{N}\}, U_1 \cap X_1 = \{b\}$ and $U_{n+1} \subseteq U_n, U_n\backslash U_{n+1} \neq \emptyset$ for any $n \in \omega$. The set $H_n = U_n\backslash U_{n+1}$ is finite. We fix a point $b_n \in U_n\backslash U_{n+1}$. Then $b = \lim b_n$. We put $V = \cup\{U_{2n}\backslash U_{2n+1}|n \in \mathbb{N}\}$ and $W = \cup\{U_{2n-1}\backslash U_{2n}|n \in \mathbb{N}\}$. Then $X\backslash Y = V \sqcup W$. If $I = M_Y$ and $\varphi : I \to R$ is a homomorphism for which $V(\varphi) = V$ and $W(\varphi) = W$, then φ is not extendable on R. The ring R is atomic. The complete lattices $\mathcal{T}_\rho(R)$ and $\mathcal{T}_\rho^p(R)$ do not contain compact topologies.

Example 12.7. Let $C_0 = \{(x,0)|0 < x \leq 1\}, C_1 = \{(x,1)|0 \leq x < 1\}$, $X = C_0 \cup C_1$, $O(x,0,\varepsilon) = \{(x,0)\} \cup \bigcup\{\{(y,0),(y,1)\}|x - \varepsilon < y < x\}$ and $O(x,1,\varepsilon) = \{(x,1)\} \cup \bigcup\{\{(y,0),(y,1)\}|x < y < x + \varepsilon\}$.

We consider a topology on X generated by the open basis $\{O(x,i,\varepsilon)|x \in X, i \in \{0,1\}, \varepsilon > 0\}$. The space X is perfectly normal, zero-dimensional and compact. The space X is called *the two arrows space of P.S. Alexandroff and P.S. Urysohn* (see [1], [11]). Every ideal of $R = C(X)$ is countably generated. But

R is not ω-self-injective. The ring R is atomless. The sets $\mathcal{T}_\rho(R)$ and $\mathcal{T}_\rho^p(R)$ are not lattices. On X there exists a σ-additive measure μ such that $\mu(X) = 1$ and $\mu(\cup\{\{(x,0),(x,1)\}|a < x < b\}) = b - a$ provided $0 \leq a < b \leq 1$. The function $d(f,g) = \mu(\{x|x \in X, f(x) \neq g(x)\})$ is an invariant metric on R. The topology $\mathcal{T}_d$ generated by the distance d on R is a ring topology and the space $(R, \mathcal{T}_d)$ is arcwise connected. The topology $\mathcal{T}_d$ is not minimal.

Example 12.8. Let $\mathfrak{m}$ be an infinite cardinal number, $\tau = 2^{\mathfrak{m}}$ and L a dense subset of the topological space $\mathbb{F}_2^\tau$ of cardinality $\mathfrak{m}$ (see [11],Theorem 2.3.15 of Hewitt–Marczewski–Pondiczery). For every subset $A \subseteq \mathbb{F}_2^\tau$ denote by $r(A)$ the subring of the Boolean ring $\mathbb{F}_2^\tau$ generated by the set A. We consider that $r(\emptyset) = \{0,1\} \subseteq \mathbb{F}_2^\tau$. If $H \subseteq \mathbb{F}_2^\tau$ is an infinite subset, then we fix a point $a(H) \in \mathbb{F}_2^\tau$ such that $|H \cap U| = |H|$ provided U is open in $\mathbb{F}_2^\tau$ and $a(H) \in U$.

We construct the subrings $\{R_\alpha|\alpha < \tau\}$ of the ring $\mathbb{F}_2^\tau$ with the properties:

(i) $R_0 = r(L), R_\alpha \subseteq R_\beta$ for $0 \leq \alpha < \beta < \tau$.

(ii) If α is a limit ordinal, then $R_\alpha = \cup\{R_\beta|\beta < \alpha\}$.

(iii) If R_α is constructed, then $R_{\alpha+1} = r((\{a(H)|H \subseteq R_\alpha, H$ is infinite and $|H| \leq \mathfrak{m}\}) \cup R_\alpha)$.

By construction, $R = \cup\{R_\alpha|\alpha < \tau\}$ is a subring of the compact ring $\mathbb{F}_2^\tau$. Let $\mathcal{T}$ be the topology of the subspace R of the compact space $\mathbb{F}_2^\tau$.

Property 1. $|R| = \tau$ *and* $|\mathbb{F}_2^\tau| = 2^\tau$.

Property 2. *The topology $\mathcal{T}$ is $\mathfrak{m}$-compact.*

Property 3. *The topology $\mathcal{T}$ is not $\omega(\mathfrak{m})$-bounded.*

Property 4. *The ring R is atomless.*

13. Open questions

Question 1. Is it true that every minimal topological Boolean ring R is precompact?

Question 2. Under which conditions a commutative infinite ring is minimally almost periodic?

Question 3. Let τ be an infinite cardinal, $R = \mathbb{Z}\langle X\rangle$ the free associative ring over a set $X, |X| = \tau$ and $\tau^\omega > \tau$. Does R admit a pseudocompact ring topology?

Question 4. Does there exist a countably linearly compact Boolean ring which is not countably compact?

We note that every countably linearly compact Boolean ring is pseudocompact.

From Proposition 4.2 it follows that on infinite non-atomless Boolean rings no Hausdorff topology is connected. If R is a Boolean ring, $\mathcal{T} \in \mathcal{T}^p(R)$ and the space $(R, \mathcal{T})$ is connected, then $\mathcal{T} = \{\emptyset, R\}$.

Question 5. Which atomless Boolean rings admit connected ring topologies?

Question 6. Which atomless Boolean rings admit arcwise connected ring topologies?

We mention that each Hausdorff topological (Boolean) ring is a closed subring of some Hausdorff topological arcwise connected (Boolean) ring.

Let $\mathbb{Z}$ be the discrete ring of the integers.

Question 7 (see [3] for $C_p(X,\mathbb{R})$). Let X, Y be topological spaces and $\operatorname{ind} X = \operatorname{ind} Y = 0$. Determine the relations between the following assertions:

(α) The spaces X and Y are homeomorphic.
(σ) The spaces X and Y are S-equivalent.
(κ) The spaces $C_p(X,\mathbb{R})$ and $C_p(Y,\mathbb{R})$ are homeomorphic, i.e., the spaces X and Y are t_p-equivalent.
(δ) The spaces $C_p(X,\mathbb{Z})$ and $C_p(Y,\mathbb{Z})$ are homeomorphic.
(γ) The topological groups $C_p(X,\mathbb{R})$ and $C_p(Y,\mathbb{R})$ are isomorphic.
(θ) The topological groups $C_p(X,\mathbb{Z})$ and $C_p(Y,\mathbb{Z})$ are isomorphic.
(λ) The spaces $C_p(X,\mathbb{R})$ and $C_p(Y,\mathbb{R})$ are linear homeomorphic, i.e., the spaces X and Y are l_p-equivalent.

References

[1] P.S. Alexandroff et P.S. Urysohn, *Mémoire sur les Espaces Topologiques Compacts.* – Verhandelingen Kon. Akad., van Wetenschappen. Amsterdam 14, 1929, 1–96.

[2] A.V. Arhangel'skii, *Topological Invariants in Algebraic Environment.* – In: Recent Progress in General Topology II. North Holland. Elsevier Science Publ., 2002, 1–57.

[3] A.V. Arhangel'skii, *Topological Function Spaces.* – Kluwer Acad. Publ., 1992.

[4] G. Birgkhoff, *Lattice Theory.* – Providence, 1967.

[5] N. Bourbaki, *Algèbre Commutative.* – Mir, Moscow, 1971 [in Russian] French original: Algèbre Commutative, Chap. 1–7, Hermann.

[6] B. Brainerd and J. Lambek, *On the ring of quotients of a Boolean ring.* – Canad. Math. Bull. 2 (1959) 25–29.

[7] C. Chevalley, *On the theory of local rings.* – Ann. of Math. 44 (1943) 690–708.

[8] W.W. Comfort, K.H. Hofmann and D. Remus, *Topological Groups and Semigroups.* – In: Recent Progress in General Topology. – North Holland Elsevier Science Publ., 1992, 57–144.

[9] D. Dikranjan, Iv. Prodanov and L. Stoyanov, *Topological Groups: Characters, Dualities and Minimal Group Topologies.* – New York, Marcel Dekker, 1989.

[10] E.K. van Douwen, *The product of two countably compact topological groups.* – Trans. Am. Math. Soc. 262 (1980) 417–427.

[11] R. Engelking, *General Topology.* – Warszawa, PWN, 1977.

[12] N.J. Fine, L. Gillman and J. Lambek, *Rings of Quotients of Rings of Functions.* Montreal, 1966.

[13] L. Gillman and M. Jerison, *Rings of Continuous Functions.* – Princeton, 1960.

[14] P. Holm, *On the Bohr compactification.* – Math. Ann. 156 (1964) 34–46.

[15] K. Iwasawa, *On the rings of valuation vectors.* – Ann. of Math. (2) 57 (1953) 331–356.

[16] J. Lambek, *Lectures on Rings and Modules.* – Waltham-London, Blaisdell Publ. Company, 1966.

[17] H. Leptin, *Linear kompakte Moduln und Ringe.* – Math. Z. 63 (1955) 241–267.

[18] H. Leptin, *Linear kompakte Moduln und Ringe, II.* – Math. Z. 66 (1957) 289–327.

[19] L. Nachbin, *On strictly minimal topological division rings.* – Bull. Amer. Math. Soc. 55 (1949) 1128–1136.

[20] Z. Semadeni, *Banach Spaces of Continuous Functions.* – Warszawa, PWN, 1971.

[21] R. Sikorski, *Boolean Algebras.* – Mir. Moscow. 1969 [In Russian] English original: Boolean algebras, Berlin, Springer-Verlag, 1964.

[22] M.H. Stone, *The theory of representations for Boolean algebras.* – Trans. Amer. Math. Soc. 40 (1936) 375–481.

[23] M.H. Stone, *Applications of the theory of Boolean rings to general topology.* – Trans. Amer. Math. Soc. 41 (1937) 37–111.

[24] M.H. Stone, *Algebraic characterization of special Boolean rings.* – Fund. Math. 29 (1937) 223–303.

[25] M.I. Ursul, *Topological Rings Satisfying Compactness Conditions.* – Kluwer Acad. Publ., 2002.

[26] M.I. Ursul, *Compact Rings and their Generalizations.* – Kishinev, Stiinta, 1991.

[27] M.I. Ursul and A. Tripe, *Totally bounded rings and their groups of units.* – Buletinul Acad. de Stiinte a Rep. Moldova. Matematica 1 (2004) 93–97.

[28] D. Zelinsky, *Linearly compact modules and rings.* – Amer. J. Math. 75 (1953) 79–90.

Mitrofan M. Choban
Tiraspol State University
str. Gh. Iablocichin 5
MD-2069 Chisinau, Republic of Moldova
e-mail: `mmchoban@mail.md`

Mihail I.Ursul
University of Oradea
str. Universitatii 1
RO-410087 Oradea, Romania
e-mail: `ursul@uoradea.ro`
`mihail.ursul@gmail.com`

Advances in Ring Theory
Trends in Mathematics, 113–129

Over Rings and Functors

John Dauns

Abstract. For any ring R, let $\mathcal{J}(R)$ be the unique Boolean lattice of two sided ideals which is isomorphic to the lattice of natural classes of non-singular right R-modules $\mathcal{N}_f(R)$. Let $1 = 1_R = 1_Q \in R \subset Q$ be rings with $R \subset Q$ an essential extension of right R-modules. Under some appropriate assumptions it is shown that there is an isomorphism of Boolean lattices $\Psi : \mathcal{J}(R) \longrightarrow \mathcal{J}(Q)$. The natural inclusion map $\phi : R \longrightarrow Q$, induces a natural order preserving map $\phi^* : \mathcal{N}_f(Q) \longrightarrow \mathcal{N}_f(R)$ of the Boolean lattices of natural classes of Q and R. It is shown that ϕ^* is essentially the inverse of Ψ.

Mathematics Subject Classification (2000). Primary 16D40; Secondary 16P50, 16D99.

Keywords. Rational extension of modules, complement submodule, complement closure of a submodule, natural class, Boolean lattice, second singular submodule.

Introduction

It is known that the set $\mathcal{N}(R)$ of natural classes of right R-modules is a lattice direct sum $\mathcal{N}(R) = \mathcal{N}_t(R) \oplus \mathcal{N}_f(R)$ of complete Boolean sublattices, where $\mathcal{N}_f(R)$ consists of all non-singular (called **torsion free**) classes of right R-modules. Every associative ring with identity contains a unique lattice of (two-sided) ideals $\mathcal{J}(R) \cong \mathcal{N}_f(R)$. (See [11; Thm. 6.6.6, p. 202].)

If T is a regular right self injective ring, then it was shown in [13 ; Prop. 4.1, p. 25] that the set $B(T)$ of central idempotents of T can be made into a complete Boolean lattice. The lattice operations in $B(T)$ are not the ring operations in $B(T)$: $e \wedge f = ef$, $e \vee f = e + f - ef$, where $e, f \in T$. More generally, for any ring R with identity with $Z_2(R) \leq R$ the second right singular submodule the right R-injective hull $T = E[R/(Z_2(R))]$ is such a ring as described above, and in [4; 5.11, p. 74] it was shown that $\mathcal{J}(R) \cong B(E[R/(Z_2(R))]) = B(T)$.

If $R = Z_2(R)$, then $\mathcal{J}(R) = \{0\}$ is a singleton. If R is right non-singular, then the maximal right ring of quotients of R is its right R-injective hull $ER = E(R_R)$. It was shown recently ([11 ; Corollary 6.6.7, p. 203]) that in this case $\mathcal{J}(R) \cong \mathcal{J}(ER)$. This note is just the beginning introduction of a larger project described below.

Let $1_R = 1_Q = 1 \in R \subseteq Q$ be rings such that $R \subseteq Q_R$ is an essential extension of right R- modules. This note begins with determining how the essential right R-ideals of R are related to the essential right Q-ideals of Q (Theorem 3.4). Let $\mathcal{L}(R)$ denote the essential right ideals of R, and $\mathcal{L}(Q)$ the Q-essential right ideals of Q. Then under two additional hypotheses (H1) and (H2) on the essential extension $R \subseteq Q_R$, it is shown that $\mathcal{J}(R) \cong \mathcal{J}(Q)$. If $R \subseteq Q$ is a rational extension of right R-modules, that is Q is a right ring of quotients of R, then (H1), (H2) hold. In order to prove that $\mathcal{J}(R) \cong \mathcal{J}(Q)$ it seems that there must be one to one functions $\mathcal{L}(R) \longrightarrow \mathcal{L}(Q)$, and $\mathcal{L}(Q) \longrightarrow \mathcal{L}(R)$. The hypotheses (H1), (H2) seem to be the minimal hypotheses that guarantee this.

Recently it has been shown that there exist subrings $R \subseteq Q$, where $Q \subseteq E(R)$ is an essential right R-submodule such that Q carries several non isomorphic multiplicative ring structures extending the multiplication on R. Examples in this area are hard to come by. In [1] and [2], all the examples of rings R are upper triangular matrix rings, in which case $\mathcal{J}(R) = \mathcal{J}(Q) = \{0\}$ because $R = Z_2(R)$. The same applies to the familiar first example in [16]. For the other examples in [16] and [17], the author has not been able to compute $Z_2(R)$. The results of Section 3 might be useful in answering questions like the next one. Under what conditions on the correspondence between $\mathcal{L}(R) \longleftrightarrow \mathcal{L}(Q)$ is there up to isomorphism over R a unique ring structure on Q compatible with the right R-module structure? Are the conditions (H1) and (H2) necessary and sufficient for this? All of the above is in Sections 1–4. For those readers not interested in natural classes, their use has been avoided in Sections 1–4.

This note is a part of an on going larger project. Section 5 applies the results of Section 4 to what should be functors. If $\phi : R \longrightarrow S$ is an identity preserving homomorphism of rings, then there always is an order preserving induced map $\phi^* : \mathcal{N}_f(S) \longrightarrow \mathcal{N}_f(R)$. If only surjective ring homomorphisms ϕ whose kernels are closed as right ideals are used, then it has been shown $\mathcal{N}(\ )$, $\mathcal{N}_t(\ )$, $\mathcal{N}_f(\ )$ are functors ([11; Theorem 6.5.14, p. 195], [9; 5.13, p. 538]), and they have been studied and used a lot in some form or other ([11], [10], [20], [9], and also in [4, 5, 6, 7, 8]). In this special case $\phi^* = \mathcal{N}_f(\phi) : \mathcal{N}_f(S) \longrightarrow \mathcal{N}_f(R)$ are lattice monomorphisms. The minimum condition needed to make $\mathcal{N}_f(\ )$ into a functor is that $\phi^*(\mathcal{N}_f(S)) \subseteq \mathcal{N}_f(R)$. This note is the first step in extending some of the previous results (where ϕ had to be surjective with a right closed kernel) to the case when $S = Q$, $1 \in R \subseteq Q_R$ is essential, and $\phi : R \longrightarrow Q$ is the natural inclusion. Here it is also shown that $\phi^*(\mathcal{N}_f(Q)) \subseteq \mathcal{N}_f(R)$ in 5.3(iv). One possible next step which is beyond the scope of this note is to create a category of rings and ring homomorphisms which includes all maps of the above-described kind $\phi : R \longrightarrow Q$. Clearly, this is very far from allowing $\phi : R \longrightarrow S$ to be an arbitrary ring homomorphism, which is a project that is beyond the scope of this note. Y. Zhou invented and studied M-natural classes in [19] and pre-natural classes in [20], where he showed that the set $\mathcal{N}^p(R)$ of all pre-natural classes of right R-modules is a lattice. Another part of the on going larger project is to replace $\mathcal{N}(R)$ with $\mathcal{N}^p(R)$.

1. Preliminaries

Notation 1.1. The categories of right and left unital modules over an arbitrary ring R are denoted by Mod-R and R-Mod. The symbols $<$, $\leq$ denote right R-modules, and only right R-modules, while $<_e$, $\leq_e$, $<_r$, $\leq_r$ denote essential and rational extensions of right R-modules. For $K \leq N \in$Mod-R, and $x \in N$, define $x^{\perp} = \{r \in R \mid xr = 0\} \leq R$, and $x^{-1}K = (x+K)^{\perp} = \{r \in R \mid xr \in K\}$. (Note that if $L_x : R_R \longrightarrow M$, $r \longrightarrow L_x(r) = xr$, then $x^{-1}K = L_x^{-1}K$.) Set $N^{\perp} = \{r \in R \mid Nr = 0\} \leq R$. Let '$\lhd$' denote two sided ideals in any ring. Thus $N^{\perp} \lhd R$. Right R-injective hulls are denoted by both '$\widehat{\ }$' and 'E' as $\widehat{N} = EN = E(N)$, where the latter is used if N is given by a complex formula.

For a module M its singular submodule $ZM = Z(M) \leq M$ is $ZM = \{m \in M \mid m^{\perp} \leq_e R\}$, while the second singular submodule $ZM \leq_e Z_2(M) = Z_2M$ is defined by $Z[M/Z(M)] = Z_2M/ZM \leq M/ZM$. Right R-modules M_1, M_2 are orthogonal (=perpendicular), denoted by $M_1 \perp M_2$, if there do **not** exist $0 \neq V_i \leq M_i$ such that $V_1 \cong V_2$.

A class $\mathcal{K}$ of right R-modules is a **natural class** if it is closed under isomorphic copies, submodules, arbitrary direct sums, and injective hulls. Let $\mathcal{N}(R)$ denote the set of all natural classes of right R-modules. It is well known that $\mathcal{N}(R)$ is a complete Boolean lattice, where the partial order is simply class inclusion of natural classes. The Boolean complement of $\mathcal{K}$ is $\mathcal{K}^{\perp} = \{N_R \mid \forall M \in \mathcal{K},\ N \perp M\}$.

For simplicity, a module M is said to be **torsion** if $M = Z_2M$, and **torsion free** (abbreviate **t.f.**) if $ZM = 0$ ($\Longrightarrow Z_2M = 0$). A class $\mathcal{K}$ is **torsion free** if every module in $\mathcal{K}$ is torsion free, and $\mathcal{N}_f(R)$ denotes the set of all torsion free natural classes $\mathcal{N}_f(R) = \{\mathcal{K} \in \mathcal{N}(R) \mid \mathcal{K}\ is\ \mathbf{t.f.}\}$. The applicant has proved that $\mathcal{N}(R) = \mathcal{N}_t(R) \oplus \mathcal{N}_f(R)$ is a lattice direct sum, where $\mathcal{N}_t(R)$ are the torsion natural classes which are defined in a similar way.

For any subclass $\mathcal{F}$ of right R-modules, since natural classes are closed under arbitrary intersections, $d(\mathcal{F}) \in \mathcal{N}(R)$ denotes the natural class generated by $\mathcal{F}$. Very explicitly,

$$d(\mathcal{F}) = \{\ N = N_R \mid \forall\ 0 \neq W \leq N,\ \exists\ 0 \neq V \leq W,\ \text{and}\ \exists\ V \hookrightarrow A,\ \text{some}\ A \in \mathcal{F}\}.$$

Note that this just says that $N \in d(\mathcal{F})$ if and only if there exists an essential direct sum $\oplus_{\alpha} x_{\alpha} R \leq_e N$ of cyclics $x_{\alpha}R \hookrightarrow A_{\alpha} \in \mathcal{F}$ for some $A_{\alpha} \in \mathcal{F}$.

In fact, if Y is a right R-module, then $d(Y)$ consists of all right R-modules which can be embedded in the injective hull of some direct sum of Y's. Also, $d(Y)^{\perp} = \{\ M_R \mid M \perp Y\ \}$.

Terminology 1.2. As usual, R being a unital subring of Q means that $1 = 1_R = 1_Q \in R \subseteq Q$ are rings. If in addition also $R \leq_e Q_R$ is an essential extension of right R-modules, then Q will be said to be a **right over ring** of R. Left over rings are defined similarly. If for a unital subring $R \subseteq Q$, $R \leq_r Q_R$ is a rational extension of right R-modules, then as usual, Q will be said to be a **right ring of quotients** of R.

Notation 1.3. The same objets $\mathcal{K}$, Z , Z_2 , E, $\leq$, $\leq_e$, $\leq_r$, and $d(\cdot)$ as defined above for R, over another ring S will be denoted by superscripts as $\mathcal{K}^S$, Z^S , Z_2^S , E^S, $\subseteq$, $\subseteq^e$, $\subseteq^r$, and $d^S(\cdot)$. For $y \in N_S$, define $\text{ann}(y) = \text{ann}_S(y) = \{s \in S \mid ys = 0\}$.

Functors 1.4. In order to investigate the functors $\mathcal{N}(\cdot)$, $\mathcal{N}(\cdot)_t$, and under certain conditions $\mathcal{N}_f(\cdot)$, let $\phi : R \longrightarrow S$ be any ring homomorphism preserving identities. The above map induces a map Mod-$S \longrightarrow$ Mod-R, $N = N_S \longrightarrow N_\phi \in$Mod-$R$, where for $y \in N$ and $r \in R$, $y \cdot r = y(\phi r)$. For $\mathcal{K}^S \in \mathcal{N}(S)$, define $\mathcal{K}^S_\phi = \{N_\phi : N \in \mathcal{K}^S\}$. The correspondence $N \to N_\phi$ induces a covariant functor $\phi^\# :$ Mod-$S \longrightarrow$ Mod-R which on morphisms, is the identity. This functor $\phi^\#$ induces a map $\phi^* : \mathcal{N}(S) \longrightarrow \mathcal{N}(R)$ by $\phi^*(\mathcal{K}) = d(\mathcal{K}^S_\phi)$. For any ring with identity R, the ring contains a unique lattice $\mathcal{J}(R)$ of two sided ideals, defined as $\mathcal{J}(R) = \{ I \leq R_R$ is a right complement $| Z_2(R) \leq I \leq R_R,\ E(I) \leq E(R)$ is fully invariant$\}$.

For any $I \in \mathcal{J}(R)$, for any $C \leq R_R$ such that $I \oplus C \leq_e R_R$, necessarily $I \perp C$.

Always, there is an isomorphism of complete Boolean lattices $\eta_R : \mathcal{J}(R) \longrightarrow \mathcal{N}_f(R)$ given by $\eta_R(I) = d(I/Z_2(R))$. (See [11; Thm. 6.6.6, p. 202], [5; Thm. 2.6, p. 106], [6; Thm. 2.6, p. 333].)

2. Large right ideals

We review and develop some facts about complement submodule which we need and which may be of independent interest.

Complement closure 2.1. Let $ZM \subseteq K < M$ be right R-modules. Then K has in M a complement closure $K \leq_e \overline{K} \leq M$ satisfying the following:

(1) $\overline{K} = \{ x \in M \mid x^{-1}K <_e R \}$;
(2) $\overline{K} = \{ x \in M \mid K \leq_e K + xR \}$;
(3) $\overline{K} = \bigcap \{ C \mid K \leq C,\ C \leq M$ is a complement $\}$.
(4) $\overline{K}$ is the unique smallest complement closure of K in M.
(5) If $L \leq_e K$, then $\overline{L} = \overline{K}$.
(6) $\overline{K}/Z_2M \leq M/Z_2M$ is a complement.

Proof. Conclusions (1)–(4) are in [4; p. 53, Prop. 1.3]. (5) By (3) and (4), $\overline{L} \subseteq \overline{K}$. From $L <_e K \leq_e \overline{K}$, we get $\overline{L} \leq_e \overline{K}$, and hence $\overline{L} = \overline{K}$. (6) Since by 2.1 (1)$\Longleftrightarrow$ (2), for $x \in M$, $(x + Z_2M)^{-1}(\overline{K}/Z_2M)) = \{ r \in R \mid xr \in \overline{K} \}$ is a large right ideal if and only if $\overline{K} \leq_e \overline{K} + xR = \overline{K}$. Thus $\overline{\overline{K}/Z_2M} = \overline{K}/(Z_2M)$. □

The following lemma circumvents the difficulties posed by the possibility that $Z_2^Q Q \setminus JQ \neq \emptyset$ for $J \in \mathcal{J}(R)$, or equivalently, $Z_2^Q Q \setminus (Z_2R)Q \neq \emptyset$.

Lemma 2.2. *For any ring R, and any right R-modules $A <_e B \leq M$, and any $x \in M$,*

$$x^{-1}B \leq_e R \iff x^{-1}A \leq_e R.$$

Proof. $\Longrightarrow$ If not, then $(x^{-1}A) \oplus C \leq R$ for some $C \neq 0$. There exists $0 \neq c_0 \in C \cap x^{-1}B$. If $xc_0 = 0$, then $0 \neq c_0 \in (x^{-1}A) \cap C = 0$, a contradiction. So $0 \neq xc_0R \leq B$. But $A \leq_e B$, and hence also $0 \neq xc_0R \cap A \subseteq xC \cap A = 0$, since $xC + A = xC \oplus A$. Thus $C = 0$. □

3. Mod-R and mod-Q for $R \subset Q$

Throughout this section, unless otherwise stated, it is assumed that $1 = 1_Q = 1_R \in R <_e Q$ are rings, i.e., Q is a right over ring of R. If $A, B \in$ Mod-R, with $A \cup B \subseteq N$ where $N \in$ Mod-Q, then $(AQ + BQ) \subseteq N_Q$. This section considers when such binary module properties as $A \cap B = 0$, $A <_e B$, $A \perp B$ do or do not transfer to $AQ \cap BQ = 0$, $AQ \subseteq^e BQ$, $AQ \perp^Q BQ$.

In (i) of the next lemma, $R <_e Q$ is not required. It is worth stating explicitly in words, that below the converse of 3.1(i) holds either if Y is R-nonsingular, or if Q is a ring of right quotients of R.

Lemma 3.1. *If $X \subset Y$ are right Q-modules, then*

(i) $X <_e Y \implies X \subset^e Y$;

(ii) $ZX = 0,\ X \subset^e Y \implies X <_e Y$;

(iii) $R <_r Q,\ X \subset^e Y \implies X <_e Y$.

Proof. (i) is clear. (ii) For any $0 \neq y \in Y$, there is a $q \in Q$ with $0 \neq yq \in X$. But then also $0 \neq (yq)q^{-1}R \subseteq yR \cap X$, since $0 \neq yq \notin ZX$. So $X \subset^e Y$.

(iii) For $0 \neq y \in Y$, again let $0 \neq yq \in X$, $q \in Q$. Since $R <_r Q$, and $yq,\ q \in Q$, there is an $r \in R$ with $yqr \neq 0$ and $qr \in R$. Thus $0 \neq yqr \in yR \cap X$. Hence $X <_e Y$. □

Lemma 3.2. *For $X,\ Y \in$ Mod-Q, $Hom_Q(X,Y)$ =$Hom_R(X,Y)$ if one of* (i), (ii), *or* (iii) *holds:*

(i) $ZY = 0$;

(ii) $R <_r Q$, *and* $Y \subseteq Q$;

(iii) $X \leq \widehat{R},\ Y \leq \widehat{R};\ X, Y \in$ *Mod-Q*, $R <_r Q$.

Proof. For $f \in \mathrm{Hom}_R(X,Y)$ and any $x \in X,\ q \in Y$, define $z = f(xq) - (fx)q$. (i) For any $r \in q^{-1}R \leq_e R$, $zr = f((xq)r) - f(x(qr)) = 0$. Thus $z \in ZY = 0$.

(ii) Since $R <_r Q$ and $z \in Y \subseteq Q$, if $z \neq 0$, we can find an $r \in R$ with $zr \neq 0$ and $qr \in R$. Thus $r \in q^{-1}R$, and as above in (i) we have $zr = 0$, a contradiction. Thus $z = 0$.

(iii) By [15; p.95, Prop.2], $\mathrm{Hom}_R(\widehat{R},\widehat{R})$=$\mathrm{Hom}_Q(\widehat{R},\widehat{R})$. Extend f to $\widetilde{f} \in \mathrm{Hom}_R(\widehat{R},\widehat{R})$. Then $z = \widetilde{f}(xq) - (\widetilde{f}x)q = 0$, and $f \in \mathrm{Hom}_Q(X,Y)$. □

The notation $X_Q \perp Y_Q$ or $X \perp^Q Y$ denotes orthogonality of Q-modules.

Lemma 3.3. *For rings $1 = 1_Q = 1_R \in R <_e Q$, let $\phi : R \longrightarrow Q$ be the inclusion map. Let $A, B \in$ Mod-R, $N \in$ Mod-Q with $A \leq N_\phi$, $B \leq N_\phi$ and let $X, Y \in$ Mod-Q. Then the following hold:*

(i) $A + B = A_R \oplus B_R \le N_R,\ Z(A \oplus B) = 0 \Longrightarrow AQ \oplus BQ \le N_Q$;
(ii) $A_R \perp B_R \implies (AQ) \perp (BQ)$, *and* $(AQ) \perp^Q (BQ)$;
(iii) $X_\phi \perp Y_\phi \implies X_Q \perp Y_Q$;
(iv) $ZX_\phi = 0,\ ZY_\phi = 0;\ X_Q \perp Y_Q \Longrightarrow X_\phi \perp Y_\phi$;
(v) $Z[A(Q)_\phi] \subseteq A \implies A \le_e AQ$.

Proof. (i) If $0 \ne \xi = \sum_{i=1}^n a_i p_j = \sum_{j=1}^m b_j q_j \in AQ \cap BQ,\ a_i \in A,\ b_j \in B;\ p_i, q_j \in Q$, and $L = [\cap_1^n p_i^{-1} R] \cap [\cap_1^m q_j^{-1} R] \le_e R$, then $\xi L \subseteq A \cap B = (0)$, and hence $\xi \in Z(AQ \cap BQ) = 0$, a contradiction.

(ii) If $0 \ne \alpha Q \cong \beta Q,\ \alpha \in AQ,\ \beta \in BQ$ as Q-modules and with $\operatorname{ann}_Q \alpha = \operatorname{ann}_Q \beta$, then as in (i), there is an $L \le_e R$ with $\alpha L \subseteq A,\ \beta L \subseteq B$. Since $ZB = 0$, there is a $\lambda \in L$ with $0 \ne \beta\lambda \in B$. Since $\alpha^\perp = R\ \cap \operatorname{ann}_Q \alpha = \beta^\perp = R\ \cap \operatorname{ann}_Q\ \beta = \beta^\perp$, also $0 \ne \alpha\lambda$, and $(\alpha\lambda)^\perp = \lambda^{-1}\beta^\perp = (\beta\lambda)^\perp$. Thus $\alpha\lambda R \cong \beta\lambda R$ is a contradiction.

(iii) If $0 \ne xQ \cong yQ,\ x \in X,\ y \in Y$ with $\operatorname{ann}_Q(x) = \operatorname{ann}_Q(y) \subset Q$, then $x^\perp = R\ \cap \operatorname{ann}_Q(y) = y^\perp$. Thus $xR \cong yR$ is a contradiction.

(iv) If not, let $x \in X,\ y \in Y,\ x^\perp = y^\perp$. Now define $\psi : xQ \longrightarrow yQ$ by $\psi xq = yq$ for $q \in Q$. This is well defined, for if $xq = 0$, then $xR \cong yR$ implies that $0 = x[q(q^{-1}R] \cong y[q(q^{-1}R]$, and hence that $yq \in ZY = 0$, and $xq \in ZX = 0$. Thus not only ψ is well defined, but it is also a Q-isomorphism, which is a contradiction.

(v) For any $0 \ne \eta \in AQ \setminus A,\ \eta = \sum_1^n a_i q_i,\ a_i \in A,\ q_i \in Q$, since $L = \cap_1^n q_i^{-1} R \le_e R$ and since $\eta \notin Z(AQ)$, it follows that $0 \ne \eta L \subseteq A$. Thus $A \le_e AQ$. □

Theorem 3.4. *Let $1 = 1_Q = 1_R \in R <_e Q$ be rings, and $I \le R,\ J \le R,\ X \subseteq Q_Q,\ Y \subseteq Q_Q$ be any right ideals.*

(1) *If $R <_e Q$, then*
  (i) $X \cap R \le_e R \implies X \subseteq^e Q$;
  (ii) $I \le_e R \implies IQ \subseteq^e Q$.
(2) *If $R <_r Q$, then the converses hold*
  (i) $X \subseteq^e Q \implies X \cap R \le_e R$;
  (ii) $IQ \subseteq^e Q \implies I \le_e R$.

Proof. (1)(i). First we show that $(X \cap R)Q \le_e Q_R$. For any $0 \ne q \in Q$, there is an $r \in R$ with $0 \ne qr \in R$. Since $X \cap R \le_e R$, there is an $s \in R$ with $0 \ne (qr)s \in X \cap R \subseteq (X \cap R)Q$. Hence $(X \cap R)Q \le_e Q_R$, and by Lemma 3.1(i), hence also $(X \cap R)Q \subseteq^e Q_Q$.

(1)(ii). For $0 \ne q \in Q$, it suffices to show that $0 \ne qR \cap I$, since $qR \cap I \subseteq qQ \cap IQ$. First, $0 \ne qr \in R$ for some $r \in R$. Then $0 \ne qrs \in I$.

(2)(i). It suffices to show that $X_R \le_e Q_R$, for then also $X \cap R \le_e R = Q \cap R$. For any $0 \ne y \in Q$, $0 \ne yq \in X$ for some $q \in Q$. Since $R <_r Q$, for $0 \ne yq,\ q \in Q$, there exists an $r \in R$ with $0 \ne yqr$ and $qr \in R$, or $0 \ne yqr \in X$ with $qr \in R$. Hence $0 \ne yqr \in yR \cap X$, and $X_R \le_e Q_R$, and hence $X_R \le_e Q_R$.

(2)(ii). It suffices to show that for any $0 \neq y \in Q$, $yR \cap I \neq 0$. By hypothesis we have for some $0 \neq q_0 \in Q$, $0 \neq yq_0 \in IQ \subseteq^e Q$, where $yq_0 = \sum_{k=1}^{m} y_k q_k$ for some $y_k \in I$, $q_k \in Q$.

Since $R <_r Q$, for $0 \neq yq_0$, $q_0 \in Q$, there is an $r_0 \in R$, with $yq_0r_0 \neq 0$, and $q_0r_0 \in R$. Thus $yq_0r_0 = \sum_{k=1}^{m} y_k q_k r_0$. Next, for $0 \neq yq_0r_0$, $q_1r_1 \in Q$, there is an $r_1 \in R$ with $Q_1r_0r_1 \in R$, and $0 \neq yq_0r_0r_1 = \sum_{k=1}^{m} y_k q_k r_0 r_1$. There is an $r_2 \in R$ with $0 \neq yq_0r_0r_1r_2 = \sum_{k=1}^{m} y_k q_k r_0 r_1 r_2$, where $q_2r_0r_1r_2 \in R$, as well the previous $q_1r_0r_1r_2 \in R$. Continuing this way we get $r_0, r_1, \ldots, r_n \in R$ such that

$$0 \neq yq_0r_0r_1 \ldots r_n = \sum_{k=1}^{m} y_k q_k r_0 r_1 \ldots r_n \in yR \cap I, \; q_k r_0 r_1 \ldots r_k \in R.$$

Thus $I \leq_e R$. □

Corollary 3.5. *If above $R <_r Q$, then $\forall\, Y \subseteq^e Q$, $\exists\, J \leq_e R$, $Y \supseteq JQ \subseteq^e Q$.*

Proof. From $Y \subseteq^e Q$ it follows by 3.4 (2)(i) that $Y \cap R \leq_e R$. Then for $J = Y \cap R \leq_e R$, by 3.4 (1)(ii), $JQ \subseteq^e Q$. □

It is beneficial for later arguments to visualize the previous facts as set containment relations in the lattices $\mathcal{L}(R)$, and $\mathcal{L}(Q)$ of all large right R-ideals and all large Q-ideals.

Corollary 3.6. *Let $1 = 1_R = 1_Q \in R \leq_e Q$ be any rings, and let $\mathcal{L}(R)$ and $\mathcal{L}(Q)$ denote the set of all large right ideals of R and Q. Then the following hold:*

(1) (i) $\mathcal{L}(Q) \supseteq \{\, X \subseteq Q \mid X \cap R \in \mathcal{L}(R) \,\}$;
(ii) $\mathcal{L}(Q) \supseteq \{\, IQ \mid I \in \mathcal{L}(R) \,\}$;

(2) *If in addition, $R <_r Q$, then*
(i) $\mathcal{L}(R) \supseteq \{\, X \cap R \mid X \subseteq^e Q \,\}$;
(ii) $\mathcal{L}(R) \supseteq \{\, I \leq R \mid IQ \subseteq^e Q \,\}$.

Below in (1) and (2), $\subseteq$ refers to R-submodules, but not Q-submodules.

Proposition 3.7. *Let $1 = 1_Q = 1_R \in R <_e Q$ be an over ring of R and let $\phi : R \longrightarrow Q$ be the inclusion map. Then for any $N \in$Mod-Q, the following hold:*

(1) $ZN \subseteq Z^Q N$;
(2) $Z_2 N \subseteq Z_2^Q N$.

If in addition , $R <_r Q$, then

(3) $ZN = Z^Q N$;
(4) $Z_2 N = Z_2^Q N$.

Proof. (1) If $\xi \in ZN$, then $\xi^\perp \leq_e R \leq_e Q$, implies that $\xi^\perp \leq (\xi^\perp Q)_R \leq_e Q$. By Lemma 3.1(i), $\xi^\perp Q \subseteq^e Q$. But since $\xi(\xi^\perp Q) = 0$, it follows that $\xi^\perp Q \subseteq \mathrm{ann}_Q\, \xi \subseteq^e Q$. Thus $\xi \in Z^Q N$, and $ZN \subseteq Z^Q N$.

(2) Note that $Z^Q N$ by definition is a Q-module, while ZN may not be, and that $[N/Z^Q N]_\phi = N_\phi/(Z^Q N)_\phi$. Let π; $N/ZN \longrightarrow N/Z^Q N$ be the projection induced by the inclusion of right R -modules in (1), and that π is a homomorphism

of R-modules. Then since Z is a subfunctor of the identity functor, it follows that on objects $\pi Z[\cdot] \subseteq Z\pi[\cdot]$. Hence

(i) $$\pi Z \frac{N}{ZN} \subseteq Z\left(\pi\left[\frac{N}{ZN}\right]\right) = Z\left(\pi\left[\frac{N}{Z^Q N}\right]_\phi\right) = Z\left[\frac{N}{ZN}\right].$$

By (1) applied to $N/Z^Q N \in$ Mod-Q, we get

(ii) $$Z\left[\frac{N}{Z^Q N}\right] \subseteq Z^Q\left[\frac{N}{Z^Q N}\right] = \frac{Z_2^Q N}{Z^Q N}.$$

But

(iii) $$\pi\left(Z\left[\frac{N}{ZN}\right]\right) = \pi\left(\frac{Z_2 N}{ZN}\right) = \frac{Z_2 N + Z^Q N}{Z^Q N}.$$

Thus

(iv) $$\frac{Z_2 N + Z^Q N}{Z^Q N} \subseteq Z\left[\frac{Z_2^Q N}{Z^Q N}\right], \; Z_2 N \subseteq Z_2^Q N.$$

Note that if in equation (ii), the "$\subseteq$" was "$=$", then in (iv) we would get that $Z_2 N + Z^Q N = Z_2 N$. Now if in addition it was known that $Z^Q N = ZN$, then we could conclude that $Z_2 N = Z_2^Q N$.

(3) Since $ZN \subseteq Z^Q N$ by (1), let $0 \neq x \in Z^Q N$, and we will show that $x \in ZN$. Since $\text{ann}_Q(x) \subseteq^e Q_Q$, by Theorem 3.4, (2)(i) we get that $x^\perp = \text{ann}_Q(x) \leq_e R$. Hence $x \in ZN$, hence $Z^Q N \subseteq ZN$, and by (2) above, $Z^Q N = ZN$. (4). The remark at the end of the proof of (2) above shows that $Z_2 N = Z_2^Q N$. □

Corollary 3.8. *If $R <_e Q$, then*

(i) $ZR \subseteq ZRQ \subseteq ZQ \subseteq Z^Q Q$;

(ii) $Z_2 R \subseteq Z_2 RQ \subseteq Z_2 Q \subseteq Z_2^Q Q$.

There is no reason to expect that the above inclusions in 3.8(ii) are essential extensions of R-modules.

Proposition 3.9. *If $R <_r Q$, then*

(i) $ZR \leq_e ZQ = Z^Q Q$;

(ii) $Z^Q(Q) \leq_e Z_2^Q(Q)$; *hence*

(iii) $ZR \leq_e Z_2 R \leq_e Z_2^Q(Q)$.

Proof. (i) If $0 \neq \eta \in ZQ$ then for some $r \in R$, $0 \neq \eta r \in R$. Then $(\eta r)^\perp = \text{ann}_Q(\eta r) \leq_e R$ by Theorem 3.4 (2)(i). Thus $0 \neq (\eta r) \in ZR$, and $ZR \leq_e ZQ$. Lastly, by Proposition 3.7 (3), $ZQ = Z^Q Q$. (ii) This follows by Lemma 3.1(iii). Finally, upon combining (i) and (ii) we get conclusion (iii). □

Proposition 3.10. *If $R <_r Q$, consider $0 \neq A \leq R$, and $0 \neq B \leq R$, with $ZB = 0$. Then*

(i) $A \leq_e AQ$; *and*

(ii) $B \leq_e BQ$, *and* $Z_2^Q(Q) + BQ = Z_2^Q(Q) \oplus BQ$.

Proof. (i) Let $0 \neq \eta = \sum_1^n a_i q_i \in AQ$ with $a_i \in A$, $q_i \in Q$. Since $R <_e Q$, $0 \neq \eta r_0 \in R$. Since $R \leq_r Q$, there is an $r_1 \in R$, $0 \neq \eta r_0 \in R$, and $q_1 r_1 \in R$. Continuing this way we get $r = r_0 r_1 \ldots r_n \in R$ and $0 \neq \eta \sum_1^n a_i q_i \in AQ$.

(ii) Suppose that $0 \neq \eta = \sum_1^n b_i q_i \in ZR \cap BQ$. Since $R <_r Q$, there is an $r_1 \in R$, $\eta r_1 \neq 0$, and $q_1 r_1 \in R$. Continuing this way we get $r = r_1 \ldots r_n \in R$, and get the contradiction $0 \neq \eta r \in B \cap ZR = 0$. Thus $ZR \oplus BQ \leq_e Z_2^Q(Q) \oplus BQ$ by the previous Proposition 3.8(iii). □

Hypotheses 3.11. Let $1 = 1_R = 1_Q \in R \leq_e Q_R = Q$ be rings, and $\phi : R \hookrightarrow Q$, the inclusion map. Define (H1) and (H2) to be the following hypotheses.

(H1) $\forall\, B \leq R,\ B \leq_e BQ$.

(H2) $ZR \leq_e Z_2^Q(Q)$. $(ZR = 0 \Longrightarrow\ Z_2^Q(Q) = 0)$.

If $R <_r Q$, all of the above hold.

The next theorem allows us to transfer properties of right ideals between R and Q.

Theorem 3.12. *Let $R <_e Q$ be rings, and let (H1) and (H2) hold. Let X, Y, U be any right Q ideals, where $Z^Q Q \subseteq Y \subseteq^e Q_Q$; $Z^Q Q \subseteq X \subseteq X \oplus U \subseteq^e Q_Q$. Then the following hold.*

(i) *$Y \cap R \leq_e R$; in particular $(X \cap R) \oplus (U \cap R) \leq_e R$.*

(ii) *$X \perp^Q U \iff (X \cap R) \perp (U \cap R)$.*

(iii) *$ZR \leq_e Z^Q Q_R$ and $ZR \leq I \leq I \oplus V \leq R$, $\Longrightarrow IQ + VQ = IQ \oplus VQ \subseteq Q_Q$.*

(iv) *$ZR \subseteq I \subseteq I \oplus V \leq_e R$, $\Longrightarrow IQ \oplus VQ \subseteq_e Q_Q$.*

(v) *$I \perp V \iff IQ \perp^Q VQ$.*

Proof. If not for some $V \neq 0$, we have $(Y \cap R) \oplus V \leq R$. By (H1), $V \leq_e VQ$, and hence $(Y \cap R) \oplus VQ \leq Q_R$. Also, $ZR \oplus VQ \leq Q_R$. Since $Y \subseteq^e Q$, there exists $0 \neq \xi = \sum_1^n v_i q_i \in Y \cap VQ$. Let $L = \cap_1^n q_i^{-1} R \leq_e R$. Then $0 \neq \xi L \subseteq V \cap X = V \cap Y \cap R = 0$, a contradiction. Thus $V = 0$ and $Y \cap R \leq_e R$.

(ii) Always, (ii) $\Longleftarrow$ by Lemma 3.3 (ii). $\Longrightarrow$: Since $ZR \leq X \cap R \leq_e X$ and $U \cap R \leq_e U$, by Lemma 3.2 (i), this holds.

(iii) Since $ZR \leq I \leq_e IQ$, and $V \leq_e VQ$, we have $Z[(IQ) \cap (VQ)] = 0$. Let $0 \neq \xi = \sum_1^n y_i q_i = \sum_1^m v_j q_j \in [(IQ) \cap (VQ)]$, and set $L = [\cap_1^n q_i^{-1} R] \cap [\cap_1^m q_j^{-1} R]$. Then $\xi L \subseteq I \cap V = 0$ gives the contradiction that $0 \neq \xi \in Z[(IQ) \cap (VQ)] = 0$. Always $<_e \Longrightarrow \subseteq^e$.

(iv) Since $I \oplus V \leq_e R \leq_e Q$, and since $I \oplus V \subseteq IQ \oplus VQ \leq Q_R$, it follows that $IQ \oplus VQ \leq_e Q_R$. But by Lemma 3.1, for Q-modules, always $\leq_e \Longrightarrow \subseteq^e$, and thus (iv) follows.

(v)$\Longleftarrow$ always holds. $\Longrightarrow$: Since $ZR \leq I \leq_e IQ$, and $V \leq_e VQ$, $ZVQ = 0$, and the rest follows from Lemma 3.2 (i). □

4. Over rings

Throughout this section, many of the submodules used are so called type submodules, even when this is not stated. Type submodules are defined below, and the connection with natural classes clarified.

Type submodules 4.1. For any right R-module M, a submodule $N \leq M$ is a **type submodule** if $N \leq M$ is a complement, and if $\exists\, P \leq M$ such that $N \oplus P \leq_e M$ is an essential extension, and $N \perp P$ (see 1.1 for '$\perp$'). If above '$\exists$' is replaced by '$\forall$', we get the same definition. Another equivalent definition is the following. A submodule $N \leq M$ is a **type submodule** if there exists a natural class $\mathcal{K} \in \mathcal{N}(R)$ such that $N \in \mathcal{K}$, but for any $N < L \leq M$, $L \notin \mathcal{K}$. In other words, $N \leq M$ is a **maximal** $\mathcal{K}$ **submodule**. Such a submodule N is also called a **type** $\mathcal{K}$ type submodule of M.

A sort of converse of the above is that for any natural class $\mathcal{K}$ and any right R-module M whatever, M always has at least one maximal $\mathcal{K}$-submodule.

Notation 4.2. Recall that by definition $\mathcal{J}(R) = \{I \leq R_R$ is a right complement $| Z_2(R) \leq I \leq R_R, E(I) \leq E(R)$ is fully invariant$\}$. Now let $J \in \mathcal{J}(R)$. Then firstly [11; Lemma 6.6.5, p. 202] states that $J/Z_2R \leq R/Z_2R$ is the unique type submodule of the right R-module R/Z_2R of type $d(J/Z_2R)$. Secondly, [11; Proposition 6.6.4 part (1), p. 201] says that $J \leq R_R$ is also a type submodule.

Let $Z_2(R) \leq J \leq$R. Then $J \leq R$ being a right complement is equivalent to $J/Z_2R \leq R/Z_2R$ being a right R- complement. (See first [11; Lemma 6.6.3, p. 201], and secondly observe the fact [12; Exercise 15, p. 20] that for any R-modules $A \leq B \leq C$, if B is closed in C, then $B/A \leq C/A$ is closed in C/A.)

Elements of $\mathcal{J}(R)$ are complement right ideals of the form $J \leq R$ with $Z_2R \leq J$, and such that $J \perp C$ for any $J \oplus C \leq_e R$. So given $J \in \mathcal{J}(R)$, select any right ideals $B, C \leq R$, where

$$Z_2R \oplus B \leq_e J \leq J \oplus C \leq_e R,\ Z_2R \oplus B \oplus C \leq_e R,\ B \perp C.$$

Furthermore note that it follows that $J \lhd R$ where '$\lhd$' denotes two sided ideals in an ring.

Next 3.1, 3.3, 3.4, and 3.11 allow us to transfer essentiality, direct sums, orthogonality of right ideals from R to Q, and vice versa. We do this in Section 4 without explicitly quoting the appropriate justifying result from Section 3.

Lemma 4.3. *With the above notation from* 4.2, *if* 3.11 (H1) *and* (H2) *hold, then*

$$Z_2^Q Q \oplus BQ \oplus CQ \subseteq^e Q,\ BQ \perp^Q CQ,\ \text{and } (BQ)_R \perp (CQ)_R.$$

Lemma 4.4. *Let* $Z_2R \oplus B \leq_e J \leq J \oplus C \leq_e R$ *be as in* 4.2, *and define* $\widetilde{J}$ *to be* $\widetilde{J} = \{\ \zeta \in Q \mid \{\ q \in Q \mid \zeta q \in Z_2^Q Q \oplus BQ\ \} \subseteq^e Q\ \}$. *Then the following hold.*

(i) $\widetilde{J}$ *is the unique right* Q*-complement closure of* $Z_2^Q Q \oplus BQ \subseteq^e Q$ *in* Q*; it is independent of the choice of* B *in* 4.2.

(ii) $\widetilde{J} \oplus CQ \subseteq^e Q$*;*

(iii) $JQ \subseteq \widetilde{J}$.

Lemma 4.5. *In the notation of* 4.3, 4.4*:*

(i) $Z_2^Q Q$*,* BQ*, and* CQ *are pairwise orthogonal as* Q*-modules;*
(ii) $\widetilde{J} \perp^Q CQ$.
(iii) $\widetilde{J} \in \mathcal{J}(Q)$*;*
(iv) $BQ \hookrightarrow^e \widetilde{J}/Z_2^Q(Q)$.

Construction 4.6. Let $J_1,\ J_2 \in \mathcal{J}(R)$. Define $U_1,\ U_2$ as any right ideals such that $Z_2R \oplus (J_1 \cap J_2) \oplus U_1 \leq_e J_1$ and $Z_2R \oplus (J_1 \cap J_2) \oplus U_2 \leq_e J_2$. Then

(i) $U_1 \perp J_2,\ U_2 \perp J_1$ and in particular $U_1 \perp U_2$;
(ii) $U_1Q \perp J_2Q,\ U_2Q \perp J_1Q$, and $U_1Q \perp U_2Q$;
(iii) $\mathcal{J}(R) \longrightarrow \mathcal{J}(Q),\ J \longrightarrow \widetilde{J}$ is one to one.

Aside from showing that the function below is well defined and one to one, the next proposition gives a wealth of information about how the ideal structure of R relates to that of Q.

Proposition 4.7. *Assume* (H1) *and* (H2)*, and let* $J \in \mathcal{J}(R),\ B,\ \widetilde{J},\ C$ *be as above. Then the following hold.*

(i) $\widetilde{J} \cap CQ = 0$.
(ii) $\widetilde{J} \oplus CQ \subseteq^e Q$.
(iii) $\widetilde{J} \perp^Q CQ$.
(iv) $\widetilde{J} \in \mathcal{J}(Q)$.
(v) $\widetilde{J}$ *is independent of the choice of* B.
(vi) $J \longrightarrow \widetilde{J}$ *is a well-defined map* $\Psi :\ \mathcal{J}(R) \longrightarrow \mathcal{J}(Q)$.
(vii) *The above map* Ψ *is one to one.*

Proposition 4.8. *If as before* $R <_e Q$ *and* (H1) *and* (H2) *hold, then the map* Ψ *in* 4.7 *is an isomorphism. Both* Ψ *and its inverse are order preserving maps.*

Proof. To show that the map is onto take $Y \in \mathcal{J}(Q)$, and choose any V and U such that $Z_2^Q Q \oplus V \subseteq^e Y \subseteq Y \oplus U \subseteq^e Q$ with $Y \subset Q_Q$ Q-closed. Then since Y is a type submodule we have $Y \perp^Q U$.

Since (H1) and (H2) guarantee that $Z_2R \leq_e Z_2^Q Q \cap R$, and that $\subseteq^e \Longrightarrow \leq_e$, we conclude that

$$Z_2^Q Q \oplus V \oplus U \subseteq^e Q \implies Z_2R \oplus V \cap R \oplus U \cap R \leq_e R.$$

Next define I to be the unique complement closure of $Z_2R \oplus V \cap R \leq_e I \leq R$.

Then the following five steps finish the proof.

(i) $I \oplus U \cap R \leq_e R$;
(ii) $I \perp U \cap R$;
(iii) $I \in \mathcal{J}(R)$ and $\forall I \in \mathcal{J}(R),\ \widetilde{I} \in \mathcal{J}(Q)$ as defined in 4.4 is a Q-complement with $Z_2^Q Q \oplus [(V \cap R)Q] \subseteq^e \widetilde{I} \subseteq Q_Q$.
(iv) Since $[(V \cap R)Q] \subseteq^e V,\ Z_2^Q Q \oplus [(V \cap R)Q] \subseteq^e Y$. Thus
(v) $Y = \widetilde{I}$, and Ψ is onto, and hence an isomorphism. □

Since both Ψ and its inverse, almost by definition preserve the partial order $\subset \subseteq$, Ψ is an order isomorphism, and hence a complete isomorphism (i.e., preserving arbitrary infinite joins and meets in both directions) of the complete Boolean lattices $\mathcal{J}(R) \cong \mathcal{J}(Q)$.

The next theorem is formulated in terms of lattices of ideals of the rings R and Q, and does not require a knowledge of natural classes and the functor $\mathcal{N}(\cdot)$.

Theorem 4.11. *Let $1 = 1_R = 1_Q \in R \leq_e Q_R = Q$ be rings, assume* (H1) *and* (H2), *and let Ψ be the map $\Psi : \mathcal{J}(R) \longrightarrow \mathcal{J}(Q)$, $J \in \mathcal{J}(R)$, $\Psi(J) = \widetilde{J}$, where $\widetilde{J}$ is as defined in Lemma 4.4. Then Ψ is an isomorphism, and both Ψ and its inverse preserve the order.*

5. Lattices

For any identity preserving homomorphism $\phi : R \longrightarrow Q$ of associative rings R, Q, there is an order preserving induced function $\phi^* = \mathcal{N}(\phi) : \mathcal{N}(R) \longrightarrow \mathcal{N}(Q)$. And there is always a lattice direct sum $\mathcal{N}(R) = \mathcal{N}_t(R) \oplus \mathcal{N}_f(R)$ of complete Boolean lattices, and similarly for Q.

The last section showed that Ψ is an isomorphism. This section contains new additional information that the restriction and co-restriction of ϕ^* to $\phi^* : \mathcal{N}_f(R) \longrightarrow \mathcal{N}_f(Q)$ is not only an isomorphism, but essentially the inverse of Ψ, in view of the isomorphisms $\mathcal{N}_f(R) \cong \mathcal{J}(R)$ for all R.

Since the above map ϕ^* has been used often in many articles for a long period of time, it is of interest to see what happens under the assumptions (H1), (H2) on the rings $R <_e Q$, and how the map Ψ relates to the map ϕ^*.

Recall that in general, a one to one and onto order preserving map of partially ordered sets is not an isomorphism of ordered sets, for the inverse map need not preserve order.

Corollary 5.1. *When $R <_e Q$ and* (H1), (H2) *hold, then Ψ induces a lattice isomorphism $\widetilde{\Psi} = \eta_Q \Psi \eta_R^{-1} : \mathcal{N}_f(R) \longrightarrow \mathcal{N}_f(Q)$, where η_R, η_Q are as in 1.4.*

Proof. For any ring, $\mathcal{J}(R) \cong \mathcal{N}_f(R)$ under the isomorphism $J \longrightarrow d(J/Z_2R) \in \mathcal{N}_f(R)$. □

Observation 5.2. The following is valid for any identity preserving homomorphism $\phi : R \longrightarrow Q$ of any rings R and Q without any special assumptions. Every element of $\mathcal{N}(Q)$ is of the form $d^Q(N)$, $N = N_Q$. Then $\phi^*(d^Q(N)) = \phi^*\{ V_\phi \mid V_Q \in d^Q(N)\} = d(N_\phi)$.

Proposition 5.3. *Let $\phi : R \longrightarrow Q$ be the inclusion map of $R \leq_e Q$, and let $\phi^* : \mathcal{N}(Q) \longrightarrow \mathcal{N}(R)$ and $\phi^*(\mathcal{K}^Q) = d(\mathcal{K}^Q_\phi)$ be the induced map. Let $\mathcal{N}(R) = \mathcal{N}_t(R) \oplus \mathcal{N}_f(R)$, and similarly for Q. Assume* (H1) *and* H(2) *and let $Y \in \mathcal{J}(Q)$, and let*

$$Z_2^Q(Q) \subseteq Z_2^Q(Q) \oplus V \subseteq^e Y \subseteq Y \oplus U \subseteq^e Q.$$

Then define I to be the unique complement closure of $Z_2(R) \oplus (V \cap R) \leq_e (Y \cap R) \leq_e I$.

Then the following hold:

(i) $\eta_Q(Y) = d^Q(V_Q)$; $\eta_R(I) = d(V \cap R)$.
(ii) $Z_2R \oplus (V \cap R) \leq_e (Y \cap R) \leq (Y \cap R) \oplus (U \cap R) \leq_e R$.
(iii) $(Y \cap R) \oplus (U \cap R) \leq_e I \oplus (U \cap R)$; $I \in \mathcal{J}(R)$; $\Psi(I) = Y$.
(iv) $\phi^*[d^Q(V)] = d((V \cap R)_R) = d[(Y \cap R)/Z_2R]$; $\phi^*(\mathcal{N}_f(Q)) \subseteq \mathcal{N}(R)_f$.
(v) $\widetilde{\Psi}\phi^* = \mathbf{1}_{\mathcal{N}_f(Q)}$, $\phi^*\widetilde{\Psi} = \mathbf{1}_{\mathcal{N}_f(R)}$, $\widetilde{\Psi}^{-1} = \phi^*$.
(vi) *In particular, the restriction and co-restriction $\phi^* : \mathcal{N}_f(Q) \longrightarrow \mathcal{N}_f(R)$ is a lattice isomorphism.*

Proof. (i) Since $V_Q \hookrightarrow_e Y/Z_2^Q(Q)$, and $V \cap R \hookrightarrow_e I/Z_2(R)$ are essential submodules, (i) now follows by the definitions of $\eta_R(I) = d(I/Z_2(R))$ and $\eta_Q(Y) = d(Y/Z_2^Q(Q))$.

(ii) This follows by intersecting $Z_2^Q(Q) \subseteq Z_2^Q(Q) \oplus X \subseteq^e Y \subseteq Y \oplus U \subseteq^e Q$ with R.

(iii) Since $Y \perp^Q U$, by 3.11 $Y \cap R \perp U \cap R$. But then $(Y \cap R) \perp^Q (U \cap R)Q$. Since $(Y \cap R)Q \subseteq^e Y$, and similarly $(U \cap R)Q \subseteq^e U$, it follows that $IQ \perp^Q U$. By 2.1 (4), $IQ \subseteq^e Y$, and this means that $\Psi(I) = Y$.

(iv) By definition of ϕ^*, $\phi^*(d^Q(V)) = d(\{\ V_\phi \mid V \in d^Q(V)\ \}) = d(V_R) = d((V \cap R)_R)$. Since $Z_2(R) \leq R$ is a complement submodule, by (**) $V \cap R$ embeds as an essential submodule of $(Y \cap R)/Z_2R$. Since an arbitrary element of $\mathcal{N}_f(Q)$ is of the form $d^Q(V_Q)$, and since $Z(V_R) = 0$, $d(V \cap R) \in \mathcal{N}_f(R)$, (iv) follows.

(v) An arbitrary element of $\mathcal{N}(Q)_f$ is of the form $d^Q(V)$ as in (i). Then $\widetilde{\Psi}[\phi^* d^Q(Q)] = \eta_Q\Psi(\eta_R^{-1}[d(V \cap R)]) = \eta_Q\Psi(I)\eta_Q(Y) = d^Q(Y/Z_2^Q(Q)) = d^Q(V)$. Thus $\widetilde{\Psi}\phi^* = \mathbf{1}$.

An arbitrary element of $\mathcal{N}(R)_f$ is of the form $d(B)$, where $Z_2R \oplus B \leq_e J \in \mathcal{J}(R)$ is as in 4.2. Let $\widetilde{J} \in \mathcal{J}(Q)$ be as in 4.4. Then $\phi^*\widetilde{\Psi}d(B) = \phi^*\eta_Q\Psi(\eta_R^{-1}[d(B)]) = \phi^*\eta_Q\Psi(J) = \phi^* d^Q[\widetilde{J}/(Z_2^Q(Q)] = \phi^* d^Q(BQ)$ by Lemma 4.5 (iv). In view of (H1), and (H2), since $B \leq_e BQ$, $\phi^*\widetilde{\Psi}d(B) = \phi^* d^(BQ) = d(B)$. Hence $\phi * \widetilde{\Psi} = \mathbf{1}$. □

Recall that $\mathcal{N}(R)_f \cong \mathcal{N}(Q)_f$, so that the maps η_R, η_Q below are isomorphisms of partially ordered sets.

Finally, we summarize the main results of Section 5 in the next theorem.

Theorem 5.4. *Let $1 = 1_R = 1_Q \in R \leq_e Q_R = Q$ be rings, assume* (H1) *and* (H2), *and let Ψ be the map $\Psi : \mathcal{J}(R) \longrightarrow \mathcal{J}(Q)$, $J \in \mathcal{J}(R)$, $\Psi(J) = \widetilde{J}$, where $\widetilde{J}$ is as defined in Lemma 4.4. Let ϕ^*, η_R, η_Q be as in 1.4. Then there is a commutative* *diagram of isomorphisms of complete Boolean lattices. In particular,*

$$\begin{array}{ccc} \mathcal{J}(R) & \xrightarrow{\Psi} & \mathcal{J}(Q) \\ \eta_R \big\downarrow & & \eta_Q \big\downarrow \\ \mathcal{N}_f(R) & \xrightarrow{\phi^*} & \mathcal{N}_f(Q) \end{array}$$

(i) Ψ , ϕ^* *are isomorphisms, and* Ψ , ϕ^* *and their inverses preserve the order.*
(ii) $\widetilde{\Psi} = \eta_Q \phi^* \eta_R^{-1} : \mathcal{N}_f(R) \longrightarrow \mathcal{N}_f(Q)$ *is an isomorphism.*
(iii) $\widetilde{\Psi}^{-1} = \phi^*$.

6. Examples

The earlier proof that $\mathcal{J}(R) \cong \mathcal{J}(ER)$ when $ZR = 0$ was more of an existence type of proof, while here an explicit way of evaluating the map $\Psi : \mathcal{J}(R) \longrightarrow \mathcal{J}(Q)$ is given. Even in cases when $ZR = 0$ and $Q = ER$, examples are of interest. Examples are hard to construct, since if $R = Z_2(R)$, then $\mathcal{J}(R) = \{0\}$ is degenerate.

Example 6.1. For $p \in \mathbb{Z}$, set $Z_p = \mathbb{Z}/p\mathbb{Z}$, $Z_{(p)} = \{ a/b \mid a, b \in \mathbb{Z},\ \gcd(b,p) = 1\}$, and set e_{ij} to be the usual matrix units, $\overline{n} = n + (p) \in \mathbb{Z}_p$, and $\overline{e_{ij}} = \overline{1} e_{ij}$. Define

$$Z_2(R) = \begin{bmatrix} 0 & 0 & 0 \\ \mathbb{Z}_p & \mathbb{Z}_p & 0 \\ 0 & 0 & 0 \end{bmatrix} < Z_2(R) \oplus B \oplus C$$
$$= R = \begin{bmatrix} \mathbb{Z} & 0 & 0 \\ \mathbb{Z}_p & \mathbb{Z}_p & 0 \\ \mathbb{Z} & 0 & \mathbb{Z} \end{bmatrix} \subset Q = \begin{bmatrix} \mathbb{Z}_{(p)} & 0 & 0 \\ \mathbb{Z}_p & \mathbb{Z}_p & 0 \\ \mathbb{Z} & 0 & \mathbb{Z} \end{bmatrix}.$$

Recall that '$\lhd$' is used to denote two-sided ideals in both of the rings R as well as Q. Above $B = \mathbb{Z}e_{11} < R$, $C = \mathbb{Z}e_{31} + \mathbb{Z}\overline{e_{33}} \lhd R_R$, $BC = 0$, $CB \neq 0$, $Z(R) = \mathbb{Z}_p\overline{e_{21}} < Z_2(R) \lhd R$. Then $\mathcal{J}(R) = \{ 0,\ I,\ J,\ R \}$, where $I = Z_2(R) \oplus B \lhd R$, $J = Z_2(R) \oplus C \lhd R$, $\Psi(I) = \widetilde{I} = Z_2^Q(Q) \oplus \mathbb{Z}_{(p)}e_{11} \lhd Q \in \mathcal{J}(Q)$, where $Z_2^Q(Q) = Z_2(R)$; and similarly $\Psi(J) = \widetilde{J} = J \lhd Q$. Also $\Psi(R) = Q$ and $\Psi(0) = 0$. Note that $\widetilde{I} = Z_2(R) \oplus BQ$, $BQ \lhd Q$.

Here $\mathcal{N}_f(R) = \{ d(B) = d(I/Z_2(R)),\ d(C) = d(J/Z_2(R)),\ d(B \oplus C) = d(R/Z_2(R)), 0\}$, and similarly

$$\mathcal{J}(Q) = \{d^Q(BQ) = d^Q(\widetilde{I}/Z_2^Q(Q)),\ d^Q(CQ),\ d^Q(BQ \oplus CQ) = Q/Z_2^Q(Q), 0\}.$$

Recall that the isomorphism $\eta_R : \mathcal{J}(R) \longrightarrow \mathcal{N}_f(R)$ is defined by

$$\eta_R(I) = d(I/Z_2(R)) = d(B),$$

and induces an isomorphism $\widetilde{\Psi}^{-1} = \eta_R \Psi^{-1} \eta_Q^{-1} : \mathcal{N}_f(Q) \longrightarrow \mathcal{N}_f(R)$, which is essentially the inverse of Ψ. We now verify that $\widetilde{\Psi}^{-1} = \phi^*$, where the latter is the restriction and co-restriction to ϕ^*; $\mathcal{N}_f(Q) \longrightarrow \mathcal{N}_f(R)$, which is an order isomorphism.

First,

$$\widetilde{\Psi}^{-1}(d^Q(\widetilde{I}/Z_2^Q(Q))) = \eta_R\Psi^{-1}\eta_Q^{-1}(d^Q(\widetilde{I}/Z_2^Q(Q)))$$
$$= \eta_R\Psi^{-1}(\widetilde{I}) = \eta_R(I) = d(I/Z_2(R)) = d(B).$$

Next,

$$\phi^*(d^Q(\widetilde{I}/Z_2^Q(Q))) = \phi^*(d^Q(BQ)) = d((BQ)_R) = d(B).$$

Thus

$$\widetilde{\Psi}^{-1} = \phi^*, \quad \text{or} \quad \widetilde{\Psi} = (\phi^*|_{\mathcal{N}_f(Q)})^{-1} = (\phi^*)^{-1}.$$

In the previous example all the rings in each row of the matrix had to be the same (i.e., $\mathbb{Z}$, $\mathbb{Z}_p$, and $\mathbb{Z}$). Here in the next example the fact that the ring R has a well-defined multiplication hinges on the fact that $p\mathbb{Z}\cdot\mathbb{Z}_p = 0$. The notation of the previous example is continued.

Example 6.2. Let

$$Z_2(R) = \begin{bmatrix} 0 & 0 & 0 \\ \mathbb{Z}_p & \mathbb{Z} & 0 \\ 0 & 0 & 0 \end{bmatrix} \lhd R = Z_2(R) \oplus B \oplus C = \begin{bmatrix} \mathbb{Z} & p\cdot\mathbb{Z} & 0 \\ \mathbb{Z}_p & \mathbb{Z} & 0 \\ \mathbb{Z}_p & 0 & \mathbb{Z} \end{bmatrix}$$
$$<_e \mathbb{Q} = \begin{bmatrix} \mathbb{Z}_{(p)} & p\cdot\mathbb{Z}_{(p)} & 0 \\ \mathbb{Z}_p & \mathbb{Z}_{(p)} & 0 \\ \mathbb{Z}_p & 0 & \mathbb{Z}_{(p)} \end{bmatrix},$$

where $B = \mathbb{Z}e_{11} + p\cdot\mathbb{Z}e_{12} < R$, $C = \mathbb{Z}_p\overline{e_{31}} + \mathbb{Z}e_{33} \lhd R_R$. Then $BC = 0$, $CB \neq 0$, $Z(R) = \mathbb{Z}_p\overline{e_{21}} < Z_2(R) = Z(R) + \mathbb{Z}e_{22} \lhd R$. And $\mathbb{Z}_2^Q(Q) = \mathbb{Z}_p\overline{e_{21}} + \mathbb{Z}_{(p)}e_{21}$. Set

$$I = \mathbb{Z}_2R \oplus B \subset \widetilde{I} = \mathbb{Z}_2^Q(Q) \oplus BQ,$$

or

$$I = \mathbb{Z}_2(R) + \mathbb{Z}e_{11} + p\cdot\mathbb{Z}e_{12} \subset \widetilde{I} = \mathbb{Z}_2^Q(Q) + \mathbb{Z}_{(p)}e_{11} + p\cdot\mathbb{Z}_{(p)}e_{12}.$$

Similarly set

$$J = \mathbb{Z}_2R \oplus C \subset \widetilde{J} = \mathbb{Z}_2^Q(Q) \oplus CQ,$$

that is

$$J = \mathbb{Z}_2(R) + \mathbb{Z}_p\overline{e_{31}} + \mathbb{Z}e_{33} \subset \widetilde{J} = \mathbb{Z}_2^Q(Q) + \mathbb{Z}_p\overline{e_{31}} + \mathbb{Z}_{(p)}e_{33}.$$

Then

$$\mathcal{J}(R) = \{\ 0,\ I,\ J,\ R\ \},$$

while

$$\mathcal{J}(Q) = \{\ 0,\ \widetilde{I},\ \widetilde{J},\ R\ \}.$$

Again

$$\Psi : \mathcal{J}(R) \longrightarrow \mathcal{J}(Q)$$

by

$$I \longrightarrow \widetilde{I},\ J \longrightarrow \widetilde{J},\ R \longrightarrow Q,\ 0 \longrightarrow 0.$$

Here

$$\mathcal{N}_f(R) = \{\ d(B),\ d(C),\ d(B\oplus C) = d(R/Z_2(R)), 0\ \},$$

and similarly

$$\mathcal{J}(Q) = \{\ d^Q(BQ),\ d^Q(CQ),\ d^Q(BQ \oplus CQ) = Q/Z_2^Q(Q),\ 0\}.$$

Thus $\widetilde{\Psi} : \mathcal{N}_f(R) \longrightarrow \mathcal{N}_f(Q)$ by

$$d(I/Z_2(R)) \longrightarrow d^Q(\widetilde{I}/Z_2^Q(Q)), d(J/Z_2(R)) \longrightarrow d^Q(\widetilde{J}/Z_2^Q(Q)),$$
$$d(R/Z_2(R)) \longrightarrow d^Q(Q/Z_2(Q)),\ 0 \longrightarrow 0.$$

Here

$$\phi^*[\widetilde{I}/Z_2^Q(Q)] = [d^Q(BQ)] = d[(BQ)_R] = d[B] = (\widetilde{\Psi})^{-1}\{d^Q(BQ)\} = d(B),$$

and similarly for the others. Thus again $(\phi^*)^{-1} = \widetilde{\Psi}$ as previously.

It was only due to the uncomplicated nature of the last two examples that in both $R/Z_2(R) = (I/Z_2(R)) \oplus (J/Z_2(R))$ is a direct sum, usually it is only essential in $R/Z_2(R)$, as is illustrated in the next example from T.Y. Lam ([14; p. 372] and [14; p. 381, Ex. 14]).

Example 6.3. For any field or division ring F, let

$$B \oplus C <_e R = \begin{bmatrix} F & F & F \\ 0 & F & 0 \\ 0 & 0 & F \end{bmatrix},\ B = \begin{bmatrix} 0 & F & 0 \\ 0 & F & 0 \\ 0 & 0 & 0 \end{bmatrix},\ C = \begin{bmatrix} 0 & 0 & F \\ 0 & 0 & 0 \\ 0 & 0 & F \end{bmatrix},\ Z(R) = 0,\ B \perp C.$$

Then $\mathcal{J}(R) = \{\ 0,\ B,\ C,\ R\ \}$, and $\mathcal{N}_f(R) = \{\ d(B),\ d(C),\ d(B \oplus C) = d(R),\ 0\ \}$. Here the maximal right ring of quotients of R is the direct product of the full two by two matrix ring $Q = E(R_R) = M_2(F) \times M_2(F)$, where $\phi : R \longrightarrow Q$ is given by

$$r = \begin{bmatrix} a & b & c \\ 0 & d & 0 \\ 0 & 0 & e \end{bmatrix} \longrightarrow \phi(r) = \left(\begin{bmatrix} a & b \\ 0 & d \end{bmatrix}, \begin{bmatrix} a & c \\ 0 & e \end{bmatrix} \right).$$

Now $\mathcal{J}(Q) = \{\ 0, M_2(F) \times (0),\ (0) \times M_2(F),\ Q\}$; $\Psi : \mathcal{J}(R) \longrightarrow \mathcal{J}(Q)$ is $B \longrightarrow M_2(F) \times (0),\ C \longrightarrow (0) \times M_2(F),\ 0 \longrightarrow 0,\ R \longrightarrow Q$. Thus $\widetilde{\Psi}(d(B)) = d^Q(M_2(F)) \times (0)),\ \widetilde{\Psi}(d(C)) = d^Q((0) \times M_2(F)), 0 \longrightarrow 0,\ d^Q(Q) \longrightarrow d(R)$. Note that $d(R) =$Mod-$R = \mathbf{1} \in \mathcal{N}_f(R)$ and similarly for $d^Q(Q)$.

Recall that the map $\phi^* : \mathcal{N}_f(Q) \longrightarrow \mathcal{N}_f(R)$ is given by $\phi^*[d^Q(M_2(F) \times (0)] = d[(M_2(F) \times (0))_\phi] = d(B)$, where the last step holds because $B \leq_e (M_2(F) \times (0))_R$. The latter is of course an R-module via the homomorphism ϕ. Thus $\phi^*[\widetilde{\Psi}(d(B))] = \phi^*[d^Q(M_2(F) \times (0))] = d(B)$, and similarly for C. Thus $\phi^*\widetilde{\Psi} = \mathbf{1}$. In order to prove that $\widetilde{\Psi}\phi^* = \mathbf{1}$, we use the steps in the last two computations to get that $\widetilde{\Psi}\phi^*[d^Q(M_2(F) \times (0))] = \widetilde{\Psi}[d(B)] = d^Q(M_2(F)) \times (0))$ and similarly for C. Thus $\widetilde{\Psi}^{-1} = \phi^*$.

Acknowledgement

The author thanks the referee for a careful reading of the article and for correcting many mistakes.

References

[1] G.F. Birkenmeier, J.K. Park and S.T. Rizvi, An essential extension with non isomorphic ring structures, Algebra and its Applications, pp. 29–48, *Contemp. Math.*, **419**, Amer. Math. Society, Providence, RI, 2006.

[2] G.F. Birkenmeier, J.K. Park and S.T. Rizvi, An essential extension with non isomorphic ring structures, II. Comm. Algebra **35**, no. 12, (2007), 3986–4004.

[3] G.F. Birkenmeier, B.L. Osofsky, J.K. Park, and S.T. Rizvi, Injective hulls with distinct ring structures, Journal of Pure and Applied Algebra **213** (2009), 732–736.

[4] J. Dauns, Torsion free modules, *Ann. Mat. Pure Appl.* **154**(4)(1989), 49–81.

[5] J. Dauns, Torsion free types, *Fund. Math.* **139**(1991), 99–117.

[6] J. Dauns, Classes of modules, *Forum Math.* **3**(1991), 327–338.

[7] J. Dauns, Modules classifying functors, *Czechoslovak Math. J.* **42**(117)(1992), 741–756.

[8] J. Dauns, Functors and Σ-products, pp. 149–171, in: Ring Theory (S.K. Jain and S.T. Rizvi, eds.), World Sci. Pub., Singapore, 1993.

[9] J. Dauns, Module types, *Rocky Mountain J. Math.* **27**(2)(1997), 503–557.

[10] J. Dauns and Y. Zhou, Sublattices of the lattice of pre-natural classes, *J. Algebra* **231**(2000), 138–162.

[11] J. Dauns and Y. Zhou, Classes of Modules, Pure and Applied Math., a Series of Math. Monographs and Textbooks, No. 281 (previously known as Marcel Dekker), pp. 1–218. Chapman-Hall CRC Press (Taylor and Francis Group), New York, June 2006.

[12] K.R. Goodearl, Ring Theory, pp. 1–206, Marcel Dekker, NY, 1976.

[13] K.R. Goodearl and A.K. Boyle, Dimension theory for nonsingular injective modules, Amer. Math. Society Memoir Vol. 7, No. 177, pp. 1–112, Providence, RI, 1976.

[14] T.Y. Lam, Lectures on Rings and Modules, Graduate Texts in Math. 189, Springer, New York, 1999.

[15] J. Lambek, Lectures on Rings and Modules, Blaisdell Publishing Co., Waltham, Massachusetts, 1966.

[16] B.L. Osofsky, On ring properties of injective hulls, *Canad. Math. Bull.* **7**(1964), 405–413.

[17] B.L. Osofsky, A non-trivial ring with non-rational injective hull, *Canad. Math. Bull.* **10**(1967), 275–282.

[18] Y. Zhou, The lattice of natural classes of modules, *Comm. Algebra* **24**(5)(1996), 1637–1648.

[19] Y. Zhou, Direct sums of M-injective modules and module classes, *Comm. Algebra* **23**(1995), 927–940.

[20] Y. Zhou, The lattice of pre-natural classes of modules, *J. Pure Appl. Algebra* **140**(2)(1999), 191–207.

John Dauns
Department of Mathematics
Tulane University
New Orleans, LA 70118, USA

Advances in Ring Theory
Trends in Mathematics, 131–147

On Some Classes of Repeated-root Constacyclic Codes of Length a Power of 2 over Galois Rings

Hai Q. Dinh

Dedicated to Professor S.K. Jain on the occasion of his seventieth birthday.

Abstract. Negacyclic codes of length 2^s over the Galois ring $\mathrm{GR}(2^a, m)$ are ideals of the chain ring $\frac{\mathrm{GR}(2^a,m)[x]}{\langle x^{2^s}+1\rangle}$. This structure is used to provide the Hamming and homogeneous distances of all such negacyclic codes. The technique is then generalized to obtain the structure and Hamming and homogeneous distances of all γ-constacyclic codes of length 2^s over $\mathrm{GR}(2^a, m)$, where γ is any unit of the ring $\mathrm{GR}(2^a, m)$ that has the form $\gamma = (4k_0 - 1) + 4k_1\xi + \cdots + 4k_{m-1}\xi^{m-1}$, for integers $k_0, k_1, \ldots, k_{m-1}$. Among other results, duals of such γ-constacyclic codes are studied, and necessary and sufficient conditions for the existence of a self-dual γ-constacyclic code are established.

Mathematics Subject Classification (2000). Primary 94B15, 94B05; Secondary 11T71.

Keywords. Negacyclic codes, cyclic codes, constacyclic codes, repeated-root codes, dual codes, codes over rings, Hamming distance, homogeneous distance, Galois extension, chain rings, Galois rings.

1. Introduction

Constacyclic codes over finite fields play a very significant role in algebraic coding theory. The most important class of these codes is the class of cyclic codes, which has been well studied since the late 1950's [47, 48, 49, 50]. However, most of the research is concentrated on the situation when the code length n is relatively prime to the characteristic of the field F. In this case, cyclic codes of length n are classified as ideals $\langle f(x)\rangle$ of $\frac{F[x]}{\langle x^n-1\rangle}$, where $f(x)$ is a divisor of $x^n - 1$. The case when the code length n is divisible by the characteristic p of the field yields the so-called repeated-root codes, which were first studied since 1967 by Berman [6], and then in the 1970's and 1980's by several authors such as Massey *et al.* [41],

Falkner *et al.* [26], Roth and Seroussi [52]. However, repeated-root codes over finite fields were investigated in the most generality in the 1990's by Castagnoli *et al.* [15], and van Lint [55], where they showed that repeated-root cyclic codes have a concatenated construction, and are asymptotically bad. Nevertheless, such codes are optimal in a few cases, that motivates researchers to further study this class of codes (see, for example, [54, 44]).

In the early 1990's, Nechaev [43], and Hammons *et al.* [13, 29] established the celebrated result that many well-known seemingly nonlinear codes over finite fields such as Kerdock and Preparata codes are actually closely related to linear codes over the ring $\mathbb{Z}_4$. Since then, codes over $\mathbb{Z}_4$ in particular, and codes over finite rings in general, have proved their importance, and they have received a great deal of attention.

The Galois ring of characteristic p^a and dimension m, denoted by $\mathrm{GR}(p^a, m)$, is the Galois extension of degree m of the ring $\mathbb{Z}_{p^a}$, for some prime number p. In particular, rings of the form $\mathbb{Z}_{p^a}$ such as $\mathbb{Z}_4$ are Galois rings. The class of Galois rings has been used widely as an alphabet for cyclic and negacyclic codes, for instance [14, 56, 24, 3, 7, 8, 34, 10]. Various decoding schemes for codes over Galois rings have also been addressed [9, 10, 11, 12].

In recent years, we have been working on the description of several classes of constacyclic codes, such as cyclic and negacyclic codes, over various types of Galois rings. In 2004, the structure of negacyclic codes of length 2^s over $\mathbb{Z}_{2^a}$ was obtained [24]. In 2005 [19], we investigated negacyclic codes of length 2^s over the Galois ring $\mathrm{GR}(2^a, m)$. We showed that the ring $\frac{\mathrm{GR}(2^a,m)[x]}{\langle x^{2^s}+1\rangle}$ is indeed a chain ring, and the negacyclic codes of length 2^s over $\mathrm{GR}(2^a, m)$ are precisely the ideals generated by $(x+1)^i$ of this chain ring, for $i = 0, 1, \ldots, 2^s a$. Using this structure, Hamming distances of such negacyclic codes were addressed. We were able to provide Hamming distances of such negacyclic codes $\langle (x+1)^i \rangle$ for $0 \le i \le 2^s(a-1)+2^{s-1}$. The computation technique in [19] was used by other authors [59] to give Hamming distances of such negacyclic codes for all i. In 2007, we computed the Hamming distances of all those negacyclic codes for the case when the alphabet is $\mathbb{Z}_{2^a}$ [22], i.e., the Galois ring $\mathrm{GR}(2^a, m)$ with dimension $m = 1$. We also provided the Lee, homogeneous, and Euclidean distances of all such codes. Since 2003, special classes of repeated-root constacyclic codes over certain classes of finite chain rings and Galois rings have been studied by numerous other authors (see, for example, [1, 7, 8, 37, 45, 53]).

The purpose of this paper is to complete the computation of Hamming distances and furthermore provide the homogeneous distances of negacyclic codes of length 2^s over $\mathrm{GR}(2^a, m)$ that started in [19], and then extend such structure and distances to other more general classes of constacyclic codes. Although our technique in [19] was used recently in [59] to give Hamming distances of such negacyclic codes, our computation here is simpler with the use of the newly obtained Hamming distances of 2^m-ary cyclic codes [23], and more importantly, it is applicable to the more general classes of γ-constacyclic codes considered in Section 4. In Section 3, we first use the structure of negacyclic codes of length 2^s over $\mathrm{GR}(2^a, m)$, and the Hamming distance of 2^m-ary cyclic codes of length 2^s, that we obtained in [19], and [23], to provide the Hamming distances of the remaining negacyclic codes from [19] (i.e., the codes $\langle (x+1)^i \rangle$ for $2^s(a-1)+2^{s-1}+1 \le i \le 2^s a-1$.) Section 3 includes a different proof of the structure of the ring $\frac{\mathrm{GR}(2^a,m)[x]}{\langle x^{2^s}+1\rangle}$, which makes it easier to carry our results over to the setting of γ-constacyclic codes in Section 4. We also establish the homogeneous distances of all such negacyclic codes. In Section 4, this structure and computation technique are then extended to give the structure and Hamming and homogeneous distances of all codes of much larger classes of constacyclic codes over $\mathrm{GR}(2^a, m)$, namely, the classes of γ-constacyclic codes of length 2^s over $\mathrm{GR}(2^a, m)$, where γ is a unit of the Galois ring $\mathrm{GR}(2^a, m)$ with the form $\gamma = (4k_0 - 1) + 4k_1\xi + \cdots + 4k_{m-1}\xi^{m-1}$, for integers $k_0, k_1, \ldots, k_{m-1}$. These γ-constacyclic codes include as particular cases many classes of constacyclic codes that were investigated, such as negacyclic codes [24, 19, 22], $(2^\theta - 1)$-constacyclic codes [20], $(4k-1)$-constacyclic codes [21, 22]. Among other results, we give the duals of all such γ-constacyclic code, and provide necessary and sufficient conditions for the existance of a self-dual γ-constacyclic code.

2. Chain rings, Galois rings, and constacyclic codes

In this paper, all rings under consideration are associative rings with identity. An ideal I of a ring R is called *principal* if it is generated by one element. A ring R is a *principal ideal ring* if its ideals are principal. R is said to be *local* if it has a unique maximal right (left) ideal. Furthermore, a ring R is called a *chain ring* if the set of all right (left) ideals of R is linearly ordered under set-theoretic inclusion.

While we only consider finite commutative chain rings in this paper, it is worth mentioning that a finite chain ring need not be commutative. The smallest noncommutative chain ring has order 16 [35], that can be represented as $R = \mathrm{GF}(4) \oplus \mathrm{GF}(4)$, where the operations $+, \cdot$ are defined as

$$(a_1, b_1) + (a_2, b_2) = (a_1 + a_2, b_1 + b_2),$$
$$(a_1, b_1) \cdot (a_2, b_2) = (a_1 a_2, a_1 b_2 + b_1 a_2^2).$$

The following equivalent conditions are known for the class of finite commutative chain rings (see for example [24, Prop. 2.1]).

Proposition 2.1. *Let R be a finite commutative ring, then the following conditions are equivalent:*

(i) *R is a local ring and the maximal ideal M of R is principal,*
(ii) *R is a local principal ideal ring,*
(iii) *R is a chain ring.*

The following properties of chain rings are well known.

Proposition 2.2. *Let R be a finite commutative chain ring, with maximal ideal $M = \langle r \rangle$. Denote $\bar{R} = \frac{R}{M}$, and let β be the nilpotency of r. Then*

(a) *There is some prime p and positive integers k, l with $k \geq l$ such that $|R| = p^k, |\bar{R}| = p^l$, and the characteristic of R and $\bar{R}$ are powers of p,*

(b) *The ideals of R are $\langle r^i \rangle$, where $i = 0, 1, \ldots, \beta$, and they are strictly inclusive:*

$$R = \langle r^0 \rangle \supsetneq \langle r^1 \rangle \supsetneq \cdots \supsetneq \langle r^{\beta-1} \rangle \supsetneq \langle r^\beta \rangle = \langle 0 \rangle.$$

(c) *For $i = 0, \ldots, \beta$, $|\langle r^i \rangle| = |\bar{R}|^{\beta-i}$. In particular, $|R| = |\bar{R}|^\beta$, i.e., $k = l\beta$.*

A polynomial in $\mathbb{Z}_{p^a}[x]$ is called a *basic irreducible polynomial* if its reduction modulo p is irreducible in $\mathbb{Z}_p[x]$. The *Galois ring of characteristic p^a and dimension m*, denoted by $\mathrm{GR}(p^a, m)$, is the Galois extension of degree m of the ring $\mathbb{Z}_{p^a}$. Equivalently,

$$\mathrm{GR}(p^a, m) = \frac{\mathbb{Z}_{p^a}[u]}{\langle h(u) \rangle},$$

where $h(u)$ is a monic basic irreducible polynomial of degree m in $\mathbb{Z}_{p^a}[u]$. Krull [36] initiated the study of Galois rings in 1924, and later these rings were rediscovered independently by Janusz [33] in 1966, and Raghavendran [51] in 1969. Since then, Galois rings have been proven to be very applicable in many branches of mathematics such as combinatorics and coding theory. Note that if $a = 1$, then $\mathrm{GR}(p, m) = \mathrm{GF}(p^m)$, and if $m = 1$, then $\mathrm{GR}(p^a, 1) = \mathbb{Z}_{p^a}$. We list some well-known facts about Galois rings (cf. [42, 29, 32, 46]), which will be used freely throughout this paper.

Proposition 2.3. *Let $\mathrm{GR}(p^a, m) = \frac{\mathbb{Z}_{p^a}[u]}{\langle h(u) \rangle}$ be a Galois ring, then the following conditions hold:*

(i) *Each ideal of $\mathrm{GR}(p^a, m)$ is of the form $\langle p^k \rangle = p^k\,\mathrm{GR}(p^a, m)$, for $0 \leqslant k \leqslant a$. In particular, $\mathrm{GR}(p^a, m)$ is a chain ring with maximal ideal $\langle p \rangle = p\,\mathrm{GR}(p^a, m)$, and residue field $\mathrm{GF}(p^m)$.*

(ii) *For $0 \leqslant i \leqslant a$, $|p^i\,\mathrm{GR}(p^a, m)| = p^{m(a-i)}$.*

(iii) *Each element of $\mathrm{GR}(p^a, m)$ can be represented as vp^k, where v is a unit and $0 \leqslant k \leqslant a$, in this representation k is unique and v is unique modulo $\langle p^{n-k} \rangle$*

(iv) *$h(u)$ has a root ζ in $\mathrm{GR}(p^a, m)$, which is also a primitive $(p^m - 1)$th root of unity. The set*

$$\mathcal{T}(p, m) = \{0, 1, \zeta, \zeta^2, \ldots, \zeta^{p^m-2}\}$$

is a complete set of representatives of the cosets $\frac{\mathrm{GR}(p^a, m)}{p\,\mathrm{GR}(p^a, m)} = \mathrm{GF}(p^m)$ in $\mathrm{GR}(p^a, m)$. Each element $r \in \mathrm{GR}(p^a, m)$ can be written uniquely as

$$r = \zeta_0 + \zeta_1 p + \cdots + \zeta_{a-1} p^{a-1},$$

with $\zeta_i \in \mathcal{T}(p, m)$, $0 \leqslant i \leqslant a - 1$.

(v) *For each positive integer d, there is a natural injective ring homomorphism $\mathrm{GR}(p^a, m) \to \mathrm{GR}(p^a, md)$.*

(vi) *In $\mathrm{GR}(p^a, m) = \frac{\mathbb{Z}_{p^a}[u]}{\langle h(u) \rangle}$, let $\xi = u + \langle h(u) \rangle$, then $h(\xi) = 0$, and ξ is in fact a primitive element of $\mathrm{GR}(p^a, m)$. The Galois ring $\mathrm{GR}(p^a, m)$ can be viewed* *as* $\mathrm{GR}(p^a, m) = \mathbb{Z}_{p^a}[\xi]$. *Every element* $r \in \mathrm{GR}(p^a, m)$ *can be expressed as*

$$r = r_0 + r_1\xi + \cdots + r_{m-1}\xi^{m-1},$$

where $r_0, r_1, \ldots, r_{m-1} \in \mathbb{Z}_{p^a}$.

(vii) *There is a natural surjective ring homomorphism* $\mathrm{GR}(p^a, m) \to \mathrm{GR}(p^{a-1}, m)$ *with kernel* $\langle p^{a-1} \rangle$.

(viii) *Each subring of* $\mathrm{GR}(p^a, m)$ *is a Galois ring of the form* $\mathrm{GR}(p^a, l)$, *where* l *divides* m. *Conversely, if* l *divides* m *then* $\mathrm{GR}(p^a, m)$ *contains a unique copy of* $\mathrm{GR}(p^a, l)$. *That means, the number of subrings of* $\mathrm{GR}(p^a, m)$ *is the number of positive divisors of* m.

For a finite ring R, consider the set R^n of n-tuples of elements from R as a module over R. Any subset $C \subseteq R^n$ is called a *code of length* n over R, the code C is *linear* if in addition, C is an R-submodule of R^n. Given an n-tuple $(x_0, x_1, \ldots, x_{n-1}) \in R^n$, the *cyclic shift* τ and *negashift* ν on R^n are defined as usual, i.e.,

$$\tau(x_0, x_1, \ldots, x_{n-1}) = (x_{n-1}, x_0, x_1, \cdots, x_{n-2}),$$

and

$$\nu(x_0, x_1, \ldots, x_{n-1}) = (-x_{n-1}, x_0, x_1, \cdots, x_{n-2}).$$

A code C is called *cyclic* if $\tau(C) = C$, and C is called *negacyclic* if $\nu(C) = C$. Cyclic codes over finite fields, and more generally, over finite rings, are well studied, while negacyclic codes over finite fields were initiated by Berlekamp in the late 1960's (cf. [4, 5]), and since 1999, repeated-root negacyclic codes over finite rings have been brought to attention by Wolfmann [57], Blackford [7], and Dinh and López-Permouth [24], among others.

More generally, if λ is a unit of the ring R, then the λ-*constacyclic* (λ-twisted) *shift* τ_λ on R^n is the shift

$$\tau_\lambda(x_0, x_1, \ldots, x_{n-1}) = (\lambda x_{n-1}, x_0, x_1, \cdots, x_{n-2}),$$

and a code C is said to be λ-*constacyclic* if $\tau_\lambda(C) = C$, i.e., if C is closed under the λ-constacyclic shift τ_λ. In light of this definition, when $\lambda = 1$, λ-constacyclic codes are cyclic codes, and when $\lambda = -1$, λ-constacyclic codes are just negacyclic codes.

Each codeword $c = (c_0, c_1, \ldots, c_{n-1})$ is customarily identified with its polynomial representation $c(x) = c_0 + c_1x + \cdots + c_{n-1}x^{n-1}$, and the code C is in turn identified with the set of all polynomial representations of its codewords. Then in the ring $\frac{R[x]}{\langle x^n - \lambda \rangle}$, $xc(x)$ corresponds to a λ-constacyclic shift of $c(x)$. From that, the following fact is well known and straightforward:

Proposition 2.4. *A linear code* C *of length* n *is* λ-*constacyclic over* R *if and only if* C *is an ideal of* $\frac{R[x]}{\langle x^n - \lambda \rangle}$.

Given n-tuples $x = (x_0, x_1, \ldots, x_{n-1}), y = (y_0, y_1, \ldots, y_{n-1}) \in R^n$, their *inner product* or *dot product* is defined as usual

$$x \cdot y = x_0y_0 + x_1y_1 + \cdots + x_{n-1}y_{n-1}$$

(evaluated in R). Two n-tuples x, y are called *orthogonal* if $x \cdot y = 0$. For a linear code C over R, its *dual code* $C^\perp$ is the set of n-tuples over R that are orthogonal to all codewords of C, i.e.,

$$C^\perp = \{x \mid x \cdot y = 0, \forall y \in C\}.$$

A code C is called *self-orthogonal* if $C \subseteq C^\perp$, and it is called *self-dual* if $C = C^\perp$. The following result is well known (cf. [14, 24]).

Proposition 2.5. *Let R be a finite chain ring of size p^α. The number of codewords in any linear code C of length n over R is p^k, for some integer k, $0 \le k \le \alpha n$. Moreover, the dual code $C^\perp$ has $p^{\alpha n - k}$ codewords, so that $|C| \cdot |C^\perp| = |R|^n$.*

The dual of a cyclic code is a cyclic code, and the dual of a negacyclic code is a negacyclic code. In general, we have the following implication of the dual of a λ-constacyclic code.

Proposition 2.6. *The dual of a λ-constacyclic code is a λ^{-1}-constacyclic code.*

Proof. Let C be a λ-constacyclic code length n over R. Consider arbitrary elements $x \in C^\perp$, and $y \in C$. Since C is λ-constacyclic, $\tau_\lambda^{n-1}(y) \in C$. Thus,

$$0 = x \cdot \tau_\lambda^{n-1}(y) = \lambda \tau_{\lambda^{-1}}(x) \cdot y = \tau_{\lambda^{-1}}(x) \cdot y.$$

That means $\tau_{\lambda^{-1}}(x) \in C^\perp$. Therefore, $C^\perp$ is closed under the $\tau_{\lambda^{-1}}$-shift, i.e., $C^\perp$ is a λ^{-1}-constacyclic code. □

3. Negacyclic codes of length 2^s over $\mathrm{GR}(2^a, m)$

As mentioned in Proposition 2.4, negacyclic codes of length 2^s over a Galois ring $\mathrm{GR}(2^a, m)$ are precisely the ideals of ring

$$\mathcal{R}(a, m) = \frac{\mathrm{GR}(2^a, m)[x]}{\langle x^{2^s} + 1 \rangle}.$$

The following fact is useful in expressing $(x+1)^{2^n}$ in $\mathcal{R}(a, m)$, as well as relating the roles of the elements $x + 1$ and 2 in $\mathcal{R}(a, m)$:

Proposition 3.1. (cf. [19, Lemma 3.1, Proposition 3.2]) *For any positive integer n, there exists a polynomial $\alpha_n(x) \in \mathbb{Z}[x]$ such that $(x+1)^{2^n} = x^{2^n} + 1 + 2\alpha_n(x)$, and $\alpha_n(x)$ is a unit in $\mathcal{R}(a, m)$. In particular, $\langle (x+1)^{2^s} \rangle = \langle 2 \rangle$ in $\mathcal{R}(a, m)$, and the element $x + 1$ is nilpotent in $\mathcal{R}(a, m)$ with nilpotency $2^s a$.*

We proved in [19, Proposition 3.3, Theorem 3.6] that $\mathcal{R}(a, m)$ is a chain ring, and from that derived the structure of its ideals as follows:

Theorem 3.2. *The ring $\mathcal{R}(a, m)$ is a chain ring with maximal ideal $\langle x + 1 \rangle$ and residue field $\mathrm{GF}(2^m)$. Negacyclic codes of length 2^s over the Galois ring $\mathrm{GR}(2^a, m)$ are precisely the ideals $\langle (x+1)^i \rangle$, $0 \le i \le 2^s a$, of $\mathcal{R}(a, m)$. Each negacyclic code $\langle (x+1)^i \rangle$ has $2^{m(2^s a - i)}$ codewords, its dual is the negacyclic code $\langle (x+1)^{2^s a - i} \rangle$, which contains 2^{mi} codewords.*

Here, we provide a different proof, using polynomial representation, which makes it easier to extend to a more general setting in Section 4.

Proof. Let $f(x) \in \mathcal{R}(a,m)$, then $f(x)$ can be expressed as

$$f(x) = b_0 + b_1(x+1) + \cdots + b_{2^s-1}(x+1)^{2^s-1},$$

where $b_i \in \mathrm{GR}(2^a,m)$. If b_0 is in the maximal ideal $2\,\mathrm{GR}(2^a,m)$, then $f(x)$ is nilpotent, and so it is not invertible. On the other hand, assume that $f(x)$ is not invertible. If $b_0 \notin 2\,\mathrm{GR}(2^a,m)$, then b_0 is a unit in $\mathrm{GR}(2^a,m)$. In $\mathcal{R}(a,m)$, by Proposition 3.1, $x+1$ is nilpotent. So there is an odd integer k such that $(x+1)^k = 0$, and thus, $g(x)^k = 0$, where $g(x) = b_1(x+1) + \cdots + b_{2^s-1}(x+1)^{2^s-1}$. Now, $f(x) = b_0 + g(x)$, and thus,

$$1 = 1 + [g(x)b_0^{-1}]^k = [1 + g(x)b_0^{-1}]h(x) = f(x)b_0^{-1}h(x),$$

which contradicts the assumption that $f(x)$ is not invertible. Therefore, b_0 must be in $2\,\mathrm{GR}(2^a,m)$. That means that $f(x)$ is not invertible if and only if $b_0 \in 2\,\mathrm{GR}(2^a,m)$. Moreover, in light of Proposition 3.1, $2 \in \langle (x+1)^{2^s} \rangle \subseteq \langle x+1 \rangle$. Thus, $\langle x+1 \rangle$ is the set of noninvertible elements of $\mathcal{R}(a,m)$, proving that $\mathcal{R}(a,m)$ is a chain ring whose maximal ideal is $\langle x+1 \rangle$ (cf. Proposition 2.1). By Proposition 3.1, the nilpotency of $x+1$ is $2^s a$, so the ideals of $\mathcal{R}(a,m)$ are $\langle (x+1)^i \rangle$, $0 \le i \le 2^s a$ (cf. Proposition 2.2). The rest of the theorem follows readily from the fact that negacyclic codes of length 2^s over $\mathrm{GR}(2^a,m)$ are ideals of this chain ring $\mathcal{R}(a,m)$. □

Using this structure, we obtained in [19] the Hamming distances of a large part of the class of such negacyclic codes. Indeed, the Hamming distances of the codes $\langle (x+1)^i \rangle$, where $0 \le i \le 2^s(a-1) + 2^{s-1}$, were established:

Proposition 3.3. (cf. [19, Theorem 4.1]) *Let C be a negacyclic code of length 2^s over $\mathrm{GR}(2^a,m)$, i.e., $C = \langle (x+1)^i \rangle \subseteq \mathcal{R}(a,m)$, for some integer $i \in \{0,1,\dots,2^s a\}$, and let $d(C)$ denote the Hamming distance of C. Then*

$$d(C) \begin{cases} = 0 & \text{if } \; i = 2^s a, \\ = 1 & \text{if } \; 0 \le i \le 2^s(a-1), \\ = 2 & \text{if } \; 2^s(a-1)+1 \le i \le 2^s(a-1)+2^{s-1}, \\ \ge 2 & \text{if } \; 2^s(a-1)+2^{s-1}+1 \le i \le 2^s a - 1. \end{cases}$$

We now obtain the Hamming distances of the remaining negacyclic codes, i.e., for $2^s(a-1) + 2^{s-1} + 1 \le i \le 2^s a - 1$. Although our technique in [19] was used recently in [59] to compute Hamming distances of such negacyclic codes, our computation here is simpler, and more importantly, it is applicable also for the more general classes of γ-constacyclic codes in the next section. The main tool is the Hamming distances of all 2^m-ary cyclic codes of length 2^s, that we gave in [23]. Note that, over $\mathbb{F}_{2^m}$, cyclic codes and negacyclic codes coincide. Their Hamming distances are as follows:

Theorem 3.4. (cf. [23, Corollary 4.12]) *Let C be a (nega)cyclic code of length 2^s over $\mathbb{F}_{2^m}$, then $C = \langle (x+1)^i \rangle \subseteq \frac{\mathbb{F}_{2^m}[x]}{\langle x^{2^s}+1 \rangle}$, for $i \in \{0, 1, \ldots, 2^s\}$. The Hamming distance $d(C)$ of C is determined by:*

$$d(C) = \begin{cases} 1, & \text{if } i = 0, \\ 2, & \text{if } 1 \le i \le 2^{s-1}, \\ 2^{k+1}, & \text{if } 2^s - 2^{s-k} + 1 \le i \le 2^s - 2^{s-k} + 2^{s-k-1}, \\ & \quad \text{i.e., } 1 + \sum_{l=1}^{k} 2^{s-l} \le i \le \sum_{l=1}^{k+1} 2^{s-l}, \\ & \quad \text{where } 1 \le k \le s-1, \\ 0, & \text{if } i = 2^s. \end{cases}$$

Proposition 3.5. *Let j be an integer with $1 \le j \le s-1$. Then in $\mathcal{R}(a,m)$,*

(i) *the Hamming distance of $\left\langle (x+1)^{2^s(a-1)+\sum_{l=1}^{j} 2^{s-l}} \right\rangle$ is 2^j,*

(ii) *the Hamming distance of $\left\langle (x+1)^{2^s(a-1)+1+\sum_{l=1}^{j} 2^{s-l}} \right\rangle$ is 2^{j+1},*

(iii) *the Hamming distance of $\langle (x+1)^i \rangle$ is 2^{k+1}, for any integer i such that*

$$2^s(a-1) + 2^s - 2^{s-k} + 1 \le i \le 2^s(a-1) + 2^s - 2^{s-k} + 2^{s-k-1},$$

i.e.,

$$2^s(a-1) + 1 + \sum_{l=1}^{k} 2^{s-l} \le i \le 2^s(a-1) + \sum_{l=1}^{k+1} 2^{s-l},$$

where $1 \le k \le s-1$.

Proof. By Proposition 3.1, $\langle (x+1)^{2^s} \rangle = \langle 2 \rangle$ in $\mathcal{R}(a,m)$. Therefore,

$$\left\langle (x+1)^{2^s(a-1)+\sum_{l=1}^{j} 2^{s-l}} \right\rangle = \left\langle 2^{a-1} (x+1)^{\sum_{l=1}^{j} 2^{s-l}} \right\rangle,$$

and

$$\left\langle (x+1)^{2^s(a-1)+1+\sum_{l=1}^{j} 2^{s-l}} \right\rangle = \left\langle 2^{a-1} (x+1)^{1+\sum_{l=1}^{j} 2^{s-l}} \right\rangle.$$

Now, the ideals

$$\left\langle 2^{a-1} (x+1)^{\sum_{l=1}^{j} 2^{s-l}} \right\rangle \quad \text{and} \quad \left\langle 2^{a-1} (x+1)^{1+\sum_{l=1}^{j} 2^{s-l}} \right\rangle$$

of $\mathcal{R}(a,m)$ are indeed the sets of elements from the ideals $\left\langle (x+1)^{\sum_{l=1}^{j} 2^{s-l}} \right\rangle$ and $\left\langle (x+1)^{1+\sum_{l=1}^{j} 2^{s-l}} \right\rangle$ of $\frac{\mathbb{F}_{2^m}[x]}{\langle x^{2^s}+1 \rangle}$ multiplied with 2^{a-1}. Hence, (i) and (ii) follow from Theorem 3.4.

Now, using (i) and (ii), we get that the Hamming distance of both negacyclic codes $\left\langle (x+1)^{2^s(a-1)+1+\sum_{l=1}^{k} 2^{s-l}} \right\rangle$, and $\left\langle (x+1)^{2^s(a-1)+\sum_{l=1}^{k+1} 2^{s-l}} \right\rangle$ are 2^{k+1}. Since the ideals of $\mathcal{R}(a,m)$ are strictly inclusive, we get (iii). □

From Propositions 3.3 and 3.5, we now have the Hamming distances of all negacyclic codes of length 2^s over $\mathrm{GR}(2^a, m)$:

Theorem 3.6. *Let C be a negacyclic code of length 2^s over $\mathrm{GR}(2^a, m)$, i.e., $C = \langle (x+1)^i \rangle \subseteq \mathcal{R}(a,m)$, for some integer $i \in \{0, 1, \dots, 2^s a\}$. Then the Hamming distance of C can be completely determined as follows:*

$$d(C) = \begin{cases} 0 & \text{if} \quad i = 2^s a, \\ 1 & \text{if} \quad 0 \le i \le 2^s(a-1), \\ 2 & \text{if} \quad 2^s(a-1) + 1 \le i \le 2^s(a-1) + 2^{s-1}, \\ 2^{k+1} & \text{if} \quad 2^s(a-1) + 2^s - 2^{s-k} + 1 \le i \\ & \qquad \le 2^s(a-1) + 2^s - 2^{s-k} + 2^{s-k-1}, \\ & \quad \text{i.e., } 2^s(a-1) + 1 + \sum_{l=1}^{k} 2^{s-l} \le i \\ & \qquad \le 2^s(a-1) + \sum_{l=1}^{k+1} 2^{s-l}, \\ & \quad \text{where } 1 \le k \le s-1. \end{cases}$$

In [22, Theorem 4.4], we derived the Hamming distances of all negacyclic codes of length 2^s over the Galois ring $\mathbb{Z}_{2^a}$, i.e., $\mathrm{GR}(2^a, 1)$. That gave an affirmative answer to our conjecture in [19, Conjecture 4.16]. Theorem 3.6 shows that the same formula can be used to determine the Hamming distances of all negacyclic codes of length 2^s over $\mathrm{GR}(2^a, m)$, for any m.

We now establish the other kind of distance of all negacyclic codes, namely, the homogeneous distance. The homogeneous weight was first introduced in [16] (see also [17, 18]) over integer residue rings, and later over finite Frobenius rings. This weight has numerous applications for codes over finite rings, such as constructing extensions of the Gray isometry to finite chain rings [31, 30, 27], or providing a combinatorial approach to MacWilliams equivalence theorems (cf. [38, 39, 58]) for codes over finite Frobenius rings [28].

Let $a \ge 2$, the *homogeneous weight* on $\mathrm{GR}(2^a, m)$ is a weight function on $\mathrm{GR}(2^a, m)$ given as

$$\mathrm{wt_h} : \mathrm{GR}(2^a, m) \longrightarrow \mathbb{N},$$
$$r \mapsto \begin{cases} 0, & \text{if } r = 0 \\ (2^m - 1)\, 2^{m(a-2)}, & \text{if } r \in \mathrm{GR}(2^a, m) \setminus 2^{a-1}\,\mathrm{GR}(2^a, m) \\ 2^{m(a-1)}, & \text{if } r \in 2^{a-1}\,\mathrm{GR}(2^a, m) \setminus \{0\}. \end{cases}$$

The homogeneous weight of a codeword $(c_0, c_1, \dots, c_{n-1})$ of length n over $\mathrm{GR}(2^a, m)$ is the rational sum of the homogeneous weights of its components, i.e.,

$$\mathrm{wt_h}(c_0, c_1, \dots, c_{n-1}) = \mathrm{wt_h}(c_0) + \mathrm{wt_h}(c_1) + \cdots + \mathrm{wt_h}(c_{n-1}).$$

The *homogeneous distance* (or minimum homogeneous weight) d_{h} of a linear code C is the minimum homogeneous weight of nonzero codewords of C:

$$d_{\mathrm{h}}(C) = \min\{\mathrm{wt_h}(x - y) : x, y \in C,\ x \ne y\} = \min\{\mathrm{wt_h}(c) : c \in C,\ c \ne 0\}.$$

Theorem 3.7. *Let C be a negacyclic code of length 2^s over $\mathrm{GR}(2^a, m)$, i.e., $C = \langle (x+1)^i \rangle \subseteq \mathcal{R}(a,m)$, for some integer $i \in \{0, 1, \dots, 2^s a\}$. Then the homogeneous* *distance* $d_{\mathrm{h}}(C)$ *of* C *can be completely determined as follows:*

$$d_{\mathrm{h}}(C)=\begin{cases} 0 & \textit{if} \quad i=2^s a,\\ (2^m-1)\,2^{m(a-2)} & \textit{if} \quad 0\le i\le 2^s(a-2),\\ 2^{m(a-1)} & \textit{if} \quad 2^s(a-2)+1\le i\le 2^s(a-1),\\ 2^{m(a-1)+1} & \textit{if} \quad 2^s(a-1)+1\le i\le 2^s(a-1)+2^{s-1},\\ 2^{m(a-1)+k+1} & \textit{if} \quad 2^s(a-1)+2^s-2^{s-k}+1\le i\\ & \qquad \le 2^s(a-1)+2^s-2^{s-k}+2^{s-k-1},\\ & \quad \textit{i.e.,}\ 2^s(a-1)+1+\sum_{l=1}^{k}2^{s-l}\le i\\ & \qquad \le 2^s(a-1)+\sum_{l=1}^{k+1}2^{s-l},\\ & \quad \textit{where}\ 1\le k\le s-1. \end{cases}$$

Proof. In $\mathcal{R}(a,m)$, note that, by Proposition 3.1, $\langle (x+1)^{2^s}\rangle = \langle 2\rangle$, therefore $\langle (x+1)^{2^s j+t}\rangle = \langle 2^j(x+1)^t\rangle$.

If $0\le i\le 2^s(a-2)$, we get $\langle 1\rangle \supseteq C\supseteq \langle 2^{a-2}\rangle$. Since $d_{\mathrm{h}}(\langle 1\rangle)=d_{\mathrm{h}}(\langle 2^{a-2}\rangle)=(2^m-1)\,2^{m(a-2)}$, $d_{\mathrm{h}}(C)=(2^m-1)\,2^{m(a-2)}$.

If $2^s(a-2)+1\le i\le 2^s(a-1)$, then $\langle 2^{a-2}(x+1)\rangle\supseteq C\supseteq\langle 2^{a-1}\rangle$. Clearly, $d_{\mathrm{h}}(\langle 2^{a-1}\rangle)=2^{m(a-1)}$, and $d_{\mathrm{h}}(\langle 2^{a-2}(x+1)\rangle)\ge 2(2^m-1)2^{m(a-2)}\ge 2^{m(a-1)}$. Thus,

$$2^{m(a-1)}\le d_{\mathrm{h}}(\langle 2^{m(a-1)}\rangle)\le d_{\mathrm{h}}(C)\le d_{\mathrm{h}}(\langle 2^{a-1}\rangle)=2^{m(a-1)},$$

implying $d_{\mathrm{h}}(C)=2^{m(a-1)}$.

If $2^s(a-1)+1\le i\le 2^s(a-1)+2^{s-1}$, then $\langle 2^{a-1}(x+1)\rangle\supseteq C\supseteq\langle 2^{a-1}(x+1)^{2^{s-1}}\rangle$. By Theorem 3.6, the Hamming distances of both $\langle 2^{a-1}(x+1)\rangle$ and $\langle 2^{a-1}(x+1)^{2^{s-1}}\rangle$ are 2, thus their homogeneous distances are $2^{m(a-1)+1}$. Hence, $d_{\mathrm{h}}(C)=2^{m(a-1)+1}$.

For $1\le k\le s-1$, if $2^s(a-1)+1+\sum\limits_{l=1}^{k}2^{s-l}\le i\le 2^s(a-1)+\sum\limits_{l=1}^{k+1}2^{s-l}$, then

$$\left\langle 2^{a-1}(x+1)^{1+\sum\limits_{l=1}^{k}2^{s-l}}\right\rangle\supseteq C\supseteq\left\langle 2^{a-1}(x+1)^{\sum\limits_{l=1}^{k+1}2^{s-l}}\right\rangle.$$

Using Theorem 3.6, we have the Hamming distances of both

$$\left\langle 2^{a-1}(x+1)^{1+\sum\limits_{l=1}^{k}2^{s-l}}\right\rangle \quad\text{and}\quad \left\langle 2^{a-1}(x+1)^{\sum\limits_{l=1}^{k+1}2^{s-l}}\right\rangle$$

are 2^{k+1}. Thus, their homogeneous distances are $2^{k+1}\cdot 2^{m(a-1)}=2^{m(a-1)+k+1}$. Therefore, $d_{\mathrm{h}}(C)=2^{m(a-1)+k+1}$. □

4. Some classes of constacyclic codes of length 2^s over $\mathrm{GR}(2^a,m)$

As mentioned in Proposition 2.3, since the Galois ring $\mathrm{GR}(2^a,m)$ is a Galois extension of $\mathbb{Z}_{2^a}$, there exists a primitive element ξ such that $\mathrm{GR}(2^a,m)=\mathbb{Z}_{2^a}[\xi]$.

Each element $r \in \mathrm{GR}(2^a, m)$ is uniquely expressed as:

$$r = r_0 + r_1\xi + \cdots + r_{m-1}\xi^{m-1},$$

where $r_i \in \mathbb{Z}_{2^a}$. To simplify notations, we will say that an element γ of $\mathrm{GR}(2^a, m)$ is of Type $(*^-)$ if it has the form $\gamma = (4k_0 - 1) + 4k_1\xi + \cdots + 4k_{m-1}\xi^{m-1}$, and γ is called to be of Type $(*^+)$ if it has of the form $\gamma = (4k_0+1)+4k_1\xi+\cdots+4k_{m-1}\xi^{m-1}$, for integers $k_0, k_1, \ldots, k_{m-1}$. Clearly, elements of Type $(*^-)$ or $(*^+)$ are invertible in $\mathrm{GR}(2^a, m)$. Furthermore, the sets of Type $(*^-)$ and Type $(*^+)$ elements are disjoint when $a \geq 2$, they coincide if $a = 1$.

We now extend our technique in Section 3 to consider γ-constacyclic codes of length 2^s over $\mathrm{GR}(2^a, m)$ where γ is of Type $(*^-)$. Such γ-constacyclic codes are ideals of the ring

$$\mathcal{R}(a, m, \gamma) = \frac{\mathrm{GR}(2^a, m)}{\langle x^{2^s} - \gamma \rangle}.$$

Proposition 3.1 holds true when we replace the ring $\mathcal{R}(a, m)$ by $\mathcal{R}(a, m, \gamma)$.

Proposition 4.1. *Let $\gamma \in \mathrm{GR}(2^a, m)$ be of Type $(*^-)$. For any positive integer n, there exists a polynomial $\alpha_n(x) \in \mathbb{Z}[x]$ such that $(x+1)^{2^n} = x^{2^n} + 1 + 2\alpha_n(x)$, and $\alpha_n(x)$ is a unit in $\mathcal{R}(a, m, \gamma)$. In $\mathcal{R}(a, m, \gamma)$, $\langle (x + 1)^{2^s} \rangle = \langle 2 \rangle$, and the element $x + 1$ is nilpotent with nilpotency $2^s a$.*

Proof. The first assertion can be proved using the same proof as for $\mathcal{R}(a, m)$ (cf. [19, Lemma 3.1]). For the second statement, using $n = s$, we have that there is an unit $\alpha_s(x)$ in $\mathcal{R}(a, m, \gamma)$ such that:

$$\begin{aligned}
(x+1)^{2^s} &= x^{2^s} + 1 + 2\alpha_s(x) \\
&= \gamma + 1 + 2\alpha_s(x) \\
&= 4k_0 + 4k_1\xi + \cdots + 4k_{m-1}\xi^{m-1} + 2\alpha_s(x) \\
&= 2\alpha_s(x)(1 + y),
\end{aligned}$$

where $y = 2(k_0+k_1\xi+\cdots+k_{m-1}\xi^{m-1})\alpha_s^{-1}(x)$ is a nilpotent element in $\mathcal{R}(a, m, \gamma)$. Thus, in $\mathcal{R}(a, m, \gamma)$, $1+y$ is invertible, and hence, $\langle (x+1)^{2^s} \rangle = \langle 2 \rangle$, and $x+1$ has nilpotency $2^s a$. □

Using Proposition 4.1, the same proofs as in Section 3 can be extended from $\mathcal{R}(a, m)$ to $\mathcal{R}(a, m, \gamma)$ to provide the structure and Hamming and homogeneous distances of all γ-constacyclic codes of length 2^s over $\mathrm{GR}(2^a, m)$:

Theorem 4.2. (cf. Theorems 3.2, 3.6, 3.7) *Let $\gamma \in \mathrm{GR}(2^a, m)$ be of Type $(*^-)$. The ring $\mathcal{R}(a, m, \gamma)$ is a chain ring with maximal ideal $\langle x + 1 \rangle$ and residue field $\mathrm{GF}(2^m)$. γ-constacyclic codes of length 2^s over the Galois ring $\mathrm{GR}(2^a, m)$ are precisely the ideals $\langle (x + 1)^i \rangle$, $0 \leq i \leq 2^s a$, of $\mathcal{R}(a, m, \gamma)$. Each γ-constacyclic code $C = \langle (x+1)^i \rangle$ has $2^{m(2^s a - i)}$ codewords, and its Hamming distance $d(C)$ and* *homogeneous distances $d_{\rm h}(C)$ are completely determined as follows:*

$$d(C) = \begin{cases} 0 & \textit{if} \quad i = 2^s a, \\ 1 & \textit{if} \quad 0 \le i \le 2^s(a-1), \\ 2 & \textit{if} \quad 2^s(a-1)+1 \le i \le 2^s(a-1)+2^{s-1}, \\ 2^{k+1} & \textit{if} \quad 2^s(a-1)+2^s-2^{s-k}+1 \le i \\ & \qquad \le 2^s(a-1)+2^s-2^{s-k}+2^{s-k-1}, \\ & \quad \textit{i.e.,}\ 2^s(a-1)+1+\sum_{l=1}^{k} 2^{s-l} \le i \\ & \qquad \le 2^s(a-1)+\sum_{l=1}^{k+1} 2^{s-l}, \\ & \quad \textit{where}\ 1 \le k \le s-1. \end{cases}$$

$$d_{\rm h}(C) = \begin{cases} 0 & \textit{if} \quad i = 2^s a, \\ (2^m-1)\,2^{m(a-2)} & \textit{if} \quad 0 \le i \le 2^s(a-2), \\ 2^{m(a-1)} & \textit{if} \quad 2^s(a-2)+1 \le i \le 2^s(a-1), \\ 2^{m(a-1)+1} & \textit{if} \quad 2^s(a-1)+1 \le i \le 2^s(a-1)+2^{s-1}, \\ 2^{m(a-1)+k+1} & \textit{if} \quad 2^s(a-1)+2^s-2^{s-k}+1 \le i \\ & \qquad \le 2^s(a-1)+2^s-2^{s-k}+2^{s-k-1}, \\ & \quad \textit{i.e.,}\ 2^s(a-1)+1+\sum_{l=1}^{k} 2^{s-l} \le i \\ & \qquad \le 2^s(a-1)+\sum_{l=1}^{k+1} 2^{s-l}, \\ & \quad \textit{where}\ 1 \le k \le s-1. \end{cases}$$

Lemma 4.3 *Let γ_1, γ_2 be of Type $(*^-)$, and γ_3, γ_4 be of Type $(*^+)$. Then*

(a) *$\gamma_1\gamma_2$ is of Type $(*^+)$, i.e., the product of two elements of Type $(*^-)$ is an element of Type $(*^+)$.*
(b) *$\gamma_1\gamma_3$ is of Type $(*^-)$, i.e., the product of an element of Type $(*^-)$ and an element of Type $(*^+)$ is an element of Type $(*^-)$.*
(c) *$\gamma_3\gamma_4$ is of Type $(*^+)$, i.e., the product of two elements of Type $(*^+)$ is an element of Type $(*^+)$.*
(d) *γ_1^{-1} is of Type $(*^-)$, i.e., the inverse of an element of Type $(*^-)$ is an element of Type $(*^-)$.*
(e) *γ_3^{-1} is of Type $(*^+)$, i.e., the inverse of an element of Type $(*^+)$ is an element of Type $(*^+)$.*

Proof. Note that each element $(4k_0-1)+4k_1\xi+\cdots+4k_{m-1}\xi^{m-1}$ of Type $(*^-)$ can be expressed as $4z-1$, and any element $(4k_0'-1)+4k_1'\xi+\cdots+4k_{m-1}'\xi^{m-1}$ of Type $(*^+)$ can be expressed as $4z'+1$, for $z, z' \in \mathrm{GR}(2^a, m)$. Parts (a), (b), and (c) follow readily. For (d), write $\gamma_1 = 4z-1$. Then, observe that

$$\gamma_1 \prod_{j=0}^{a-1} \left[(4z)^{2^j}+1\right] = (4z)^{2^a} - 1 = -1,$$

therefore,

$$\gamma_1^{-1} = \gamma_1 \prod_{j=0}^{a-1} \left[(4z)^{2^j}+1\right]^2.$$

Since, for $0 \le j \le a-1$, $(4z)^{2^j}+1$ are of Type $(*^+)$, (b), (c) imply that γ^{-1} is of Type $(*^-)$, proving (d). The proof of (e) is similar to that of (d). □

Proposition 4.4. *Let $\gamma \in \mathrm{GR}(2^a, m)$ be of Type $(*^-)$. Assume that C is a γ-constacyclic code of length 2^s over $\mathrm{GR}(2^a, m)$. Then $C = \langle (x+1)^i \rangle \subseteq \mathcal{R}(a, m, \gamma)$, for some $i \in \{0, 1, \dots, 2^s a\}$. Its dual is a γ^{-1}-constacyclic code of length 2^s over $\mathrm{GR}(2^a, m)$, $C^\perp = \langle (x+1)^{2^s a - i} \rangle \subseteq \mathcal{R}(a, m, \gamma^{-1})$, which contains 2^{mi} codewords.*

Proof. In light of Proposition 2.6, $C^\perp$ is a γ^{-1}-constacyclic code of length 2^s over $\mathrm{GR}(2^a, m)$. Lemma 4.3 shows that γ^{-1} is also of Type $(*^-)$. Thus, Theorem 4.2 is applicable for $C^\perp$ and $\mathcal{R}(a, m, \gamma^{-1})$. Hence, $C^\perp$ is an ideal of the form $\langle (x+1)^j \rangle$, $0 \le j \le 2^s a$, of the chain ring $\mathcal{R}(a, m, \gamma^{-1})$. On the other hand, by Proposition 2.5,

$$|C| \cdot |C^\perp| = |\,\mathrm{GR}(2^a, m)|^{2^s} = 2^{2^s am},$$

which implies,

$$|C^\perp| = \frac{2^{2^s am}}{|C|} = \frac{2^{2^s am}}{|2^{m(2^s a - i)}|} = 2^{mi}.$$

Therefore, $C^\perp$ must be the ideal $\langle (x+1)^{2^s a - i} \rangle$ of $\mathcal{R}(a, m, \gamma^{-1})$. □

Theorem 4.5. *Let $\gamma \in \mathrm{GR}(2^a, m)$ be of Type $(*^-)$, i.e., γ can be expressed as*

$$\gamma = (4k_0 - 1) + 4k_1\xi + \cdots + 4k_{m-1}\xi^{m-1},$$

for integers $k_0, k_1, \dots, k_{m-1}$. Self-dual γ-constacyclic codes of length 2^s over $\mathrm{GR}(2^a, m)$ exist if and only if $\gamma^2 = 1$, which occurs only in the following cases:

(i) $1 \le a \le 3$: *any values of $k_0, k_1, \dots, k_{m-1}$;*

(ii) $a \ge 4$: $k_i \equiv 0 \pmod{2^{a-3}}$, $0 \le i \le m-1$, *i.e., γ is of the form*

$$\gamma = (2^{a-1}l_0 - 1) + 2^{a-1}l_1\xi + \cdots + 2^{a-1}l_{m-1}\xi^{m-1},$$

for $l_i \in \{0, 1\}$.

In such case, $\langle (x+1)^{2^{s-1}a} \rangle$ is the unique self-dual γ-constacyclic code.

Proof. Let C be a γ-constacyclic code of length 2^s over $\mathrm{GR}(2^a, m)$. By Proposition 4.4, $C = \langle (x+1)^j \rangle$, for some integer $j \in \{0, 1, \dots, 2^s a\}$ and $C = C^\perp$ if and only if $\gamma = \gamma^{-1}$ and $j = 2^s a - j$. That means that C is self-dual if and only if $\gamma^2 = 1$, and in such case C must be the ideal $\langle (x+1)^{2^{s-1}a} \rangle$.

We now determine values of γ of the given form such that $\gamma^2 = 1$. Denote $z = k_0 + k_1\xi + \cdots + k_{m-1}\xi^{m-1}$, then $\gamma = 4z - 1$. We have $\gamma^2 = (4z-1)^2 = 8z(2z-1) + 1$, and so $\gamma^2 = 1$ if and only if $8z = 0$, as $2z - 1$ is invertible in $\mathrm{GR}(2^a, m)$. If $1 \le a \le 3$, $8z = 0$ in $\mathrm{GR}(2^a, m)$ for any z. When $a \ge 4$, $8z = 0$ if and only if, for $0 \le i \le m-1$, $8k_i = 0$ in $\mathbb{Z}_{2^a}$, which is equivalent to $k_i \equiv 0 \pmod{2^{a-3}}$. □

Remark 4.6.

4.6.1. When $a = 1$, there is just one class of such γ-constacyclic codes, namely, cyclic codes of length 2^s over $\mathrm{GR}(2, m) = \mathbb{F}_{2^m}$, those were studied in [23]. When $a = 2$, there is also only one class of such γ-constacyclic codes, namely, negacyclic codes of length 2^s over $\mathrm{GR}(4, m)$, those were investigated by many researchers in

TABLE 1. γ-constacyclic codes of length 2^s over $\mathrm{GR}(2^a, m)$ for any unit γ of the form $\gamma = (4k_0 - 1) + 4k_1\xi + \cdots + 4k_{m-1}\xi^{m-1}$.

a	number of possible values of γ (number of classes of γ-constacyclic codes)	number of possible values of γ so that self-dual codes exist (number of classes of γ-constacyclic codes having self-dual codes)
1	1 ($\gamma = 1$)	1
2	1 ($\gamma = -1$)	1
3	2^m	2^m
≥ 4	$2^{m(a-2)}$	2^m

the general case, as well as in the special case when $m = 1$ (i.e., the alphabet is $\mathbb{Z}_4$), see, for example, [1, 2, 7, 22, 24]. When $a > 2$, there are $2^{m(a-2)}$ classes of such γ-constacyclic codes. When $a = 3$, Theorem 4.5 implies that $\gamma^2 = 1$ for any values of $k_0, k_1, \ldots, k_{m-1}$, which shows unique self-dual γ-constacyclic code over $\mathrm{GR}(2^3, m)$ always exists. Table 1 gives the number of γ-constacyclic classes, and those that have self-dual codes.

4.6.2. The class of γ-constacyclic codes with $\gamma = 4k_0 - 1$, i.e., $k_1 = \cdots = k_{m-1} = 0$, over $\mathbb{Z}_{2^a}$, i.e., the Galois ring $\mathrm{GR}(2^a, m)$ with dimension $m = 1$, was investigated in [21] and [22, Section XIII]. There are 2^{a-2} classes of such γ-constacyclic codes.

4.6.3. Cyclic codes are in general not such γ-constacyclic codes (except the case $a = 1$, when cyclic and negacyclic codes coincide). As pointed out in Proposition 2.4, cyclic codes of length 2^s over $\mathrm{GR}(2^a, m)$ are ideals of the ring $\mathcal{R}(a, m, 1) = \frac{\mathrm{GR}(2^a,m)[x]}{\langle x^{2^s} - 1\rangle}$. Using polynomial representation similar to the proof of Proposition 3.2, it can be shown that, unlike the chain ring $\mathcal{R}(a, m)$, this ring $\mathcal{R}(a, m, 1)$ is a local ring with maximal ideal $\langle 2, x + 1\rangle$, but it is not a chain ring. Indeed, [53, Theorem 3.4] showed that for any prime p, $\frac{\mathrm{GR}(p^a,m)}{\langle x^{p^s} - 1\rangle}$ is not a chain ring. While the complete structure of such cyclic codes is still unknown in general, there have been many results for the special case when the alphabet is the ring $\mathbb{Z}_4$ ($a = 2$, $m = 1$). In 2003, [1, 2] gave a structure of cyclic codes of length 2^s over $\mathbb{Z}_4$, and [8] provided a structure of cyclic codes over $\mathbb{Z}_4$ of oddly even length (length $2n$, where n is odd). In 2006, [25] established a structure of cyclic codes over $\mathbb{Z}_4$ of any length.

Acknowledgement

The author would like to sincerely thank the referee for a very meticulous reading of this manuscript, and for valuable suggestions which help to create an improved final version.

References

[1] T. Abualrub and R. Oehmke, *On the generators of* $\mathbb{Z}_4$ *cyclic codes of length* 2^e, IEEE Trans. Inform. Theory **49** (2003), 2126–2133.

[2] T. Abualrub, A. Ghrayeb, and R. Oehmke, *A mass formula and rank of* $\mathbb{Z}_4$ *cyclic codes of length* 2^e, IEEE Trans. Inform. Theory **50** (2004), 3306–3312.

[3] N.S. Babu and K.-H. Zimmermann, *Decoding of linear codes over Galois rings*, IEEE Trans. Inform. Theory **47** (2001), 1599–1603.

[4] E.R. Berlekamp, *Negacyclic Codes for the Lee Metric*, Proceedings of the Conference on Combinatorial Mathematics and Its Applications, Chapel Hill, N.C., University of North Carolina Press (1968), 298–316.

[5] E.R. Berlekamp, *Algebraic Coding Theory*, revised 1984 edition, Aegean Park Press, 1984.

[6] S.D. Berman, *Semisimple cyclic and Abelian codes. II*, Kibernetika (Kiev) **3** (1967), 21–30 (Russian). English translation: Cybernetics **3** (1967), 17–23.

[7] T. Blackford, *Negacyclic codes over* $\mathbb{Z}_4$ *of even length*, IEEE Trans. Inform. Theory **49** (2003), 1417–1424.

[8] T. Blackford, *Cyclic codes over* $\mathbb{Z}_4$ *of oddly even length*, International Workshop on Coding and Cryptography (WCC 2001) (Paris), Appl. Discr. Math. **128** (2003), 27–46.

[9] E. Byrne, *Lifting decoding schemes over a Galois ring*, Applied algebra, algebraic algorithms and Error-Correcting Codes (Melbourne, 2001), Lecture Notes in Comput. Sci. **2227**, Springer, (2001) 323–332.

[10] E. Byrne, *Decoding a class of Lee metric codes over a Galois ring*, IEEE Trans. Inform. Theory **48** (2002), 966–975.

[11] E. Byrne and P. Fitzpatrick, *Grbner bases over Galois rings with an application to decoding alternant codes*, J. Symbolic Comput. **31** (2001), 565–584.

[12] E. Byrne and P. Fitzpatrick, *Hamming metric decoding of alternant codes over Galois rings*, IEEE Trans. Inform. Theory **48** (2002), 683–694.

[13] A.R. Calderbank, A.R. Hammons, Jr., P.V. Kumar, N.J.A. Sloane, and P. Solé, *A linear construction for certain Kerdock and Preparata codes*, Bull. AMS **29** (1993), 218–222.

[14] A.R. Calderbank and N.J.A. Sloane, *Modular and p-adic codes*, Des. Codes Cryptogr **6** (1995), 21–35.

[15] G. Castagnoli, J.L. Massey, P.A. Schoeller, and N. von Seemann, *On repeated-root cyclic codes*, IEEE Trans. Inform. Theory **37** (1991), 337–342.

[16] I. Constaninescu, *Lineare Codes über Restklassenringen ganzer Zahlen und ihre Automorphismen bezüglich einer verallgemeinerten Hamming-Metrik*, Ph.D. Dissertation, Technische Universität, München, Germany, 1995.

[17] I. Constaninescu and W. Heise, *A metric for codes over residue class rings of integers*, Problemy Peredachi Informatsii **33** (1997), 22–28.

[18] I. Constaninescu, W. Heise, and T. Honold, *Monomial extensions of isometries between codes over* $\mathbb{Z}_m$, Proceedings of the 5th International Workshop on Algebraic and Combinatorial Coding Theory (ACCT'96), Unicorn Shumen (1996), 98–104.

[19] H.Q. Dinh, *Negacyclic codes of length* 2^s *over Galois rings*, IEEE Trans. Inform. Theory **51** (2005), 4252–4262.

[20] H.Q. Dinh, *Structure of some classes of repeated-root constacyclic codes over integers modulo* 2^m, Ser. Lecture Notes in Pure & Appl. Math. **248** (2006), 105–117.

[21] H.Q. Dinh, *Repeated-root constacyclic codes of length* 2^s *over* $\mathbb{Z}_{2^a}$, AMS Contemporary Mathematics **419** (2006), 95–110.

[22] H.Q. Dinh, *Complete distances of all negacyclic codes of length* 2^s *over* $\mathbb{Z}_{2^a}$, IEEE Trans. Inform. Theory **53** (2007), 147–161.

[23] H.Q. Dinh, *On the linear ordering of some classes of negacyclic and cyclic codes and their distance distributions*, Finite Fields & Appl. **14** (2008), 22–40.

[24] H.Q. Dinh and S.R. López-Permouth, *Cyclic and negacyclic codes over finite chain rings*, IEEE Trans. Inform. Theory **50** (2004), 1728–1744.

[25] S.T. Dougherty and S. Ling, *Cyclic codes over* $\mathbb{Z}_4$ *of even length*, Des. Codes Cryptogr. **39** (2006), 127–153.

[26] G. Falkner, B. Kowol, W. Heise, E. Zehendner, *On the existence of cyclic optimal codes*, Atti Sem. Mat. Fis. Univ. Modena **28** (1979), 326–341.

[27] M. Greferath and S.E. Schmidt, *Gray Isometries for Finite Chain Rings and a Nonlinear Ternary* $(36, 3^{12}, 15)$ *Code*, IEEE Trans. Inform. Theory **45** (1999), 2522–2524.

[28] M. Greferath and S.E. Schmidt, *Finite Ring Combinatorics and MacWilliams's Equivalence Theorem*, J. Combin. Theory Ser. A **92** (2000), 17–28.

[29] A.R. Hammons, Jr., P.V. Kumar, A.R. Calderbank, N.J.A. Sloane, and P. Solé, *The* $\mathbb{Z}_4$*-linearity of Kerdock, Preparata, Goethals, and related codes*, IEEE Trans. Inform. Theory **40** (1994), 301–319.

[30] W. Heise, T. Honold, and A.A. Nechaev, *Weighted modules and representations of codes*, Proceedings of the ACCT 6, Pskov, Russia (1998), 123–129.

[31] T. Honold and I. Landjev, *Linear representable codes over chain rings*, Proceedings of the ACCT 6, Pskov, Russia (1998), 135–141.

[32] W.C. Huffman and V. Pless, *Fundamentals of Error-correcting codes*, Cambridge University Press, Cambridge, 2003.

[33] G. Janusz, *Separable algebra over commutative rings*, Trans. AMS **122** (1966), 461–479.

[34] Kiran T and B.S. Rajan, *Abelian codes over Galois rings closed under certain permutations*, IEEE Trans. Inform. Theory **49** (2003), 2242–2253.

[35] E. Kleinfeld, *Finite Hjelmslev planes*, Illinois J. Math. **3** (1959), 403–407.

[36] W. Krull, *Algebraische Theorie der Ringe*, Math. Annalen **92** (1924), 183-213.

[37] S. Ling, H. Niederreiter, and P. Solé, *On the algebraic structure of quasi-cyclic codes. IV. Repeated roots*, Des. Codes Cryptogr. **38** (2006), 337–361.

[38] F.J. MacWilliams, *Error-correcting codes for multiple-level transmissions*, Bell System Tech. J. **40** (1961), 281–308.

[39] F.J. MacWilliams, *Combinatorial problems of elementary abelian groups*, PhD. Dissertation, Harvard University, Cambridge, MA, 1962.

[40] F.J. MacWilliams and N.J.A. Sloane, *The theory of error-correcting Codes*, 10th impression, North-Holland, Amsterdam, 1998.

[41] J.L. Massey, D.J. Costello, and J. Justesen, *Polynomial weights and code constructions*, IEEE Trans. Information Theory **19** (1973), 101–110.

[42] B.R. McDonald, *Finite rings with identity*, Pure and Applied Mathematics, Vol. **28**, Marcel Dekker, New York, 1974.

[43] A.A. Nechaev, *Kerdock code in a cyclic form*, (in Russian), Diskr. Math. (USSR) **1** (1989), 123–139. English translation: Discrete Math. and Appl. **1** (1991), 365–384.

[44] C.-S. Nedeloaia, *Weight distributions of cyclic self-dual codes*, IEEE Trans. Inform. Theory **49** (2003), 1582–1591.

[45] G. Norton and A. Sălăgean-Mandache, *On the structure of linear cyclic codes over finite chain rings*, Appl. Algebra Engrg. Comm. Comput. **10** (2000), 489–506.

[46] V. Pless and W.C. Huffman, *Handbook of Coding Theory*, Elsevier, Amsterdam, 1998.

[47] E. Prange, *The following technical notes issued by Air Force Cambridge Research Labs, Bedford, Mass: Cyclic Error-Correcting Codes in Two Symbols*, (September 1957), TN-57-103.

[48] E. Prange, *Some cyclic error-correcting codes with simple decoding algorithms*, (April 1958), TN-58-156.

[49] E. Prange, *The use of coset equivalence in the analysis and decoding of group codes*, (1959), TN-59-164.

[50] E. Prange, *An algorithm for factoring* $x^n - 1$ *over a finite field*, (October 1959), TN-59-175.

[51] R. Raghavendran, *Finite associative rings*, Compositio Math. **21** (1969), 55–58.

[52] R.M. Roth and G. Seroussi, *On cyclic MDS codes of length* q *over* GF(q), IEEE Trans. Inform. Theory **32** (1986), 284–285.

[53] A. Sălăgean, *Repeated-root cyclic and negacyclic codes over finite chain rings*, Discrete Appl. Math. **154** (2006), 413–419.

[54] L.-z. Tang, C.B. Soh and E. Gunawan, *A note on the* q*-ary image of a* q^m*-ary repeated-root cyclic code*, IEEE Trans. Inform. Theory **43** (1997), 732–737.

[55] J.H. van Lint, *Repeated-root cyclic codes*, IEEE Trans. Inform. Theory **37** (1991), 343–345.

[56] Z. Wan, *Cyclic Codes over Galois Rings*, Algebra Colloquium **6** (1999), 291–304.

[57] J. Wolfmann, *Negacyclic and Cyclic Codes over* $\mathbb{Z}_4$, IEEE Trans. Inform. Theory **45** (1999), 2527–2532.

[58] J.A. Wood, *Duality for modules over finite rings and applications to coding theory*, American J. of Math. **121** (1999), 555–575.

[59] S. Zhu and X. Kai, *The Hamming distances of negacyclic codes of length* 2^s *over* GR($2^a, m$), Jrl Syst Sci & Complexity **21** (2008), 60–66.

Hai Q. Dinh
Department of Mathematical Sciences
Kent State University
4314 Mahoning Avenue
Warren, OH 44483, USA
e-mail: hdinh@kent.edu

Advances in Ring Theory
Trends in Mathematics, 149–164

Couniformly Presented Modules and Dualities

Alberto Facchini and Nicola Girardi

Abstract. A module U_R is *couniform* if it has dual Goldie dimension 1, that is, it is non-zero and the sum of any two proper submodules of U_R is a proper submodule of U_R. A module M_R is *couniformly presented* if it is non-zero and there exists a short exact sequence $0 \to C_R \to P_R \to M_R \to 0$ with P_R projective and both C_R and P_R couniform modules. The endomorphism ring of a couniformly presented module has at most two maximal ideals, and a weak form of the Krull-Schmidt Theorem holds for finite direct sums of couniformly presented modules. Cokernels of morphisms between couniform projective modules are couniformly presented, provided that the morphisms are not onto. Via a suitable duality functor, finite direct sums of cokernels of morphisms between couniform projective modules correspond to finite direct sums of kernels of morphisms between uniform injective modules.

Mathematics Subject Classification (2000). 16D70.

Keywords. Krull-Schmidt Theorem, dual Goldie dimension, couniform modules.

1. Introduction

This paper is divided into two parts and devoted to two topics. In the first part, we generalize to a larger class of modules the results that were proved in [1] for cyclically presented modules over local rings. This larger class of modules consists of all *couniformly presented modules*, that is, the non-zero modules M_R for which there exists an exact sequence $0 \to C_R \to P_R \to M_R \to 0$ with P_R projective and both C_R and P_R couniform modules. Recall that a module is *couniform* if it has dual Goldie dimension one, that is, it is non-zero and the sum of any two proper submodules is a proper submodule.

During this last year, several classes of modules with a behavior very similar to the behavior described in Theorems 2.5 and 4.3 (at most two maximal ideals

---

The content of this paper is part of a thesis written by Nicola Girardi under the supervision of Alberto Facchini (University of Padova).

in the endomorphism ring and the validity of a weak form of the Krull-Schmidt Theorem) have been discovered: cyclically presented modules over local rings [1], kernels of non-zero morphisms between indecomposable injective modules [2], artinian modules whose socle is isomorphic to the direct sum of two fixed simple modules [7], and so on. All these classes mimic the behavior of uniserial modules [4], or, more generally, biuniform modules [5]. In this paper, we study the behavior of couniformly presented modules, which extend to arbitrary rings the class of cyclically presented modules over local rings.

Direct sums of modules in each of the previous classes are described by a pair of invariants: lower part and epigeny class for cyclically presented modules over local rings, upper part and monogeny class for kernels of non-zero morphisms between indecomposable injective modules, monogeny class and epigeny class for uniserial modules, or, more generally, biuniform modules. In the second part of this paper, we consider dualities that allow us to exchange, in our context, two of these notions: monogeny class and epigeny class, and lower part and upper part (Propositions 5.2 and Corollary 6.2). It was discovered in [1] that the concepts of epigeny class and lower part for cyclically presented modules over local rings are exchanged in a similar way by the Auslander-Bridger transpose.

The first author is greatly indebted to Dolors Herbera, who suggested to study the category $\mathcal{C}$ that appears in Theorem 5.1, that is, cokernels of morphisms between couniform projective modules. Our idea of studying the category of couniformly presented modules is a slight generalization of her idea. Also, some of the techniques we use in this paper were suggested by her and already appear in the paper [6].

All the rings considered in this paper will be assumed to have an identity element. For any ring R, the Jacobson radical of R will be denoted by $J(R)$, and the group of units will be denoted by $U(R)$. Thus $U(R) = R \setminus J(R)$ when R is a local ring. The modules we will consider are unitary right modules unless otherwise stated, and morphisms will be written on the left. If M_R is a module, we will write $\langle M_R \rangle$ to denote the *isomorphism class* of M_R, that is, the class of all right R-modules isomorphic to M_R.

2. Couniformly presented modules

Recall that a right module M_R over a ring R is said to be *couniform* (or *hollow*) if it has dual Goldie dimension one, that is, if it is non-zero and the sum of any two proper submodules of M_R is a proper submodule of M_R. Equivalently, a non-zero module is couniform if and only if all its proper submodules are superfluous, if and only if all its non-zero homomorphic images are indecomposable modules. For instance, every non-zero uniserial module, that is, every non-zero module whose lattice of submodules is linearly ordered under inclusion, is couniform. The following easy lemma, taken from [1, Lemma 8.7], describes the projective modules that are couniform.

Lemma 2.1. *The following conditions are equivalent for a projective right module P_R over an arbitrary ring R:*

(1) *P_R is couniform.*
(2) *P_R is the projective cover of a simple module.*
(3) *The endomorphism ring $\mathrm{End}(P_R)$ of P_R is local.*
(4) *There exists an idempotent $e \in R$ with $P_R \cong eR$ and eRe a local ring.*
(5) *P_R is a finitely generated module with a unique maximal submodule.*
(6) *P_R has a greatest proper submodule.*

Moreover, if these equivalent conditions hold, then $\mathrm{Hom}(P_R, R)$ is a couniform projective left R-module.

Notice the following interesting elementary fact:

Lemma 2.2. *If $P_R \to M_R$ is the projective cover of a couniform module M_R, then the projective module P_R is also couniform.*

Proof. Let K be the kernel of $P_R \to M_R$. Assume $P_R = A_1 + A_2$. Then $(A_1 + K)/K + (A_2 + K)/K = P_R/K \cong M_R$, so that, for $i = 1$ or $i = 2$, one has that $(A_i + K)/K = P_R/K$. Thus $A_i + K = P_R$. Now K superfluous in P_R implies $A_i = P_R$. □

By the previous two lemmas, there is a bijection between the set of all isomorphism classes $\langle P_R \rangle$, where P_R ranges in the class of all couniform projective right R-modules, and the the set of all isomorphism classes $\langle S_R \rangle$, where S_R ranges in the class of all simple right R-modules with a projective cover. It associates to each isomorphism class $\langle P_R \rangle$ the isomorphism class $\langle P_R/\mathrm{rad}(P_R) \rangle$, where $\mathrm{rad}(P_R)$ denotes the unique maximal submodule of the couniform projective module P_R. In particular, the position $\langle P_R \rangle \mapsto \langle P_R/\mathrm{rad}(P_R) \rangle$ defines a bijection between the set of isomorphism classes $\langle P_R \rangle$ of all couniform projective modules P_R and the the set of isomorphism classes $\langle S_R \rangle$ of all simple modules S_R if and only if the ring R is semiperfect [5, Theorem 3.6(d)].

We say that a module M_R is *couniformly presented* if it is non-zero and there exists an exact sequence

$$0 \to C_R \xrightarrow{\iota} P_R \to M_R \to 0$$

with P_R projective and both C_R and P_R couniform modules. In this case, we will say that $0 \to C_R \xrightarrow{\iota} P_R \to M_R \to 0$ is a *couniform presentation* of M_R. Notice that $P_R \to M_R$ is necessarily a projective cover of M_R, because every proper submodule of P_R is superfluous. Without loss of generality, we will always suppose that the monomorphism $\iota\colon C_R \to P_R$ is the inclusion. Clearly, every couniformly presented module is cyclic. Every cyclically presented module over a local ring R is either zero, or isomorphic to R, or couniformly presented. Over a right chain ring R, that is, a ring R with R_R uniserial, a right module is couniformly presented if and only if it is cyclic but not projective. In particular, couniformly presented right modules over right chain rings are uniserial.

Let R be an arbitrary ring. Given any couniformly presented right module M_R with couniform presentation $0 \to C_R \xrightarrow{\iota} P_R \to M_R \to 0$, every endomorphism f of M_R lifts to an endomorphism f_0 of the projective cover P_R, and we will denote by f_1 the restriction of f_0 to C_R. Hence we have a commutative diagram

$$\begin{array}{ccccccccc} 0 & \to & C_R & \xrightarrow{\iota} & P_R & \to & M_R & \to & 0 \\ & & {\scriptstyle f_1}\downarrow & & \downarrow{\scriptstyle f_0} & & \downarrow{\scriptstyle f} & & \\ 0 & \to & C_R & \xrightarrow{\iota} & P_R & \to & M_R & \to & 0. \end{array} \tag{1}$$

The morphisms f_0 and f_1 that complete diagram (1) are not uniquely determined by f. Nevertheless, it is easily seen that $f\colon M_R \to M_R$ is an epimorphism if and only if $f_0\colon P_R \to P_R$ is an epimorphism, if and only if f_0 is an automorphism. It follows that if we substitute f_0 and f_1 with two other morphisms f_0' and f_1' making the diagram analogous to diagram (1) commute, then $f_0\colon P_R \to P_R$ is an epimorphism if and only if $f_0'\colon P_R \to P_R$ is an epimorphism. In this notation, let us show that the same holds for C_R, i.e., that

Lemma 2.3. *$f_1\colon C_R \to C_R$ is an epimorphism if and only if $f_1'\colon C_R \to C_R$ is an epimorphism.*

Proof. The commutativity of the two diagrams (1) (one relative to f_0, f_1, the other relative to f_0', f_1') gives, by subtraction, a commutative diagram

$$\begin{array}{ccccccccc} 0 & \to & C_R & \xrightarrow{\iota} & P_R & \to & M_R & \to & 0 \\ & & {\scriptstyle f_1-f_1'}\downarrow & & \downarrow{\scriptstyle f_0-f_0'} & & \downarrow 0 & & \\ 0 & \to & C_R & \xrightarrow{\iota} & P_R & \to & M_R & \to & 0. \end{array} \tag{2}$$

Hence $(f_0 - f_0')(P_R) \subseteq C_R$. Since C_R is superfluous in P_R, it follows that $(f_0 - f_0')(C_R)$ is superfluous in $(f_0 - f_0')(P_R)$, so that $(f_0 - f_0')(C_R) = (f_1 - f_1')(C_R)$ is a proper submodule of C_R. Thus $f_1 - f_1'$ is not an epimorphism. This and the fact that C_R is couniform yields that $f_1\colon C_R \to C_R$ is an epimorphism if and only if $f_1'\colon C_R \to C_R$ is an epimorphism. □

Our proof of Lemma 2.3 is essentially the same as the proof of [6, Lemma 7.1].

Notice that, in the proof of Lemma 2.3, we have seen that, for every morphism $g\colon P_R \to C_R$ (where $P_R \supset C_R$ are couniform modules and P_R is projective), $g(C_R)$ is properly contained in C_R.

For every couniform module U_R, the endomorphism ring $\mathrm{End}(U_R)$ has a proper completely prime two-sided ideal K_{U_R} consisting of all the endomorphisms of U_R that are not surjective (see [5, Lemma 6.26]). The ring $\mathrm{End}(U_R)/K_{U_R}$ is an integral domain, but it is not a division ring in general (for instance, take as U_R the Prüfer group $\mathbb{Z}(p^\infty)$ viewed as a $\mathbb{Z}$-module.) Our proof of Lemma 2.3 also shows that for every couniformly presented right module M_R with couniform presentation $0 \to C_R \xrightarrow{\iota} P_R \to M_R \to 0$, there is a well-defined ring morphism $\mathrm{End}(M_R) \to \mathrm{End}(C_R)/K_{C_R}$, defined by $f \mapsto f_1 + K_{C_R}$.

Similarly to [6, Section 7], by Lemma 2.3, we can consider the ring morphism

$$\Phi\colon \mathrm{End}(M_R) \to \mathrm{End}(M_R)/K_{M_R} \times \mathrm{End}(C_R)/K_{C_R}$$

defined by $\Phi(f) = (f + K_{M_R}, f_1 + K_{C_R})$ for every $f \in \mathrm{End}(M_R)$. Recall that a ring morphism $\varphi\colon S \to S'$ is said to be a *local morphism* if, for every $s \in S$, $\varphi(s) \in U(S')$ implies $s \in U(S)$.

Lemma 2.4. *Let $0 \to C_R \to P_R \to M_R \to 0$ be a couniform presentation of a couniformly presented module M_R. Then the ring morphism Φ is local.*

Proof. Let $f \in \mathrm{End}(M_R)$ be an endomorphism with $\Phi(f)$ invertible. Consider the commutative diagram (1). Then $f + K_{M_R}$ and $f_1 + K_{C_R}$ are invertible in $\mathrm{End}(M_R)/K_{M_R}$ and $\mathrm{End}(C_R)/K_{C_R}$ respectively, so that, in particular, $f \notin K_{M_R}$ and $f_1 \notin K_{C_R}$, that is, the morphisms f and f_1 are epimorphisms. Thus f_0 also is an epimorphism, hence an automorphism of P_R because P_R is projective and indecomposable. By the Snake Lemma applied to diagram (1), f_0 isomorphism and f_1 epimorphism imply f monomorphism. □

The next result describes the endomorphism ring of a couniformly presented module.

Theorem 2.5. *Let $0 \to C_R \to P_R \to M_R \to 0$ be a couniform presentation of a couniformly presented module M_R. Let $K := \{ f \in \mathrm{End}(M_R) \mid f \text{ is not surjective} \}$ and $I := \{ f \in \mathrm{End}(M_R) \mid f_1\colon C_R \to C_R \text{ is not surjective} \}$. Then K and I are completely prime two-sided ideals of* $\mathrm{End}(M_R)$*, the union $K \cup I$ is the set of all non-invertible elements of* $\mathrm{End}(M_R)$*, and every proper right ideal of* $\mathrm{End}(M_R)$ *and every proper left ideal of* $\mathrm{End}(M_R)$ *is contained either in K or in I. Moreover, one of the following two conditions hold:*

(a) *Either the ideals K and I are comparable, so that* $\mathrm{End}(M_R)$ *is a local ring with maximal ideal the greatest ideal among K and I, or*
(b) *K and I are not comparable, $J(\mathrm{End}(M_R)) = K \cap I$, and*

$$\mathrm{End}(M_R)/J(\mathrm{End}(M_R))$$

is canonically isomorphic to the direct product of the two division rings $\mathrm{End}(M_R)/K$ *and* $\mathrm{End}(M_R)/I$*.*

Proof. Let π_1 and π_2 be the canonical projections of

$$\mathrm{End}(M_R)/K_{M_R} \times \mathrm{End}(C_R)/K_{C_R}$$

onto $\mathrm{End}(M_R)/K_{M_R}$ and $\mathrm{End}(C_R)/K_{C_R}$, respectively. We already know that $K = K_{M_R}$ is a completely prime ideal of $\mathrm{End}(M_R)$. Notice that I is the kernel of the composite morphism $\pi_2\Phi\colon \mathrm{End}(M_R) \to \mathrm{End}(C_R)/K_{C_R}$. As $\mathrm{End}(C_R)/K_{C_R}$ is an integral domain, it follows that I is a completely prime ideal of $\mathrm{End}(M_R)$.

As the ideals K and I are proper, it follows that $K \cup I \subseteq \mathrm{End}(M_R) \setminus U(\mathrm{End}(M_R))$. Conversely, if $f \in \mathrm{End}(M_R)$ is non-invertible, it is not an automorphism, so that it is either non-surjective or non-injective. If f is not surjective, then $f \in K$. If f is surjective but not injective, then in diagram (1) we have that f_0 is surjective, so that f_0 is an automorphism of P_R. By the Snake Lemma applied to (1), we have that f_0 automorphism of P_R and f non-injective imply f_1 non-surjective. Thus $f \in I$.

Every proper right or left ideal L of $\mathrm{End}(M_R)$ is contained in $K \cup I$. If there exist $x \in L \setminus K$ and $y \in L \setminus I$, then $x+y \in L$, $x \in I$ and $y \in K$. Hence $x+y \notin K$ and $x+y \notin I$. Thus $x+y \notin K \cup I$, so that $x+y \in L$ and is an invertible element of $\mathrm{End}(M_R)$, a contradiction. This proves that L is contained either in K or in I. In particular, the unique maximal right ideals of $\mathrm{End}(M_R)$ are at most K and I. Similarly, the unique maximal left ideals of $\mathrm{End}(M_R)$ are at most K and I.

If K and I are comparable, then (a) clearly holds. If K and I are not comparable, the ring $\mathrm{End}(M_R)$ has exactly two maximal right ideals K and I, so that $J(\mathrm{End}(M_R)) = K \cap I$, $\mathrm{End}(M_R)/K$ and $\mathrm{End}(M_R)/I$ are division rings, and there is a canonical injective ring homomorphism $\pi\colon \mathrm{End}(M_R)/J(\mathrm{End}(M_R)) \to \mathrm{End}(M_R)/K \times \mathrm{End}(M_R)/I$. But $K + I = \mathrm{End}(M_R)$ because K and I are incomparable maximal right ideals of $\mathrm{End}(M_R)$, hence π is surjective by the Chinese Remainder Theorem. □

Remark 2.6. The ideal I in the statement of Theorem 2.5 does not depend on the couniform presentation of M_R. Suppose $0 \to C_R \to P_R \to M_R \to 0$ and $0 \to C'_R \to P'_R \to M_R \to 0$ are two couniform presentations of M_R. Let f be an endomorphism of M_R, and consider a diagram (1) relative to f for each of the two couniform presentations. We need to show that f_1 is an epimorphism if and only if f'_1 is an epimorphism. Construct another diagram (1) as follows. The identity of M_R lifts to an isomorphism $g_0\colon P \to P'$ between the two projective covers of M_R, and g_0 restricts to a morphism $g_1\colon C \to C'$, which is an isomorphism as well. By Lemma 2.3, we then have that f_1 is an epimorphism if and only if $g_1^{-1} f'_1 g_1$ is an epimorphism, and this is an epimorphism if and only if f'_1 is an epimorphism.

By Theorem 2.5, couniformly presented modules have semilocal endomorphism ring, hence cancel from direct sums [5, Corollary 4.6].

Remark 2.7. When the base ring R is commutative, the endomorphism ring of the cyclic R-module $M_R \cong eR/C$ is the ring $eR/C = eRe/C$, which is a local ring with maximal ideal $eJ(R)e/C$. Hence, in this case, $K \supseteq I$. This inclusion can be proper. For instance, let R be a commutative valuation domain of Krull dimension ≥ 2, that is, a valuation domain with at least three prime ideals $0 \subset P \subset J(R)$, and consider the couniformly presented module R/P. If $r \in J(R) \setminus P$, then $r + P \in K = J(R)/P$, but $r + P \notin I$ because $rP = P$ (for every $p \in P$, $rR \supseteq P \supseteq pR$, so that $p = rs$ for some $s \in R$, and $s \in P$ because $p \in P$ and $r \notin P$.)

3. Epigeny class and lower part

Recall that if A and B are two modules, we say that A and B *have the same epigeny class*, and write $[A]_e = [B]_e$, if there exist an epimorphism $A \to B$ and an epimorphism $B \to A$; cf. [4]. If M_R and M'_R are two couniformly presented modules with couniform presentations $0 \to C_R \to P_R \to M_R \to 0$ and $0 \to C'_R \to P'_R \to M'_R \to 0$, we say that M_R and M'_R *have the same lower part*, and we write $[M_R]_\ell = [M'_R]_\ell$, if there are two homomorphisms $f_0\colon P_R \to P'_R$ and $f'_0\colon P'_R \to P_R$

such that $f_0(C_R) = C'_R$ and $f'_0(C'_R) = C_R$. In particular, if M_R and M'_R have the same lower part, then C_R and C'_R have the same epigeny class.

Notice the duality between this notion of having the same lower part, and the definition of having the same upper part given in [2]. For any right R-module A, let $E(A)$ denote the injective envelope of A. Two modules A and B are said to *have the same upper part* if there exist a homomorphism $f_0\colon E(A) \to E(B)$ and a homomorphism $f'_0\colon E(B) \to E(A)$ such that $f_0^{-1}(B) = A$ and $f_0'^{-1}(A) = B$. We write $[A]_u = [B]_u$ if A and B have the same upper part. Also notice that if M_R and M'_R are couniformly presented modules with couniform presentations $0 \to C_R \to P_R \to M_R \to 0$ and $0 \to C'_R \to P'_R \to M'_R \to 0$, then there are idempotents $e, e' \in R$ with $P_R \cong eR$ and $P'_R \cong e'R$. If we assume $P_R = eR$ and $P'_R = e'R$, C, C' right ideals of R contained in $eR, e'R$ respectively, and $M_R = eR/C, M'_R = e'R/C'$, then M_R and M'_R have the same lower part if and only if there exist $r, s \in R$ such that $rC = C'$ and $sC' = C$. Also notice that our definition of having the same lower part for arbitrary couniformly presented modules over arbitrary rings extends the definition of having the same lower part given in [1] for cyclically presented modules over local rings. Moreover, since the ideal I of $E := \mathrm{End}(M_R)$ in the statement of Theorem 2.5 does not depend on the couniform presentation of M_R (Remark 2.6), our notion of having the same lower part is well defined.

Remark 3.1. Let M_R and M'_R be couniformly presented modules. It is easily seen that M_R and M'_R have the same lower part if and only if there exists an endomorphism $f \in \mathrm{End}(M_R) \setminus I$ of M_R that factors through M'_R. Similarly, M_R and M'_R have the same epigeny class if and only if there exists an endomorphism $f \in \mathrm{End}(M_R) \setminus K$ of M_R that factors through M'_R. Here I and K are the completely prime ideals of $\mathrm{End}(M_R)$ defined in the statement of Theorem 2.5.

Lemma 3.2. *Let M_R and N_R be couniformly presented right modules over a ring R. Then $M_R \cong N_R$ if and only if $[M_R]_\ell = [N_R]_\ell$ and $[M_R]_e = [N_R]_e$.*

Proof. Let $E := \mathrm{End}(M_R)$ and let I and K be the ideals of E as in Theorem 2.5. Assume that M_R and M'_R have the same epigeny class and the same lower part. Then there exist $f \in E \setminus K$ and $g \in E \setminus I$ such that both f and g factor through M'_R. If either f or g is an automorphism, it follows that M_R is isomorphic to a non-zero direct summand of M'_R, which is indecomposable, thus $M_R \cong M'_R$. Assuming that f and g are not automorphisms, we have $f \in I \setminus K$ and $g \in K \setminus I$, hence $f + g$ is an automorphism of M_R that factors through $M'_R \oplus M'_R$. By [3, Lemma 2.3], it follows that M_R is isomorphic to a direct summand of M'_R, thus also in this case we have $M_R \cong M'_R$.

The converse is obvious. □

4. Weak Krull-Schmidt Theorem

Proposition 4.1. *Let $M, N_1, \dots, N_n$ $(n \geq 2)$ be $n+1$ couniformly presented right R-modules. Suppose that M is a direct summand of $N_1 \oplus \cdots \oplus N_n$ and that $M \not\cong N_i$ for all $i = 1, \dots, n$. Then there are two distinct indices $i, j = 1, \dots, n$ such that $[M]_\ell = [N_i]_\ell$ and $[M]_e = [N_j]_e$.*

Proof. Since M is a direct summand of $N_1 \oplus \cdots \oplus N_n$, with the obvious notation for the canonical mappings, we have $1_M = \pi_M \iota_M = \sum_{k=1}^{n} \pi_M \iota_k \pi_k \iota_M$. Let $E := \mathrm{End}(M_R)$ be the endomorphism ring of M_R and let I and K be the ideals of E as in Theorem 2.5. There exist indices i and j such that $\pi_M \iota_i \pi_i \iota_M \in E \setminus I$ and $\pi_M \iota_j \pi_j \iota_M \in E \setminus K$. This implies that $[M]_\ell = [N_i]_\ell$ and $[M]_e = [N_j]_e$. Moreover, $i \neq j$ otherwise M would be isomorphic to $N_i = N_j$, which is not. □

Lemma 4.2. *Let M, M', M'' be couniformly presented modules over an arbitrary ring R and assume $[M]_\ell = [M']_\ell$ and $[M]_e = [M'']_e$. Then*

(a) *$M \oplus D \cong M' \oplus M''$ for some R-module D;*
(b) *the module D in* (a) *is unique up to isomorphism and is couniformly presented;*
(c) *$[D]_\ell = [M'']_\ell$ and $[D]_e = [M']_e$.*

Proof. (a) Let $E = \mathrm{End}_R(M)$ and let I and K be the ideals of E as in Theorem 2.5. There exist an endomorphism $f \in E \setminus I$ which factors through M' and an endomorphism $g \in E \setminus K$ which factors through M''. If either f or g is an automorphism, then $M \cong M'$ or $M \cong M''$, thus (a) clearly holds with $D = M''$ and $D = M'$ respectively. We can thus assume $f \in K \setminus I$ and $g \in I \setminus K$. It then follows that $f + g$ is an automorphism of M which factors through $M' \oplus M''$, thus (a) holds also in this case.

(b) If $M \oplus D \cong M' \oplus M''$ and $M \oplus D' \cong M' \oplus M''$, then $M \oplus D \cong M \oplus D'$, so that $D \cong D'$ because the endomorphism ring of M is semilocal, hence M cancels from direct sums [5, Corollary 4.6]. This shows that the complement D is unique up to isomorphism.

Taking the dual Goldie dimension of both sides of $S := M \oplus D \cong M' \oplus M''$, we get that D is a couniform module. Considering the canonical projection π_D of S onto D, we have that D is a homomorphic image of M' or of M''. In fact, $D = \pi_D(S) = \pi_D(M' + M'') \subseteq \pi_D(M') + \pi_D(M'') \subseteq D$, hence $D = \pi_D(M') + \pi_D(M'')$, and the claim holds because D is couniform. Without loss of generality we can assume that D is a homomorphic image of M', thus it is a homomorphic image of the projective cover P' of M'. We then have a short exact sequence $0 \to A \to P' \to D \to 0$, which is a couniform presentation of D provided that we prove that A is couniform. With the usual notation for the couniform presentations of M, M', M'', consider the two short exact sequences $0 \to C \oplus A \to P \oplus P' \to M \oplus D \cong S \to 0$ and $0 \to C' \oplus C'' \to P' \oplus P'' \to M' \oplus M'' \cong S \to 0$. By Schanuel's Lemma we have $C \oplus A \oplus P' \oplus P'' \cong C' \oplus C'' \oplus P \oplus P'$. Taking the dual Goldie dimension of both sides, we see that A is couniform.

(c) If $D \cong M'$, then $M \cong M''$ by cancellation, so that $[D]_e = [M']_e$ and $[D]_\ell = [M']_\ell = [M]_\ell = [M'']_\ell$, as required. The case $D \cong M''$ is exactly the same. So we can assume that $D \ncong M'$ and $D \ncong M''$. By Proposition 4.1, either $[D]_\ell = [M']_\ell$ and $[D]_e = [M'']_e$ or $[D]_\ell = [M'']_\ell$ and $[D]_e = [M']_e$. In the first case, $D \cong M$ so that D, M, M', M'' are all isomorphic, which is not. Thus the second case holds, as required. □

Theorem 4.3. (Weak Krull-Schmidt Theorem) *Let $M_1, \dots, M_n, N_1, \dots, N_t$ be couniformly presented right R-modules. Then the modules $M_1 \oplus \cdots \oplus M_n$ and $N_1 \oplus \cdots \oplus N_t$ are isomorphic if and only if $n = t$ and there are two permutations σ, τ of $\{1, 2, \dots, n\}$ with $[M_i]_\ell = [N_{\sigma(i)}]_\ell$ and $[M_i]_e = [N_{\tau(i)}]_e$ for all $i = 1, \dots, n$.*

Proof. ($\Rightarrow$) Assume $M_1 \oplus \cdots \oplus M_n \cong N_1 \oplus \cdots \oplus N_t$. Comparing the dual Goldie dimension of the two sides, we get $n = t$.

We will prove by induction on n the existence of the permutations σ and τ, the case $n = 1$ being trivial. If $M_i \cong N_j$ for some indices i and j, we can cancel M_i and N_j (which is possible because their endomorphism ring is semilocal) and conclude by induction. Therefore we may assume that $M_i \ncong N_j$ for all indices i and j.

Since M_1 is isomorphic to a direct summand of $N_1 \oplus \cdots \oplus N_t$, Proposition 4.1 implies the existence of two distinct indices $i, j = 1, 2, \dots, n$ such that $[M_1]_\ell = [N_i]_\ell$ and $[M_1]_e = [N_j]_e$. For the sake of simplicity, without loss of generality, we can assume that $i = 1$ and $j = 2$. By Lemma 4.2 applied to the three couniformly presented modules M_1, N_1, N_2, we can find a couniformly presented module D, unique up to isomorphism, such that $M_1 \oplus D \cong N_1 \oplus N_2$, $[D]_\ell = [N_2]_\ell$ and $[D]_e = [N_1]_e$. Thus $M_1 \oplus \cdots \oplus M_n \cong N_1 \oplus \cdots \oplus N_n \cong M_1 \oplus D \oplus N_3 \oplus \cdots \oplus N_n$. Cancel M_1, getting that $M_2 \oplus \cdots \oplus M_n$ is isomorphic to $D \oplus N_3 \oplus \cdots \oplus N_n$. Now we deal with direct sums of $n - 1$ couniformly presented modules, so that we can again conclude by induction.

($\Leftarrow$) The statement is trivial for $n = t = 1$ by Lemma 3.2, and we proceed by induction again. Assume that $M_1, \dots, M_n, N_1, \dots, N_n$ are couniformly presented right R-modules and that there are two permutations σ, τ of $\{1, 2, \dots, n\}$ with $[M_i]_\ell = [N_{\sigma(i)}]_\ell$ and $[M_i]_e = [N_{\tau(i)}]_e$ for every $i = 1, \dots, n$. If $\sigma(1) = \tau(1)$, then $M_1 \cong N_{\sigma(1)}$. Thus σ and τ induce two bijections $\{2, 3, \dots, n\} \to \{1, 2, \dots, n\} \setminus \{\sigma(1)\}$, with the same properties as σ and τ, so that, by induction, $M_2 \oplus \cdots \oplus M_n$ is isomorphic to the direct sum of the N_j's with $j \neq \sigma(1)$, from which it clearly follows that $M_1 \oplus \cdots \oplus M_n \cong N_1 \oplus \cdots \oplus N_n$.

Thus we can suppose $\sigma(1) \neq \tau(1)$. By Lemma 4.2, there exists a couniformly presented module M_0, unique up to isomorphism, such that $M_0 \oplus M_1 \cong N_{\sigma(1)} \oplus N_{\tau(1)}$, $[M_0]_\ell = [N_{\tau(1)}]_\ell$ and $[M_0]_e = [N_{\sigma(1)}]_e$. That is, the modules M_0, M_1 and the modules $N_{\sigma(1)}, N_{\tau(1)}$ have the same lower parts and the same epigeny classes, counting multiplicities. The modules $M_0, M_1, \dots, M_n$ and the modules M_0, N_1, $\dots$, N_n have the same lower parts and the same epigeny classes as well, so that the modules M_2, M_3, $\dots$, M_n and the modules M_0, N_1, $\dots$, $\widehat{N_{\sigma(1)}}$, $\dots$, $\widehat{N_{\tau(1)}}$, $\dots$, N_n have the same lower parts and the same epigeny classes. By the inductive hypothesis, $M_2 \oplus M_3 \oplus \cdots \oplus M_n$ and the direct sum of the modules M_0 and N_j with j different from $\sigma(1)$ and $\tau(1)$ are isomorphic. Thus $M_0 \oplus N_1 \oplus \cdots \oplus N_n \cong M_2 \oplus \cdots \oplus M_n \oplus N_{\sigma(1)} \oplus N_{\tau(1)} \cong M_0 \oplus M_1 \oplus M_2 \oplus \cdots \oplus M_n$. Cancelling the module M_0, we obtain that $N_1 \oplus N_2 \oplus \cdots \oplus N_n \cong M_1 \oplus M_2 \oplus \cdots \oplus M_n$, as desired. □

5. Kernels of morphisms between indecomposable injective modules

In [2] it has been proved that theorems similar to Theorems 2.5 and 4.3 hold for kernels of morphisms between indecomposable injective modules (equivalently, uniform injective modules). More precisely, let R be an arbitrary ring. Recall that two right R-modules A and B are said to belong to the same *monogeny class* if there exist a monomorphism $A \to B$ and a monomorphism $B \to A$. In this case, we write $[A]_m = [B]_m$. If E_1, E_2, E_1', E_2' are uniform injective right R-modules and $\varphi\colon E_1 \to E_2$, $\varphi'\colon E_1' \to E_2'$ are arbitrary morphisms, then $\ker\varphi \cong \ker\varphi'$ if and only if $[\ker\varphi]_m = [\ker\varphi']_m$ and $[\ker\varphi]_u = [\ker\varphi']_u$. If $\varphi\colon E_1 \to E_2$ is a non-zero non-injective morphism, every morphism $f\colon \ker\varphi \to \ker\varphi'$ extends to a morphism $f_1\colon E_1 \to E_1'$. Any maximal ideal of $\mathrm{End}_R(\ker\varphi)$ is equal to either the completely prime ideal $\{\, f \in \mathrm{End}_R(\ker\varphi) \mid f \text{ is not injective} \,\}$ or the completely prime ideal $\{\, f \in \mathrm{End}_R(\ker\varphi) \mid f_1^{-1}(\ker\varphi) \supsetneqq \ker\varphi \,\}$. Then $\mathrm{End}_R(\ker\varphi)$ is either a local ring or has exactly two maximal ideals.

If $\varphi_i\colon E_{i,1} \to E_{i,2}$ $(i = 1, 2, \dots, n)$ and $\varphi_j'\colon E_{j,1}' \to E_{j,2}'$ $(j = 1, 2, \dots, t)$ are non-injective morphisms between uniform injective modules $E_{i,1}, E_{i,2}, E_{j,1}', E_{j,2}'$ over an arbitrary ring R, then $\oplus_{i=1}^n \ker\varphi_i \cong \oplus_{j=1}^t \ker\varphi_j'$ if and only if $n = t$ and there exist two permutations σ, τ of $\{1, 2, \dots, n\}$ such that $[\ker\varphi_i]_m = [\ker\varphi_{\sigma(i)}']_m$ and $[\ker\varphi_i]_u = [\ker\varphi_{\tau(i)}']_u$ for every $i = 1, 2, \dots, n$. The proof of all these results can be found in [2]. Let us see the relation between this theory of finite direct sums of kernels of morphisms between uniform injective modules and the the theory of couniformly presented modules developed in the previous sections.

Let R be a fixed ring. Fix a set $\{\, E_\lambda \mid \lambda \in \Lambda \,\}$ of representatives up to isomorphism of the uniform injective right R-modules. Set $E_R := E(\oplus_{\lambda\in\Lambda} E_\lambda)$ and $S := \mathrm{End}(E_R)$, so that ${}_SE_R$ turns out to be an S-R-bimodule. Consider the E-dual functors, i.e., the pair of contravariant additive functors

$$H := {}_S\mathrm{Hom}_R(-, E)\colon \text{Mod-}R \to S\text{-Mod}$$
$$H' := \mathrm{Hom}_S(-, E)_R\colon S\text{-Mod} \to \text{Mod-}R.$$

If K_R is the kernel of a morphism between uniform injective R-modules, there is an exact sequence $0 \to K_R \to E_\lambda \xrightarrow{\varphi} E_\mu$ for suitable $\lambda, \mu \in \Lambda$. Applying the exact functor H, we get an exact sequence $H(E_\mu) \xrightarrow{H(\varphi)} H(E_\lambda) \to H(K_R) \to 0$. Now for each index $\lambda \in \Lambda$ there is a direct-sum decomposition $E_R = E_\lambda \oplus E(\oplus_{\mu\neq\lambda} E_\mu)$, so that there exist a monomorphism $\iota_\lambda\colon E_\lambda \to E_R$ and an epimorphism $\pi_\lambda\colon E_R \to E_\lambda$ such that $\pi_\lambda\iota_\lambda = 1_{E_\lambda}$ and $e_\lambda := \iota_\lambda\pi_\lambda \in S$

is an idempotent endomorphism of E_R. Hence, for every $\lambda \in \Lambda$, we have that $e_\lambda E_R = \iota_\lambda E_\lambda$, and ${}_SH(E_\lambda) = \mathrm{Hom}_R(E_\lambda, E_R) \cong Se_\lambda$ is a cyclic projective left S-module whose endomorphism ring $\mathrm{End}({}_SSe_\lambda)$ is isomorphic to $e_\lambda Se_\lambda \cong \mathrm{End}(E_\lambda)$, which is a local ring. Thus ${}_SH(E_\lambda) \cong Se_\lambda$ is a couniform projective left S-module.

Now $H(\varphi)\colon {}_SH(E_\mu) \to {}_SH(E_\lambda)$ corresponds to the right multiplication $Se_\mu \to Se_\lambda$ by the endomorphism $\iota_\mu\varphi\pi_\lambda$ of E_R. Thus the finitely presented S-module ${}_SH(K_R) \cong Se_\lambda / S\iota_\mu\varphi\pi_\lambda$ is the cokernel of a morphism between couniform projective left S-modules. In particular, if $\varphi \neq 0$ and φ is non-injective, ${}_SH(K_R)$ is a couniformly presented S-module.

Conversely, let us prove that every couniform projective left S-module is isomorphic to Se_λ for some $\lambda \in \Lambda$. To see this, recall that, by Lemma 2.1, a couniform projective left S-module is isomorphic to Se for some idempotent $e \in S$ with eSe a local ring. Hence the direct summand eE_R of E_R is an injective right R-module, necessarily indecomposable because $eSe \cong \mathrm{End}_R(eE_R)$ is local. In the spectral category Spec-R of Mod-R, E_R is a semisimple object (it is the coproduct of the simple objects E_λ) and eE_R is a simple subobject of E_R [8, Ch. V, §7]. It follows that eE_R is isomorphic to E_λ, for some $\lambda \in \Lambda$, in the category Spec-R. Thus $eE_R \cong E_\lambda$ in Mod-R, so that $Se \cong Se_\lambda$ in S-Mod, which is what we wanted to prove.

If ${}_SC$ is the cokernel of a morphism between couniform projective left S-modules, there is an exact sequence $Se_\lambda \xrightarrow{f} Se_\mu \to {}_SC \to 0$ for suitable $\lambda, \mu \in \Lambda$. If we apply the left exact functor H', we get an exact sequence $0 \to H'({}_SC) \to H'(Se_\mu) \xrightarrow{H'(f)} H'(Se_\lambda)$. Now $H'(Se_\lambda)_R = \mathrm{Hom}_S(Se_\lambda, {}_SE) \cong e_\lambda E = E_\lambda$ is a uniform injective R-module. Thus $H'({}_SC)$ is the kernel of a morphism between uniform injective R-modules.

It is easily seen that $H'H(K_R) \cong K_R$ and $HH'({}_SC) \cong {}_SC$ canonically. Since H and H' are additive functors, they respect finite direct sums. It follows easily that:

Theorem 5.1. *The functors H, H' define inverse categorical dualities*

$$H\colon \mathcal{K} \to \mathcal{C} \qquad \text{and} \qquad H'\colon \mathcal{C} \to \mathcal{K}$$

between the full subcategory $\mathcal{K}$ of Mod-R *whose objects are all finite direct sums of kernels of morphisms between uniform injective right R-modules and the full subcategory $\mathcal{C}$ of* S-Mod *whose objects are all finite direct sums of cokernels of morphisms between couniform projective left S-modules.*

Proposition 5.2. *Let K_R and K'_R be kernels of non-zero non-injective morphisms between uniform injective right R-modules. Then:*

(a) $[K_R]_m = [K'_R]_m$ *if and only if* $[H(K_R)]_e = [H(K'_R)]_e$.
(b) $[K_R]_u = [K'_R]_u$ *if and only if* $[H(K_R)]_\ell = [H(K'_R)]_\ell$.

Proof. Let K_R be the kernel of the non-zero non-injective morphism $\varphi\colon E_\lambda \to E_\mu$. In view of Remark 3.1 and Theorem 5.1, it suffices to show that if f is an endomorphism of K_R, then

(a) f is in the completely prime ideal of $\mathrm{End}(K_R)$ consisting of all non-injective endomorphisms of K_R if and only if $H(f)$ is in the completely prime ideal of $\mathrm{End}(H(K_R))$ consisting of all non-surjective endomorphisms of $H(K_R)$;

(b) f is in the completely prime ideal of $\mathrm{End}(K_R)$ consisting of all the endomorphisms g of K_R with $g_1^{-1}(K_R) \supsetneqq K_R$ if and only if $H(f)$ is in the completely prime ideal I of $\mathrm{End}(H(K_R))$ consisting of all the endomorphisms h of $H(K_R)$ with $h_1\colon {}_SC \to {}_SC$ non-surjective.

(In (b) we have that $g\colon K_R \to K_R$, $g_1\colon E(K_R) \to E(K_R)$, the endomorphism h of $H(K_R)$ lifts to $h_0\colon {}_SP \to {}_SP$, where ${}_SP$ is the projective cover of $H(K_R)$, and $h_1\colon {}_SC \to {}_SC$ is the restriction of h_0 to the kernel ${}_SC$ of the epimorphism ${}_SP \to H(K_R)$.)

Now (a) follows immediately from the fact that the contravariant functor $H(-) = \mathrm{Hom}_R(-, E_R)$ is exact and E_R is an injective cogenerator. For (b), apply the contravariant exact functor H to the commutative diagram with exact rows

$$\begin{array}{ccccccc} 0 & \to & K_R & \to & E_\lambda & \to & E_\mu \\ & & {\scriptstyle f}\downarrow & & \downarrow{\scriptstyle f_1} & & \downarrow{\scriptstyle f_2} \\ 0 & \to & K_R & \to & E_\lambda & \to & E_\mu. \end{array}$$

getting a commutative diagram with exact rows

$$\begin{array}{ccccccc} H(E_\mu) = Se_\mu & \to & H(E_\lambda) = Se_\lambda = {}_SP & \to & H(K_R) & \to & 0 \\ {\scriptstyle H(f_2)}\downarrow & & \downarrow{\scriptstyle H(f_1)} & & \downarrow{\scriptstyle H(f)} & & \\ H(E_\mu) = Se_\mu & \to & H(E_\lambda) = Se_\lambda = {}_SP & \to & H(K_R) & \to & 0. \end{array}$$

Then f is in the completely prime ideal of $\mathrm{End}(K_R)$ consisting of all the endomorphisms g of K_R with $g_1^{-1}(K_R) \supsetneqq K_R$ if and only if f_2 is not injective, if and only if $H(f_2)$ is not surjective, if and only if $H(f)_1\colon {}_SC \to {}_SC$ is not surjective, if and only if $H(f)$ belongs to I. □

6. A further duality between epigeny classes and monogeny classes

In Section 5, we saw that monogeny class and epigeny class (and lower part and upper part) are related by a duality between suitable categories of modules: the category of kernels of morphisms between uniform injective modules and the category of cokernels of morphisms between couniform projective modules. In [1, Proposition 7.1] it was shown that, for cyclically presented modules over local rings, lower part and epigeny class are related by the Auslander-Bridger transpose, which also can be seen as a duality between suitable categories. In this final section, we will show that there is a similar relation between monogeny class and epigeny class in the case of suitable categories of uniserial modules.

Recall that if ${}_SA$ and ${}_SB$ are left modules over a ring S, ${}_SA$ is said to be *cogenerated by* ${}_SB$ if ${}_SA$ is isomorphic to a submodule of a direct product of copies of ${}_SB$. Equivalently, if for every non-zero $a \in {}_SA$ there exists a morphism $\varphi\colon {}_SA \to {}_SB$ such that $\varphi(a) \neq 0$.

Let R be a ring. Fix a set $\{ E_\lambda \mid \lambda \in \Lambda \}$ of representatives up to isomorphism of all injective right R-modules that are injective envelopes of non-zero uniserial R-modules. Set $E_R := E(\oplus_{\lambda\in\Lambda} E_\lambda)$ and $S := \mathrm{End}(E_R)$. Then ${}_SE_R$ is an S-R-bimodule and

$$\mathrm{Hom}(-, {}_SE_R)\colon \text{Mod-}R \to S\text{-Mod}$$

is an additive contravariant exact functor. Let $\mathcal{C}_R$ denote the full subcategory of Mod-R whose objects are all serial right R-modules of finite Goldie dimension. Let ${}_S\mathcal{C}'$ be the full subcategory of S-Mod whose objects are all finite direct sums of uniserial left S-modules with a projective cover and cogenerated by ${}_SE$. Notice that if a non-zero uniserial module U has a projective cover P, then P is a couniform module by Lemma 2.2, so that, in particular, P, hence U, are cyclic modules.

Proposition 6.1. *The functor* $\mathrm{Hom}(-, {}_SE_R)\colon \text{Mod-}R \to S\text{-Mod}$ *induces a categorical duality between* $\mathcal{C}_R$ *and* ${}_S\mathcal{C}'$.

Proof. Since the functor $\mathrm{Hom}(-, {}_SE_R)$ is additive, it respects finite direct sums. Thus it suffices to show that $\mathrm{Hom}(-, {}_SE_R)$ induces a duality between uniserial right R-modules and the uniserial left S-modules with a projective cover and cogenerated by ${}_SE$.

Let us prove that if $U_R \neq 0$ is uniserial, then $\mathrm{Hom}(U_R, {}_SE_R)$ is a uniserial left S-module with a projective cover and is cogenerated by ${}_SE$. We claim that if $\varphi, \varphi' \in \mathrm{Hom}(U_R, {}_SE_R)$ and $\ker\varphi \subseteq \ker\varphi'$, then $S\varphi \supseteq S\varphi'$. To prove the claim, assume $\ker\varphi \subseteq \ker\varphi'$. Let $\pi\colon U \to U/\ker\varphi$, $\pi'\colon U \to U/\ker\varphi'$ and $p\colon U/\ker\varphi \to U/\ker\varphi'$ be the canonical projections, so that $\pi' = p\pi$. Let $\overline{\varphi}\colon U/\ker\varphi \to E_R$ and $\overline{\varphi'}\colon U/\ker\varphi' \to E_R$ be the injective right R-module morphisms induced by φ and φ', respectively. We have $\overline{\varphi'}p = s\overline{\varphi}$ for some $s \in S$ because E_R is injective, so that $\varphi' = \overline{\varphi'}\pi' = \overline{\varphi'}p\pi = s\overline{\varphi}\pi = s\varphi$. This proves the claim. Hence, for every $\varphi, \varphi' \in \mathrm{Hom}(U_R, {}_SE_R)$, we have that $\ker\varphi \subseteq \ker\varphi'$ if and only if $S\varphi \supseteq S\varphi'$. Thus, U_R uniserial implies that $\mathrm{Hom}(U_R, {}_SE_R)$ is a uniserial left S-module.

We will now determine the projective cover of $\mathrm{Hom}(U_R, {}_SE_R)$. Since $U_R \neq 0$, there exists a unique $\lambda \in \Lambda$ with E_λ an injective envelope of U_R. Let $\iota\colon U_R \to E_\lambda$ be a fixed essential monomorphism. Applying the functor $\mathrm{Hom}(-, {}_SE_R)$, we get an S-module epimorphism ${}_S\mathrm{Hom}(E_\lambda, E_R) \overset{{}_S\mathrm{Hom}(\iota, E_R)}{\longrightarrow} {}_S\mathrm{Hom}(U_R, E_R)$. Since $E_R \cong E_\lambda \oplus E(\oplus_{\mu\neq\lambda} E_\mu)$, there exists a monomorphism $\iota_\lambda\colon E_\lambda \to E_R$ and an epimorphism $\pi_\lambda\colon E_R \to E_\lambda$ such that $\pi_\lambda\iota_\lambda = 1_{E_\lambda}$ and $e_\lambda := \iota_\lambda\pi_\lambda \in S$ is an idempotent endomorphism of E_R. Hence we have an isomorphism ${}_S\mathrm{Hom}(E_\lambda, E_R) \to Se_\lambda$ given by $f \mapsto f\pi_\lambda$. Moreover, we have an isomorphism $\mathrm{End}_R(E_\lambda) \to e_\lambda Se_\lambda$ given by $f \mapsto \iota_\lambda f\pi_\lambda$, so that $e_\lambda Se_\lambda \cong \mathrm{End}_S(Se_\lambda)$ is a local ring. Thus Se_λ is a couniform projective left S-module by Lemma 2.1. Finally, $\mathrm{Hom}(U_R, {}_SE_R) \neq 0$ because it contains the embedding $\iota_\lambda\iota\colon U_R \to E_R$, and ${}_S\mathrm{Hom}(\iota, E_R)$ is a projective cover of $\mathrm{Hom}(U_R, {}_SE_R)$.

Finally, $\mathrm{Hom}(U_R, {}_SE_R)$ is cogenerated by ${}_SE$ because it is an S-submodule of ${}_SE^U$, the direct product of a family of copies of ${}_SE$ indexed in the set U.

Conversely, let us prove that every uniserial left S-module with a projective cover and cogenerated by ${}_SE$ is isomorphic to $\mathrm{Hom}(U_R, {}_SE_R)$ for some uniserial module U_R.

To this end, we claim that any couniform projective left S-module is isomorphic to Se_λ for some $\lambda \in \Lambda$. To see this, notice that, by Lemma 2.1, a couniform projective left S-module is isomorphic to Se for some idempotent $e \in S$ with eSe a local ring. Hence the direct summand eE_R of E_R is an injective right R-module, necessarily indecomposable because $eSe \cong \mathrm{End}_R(eE_R)$ is local. In the spectral category Spec-R of Mod-R, E_R is a semisimple object, namely the coproduct of the simple objects E_λ, and eE_R is a simple subobject of E_R [8, Ch. V, §7]. It follows that eE_R is isomorphic to E_λ, for some $\lambda \in \Lambda$, in the category Spec-R. Thus $eE_R \cong E_\lambda$ in Mod-R, so that $Se \cong Se_\lambda$ in S-Mod, which is what we wanted to show.

Hence couniform left S-modules with a projective cover are isomorphic to $Se_\lambda/{}_ST$ for some left ideal ${}_ST$ of S properly contained in Se_λ. Thus, assume we have a non-zero uniserial module $Se_\lambda/{}_ST$ cogenerated by ${}_SE$ and let us prove that there exists a uniserial module U_R with $\mathrm{Hom}(U_R, {}_SE_R) \cong Se_\lambda/{}_ST$. We will show that the required uniserial module is $U_R := \{\, x \in \iota_\lambda(E_\lambda) \mid tx = 0 \text{ for every } t \in T\,\}$ (notice that $T \subseteq S$ and $\iota_\lambda(E_\lambda) \subseteq E_R$, so that the product tx is defined).

Let us first prove that the submodule U_R of $\iota_\lambda(E_\lambda)$ is uniserial. Let x, y be any two elements of U_R. Then $\mathrm{l.ann}_S(x)$ and $\mathrm{l.ann}_S(y)$ are two left ideals of S that contain $1-e_\lambda$ and T. Thus $\mathrm{l.ann}_S(x)/(S(1-e_\lambda)+T)$ and $\mathrm{l.ann}_S(y)/(S(1-e_\lambda)+T)$ are two submodules of $S/(S(1-e_\lambda)+T) \cong Se_\lambda/{}_ST$, which is a uniserial S-module. Hence $\mathrm{l.ann}_S(x)/(S(1-e_\lambda)+T)$ and $\mathrm{l.ann}_S(y)/(S(1-e_\lambda)+T)$ are comparable, so that $\mathrm{l.ann}_S(x)$ and $\mathrm{l.ann}_S(y)$ are comparable as well. Without loss of generality, we can suppose $\mathrm{l.ann}_S(x) \subseteq \mathrm{l.ann}_S(y)$. Let us prove that this implies $yR \subseteq xR$. Assume the contrary, that is, assume $yR \not\subseteq xR$. Then $(yR + xR)/xR$ is a non-zero module. Now E_R is an injective cogenerator because simple modules are non-zero uniserial modules, so that there exists a morphism $\varphi\colon yR + xR \to E_R$ with $\varphi(x) = 0$ and $\varphi(y) \neq 0$. The R-module morphism φ extends to an element $s \in S$, and $sx = 0$, $sy \neq 0$. This contradicts $\mathrm{l.ann}_S(x) \subseteq \mathrm{l.ann}_S(y)$. Hence U_R is a uniserial submodule of $\iota_\lambda(E_\lambda)$.

We finally prove that $\mathrm{Hom}_R(U_R, {}_SE_R) \cong Se_\lambda/{}_ST$. We have a restriction morphism $\rho\colon Se_\lambda \to \mathrm{Hom}(U_R, {}_SE_R)$ because $U_R \subseteq \iota_\lambda(E_\lambda) \subseteq E_R$, and ρ is an epimorphism because E_R is injective. It remains to show that $\ker\rho = T$. From $TU_R = 0$, it follows that $\ker\rho = \mathrm{l.ann}_S(U_R) \cap Se_\lambda \supseteq T$. Conversely, if $se_\lambda \in Se_\lambda$ but $se_\lambda \notin T$, then $se_\lambda + {}_ST \neq 0$. As $Se_\lambda/{}_ST$ is cogenerated by ${}_SE$, there is an S-module morphism $Se_\lambda \to {}_SE$ that maps ${}_ST$ to 0 and se_λ to a non-zero element of ${}_SE$. S-module morphisms $Se_\lambda \to {}_SE$ are given by right multiplication by elements of $e_\lambda E = \iota_\lambda(E_\lambda)$. Hence there exists $x \in U_R$ with $se_\lambda x \neq 0$, that is, $\rho(se_\lambda)(x) \neq 0$. Thus $se_\lambda \notin \ker\rho$, which proves that $\ker\rho = T$.

We now show that the functor $\mathrm{Hom}(-, {}_SE_R)$ is full and faithful. Let U_1 and U_2 be non-zero uniserial modules. For each $i = 1, 2$, there exist an essential monomorphism $\iota_i\colon U_i \to E_{\lambda_i}$ and an epimorphism $\rho_i\colon Se_{\lambda_i} \to \mathrm{Hom}(U_i, E_R)$.

Notice that ρ_1 and ρ_2 are projective covers. Any S-module morphism $f\colon {}_S\mathrm{Hom}(U_1, E_R) \to {}_S\mathrm{Hom}(U_2, E_R)$ lifts to an S-module morphism g between the projective covers:

$$\begin{array}{ccccccccc} 0 & \to & \ker\rho_1 & \longrightarrow & Se_{\lambda_1} & \xrightarrow{\rho_1} & {}_S\mathrm{Hom}(U_1,E) & \to & 0 \\ & & \downarrow & & \downarrow g & & \downarrow f & & \\ 0 & \to & \ker\rho_2 & \longrightarrow & Se_{\lambda_2} & \xrightarrow{\rho_2} & {}_S\mathrm{Hom}(U_2,E) & \to & 0. \end{array} \tag{3}$$

Now g is right multiplication by some $e_{\lambda_1} s e_{\lambda_2} \in e_{\lambda_1} S e_{\lambda_2}$. Set $\alpha := \pi_{\lambda_1} s \iota_{\lambda_2} : E_{\lambda_2} \to E_{\lambda_1}$. Let us prove that $\alpha(\iota_2(U_2)) \subseteq \iota_1(U_1)$. Suppose not. Then there exists $y \in \iota_2(U_2)$ with $\alpha(y) \notin \iota_1(U_1)$. Since E_R is an injective cogenerator, there exists $t\colon E_{\lambda_1} \to E_R$ such that $t(\iota_1(U_1)) = 0$ and $t(\alpha(y)) \neq 0$, that is, $t \in Se_{\lambda_1}$, $t \in \ker\rho_1$ and $t\pi_{\lambda_1} s\iota_{\lambda_2}(y) \neq 0$. Hence $t\pi_{\lambda_1} s \iota_{\lambda_2} = g(t) \neq 0$. Thus $t \in \ker\rho_1$ and $g(t) \notin \ker\rho_2$, which contradicts the commutativity of diagram (3). Thus $\alpha(\iota_2(U_2)) \subseteq \iota_1(U_1)$, there exists $\beta\colon U_2 \to U_1$ with $\iota_1\beta = \alpha\iota_2$, and $\mathrm{Hom}(\beta, {}_SE_R) = f$, which proves that the functor $\mathrm{Hom}(-, {}_SE_R)$ is full.

It is also faithful, because if $\beta\colon U_2 \to U_1$ is a non-zero R-module morphism, there exists $u_2 \in U_2$ with $\beta(u_2) \neq 0$. But E_R is an injective cogenerator, so that there exists $\varphi\colon U_1 \to E_R$ with $\varphi(\beta(u_2)) \neq 0$. Then $\mathrm{Hom}(\beta, {}_SE_R)(\varphi) = \varphi\beta \neq 0$, which proves that $\mathrm{Hom}(-, {}_SE_R)$ is faithful. □

Corollary 6.2. *If U_R, U'_R are uniserial right R-modules, then:*

(a) *U_R and U'_R are in the same monogeny class if and only if the uniserial left S-modules $\mathrm{Hom}(U_R, {}_SE_R)$ and $\mathrm{Hom}(U'_R, {}_SE_R)$ are in the same epigeny class.*

(b) *U_R and U'_R are in the same epigeny class if and only if the uniserial left S-modules $\mathrm{Hom}(U_R, {}_SE_R)$ and $\mathrm{Hom}(U'_R, {}_SE_R)$ are in the same monogeny class.*

Proof. (a) The implication ($\Rightarrow$) follows from the fact that $\mathrm{Hom}(-, E_R)$ is an exact contravariant functor. For ($\Leftarrow$), every epimorphism $\mathrm{Hom}(U_R, E_R) \to \mathrm{Hom}(U'_R, E_R)$ is of the form $\mathrm{Hom}(\varphi, E_R)$ for some $\varphi\colon U'_R \to U_R$ by Proposition 6.1. Applying the exact functor $\mathrm{Hom}(-, E_R)$ to the exact sequence $0 \to \ker\varphi \to U'_R \xrightarrow{\varphi} U_R$, we see that $\mathrm{Hom}(\ker\varphi, E_R) = 0$, so that $\ker\varphi = 0$ because E_R is a cogenerator. Thus $\varphi\colon U'_R \to U_R$ is a monomorphism.

The proof for (b) is similar. □

References

[1] B. Amini, A. Amini and A. Facchini, *Equivalence of diagonal matrices over local rings*, J. Algebra **320** (2008), 1288–1310.

[2] A. Facchini, Ş. Ecevit, M.T. Koşan and T. Özdin, *Kernels of morphisms between indecomposable injective modules,* to appear in Glasg. Math. J., 2010.

[3] N.V. Dung and A. Facchini, *Direct summands of serial modules*, J. Pure Appl. Algebra **133** (1998), 93–106.

[4] A. Facchini, *Krull-Schmidt fails for serial modules,* Trans. Amer. Math. Soc. **348** (1996), 4561–4575.

[5] A. Facchini, "Module Theory. Endomorphism rings and direct sum decompositions in some classes of modules", Progress in Math. **167**, Birkhäuser Verlag, Basel, 1998.

[6] A. Facchini and D. Herbera, *Local morphisms and modules with a semilocal endomorphism ring,* Algebr. Represent. Theory **9** (2006), 403–422.

[7] A. Facchini and P. Příhoda, *Endomorphism rings with finitely many maximal right ideals*, to appear in Comm. Algebra, 2010.

[8] B. Stenström, *Rings of quotients*, Die Grundlehren der Mathematischen Wissenschaften, Band 217, Springer-Verlag, New York-Heidelberg, 1975.

Alberto Facchini and Nicola Girardi
Dipartimento di Matematica Pura e Applicata
Università di Padova
I-35131 Padova, Italy

Advances in Ring Theory
Trends in Mathematics, 165–204

Semiclassical Limits of Quantized Coordinate Rings

K.R. Goodearl

Dedicated to S.K. Jain on the occasion of his 70th birthday

Abstract. This paper offers an expository account of some ideas, methods, and conjectures concerning quantized coordinate rings and their semiclassical limits, with a particular focus on primitive ideal spaces. The semiclassical limit of a family of quantized coordinate rings of an affine algebraic variety V consists of the classical coordinate ring $\mathcal{O}(V)$ equipped with an associated Poisson structure. Conjectured relationships between primitive ideals of a generic quantized coordinate ring A and symplectic leaves in V (relative to a semiclassical limit Poisson structure on $\mathcal{O}(V)$) are discussed, as are breakdowns in the connections when the symplectic leaves are not algebraic. This prompts replacement of the differential-geometric concept of symplectic leaves with the algebraic concept of symplectic cores, and a reformulated conjecture is proposed: The primitive spectrum of A should be homeomorphic to the space of symplectic cores in V, and to the Poisson-primitive spectrum of $\mathcal{O}(V)$. Various examples, including both quantized coordinate rings and enveloping algebras of solvable Lie algebras, are analyzed to support the choice of symplectic cores to replace symplectic leaves.

Mathematics Subject Classification (2000). 16W35; 16D60, 17B63, 20G42.
Keywords. Quantized coordinate ring, semiclassical limit, Poisson algebra, symplectic leaf, symplectic core, Dixmier map.

0. Introduction

By now, the "Cheshire cat" description of quantum groups is well known – a quantum group is not a group at all, but something that remains when a group has faded away, leaving an algebra of functions behind. The appropriate functions depend on which category of group is under investigation. We concentrate here

This research was partially supported by National Science Foundation grant DMS-0800948.

on (affine) algebraic groups G, on which the natural functions of interest are the polynomial functions. These constitute the classical coordinate ring of G, which we denote $\mathcal{O}(G)$. (The group structure on G induces a Hopf algebra structure on $\mathcal{O}(G)$, but we shall not make use of that.) A *quantized coordinate ring* of G is, informally, a deformation of $\mathcal{O}(G)$, in the sense that it is an algebra with a set of generators patterned after those in $\mathcal{O}(G)$, but with a new multiplication that is typically noncommutative. Examples and references will be given in Section 1. We do not address the question of what properties are required to qualify an algebra as a quantized coordinate ring – this remains a fundamental open problem. Quantized coordinate rings have also been defined for a number of algebraic varieties other than algebraic groups, and our discussion will incorporate them as well.

Many parallels have been found between the structures of quantized and classical coordinate rings, and general principles for organizing and predicting such parallels are needed. The present paper concentrates on a circle of ideas and results focussed on ideal structure, particularly spaces of prime or primitive ideals. The theme/principle we follow, based on much previous work, can be stated this way:

> *The primitive ideals of a suitably generic quantized coordinate ring of an algebraic variety V should match subsets of V in some partition defined through the geometry of V and a Poisson structure obtained from a semiclassical limit process.*

Many of the terms just mentioned require explanations, which we will give over the course of the paper. Here we just mention that, in the above statement, "generic" refers to the assumption that suitable parameters in the construction of the quantized coordinate ring should be non-roots of unity.

To begin the story (omitting many definitions and details), we refer to the results of Soibelman and Vaksman [51, 45, 46], who studied the "standard" generic quantized coordinate rings of simple compact Lie groups K. They established a bijection between the irreducible $*$-representations of K (on Hilbert spaces) and the symplectic leaves in K (relative to a Poisson structure arising from the quantization). This amounts to a linkage between primitive ideals and symplectic leaves, a relationship which is a key ingredient of the Orbit Method from Lie theory. Informed by this principle, and inspired by the work of Soibelman and Vaksman, Hodges and Levasseur conjectured that similar bijections should exist for semisimple complex algebraic groups [22]. The case of $SL_2(\mathbb{C})$ being easy [22, Appendix], they first verified the conjecture for $SL_3(\mathbb{C})$ [op. cit.], and then for $SL_n(\mathbb{C})$ [23]. It was established for connected semisimple groups by Joseph [27] and by Hodges, Levasseur, and Toro [24]. In light of these achievements, it is natural to pose this conjecture for other classes of generic quantized coordinate rings. (It is easily seen that the above conjecture cannot hold for non-generic quantized coordinate rings. In such cases, the quantized coordinate rings are usually finitely generated modules over their centers, and they have far more primitive ideals than can be matched to symplectic leaves.)

In the specific cases just mentioned, the symplectic leaves turn out to be algebraic, in the sense that they are locally closed in the Zariski topology. Hodges, Levasseur, and Toro pointed out in [24] that symplectic leaves need not be algebraic for Poisson structures arising from multiparameter quantizations, and that the above conjecture cannot be expected to hold in such cases. We argue that it should not be surprising that the concept of symplectic leaves, which comes from differential geometry, is not always well suited for algebraic problems. Thus, symplectic leaves should be replaced by more algebraically defined objects. The notion of symplectic cores introduced by Brown and Gordon [6] fills the role well, up to the present state of knowledge; we will give evidence to buttress this statement.

Our aim here is to present an account of the above story, with introductions to and discussions of the relevant concepts. In particular, the tour will pass through way stations such as *quantized coordinate rings*, *semiclassical limits*, *Poisson structures*, *symplectic leaves*, the *Orbit Method*, *symplectic cores*, and the *Dixmier map*. By the end of the tour, we will be in purely algebraic territory, where we can formulate a conjecture that does not require any differential geometry (i.e., symplectic leaves). Namely:

> *If A is a generic quantized coordinate ring of an affine algebraic variety V over an algebraically closed field of characteristic zero, and if V is given the Poisson structure arising from an appropriate semiclassical limit, then the spaces of primitive ideals in A and symplectic cores in V, with their respective Zariski topologies, are homeomorphic.*

A parallel conjecture relates the prime and primitive spectra of A to the spaces of Poisson prime and Poisson-primitive ideals in $\mathcal{O}(V)$.

Fix a base field k throughout the paper; all algebras mentioned will be unital k-algebras. This field can be general at first, but then we will require it to have characteristic zero, and/or be algebraically closed. When discussing symplectic leaves, we restrict k to $\mathbb{R}$ or $\mathbb{C}$.

1. Quantized coordinate rings

We begin by recalling two basic examples, to clarify the idea that a quantized coordinate ring of an algebraic group (or variety) is, loosely speaking, a deformation of the classical coordinate ring. References to many other examples are given in §§1.2, 1.4, 1.5.

1.1. Quantum SL_2. Recall that the group $SL_2(k)$ is a closed subvariety of the variety of 2×2 matrices over k, defined by the single equation "determinant $=$ 1". The coordinate ring of the matrix variety is naturally realized as a polynomial ring in four variables X_{ij}, corresponding to the functions that pick out the four entries of the matrices. The coordinate ring of $SL_2(k)$ can thus be described as follows:

$$\mathcal{O}(SL_2(k)) = k[X_{11}, X_{12}, X_{21}, X_{22}]/\langle X_{11}X_{22} - X_{12}X_{21} - 1\rangle.$$

To "quantize" this coordinate ring, we replace the commutative multiplication by a noncommutative one, parametrized by a nonzero scalar q, as below. The reasons for this particular choice of relations will not be given here; see [4, §§I.1.6, I.1.8], for instance, for a discussion.

Given a choice of scalar $q \in k^\times$, the "standard" one-parameter *quantized coordinate ring of* $SL_2(k)$ is the k-algebra $\mathcal{O}_q(SL_2(k))$ presented by generators X_{11}, X_{12}, X_{21}, X_{22} and the following relations:

$$X_{11}X_{12} = qX_{12}X_{11} \qquad X_{11}X_{21} = qX_{21}X_{11}$$
$$X_{12}X_{22} = qX_{22}X_{12} \qquad X_{21}X_{22} = qX_{22}X_{21}$$
$$X_{12}X_{21} = X_{21}X_{12} \qquad X_{11}X_{22} - X_{22}X_{11} = (q - q^{-1})X_{12}X_{21}$$
$$X_{11}X_{22} - qX_{12}X_{21} = 1\,.$$

The case when $q = 1$ is special: The first six relations then reduce to saying that the generators X_{ij} commute with each other, the last reduces to the defining relation for the variety $SL_2(k)$, and so the algebra $\mathcal{O}_1(SL_2(k))$ is just the classical coordinate ring. We write this, very informally, as

$$\mathcal{O}(SL_2(k)) = \lim_{q \to 1} \mathcal{O}_q(SL_2(k));$$

it is our first example of a "semiclassical limit".

1.2. Quantum matrices, quantum SL_n and GL_n. The pattern indicated in §1.1 extends to definitions of "standard" single parameter quantized coordinate rings $\mathcal{O}_q(M_n(k))$, $\mathcal{O}_q(SL_n(k))$, and $\mathcal{O}_q(GL_n(k))$ for all positive integers n. Multiparameter versions, which we label in the form $\mathcal{O}_{\lambda,\boldsymbol{p}}(-)$, have also been defined. Generators and relations for these algebras may be found, for instance, in [12, §§ 1.2–1.4], [4, §§ I.2.2–I.2.4].

1.3. Quantum affine spaces. The coordinate ring of affine n-space over k is the polynomial algebra in n indeterminates, and the most basic quantization is obtained by replacing commutativity ($xy = yx$) with *q-commutativity*: $xy = qyx$. Thus, the "standard" one-parameter *quantized coordinate ring of k^n*, relative to a choice of scalar $q \in k^\times$, is the k-algebra

$$\mathcal{O}_q(k^n) = k\langle x_1, \dots, x_n \mid x_i x_j = q x_j x_i \text{ for } 1 \le i < j \le n\rangle.$$

The multiparameter version of this algebra requires an $n \times n$ matrix of nonzero scalars, $\boldsymbol{q} = (q_{ij})$, which is *multiplicatively antisymmetric* in the sense that $q_{ii} = 1$ and $q_{ji} = q_{ij}^{-1}$ for all i, j. The *multiparameter quantized coordinate ring of k^n* corresponding to a choice of $\boldsymbol{q}$ is the k-algebra

$$\mathcal{O}_{\boldsymbol{q}}(k^n) = k\langle x_1, \dots, x_n \mid x_i x_j = q_{ij} x_j x_i \text{ for all } i, j\rangle.$$

In the one-parameter case, we can write $\mathcal{O}(k^n) = \lim_{q\to 1} \mathcal{O}_q(k^n)$ in the same sense as above. For the multiparameter case, we imagine a limit in which all $q_{ij} \to 1$.

1.4. Quantized coordinate rings of semisimple groups. The single parameter versions of these Hopf algebras, which we denote $\mathcal{O}_q(G)$, were first defined for semisimple algebraic groups G of classical type (types A, B, C, D) via generators and relations, by Faddeev, Reshetikhin, and Takhtadjan [43] and Takeuchi [48]. A detailed development (done for $k = \mathbb{C}$, but the pattern is the same over other fields) can be found in [32, Chapter 9]. In most of the more recent literature, $\mathcal{O}_q(G)$ is defined as a restricted Hopf dual of the quantized enveloping algebra of the Lie algebra of G (e.g., see [4, Chapter I.7]). This is a more uniform approach, which also covers groups of exceptional type. That the two approaches yield the same Hopf algebras in the classical cases was established by Hayashi [20] and Takeuchi [48] (see [32, Theorem 11.22]).

The single parameter algebras $\mathcal{O}_q(G)$ constitute the "standard" quantized coordinate rings of semisimple groups. Multiparameter versions, which we label $\mathcal{O}_{q,p}(G)$, were introduced by Hodges, Levasseur, and Toro [24].

1.5. Additional examples. Quantized coordinate rings, both single- and multiparameter, have been defined for many algebraic varieties, such as algebraic tori, toric varieties, and versions of affine spaces related to classical groups of types B, C, D. For a general survey, see [12, Section 1]. Quantized toric varieties were introduced in [26] (see also [17, 16]). A family of iterated skew polynomial algebras covering multiparameter quantized Euclidean and symplectic spaces was introduced by Oh [38] and extended by Horton [25] (see also [14, § 2.5] for the odd-dimensional Euclidean case). Among other algebras that have been studied in the literature, we mention quantized coordinate rings for varieties of antisymmetric matrices [47] and varieties of symmetric matrices [37, 28].

1.6. Limits of families of algebras. The semiclassical limits informally introduced in §§1.1, 1.3 are more properly viewed in the framework of families of algebras. For example, the algebras $\mathcal{O}_q(SL_2(k))$ are quotients of a single algebra over a Laurent polynomial ring $k[t^{\pm 1}]$, namely the algebra A given by generators X_{11}, X_{12}, X_{21}, X_{22} and relations as in §1.1, but with q replaced by t:

$$\begin{aligned} &X_{11}X_{12} = tX_{12}X_{11} && X_{11}X_{21} = tX_{21}X_{11} \\ &X_{12}X_{22} = tX_{22}X_{12} && X_{21}X_{22} = tX_{22}X_{21} \\ &X_{12}X_{21} = X_{21}X_{12} && X_{11}X_{22} - X_{22}X_{11} = (t - t^{-1})X_{12}X_{21} \\ &X_{11}X_{22} - tX_{12}X_{21} = 1\,. \end{aligned} \tag{1.6a}$$

For each $q \in k^{\times}$, there is a natural identification $A/(t-q)A \equiv \mathcal{O}_q(SL_2(k))$. The "limit as $q \to 1$" is then simply the case $q = 1$ of these identifications: $A/(t-1)A \equiv \mathcal{O}(SL_2(k))$.

Similarly, if we take

$$B = k[t^{\pm 1}]\langle x_1, \ldots, x_n \mid x_i x_j = t x_j x_i \text{ for } 1 \le i < j \le n \rangle, \tag{1.6b}$$

then $B/(t-q)B \equiv \mathcal{O}_q(k^n)$ for all $q \in k^{\times}$, and

$$\lim_{q \to 1} \mathcal{O}_q(k^n) = B/(t-1)B \equiv \mathcal{O}(k^n).$$

The multiparameter algebras $\mathcal{O}_{\mathbf{q}}(k^n)$ can, likewise, be set up as common quotients of an algebra over a Laurent polynomial ring $k[t_{ij}^{\pm 1} \mid 1 \leq i < j \leq n]$. However, for purposes such as obtaining Poisson structures on semiclassical limits, we need to be able to exhibit the $\mathcal{O}_{\mathbf{q}}(k^n)$ as quotients of $k[t^{\pm 1}]$-algebras. There are many ways to do this; we will discuss some in §2.3.

1.7. An older example: the Weyl algebra. Weyl defined the algebra we now call the *first Weyl algebra* as

$$\mathbb{C}\langle x, y \mid xy - yx = \hbar i \rangle,$$

where $\hbar$ is Planck's constant and $i = \sqrt{-1}$. Physicists often use the term "classical limit" to denote the transition from a quantum mechanical system to a classical one by letting Planck's constant go to zero. The fact that $\lim_{\hbar \to 0}$ of the above algebra is the polynomial ring $\mathbb{C}[x, y]$ is one instance of this point of view.

To relate this semiclassical limit to the ones above, take $k = \mathbb{C}$ and take the scalar q in quantized coordinate rings to be $e^{\hbar}$. Then $\hbar \to 0$ corresponds to $q \to 1$. In many constructions, particularly the C*-algebra quantum groups corresponding to compact Lie groups, the parameter q is either written directly in the form $e^{\hbar}$ or is taken to be a nonnegative real number, with calculations involving $e^{\hbar}$ used for motivation.

2. Semiclassical limit constructions

In the context of quantized coordinate rings, semiclassical limits are constructed via quotients of algebras over Laurent polynomial rings, as in §1.6. A different version, using associated graded rings, is needed in other arenas, particularly for enveloping algebras of Lie algebras. We describe both constructions in this section.

2.1. Semiclassical limits: commutative fibre version. Let $k[h]$ be a polynomial algebra, with the indeterminate named h as a reminder of Planck's constant. Suppose that A is a torsionfree $k[h]$-algebra, and that A/hA is commutative. Since A is then a flat $k[h]$-module, the family of factor algebras $\big(A/(h-\alpha)A\big)_{\alpha \in k}$ (or, for short, A itself) is called a *flat family* of k-algebras, and A/hA is viewed as the "limit" of the family. It may happen that some of the algebras $A/(h-\alpha)A$ collapse to zero or are otherwise not desirable. If so, it is natural to treat A as an algebra over a localization of $k[h]$ (cf. Example 2.2(c), for instance). We will usually not do this explicitly.

An immediate question is, what kind of information about the algebras $A/(h-\alpha)A$ is contained in this limit? Observe that, because of the commutativity of A/hA, all additive commutators $[a, b] = ab - ba$ in A are divisible by h. Moreover, division by h is unique, since A is torsionfree as a $k[h]$-module. Hence, we obtain a well-defined binary operation $\frac{1}{h}[-,-]$ on A. This operation enjoys four key properties:

(1) Bilinearity;
(2) Antisymmetry;

(3) The Jacobi identity (thus A, equipped with $\frac{1}{h}[-,-]$, is a Lie algebra over k);
(4) The *Leibniz identities*, that is, the product rule (for derivatives) in each variable: $\frac{1}{h}[a,bc] = \big(\frac{1}{h}[a,b]\big)c + b\big(\frac{1}{h}[a,c]\big)$ for all $a,b,c \in A$, and similarly for $\frac{1}{h}[bc,a]$.

Operations satisfying properties (1)–(4) are called *Poisson brackets*.

The above Poisson bracket on A induces, uniquely, a Poisson bracket on A/hA, which we denote $\{-,-\}$. Thus, writing overbars to denote cosets modulo hA, we have

$$\{\overline{a},\overline{b}\} = \overline{\tfrac{1}{h}[a,b]}$$

for $a,b \in A$. The commutative algebra A/hA, equipped with this Poisson bracket, is called the *semiclassical limit* of the family $\big(A/(h-\alpha)A\big)_{\alpha\in k}$. Loosely speaking, the Poisson bracket on the semiclassical limit records a "first-order impression" of the commutators in A and in the algebras $A/(h-\alpha)A$.

2.2. Examples

(a) Fit the one-parameter quantum affine spaces $\mathcal{O}_q(k^n)$ into the $k[t^{\pm1}]$-algebra B of (1.6b), and set $h = t-1$. Then B represents a flat family of $k[h]$-algebras, with B/hB commutative. We identify B/hB with the polynomial ring $k[x_1,\dots,x_n]$ and compute the resulting Poisson bracket on the indeterminates as follows. For $1 \le i < j \le n$, we have $[x_i,x_j] = hx_jx_i$ in B, and hence

$$\{x_i,x_j\} = x_ix_j$$

in $k[x_1,\dots,x_n]$. Because of the Leibniz identities, the above information determines this Poisson bracket uniquely. It may be described in full as follows:

$$\{f,g\} = \sum_{1\le i<j\le n} x_ix_j \left(\frac{\partial f}{\partial x_i}\frac{\partial g}{\partial x_j} - \frac{\partial g}{\partial x_i}\frac{\partial f}{\partial x_j} \right)$$

for all $f,g \in k[x_1,\dots,x_n]$.

(b) Take $A = k[h]\langle x,y \mid xy-yx = h\rangle$. Then $A/(h-\alpha)A \cong A_1(k)$ for all nonzero $\alpha \in k$, while A/hA can be identified with the polynomial ring $k[x,y]$. In this case, the semiclassical limit Poisson bracket on $k[x,y]$ satisfies (and is determined by)

$$\{x,y\} = 1.$$

(c) The family $\big(\mathcal{O}_q(SL_2(k))\big)_{q\in k^\times}$ fits into the $k[t^{\pm1}]$-algebra A with generators X_{11}, X_{12}, X_{21}, X_{22} and relations (1.6a). This is a flat family over $k[h]$, where $h = t-1$. Since t is invertible in A, the specialization $A/(h+1)A$ is zero, corresponding to the fact that A is actually a torsionfree (even free) algebra over the localization $k[h][(h+1)^{-1}]$. Here the semiclassical limit is the classical coordinate ring $\mathcal{O}(SL_2(k))$, equipped with the Poisson bracket satisfying

$$\begin{aligned} \{X_{11},X_{12}\} &= X_{11}X_{12} & \{X_{11},X_{21}\} &= X_{11}X_{21} \\ \{X_{12},X_{22}\} &= X_{12}X_{22} & \{X_{21},X_{22}\} &= X_{21}X_{22} \\ \{X_{12},X_{21}\} &= 0 & \{X_{11},X_{22}\} &= 2X_{12}X_{21}\,. \end{aligned} \tag{2.2c}$$

(d) In parallel with (c), the family $\big(\mathcal{O}_q(M_2(k))\big)_{q\in k^\times}$ fits into a $k[t^{\pm1}]$-algebra C given by generators X_{11}, X_{12}, X_{21}, X_{22} and the first six relations of (1.6a). Its semiclassical limit is the classical coordinate ring $\mathcal{O}(M_2(k))$, equipped with the Poisson bracket satisfying the equations (2.2c).

To illustrate one piece of information reflected in this semiclassical limit, we focus on the quantum determinant in the algebra C, namely the element $D_t = X_{11}X_{22} - tX_{12}X_{21}$. It is well known that D_t is a central element. Hence, its image in $\mathcal{O}(M_2(k))$, namely the ordinary determinant $D = X_{11}X_{22} - X_{12}X_{21}$, is *Poisson central*: $\{D, f\} = 0$ for all $f \in \mathcal{O}(M_2(k))$. (This can also be checked directly, via (2.2c) and the Leibniz identities.)

2.3. Multiparameter examples. To obtain a semiclassical limit – with Poisson bracket – for a multiparameter family of algebras, we convert to a single parameter family and apply the construction of §2.1. The procedure is clear when the parameters involved are integer powers of a single parameter. For example, consider the algebras $\mathcal{O}_{\boldsymbol{q}}(k^n)$ where $\boldsymbol{q} = (q^{a_{ij}})$ for $q \in k^\times$ and an antisymmetric integer matrix (a_{ij}). Then define

$$A = k[t^{\pm1}]\langle x_1, \dots, x_n \mid x_i x_j = t^{a_{ij}} x_j x_i \text{ for all } i, j\rangle,$$

which is a torsionfree $k[t-1]$-algebra with $A/(t-1)A$ commutative. The semiclassical limit is the polynomial algebra $k[x_1, \dots, x_n]$, equipped with the Poisson bracket satisfying

$$\{x_i, x_j\} = a_{ij} x_i x_j$$

for all i, j.

More general parameters can be dealt with by various means. A simple but ad hoc method to handle any $\mathcal{O}_{\boldsymbol{q}}(k^n)$ is via the algebra

$$A = k[h]\langle x_1, \dots, x_n \mid x_i x_j = \big(1 + (q_{ij} - 1)h\big) x_j x_i \text{ for } 1 \le i < j \le n\rangle,$$

which is set up so that $A/(h-1)A \cong \mathcal{O}_{\boldsymbol{q}}(k^n)$ and $A/hA \cong k[x_1, \dots, x_n]$. This yields a Poisson bracket satisfying $\{x_i, x_j\} = (q_{ij} - 1)x_i x_j$ for all i, j.

A variant of the previous procedure, involving quadratic rather than linear polynomials in h, is used in [18] to construct semiclassical limits for which the conjecture sketched in the Introduction applies to the generic multiparameter quantum affine spaces $\mathcal{O}_{\boldsymbol{q}}(k^n)$.

Inverse to the construction of semiclassical limits is the problem of *quantization*: trying to represent a given algebra supporting a Poisson bracket as a semiclassical limit of a suitable family of algebras. We will not discuss this problem except to indicate a solution for the case of homogeneous quadratic Poisson brackets on polynomial rings. Namely, suppose we have a polynomial algebra $k[x_1, \dots, x_n]$, equipped with a Poisson bracket such that $\{x_i, x_j\} = \alpha_{ij} x_i x_j$ for all i, j, where (α_{ij}) is an antisymmetric matrix of scalars over k. In place of ad hoc procedures such as the one sketched above, it is natural, assuming that $\operatorname{char} k = 0$, to use power series. In this case, set $\exp(\alpha_{ij}) = \sum_{n=0}^{\infty} \frac{1}{n!}\alpha_{ij}^n h^n \in k[[h]]$ for all i, j, and form the $k[[h]]$-algebra

$$A = k[[h]]\langle x_1, \dots, x_n \mid x_i x_j = \exp(\alpha_{ij}) x_j x_i \text{ for all } i, j \rangle.$$

The semiclassical limit algebra is $k[x_1, \dots, x_n]$, and its Poisson bracket is the original one.

Analogous $k[[h]]$-algebra constructions are given for commonly studied families of quantized coordinate rings of skew polynomial type in [14, Section 2].

2.4. Semiclassical limits: filtered/graded version. Suppose that A is a $\mathbb{Z}$-filtered k-algebra, say with filtration $(A_n)_{n\in\mathbb{Z}}$. Thus, the A_n are k-subspaces of A, with $A_m \subseteq A_n$ when $m \leq n$, such that $A_m A_n \subseteq A_{m+n}$ for all m, n. We will assume that the filtration is *exhaustive*, that is, that $\bigcup_{n\in\mathbb{Z}} A_n = A$, and that $1 \in A_0$; thus, A_0 is a unital subalgebra of A. Finally, let $\operatorname{gr} A = \bigoplus_{n\in\mathbb{Z}} \operatorname{gr}_n A$ be the associated graded algebra, where $\operatorname{gr}_n A = A_n / A_{n-1}$.

Now assume that $\operatorname{gr} A$ is commutative. Homogeneous elements $a \in \operatorname{gr}_m A$ and $b \in \operatorname{gr}_n A$ can be lifted to elements $\widehat{a} \in A_m$ and $\widehat{b} \in A_n$, and since $\operatorname{gr} A$ is commutative, the commutator $[\widehat{a}, \widehat{b}]$ must lie in A_{m+n-1}. We then set $\{a, b\}$ equal to the coset of $[\widehat{a}, \widehat{b}]$ in $\operatorname{gr}_{m+n-1} A$. It is an easy exercise, left to the reader, to verify that $\{a, b\}$ is well defined, and that the extension of $\{-,-\}$ to sums of homogeneous elements defines a Poisson bracket on $\operatorname{gr} A$. The commutative algebra $\operatorname{gr} A$, equipped with this Poisson bracket, is called the *semiclassical limit* of A.

More generally, assume there is an integer $d < 0$ such that $[A_m, A_n] \subseteq A_{m+n+d}$ for all $m, n \in \mathbb{Z}$. This assumption of course forces $\operatorname{gr} A$ to be commutative. Modify the definition above by setting $\{a, b\}$ equal to the coset of $[\widehat{a}, \widehat{b}]$ in $\operatorname{gr}_{m+n+d} A$, for $a \in \operatorname{gr}_m A$ and $b \in \operatorname{gr}_n A$. This recipe again produces a well-defined Poisson bracket on $\operatorname{gr} A$ [34, Lemma 2.7].

2.5. Bridging the two constructions. The semiclassical limit of a $\mathbb{Z}$-filtered algebra A constructed in §2.4 can also be obtained by applying the construction of §2.1 to an auxiliary algebra, namely the *Rees ring*

$$\widetilde{A} := \sum_{n\in\mathbb{Z}} A_n h^n \subseteq A[h^{\pm 1}],$$

where $A[h^{\pm 1}]$ is a Laurent polynomial ring over A. Since $1 \in A_0$, the polynomial algebra $k[h]$ is a subalgebra of $\widetilde{A}$, and we note that $\widetilde{A}$ is a torsionfree $k[h]$-algebra. (It is not a $k[h^{\pm 1}]$-algebra unless $A_{-1} = A_0$, in which case all $A_n = A_0$.) On one hand, $\widetilde{A}/(h-1)\widetilde{A} \cong A$. On the other, $\widetilde{A}/h\widetilde{A} \cong \operatorname{gr} A$, because $h\widetilde{A} = \sum_{n\in\mathbb{Z}} A_{n-1} h^n$. Thus, if $\operatorname{gr} A$ is commutative, we have a Poisson bracket $\frac{1}{h}[-,-]$ on $\widetilde{A}$, which induces a Poisson bracket $\{-,-\}_1$ on $\operatorname{gr} A$ as in §2.1. This bracket concides with the Poisson bracket $\{-,-\}_4$ constructed in §2.4, as follows.

Start with $a \in \operatorname{gr}_m A$ and $b \in \operatorname{gr}_n A$, and lift these elements to $\widehat{a} \in A_m$ and $\widehat{b} \in A_n$. With respect to the natural epimorphism $\pi : \widetilde{A} \to \operatorname{gr} A$, the elements a

and b lift to $\widehat{a}h^m, \widehat{b}h^n \in \widetilde{A}$. Hence,

$$\{a,b\}_1 = \pi\Big(\frac{1}{h}[\widehat{a}h^m, \widehat{b}h^n]\Big) = \pi([\widehat{a},\widehat{b}]h^{m+n-1}) = [\widehat{a},\widehat{b}] + A_{m+n-2} = \{a,b\}_4 .$$

Therefore $\{-,-\}_1 = \{-,-\}_4$.

2.6. Example: enveloping algebras. Let $\mathfrak{g}$ be a finite-dimensional Lie algebra over k, and put the standard (nonnegative) filtration on the enveloping algebra $U(\mathfrak{g})$, so that $U(\mathfrak{g})_0 = k$ and $U(\mathfrak{g})_1 = k + \mathfrak{g}$, while $U(\mathfrak{g})_n = U(\mathfrak{g})_1^n$ for $n > 1$. The associated graded algebra is commutative, and is naturally identified with the symmetric algebra $S(\mathfrak{g})$ of the vector space $\mathfrak{g}$. In particular, we use the same symbol to denote an element of $\mathfrak{g}$ and its coset in $\operatorname{gr}_1 U(\mathfrak{g}) = S(\mathfrak{g})_1$. Then $S(\mathfrak{g})$ is the semiclassical limit of $U(\mathfrak{g})$, equipped with the Poisson bracket satisfying

$$\{e,f\} = [e,f]$$

for all $e, f \in \mathfrak{g}$, where $[e,f]$ denotes the Lie product in $\mathfrak{g}$. The above formula determines $\{-,-\}$ uniquely, since $\mathfrak{g}$ generates $S(\mathfrak{g})$.

Now view the dual space $\mathfrak{g}^*$ as an algebraic variety, namely the affine space $\mathbb{A}^{\dim \mathfrak{g}}$. The coordinate ring $\mathcal{O}(\mathfrak{g}^*)$ is a polynomial algebra over k in $\dim \mathfrak{g}$ indeterminates, as is $S(\mathfrak{g})$. There is a canonical isomorphism

$$\theta : S(\mathfrak{g}) \xrightarrow{\;\cong\;} \mathcal{O}(\mathfrak{g}^*) \tag{2.6}$$

which sends each $e \in \mathfrak{g}$ to the polynomial function on $\mathfrak{g}^*$ given by evaluation at e, that is, $\theta(e)(\alpha) = \alpha(e)$ for $\alpha \in \mathfrak{g}^*$. (This isomorphism is often treated as an identification of the algebras $S(\mathfrak{g})$ and $\mathcal{O}(\mathfrak{g}^*)$.) Via θ, the Poisson bracket on $S(\mathfrak{g})$ obtained from the semiclassical limit process above carries over to a Poisson bracket on $\mathcal{O}(\mathfrak{g}^*)$, known as the *Kirillov-Kostant-Souriau Poisson bracket.*

If $\{e_1, \dots, e_n\}$ is a basis for $\mathfrak{g}$, then $S(\mathfrak{g}) = k[e_1, \dots, e_n]$ and θ sends the e_i to indeterminates x_i such that $\mathcal{O}(\mathfrak{g}^*) = k[x_1, \dots, x_n]$. An explicit description of the KKS Poisson bracket on $\mathcal{O}(\mathfrak{g}^*)$ can be obtained in terms of the structure constants of $\mathfrak{g}$, as follows. These constants are scalars $c_{ij}^l \in k$ such that $[e_i, e_j] = \sum_l c_{ij}^l e_l$ for all i, j. Since $\{e_i, e_j\} = [e_i, e_j]$ in $S(\mathfrak{g})$, an application of θ yields $\{x_i, x_j\} = \sum_l c_{ij}^l x_l$ for all i, j. It follows that

$$\{p,q\} = \sum_{i,j,l} c_{ij}^l x_l \frac{\partial p}{\partial x_i} \frac{\partial q}{\partial x_j}$$

for $p, q \in \mathcal{O}(\mathfrak{g}^*)$ [7, Proposition 1.3.18]. To see this, just check that the displayed formula determines a Poisson bracket on $\mathcal{O}(\mathfrak{g}^*)$ which agrees with the KKS bracket on pairs of indeterminates.

The KKS Poisson bracket on $\mathcal{O}(\mathfrak{g}^*)$ can also be obtained by applying the method of §2.1 to the homogenization of $U(\mathfrak{g})$, that is, the $k[h]$-algebra A with generating vector space $\mathfrak{g}$ and relations $ef - fe = h[e,f]$ for $e, f \in \mathfrak{g}$ (where $[e,f]$ again denotes the Lie product in $\mathfrak{g}$). Here $A/hA \cong S(\mathfrak{g}) \cong \mathcal{O}(\mathfrak{g}^*)$ and $A/(h-\lambda)A \cong U(\mathfrak{g})$ for all $\lambda \in k^\times$.

3. Symplectic leaves

We introduce symplectic leaves first in the context of Poisson manifolds, following the original definition of Weinstein [54], and then we carry the concept over to complex affine Poisson varieties, following Brown and Gordon [6].

3.1. Poisson algebras. We reiterate the general definition from §2.1: a *Poisson bracket* on a k-algebra R is any antisymmetric bilinear map $R \times R \to R$ which satisfies the Jacobi and Leibniz identities. Unless a special notation imposes itself, we denote all Poisson brackets by curly braces: $\{-,-\}$.

A *Poisson algebra* over k is just a k-algebra R equipped with a particular Poisson bracket. We restrict our attention to commutative Poisson algebras in the present paper. As for the noncommutative case, Farkas and Letzter have shown that Poisson brackets essentially reduce to commutators [11, Theorem 1.2]: If R is a prime ring which is not commutative, any Poisson bracket on R is a multiple of the commutator bracket by an element of the extended centroid of R.

3.2. Symplectic leaves in Poisson manifolds. Let M be a smooth manifold, and let $C^\infty(M)$ denote the algebra of smooth real-valued functions on M. (Some authors replace $C^\infty(M)$ by the algebra of smooth or analytic complex-valued functions.) A *Poisson structure* on M is a choice of Poisson bracket on $C^\infty(M)$, so that $C^\infty(M)$ becomes a Poisson algebra. A smooth manifold, together with a choice of Poisson structure, is called a *Poisson manifold*.

Now assume that M is a Poisson manifold. For each $f \in C^\infty(M)$, the map $X_f = \{f,-\}$ is a derivation on $C^\infty(M)$ and thus a vector field on M. Such vector fields are called *Hamiltonian vector fields* (for the given Poisson structure), and the flows (or integral curves) of Hamiltonian vector fields are known as *Hamiltonian paths*. More specifically, a smooth path $\gamma : [0,1] \to M$ is Hamiltonian provided there is some $f \in C^\infty(M)$ such that, at each point $\gamma(t)$ along the path, the tangent vector $d\gamma/dt$ equals $X_f|_{\gamma(t)}$. Since the change from a Hamiltonian path following the flow of a vector field X_f to one following a different vector field X_g need not be smooth, one must work with *piecewise Hamiltonian paths*, i.e., finite concatenations of Hamiltonian paths.

These paths determine an equivalence relation on M, points p and p' being equivalent if and only if there is a piecewise Hamiltonian path in M running from p to p'. The resulting equivalence classes are called *symplectic leaves*, and the partition of M as the disjoint union of its symplectic leaves is known as the *symplectic foliation* of M.

3.3. Poisson bivector fields. For many purposes, it is more useful to record a Poisson structure in the form of a bivector field rather than a Poisson bracket. In particular, this allows the most direct definition of Poisson structures on non-affine algebraic varieties.

Let M be a Poisson manifold. For a point $p \in M$, let $\mathfrak{m}_p$ denote the maximal ideal of $C^\infty(M)$ consisting of those functions that vanish at p. Evaluation of Poisson brackets at p induces an antisymmetric bilinear form π_p on the cotangent space $\mathfrak{m}_p/\mathfrak{m}_p^2$, where

$$\pi_p(f+\mathfrak{m}_p^2,\, g+\mathfrak{m}_p^2) = \{f,g\}(p)$$

for $f, g \in \mathfrak{m}_p$. Now π_p acts in each variable as a linear map in the dual space of $\mathfrak{m}_p/\mathfrak{m}_p^2$, that is, as a tangent vector to M at p. Since π_p is antisymmetric, it is thus a *tangent bivector* at p, namely an element of $T_p(M) \wedge T_p(M)$. The map $\pi : p \mapsto \pi_p$ is a smooth global section of $\Lambda^2 T_M$, that is, a *tangent bivector field* on M. To recover the Poisson bracket on $C^\infty(M)$ from the bivector field π, observe that $\{f,g\}(p) = \{f - f(p),\, g - g(p)\}(p)$ for $f, g \in C^\infty(M)$ and $p \in M$, which we rewrite in the form

$$\{f,g\}(p) = \pi_p(df(p), dg(p)), \tag{3.3}$$

where $df(p) = f - f(p) + \mathfrak{m}_p^2 \in \mathfrak{m}_p/\mathfrak{m}_p^2$ and similarly for $dg(p)$.

Conversely, via (3.3) any tangent bivector field π on M induces an antisymmetric bilinear map $\{-,-\}$ on $C^\infty(M)$ satisfying the Leibniz conditions. This is a Poisson bracket exactly when the Jacobi identity is satisfied, which is equivalent to the vanishing of the *Schouten bracket* $[\pi,\pi]$ (which we will not define here; see [1, p. 44], [53, 2nd ed., Remark 2.2(3)], for instance). A *Poisson bivector field* on M is any tangent bivector field π for which $[\pi,\pi]=0$. As indicated in the sketch above, Poisson brackets on $C^\infty(M)$ correspond bijectively to Poisson bivector fields on M.

3.4. Poisson varieties. For any complex algebraic variety V, the definition of a *Poisson bivector field* on V can be copied from §3.3 – it is any tangent bivector field π on V for which $[\pi,\pi]=0$. In the context of algebraic geometry, however, the map $\pi : V \to \Lambda^2 T_V$ is required to be a regular function. Now one defines a *Poisson variety* to be a complex algebraic variety equipped with a particular Poisson bivector field. Associated concepts are defined by requiring compatibility with these bivector fields. For example, a *Poisson morphism* from a Poisson variety (V,π) to a Poisson variety (W,π') is a regular map $\phi : V \to W$ such that $(T\phi \wedge T\phi)\pi = \pi'\phi$. A *Poisson subvariety* of V is a subvariety X such that the inclusion map $X \to V$ is a Poisson morphism.

If V is an affine Poisson variety, the formula (3.3) defines a Poisson bracket on $\mathcal{O}(V)$. Conversely, any Poisson bracket on $\mathcal{O}(V)$ induces a Poisson bivector field on V as in §3.3. Thus, affine Poisson varieties can equally well be defined as complex affine varieties whose coordinate rings are Poisson algebras. This point of view can be extended to arbitrary varieties by defining a Poisson variety to be a complex algebraic variety whose sheaf of regular functions is a sheaf of Poisson algebras.

3.5. Smooth Poisson varieties as manifolds. In order to define symplectic leaves in Poisson varieties, manifold structures are needed. The fundamental result is that any smooth (i.e., nonsingular) complex variety V has a unique structure as a complex analytic manifold (e.g., [44, Chapter II, §2.3]). This allows one to view V as a smooth manifold. If V is a Poisson variety, its chosen Poisson bivector field π is necessarily smooth (because it is regular), and so V together with π becomes a Poisson manifold. One can achieve this result in the affine case with Poisson brackets as well, by showing that any Poisson bracket on $\mathcal{O}(V)$ extends uniquely to a Poisson bracket on the algebra of smooth complex functions on V; taking real parts then yields a Poisson bracket on $C^\infty(V)$.

Given a smooth Poisson variety V, we view V as a smooth manifold as above, and define the *symplectic leaves* of V to be the symplectic leaves of the manifold V, defined as in §3.2.

3.6. Symplectic leaves in singular Poisson varieties. Let V be an arbitrary complex variety, and define the sequence of closed subvarieties

$$V_0 = V \supset V_1 \supset \cdots \supset V_m = \varnothing,$$

where each V_{i+1} is the singular locus of V_i. To build this chain, recall first that the singular locus of a nonempty variety is a proper closed subvariety. Since V is a noetherian topological space, the chain must eventually reach the empty set.

If V is a Poisson variety, then V_1 is a Poisson subvariety [42, Corollary 2.4]. By induction, all the V_i are Poisson subvarieties of V. Consequently, V is (canonically) the disjoint union of smooth locally closed Poisson subvarieties $Z_i := V_{i-1} \setminus V_i$. Following [6, §3.5], we define the *symplectic leaves* of V to be the symplectic leaves of the various Z_i, defined as in §3.5.

3.7. Example. There is a known recipe, described in [22, Appendix A], for determining the symplectic leaves in a semisimple complex algebraic group G, relative to the Poisson structure arising from the "standard quantization" of G. For illustration, we present the case $G = SL_2(\mathbb{C})$; details are given in [22, Theorem B.2.1]. The Poisson bracket on $\mathcal{O}(G)$ is described in §2.2(c) above. The symplectic leaves in G are as follows:

- the singletons $\left\{\begin{bmatrix} \alpha & 0 \\ 0 & \alpha^{-1} \end{bmatrix}\right\}$, for $\alpha \in \mathbb{C}^\times$;
- the sets $\left\{\begin{bmatrix} \alpha & 0 \\ \gamma & \alpha^{-1} \end{bmatrix} \;\middle|\; \alpha, \gamma \in \mathbb{C}^\times\right\}$ and $\left\{\begin{bmatrix} \alpha & \beta \\ 0 & \alpha^{-1} \end{bmatrix} \;\middle|\; \alpha, \beta \in \mathbb{C}^\times\right\}$;
- the sets $\left\{\begin{bmatrix} \alpha & \beta \\ \gamma & \delta \end{bmatrix} \in G \;\middle|\; \beta = \lambda\gamma \neq 0\right\}$, for $\lambda \in \mathbb{C}^\times$.

3.8. Example. The standard example of a non-algebraic solvable Lie algebra is a 3-dimensional complex Lie algebra $\mathfrak{g}$ with basis $\{e_1, e_2, e_3\}$ such that

$$[e_1, e_2] = e_2 \quad [e_1, e_3] = \alpha e_3 \quad [e_2, e_3] = 0$$

for some $\alpha \in \mathbb{R} \setminus \mathbb{Q}$. Write $\mathcal{O}(\mathfrak{g}^*) = \mathbb{C}[x_1, x_2, x_3]$ following the notation of §2.6. The KKS Poisson structure on $\mathfrak{g}^*$ is given by the Poisson bracket on $\mathcal{O}(\mathfrak{g}^*)$ such that

$$\{x_1, x_2\} = x_2 \quad \{x_1, x_3\} = \alpha x_3 \quad \{x_2, x_3\} = 0.$$

As in [53, 1st ed., Example II.2.37; 2nd ed., Example II.2.43], the symplectic leaves in $\mathfrak{g}^*$ are the following sets:

- the individual points on the x_1-axis;

- the x_1x_2-plane minus the x_1-axis;
- the x_1x_3-plane minus the x_1-axis;
- the surfaces $(x_3 = \lambda x_2^\alpha \neq 0)$ for $\lambda \in \mathbb{C}^\times$.

Since α is irrational, the surfaces $(x_3 = \lambda x_2^\alpha \neq 0)$ are not algebraic – they are locally closed in the Euclidean topology but not in the Zariski topology.

4. The Orbit Method from Lie theory

4.1. The Orbit Method. This term has been applied to a whole complex of methods in the representation theory of Lie groups and Lie algebras, and extended, as a guiding principle, to many other domains. To quote Kirillov's survey article [30],

> *The idea behind the orbit method is the unification of harmonic analysis with symplectic geometry (and it can also be considered as a part of the more general idea of the unification of mathematics and physics). In fact, this is a post factum formulation. Historically, the orbit method was proposed in* [29] *for the description of the unitary dual (i.e., the set of equivalence classes of unitary irreducible representations) of nilpotent Lie groups. It turned out that not only this problem but all other principal questions of representation theory – topological structure of the unitary dual, explicit description of the restriction and induction functors, character formulae, etc. – can be naturally answered in terms of coadjoint orbits.*

In Lie theory, the relevant orbits are defined as follows. Recall that if G is a Lie group with Lie algebra $\mathfrak{g}$, then G acts on $\mathfrak{g}$ by the *adjoint action* and on $\mathfrak{g}^*$ by the *coadjoint action*. The G-orbits of these actions are called *adjoint orbits* and *coadjoint orbits*, respectively. As a particular instance, the Orbit Method suggests that the primitive ideals of the enveloping algebra of $\mathfrak{g}$, being the kernels of the irreducible representations, should be related to the coadjoint orbits in $\mathfrak{g}^*$. Kirillov's original work provided the best such relationship – a bijection – when $\mathfrak{g}$ is nilpotent. There is also a bijection in case $\mathfrak{g}$ is solvable, except that the coadjoint orbits may have to be taken with respect to a different group than a Lie group with Lie algebra $\mathfrak{g}$. We discuss this situation in Section 5.

To place the coadjoint orbits in a geometric setting, view $\mathfrak{g}^*$ as the variety $\mathbb{A}^{\dim \mathfrak{g}}$, as in §2.6. We can then ask for a geometric description of these orbits within $\mathfrak{g}^*$. The answer is a famous result discovered independently by Kirillov, Kostant, and Souriau (see, e.g., [31, §I.2.2, Theorem 2]):

4.2. Theorem. [Kirillov-Kostant-Souriau] *Let G be a Lie group and $\mathfrak{g}$ its Lie algebra. Then the coadjoint orbits of G in $\mathfrak{g}^*$ are precisely the symplectic leaves for the KKS Poisson structure.*

4.3. Example. Return to Example 3.8, and place the $x_1x_2x_3$-coordinates of points of $\mathfrak{g}^*$ in column vectors. We choose a Lie group G with Lie algebra $\mathfrak{g}$ as follows:

$$G := \left\{ \begin{bmatrix} 1 & u & v \\ 0 & t & 0 \\ 0 & 0 & t^\alpha \end{bmatrix} \;\middle|\; u, v \in \mathbb{C},\ t \in \mathbb{C}^\times \right\}.$$

The coadjoint action of G on $\mathfrak{g}^*$ can be identified with left multiplication of matrices from G on column vectors representing points in $\mathfrak{g}^*$. One easily checks that the G-orbits are exactly the symplectic leaves of $\mathfrak{g}^*$ identified in Example 3.8, as required by Theorem 4.2.

4.4. A general principle. In situations outside Lie theory, there may not be a suitable group action whose orbits play the role of coadjoint orbits. Instead, taking account of Theorem 4.2, one focusses on symplectic leaves. Restricting to the study of irreducible representations and primitive ideals, one is led to a general principle that we formulate as follows:

> *Given a noncommutative algebra A, relate the primitive ideals of A to the symplectic leaves corresponding to the Poisson structure on some associated algebraic variety arising from a semiclassical limit.*

This loose phrasing is intended to give the flavor of ideas coming out of the Orbit Method rather than to set up a precise recipe. Furthermore, this principle already requires modification in the case of enveloping algebras, and for general quantized coordinate rings.

On the other hand, the principle is right on target for the generic single parameter quantized coordinate rings $\mathcal{O}_q(G)$ of semisimple complex algebraic groups G, as conjectured by Hodges and Levasseur in [22, §2.8, Conjecture 1]: there is a bijection between the set of primitive ideals of $\mathcal{O}_q(G)$ and the set of symplectic leaves in G (for the semiclassical limit Poisson structure). They verified this conjecture for $G = SL_2(\mathbb{C})$ and $G = SL_3(\mathbb{C})$ in [22, Corollary B.2.2, Theorems 4.4.1, A.3.2], and then for $G = SL_n(\mathbb{C})$ in [23, Theorem 4.2 and following remarks]. The full conjecture was established by Joseph [27, §§10.3, A.4.5] and by Hodges, Levasseur, and Toro [24, Theorems 1.8, 4.18, Corollary 4.5] for connected semisimple complex Lie groups G. The latter results also cover the multiparameter algebra $\mathcal{O}_{q,p}(G)$ under suitable algebraicity conditions on p.

4.5. Generic versus non-generic situations. As mentioned in the introduction, the principle discussed in §4.4 does not apply to non-generic quantized coordinate rings, which typically have "too many" primitive ideals. The quantum plane provides the simplest illustration of this difficulty, and of the differences between the generic and non-generic cases. Take

$$A_q = \mathcal{O}_q(\mathbb{C}^2) = \mathbb{C}\langle x, y \mid xy = qyx \rangle,$$

where q is an arbitrary nonzero scalar in $\mathbb{C}$. By Example 2.2(a), the semiclassical limit of the family $(A_q)_{q\in\mathbb{C}^\times}$ is the polynomial ring $k[x, y]$, equipped with the

Poisson bracket such that $\{x, y\} = xy$. It is easily checked that the corresponding symplectic leaves in $\mathbb{C}^2$ consist of

- the individual points on the x- and y-axes;
- the xy-plane minus the x- and y-axes.

If q is not a root of unity, one similarly checks that the primitive ideals of A_q consist of

- the maximal ideals $\langle x - \alpha,\ y\rangle$ and $\langle x,\ y - \beta\rangle$, for $\alpha, \beta \in \mathbb{C}$;
- the zero ideal.

(See [4, Example II.7.2], for instance, for details.) In this case, there is a natural bijection between the set of primitive ideals of A_q and the set of symplectic leaves in $\mathbb{C}^2$.

On the other hand, if q is a primitive lth root of unity, the center of A_q equals the polynomial ring $\mathbb{C}[x^l, y^l]$, and A_q is a finitely generated $\mathbb{C}[x^l, y^l]$-module. In this case, the primitive ideals of A_q are maximal ideals, and they are parametrized (up to l-to-one) by the maximal ideals of $\mathbb{C}[x^l, y^l]$. While the set of primitive ideals of A_q has the same cardinality as the set of symplectic leaves in $\mathbb{C}^2$, there is no natural bijection, and certainly no homeomorphism if Zariski topologies are taken into account.

Such disparities occur in all the standard families of quantized coordinate rings, and provide just one of many distinctions between the generic and non-generic cases. We do not discuss the non-generic situation further, and concentrate on generic algebras.

5. Limitations of the Orbit Method for solvable Lie algebras

For a solvable finite dimensional complex Lie algebra $\mathfrak{g}$, the primitive ideals of the enveloping algebra $U(\mathfrak{g})$ are parametrized by means of the famous *Dixmier map*. At first glance, this is a successful instance of the Orbit Method, since the Dixmier map induces a bijection from a set of orbits in $\mathfrak{g}^*$ onto the set of primitive ideals of $U(\mathfrak{g})$. However, the relevant orbits are not, in general, those of the coadjoint action of a Lie group with Lie algebra $\mathfrak{g}$. Instead, the following group is needed.

5.1. The algebraic adjoint group. Let $\mathfrak{g}$ be a finite dimensional complex Lie algebra. Treating $\mathfrak{g}$ for a moment just as a vector space, we have the general linear group $GL(\mathfrak{g})$ on $\mathfrak{g}$, which is a complex algebraic group whose Lie algebra is the general linear Lie algebra $\mathfrak{gl}(\mathfrak{g})$. Any algebraic subgroup of $GL(\mathfrak{g})$ (i.e., any Zariski closed subgroup) has a Lie algebra which is naturally contained in $\mathfrak{gl}(\mathfrak{g})$. The *algebraic adjoint group of* $\mathfrak{g}$ is the smallest algebraic subgroup $G \subseteq GL(\mathfrak{g})$ whose Lie algebra contains $\operatorname{ad}\mathfrak{g} = \{\operatorname{ad} x \mid x \in \mathfrak{g}\}$ (cf. [2, §12.2]; [50, Definition 24.8.1]).

The natural action of $GL(\mathfrak{g})$ on $\mathfrak{g}$ by linear automorphisms restricts to an action of G on $\mathfrak{g}$, the *adjoint action*. This, in turn, induces a (left) action of G on $\mathfrak{g}^*$, the *coadjoint action*, under which

$$(g.\alpha)(x) = \alpha(g^{-1}.x)$$

for $g \in G$, $\alpha \in \mathfrak{g}^*$, and $x \in \mathfrak{g}$. The orbits of this action, the *coadjoint orbits*, are collected in the set $\mathfrak{g}^*/G$. We equip $\mathfrak{g}^*/G$ with the quotient topology induced from the Zariski topology on $\mathfrak{g}^*$, and thus refer to it as the *space* of coadjoint orbits.

5.2. Prime and primitive spectra. For any algebra A, we denote the collection of all primitive ideals of A by $\operatorname{prim} A$. This set supports a Zariski topology, under which the closed sets are the sets $V(I) := \{P \in \operatorname{prim} A \mid P \supseteq I\}$ for ideals I of A. We treat $\operatorname{prim} A$ as a topological space with this topology, and refer to it as the *primitive spectrum* of A. The analogous process, applied to the set of all prime ideals of A, results in the *prime spectrum* of A, denoted $\operatorname{spec} A$. Since primitive ideals are prime, $\operatorname{prim} A \subseteq \operatorname{spec} A$. In fact, $\operatorname{prim} A$ is a subspace of $\operatorname{spec} A$, that is, its topology coincides with the relative topology inherited from $\operatorname{spec} A$. Finally, we shall need the subspace of $\operatorname{spec} A$ consisting of all the maximal ideals of A. This is the *maximal ideal space* of A, denoted $\operatorname{maxspec} A$.

5.3. The Dixmier map. Let $\mathfrak{g}$ be a solvable finite dimensional complex Lie algebra. Following [2, §10.8], we use the name *Dixmier map* and the label Dx for the map

$$\mathrm{Dx} : \mathfrak{g}^* \longrightarrow \operatorname{prim} U(\mathfrak{g})$$

introduced by Dixmier in [9]. We do not give the definition here, but just refer to [2]. It turns out that this map is constant on G-orbits, and so it induces a *factorized Dixmier map*

$$\overline{\mathrm{Dx}} : \mathfrak{g}^*/G \longrightarrow \operatorname{prim} U(\mathfrak{g})$$

[2, §12.4]. Work of Dixmier, Conze, Duflo, and Rentschler led to the result that $\overline{\mathrm{Dx}}$ is a continuous bijection [2, Sätze 13.4, 15.1]. The conjecture that it is a homeomorphism was established later by Mathieu [35, Theorem], resulting in the following theorem:

5.4. Theorem. [Dixmier-Conze-Duflo-Rentschler-Mathieu] *Let $\mathfrak{g}$ be a solvable finite dimensional complex Lie algebra, and G its adjoint algebraic group. Then the factorized Dixmier map $\overline{\mathrm{Dx}}$ is a homeomorphism from $\mathfrak{g}^*/G$ onto* $\operatorname{prim} U(\mathfrak{g})$.

5.5. Algebraic versus non-algebraic cases. If $\mathfrak{g}$ is an *algebraic* Lie algebra, meaning that it is the Lie algebra of some algebraic group, then the adjoint algebraic group G is a Lie group, and its coadjoint orbits in $\mathfrak{g}^*$ are the symplectic leaves for the KKS Poisson structure, by Theorem 4.2. Otherwise, G is larger than the relevant Lie group, in the sense that its Lie algebra properly contains $\operatorname{ad} \mathfrak{g}$. In this case, its coadjoint orbits are larger too, typically larger than individual symplectic leaves. Our basic example illustrates this behavior.

5.6. Example. Return to Example 3.8, and again place the $x_1x_2x_3$-coordinates of points of $\mathfrak{g}^*$ in column vectors. The adjoint algebraic group G, written so as to act by left multiplication on column vectors, can be expressed as

$$G = \left\{ \begin{bmatrix} 1 & u & v \\ 0 & t & 0 \\ 0 & 0 & t' \end{bmatrix} \;\middle|\; u, v \in \mathbb{C},\ t, t' \in \mathbb{C}^\times \right\}$$

[50, §24.8.4]. The coadjoint orbits of G in $\mathfrak{g}^*$ are the following sets:

- the individual points on the x_1-axis;
- the x_1x_2-plane minus the x_1-axis;
- the x_1x_3-plane minus the x_1-axis;
- $\mathfrak{g}^*$ minus the x_1x_2- and x_1x_3-planes.

Comparing with Example 3.8, we see that the first three G-orbits are symplectic leaves, while the fourth is not. However, the fourth is at least a union of symplectic leaves.

The fourth G-orbit above is Zariski dense in $\mathfrak{g}^*$, while the others are not. Viewing these orbits as points in the orbit space $\mathfrak{g}^*/G$, we find that $\mathfrak{g}^*/G$ has a unique dense point (i.e., a unique dense singleton subset). By Theorem 5.4, the same holds for $\operatorname{prim} U(\mathfrak{g})$. (Translated into ideal theory, this means that there is one primitive ideal of $U(\mathfrak{g})$ which is contained in all other primitive ideals.) On the other hand, all the surfaces $(x_3 = \lambda x_2^\alpha \neq 0)$ are Zariski dense in $\mathfrak{g}^*$, and so the quotient topology on the space of symplectic leaves in $\mathfrak{g}^*$ has uncountably many dense points. Therefore this space of symplectic leaves cannot be homeomorphic to $\operatorname{prim} U(\mathfrak{g})$.

6. Poisson ideal theory and symplectic cores

Since the concept of symplectic leaves is differential-geometric, it should not be so surprising that it is not always suited to describe answers to algebraic problems, as seen in the previous section. Consequently, we look for an algebraic replacement. This is provided by Brown and Gordon's notion of *symplectic cores*, which is described via the ideal theory of Poisson algebras.

6.1. Poisson prime ideals. Let R be a (commutative) Poisson algebra (recall §3.1).

A *Poisson ideal* of R is any ideal I of the ring R which is also a Lie ideal relative to $\{-,-\}$, that is, $\{R, I\} \subseteq I$. Sums, products, and intersections of Poisson ideals are again Poisson ideals. Whenever I is a Poisson ideal of R, the Poisson bracket on R induces a well-defined Poisson bracket on R/I, so that R/I becomes a Poisson algebra.

The *Poisson core* of an arbitrary ideal J of R is the largest Poisson ideal contained in J. This exists and is unique, because it is the sum of all Poisson ideals contained in J. We use $\mathcal{P}(J)$ to denote the Poisson core of J.

A *Poisson-prime ideal* of R is any proper Poisson ideal P of R with the following property: whenever the product of Poisson ideals I and J of R is contained in P, one of I or J must be contained in P. Obviously any prime Poisson ideal is Poisson-prime, but the converse can fail in positive characteristic. As we shall see in a moment, (Poisson-prime) is the same as (prime Poisson) when R is noetherian and k has characteristic zero; in that case, we will drop the hyphen and speak of *Poisson prime ideals*. Note also that if Q is an arbitrary prime ideal of R, then $\mathcal{P}(Q)$ is a Poisson-prime ideal.

The *Poisson-prime spectrum* of R, denoted $\mathrm{P.spec}\, R$, is the set of all Poisson-prime ideals of R, equipped with the natural Zariski-type topology, in which the closed sets are those of the form $V_P(I) := \{P \in \mathrm{P.spec}\, R \mid P \supseteq I\}$, for ideals I of R. It suffices to consider Poisson ideals in defining closed sets, since the ideal I in the definition of a closed set can be replaced by the Poisson ideal it generates. (This observation is helpful in showing that finite unions of closed sets are closed.)

6.2. Lemma. *Let R be a Poisson k-algebra, where* $\operatorname{char} k = 0$. *Then the Poisson core of every prime ideal of R is prime, and all minimal prime ideals of R are Poisson ideals. If R is noetherian, the Poisson-prime ideals of R coincide with the prime Poisson ideals.*

Proof. Commutativity is not needed for this result. The commutative case is covered, for instance, by [13, Lemma 1.1], and the general case is proved the same way. We sketch the details for the reader's convenience.

The first conclusion is a consequence of [10, Lemma 3.3.2], and the second follows.

Now assume that R is noetherian, and let P be a Poisson-prime ideal of R. There exist prime ideals $Q_1, \dots, Q_t$ minimal over P such that $Q_1 Q_2 \cdots Q_t \subseteq P$. The minimal prime ideals Q_i/P in the Poisson algebra R/P must be Poisson ideals by what has been proved so far, and hence the Q_i are Poisson ideals of R. Poisson-primeness of P then implies that some $Q_j \subseteq P$, whence $P = Q_j$, proving that P is prime. □

6.3. Poisson-primitive ideals and symplectic cores. Let R be a (commutative) Poisson algebra.

The *Poisson-primitive ideals* of R are the Poisson cores of the maximal ideals of R. Note from §6.1 that all Poisson-primitive ideals are Poisson-prime.

This terminology is chosen to reflect the following parallel. An ideal P in an algebra A is left primitive if and only if P is the largest ideal contained in some maximal left ideal. If we view A as a (noncommutative) Poisson algebra via the commutator bracket $[-,-]$, then the ideals of A are precisely the Poisson left ideals. Thus, the left primitive ideals of A are exactly the Poisson cores of the maximal left ideals.

The *Poisson-primitive spectrum* of R, denoted $\mathrm{P.prim}\, R$, is the set of all Poisson-primitive ideals of R. This is a subset of $\mathrm{P.spec}\, R$, and we give it the relative topology.

By definition, the process of taking Poisson cores defines a surjective map

$$\operatorname{maxspec} R \longrightarrow \mathrm{P.prim}\, R,$$

and we note that this map is continuous. Its fibres, namely the sets

$$\{\mathfrak{m} \in \operatorname{maxspec} R \mid \mathcal{P}(\mathfrak{m}) = P\}$$

for $P \in \mathrm{P.prim}\, R$, are called *symplectic cores*. They determine a partition of $\operatorname{maxspec} R$.

Now suppose that $R = \mathcal{O}(V)$ is the coordinate ring of an affine variety V, and that k is algebraically closed. As in the complex case, we say that V is a *Poisson variety*. Since k is algebraically closed, there is a natural identification $V \equiv \operatorname{maxspec} R$, with which we transfer the symplectic cores from $\operatorname{maxspec} R$ to V. In other words, the *symplectic cores in V* are the sets

$$\{p \in V \mid \mathcal{P}(\mathfrak{m}_p) = P\}$$

for $P \in \operatorname{P.prim} R$, where $\mathfrak{m}_p = \{f \in R \mid f(p) = 0\}$.

6.4. Example. Return to Example 3.8, and set $R = \mathcal{O}(\mathfrak{g}^*) = \mathbb{C}[x_1, x_2, x_3]$. The Poisson-primitive ideals of R can be computed as follows:

$$\begin{aligned} \mathcal{P}(\langle x_1 - \alpha, x_2, x_3\rangle) &= \langle x_1 - \alpha, x_2, x_3\rangle && (\alpha \in \mathbb{C}) \\ \mathcal{P}(\langle x_1 - \alpha, x_2 - \beta, x_3\rangle) &= \langle x_3\rangle && (\alpha \in \mathbb{C},\ \beta \in \mathbb{C}^\times) \\ \mathcal{P}(\langle x_1 - \alpha, x_2, x_3 - \gamma\rangle) &= \langle x_2\rangle && (\alpha \in \mathbb{C},\ \gamma \in \mathbb{C}^\times) \\ \mathcal{P}(\langle x_1 - \alpha, x_2 - \beta, x_3 - \gamma\rangle) &= \langle 0\rangle && (\alpha \in \mathbb{C},\ \beta, \gamma \in \mathbb{C}^\times). \end{aligned}$$

It follows that the symplectic cores in $\mathfrak{g}^*$ are the sets

- the individual points on the x_1-axis;
- the x_1x_2-plane minus the x_1-axis;
- the x_1x_3-plane minus the x_1-axis;
- $\mathfrak{g}^*$ minus the x_1x_2- and x_1x_3-planes.

These are precisely the coadjoint orbits of the adjoint algebraic group of $\mathfrak{g}$, as we saw in Example 5.6.

7. Symplectic cores versus symplectic leaves

Symplectic cores are related to symplectic leaves by the following result of Brown and Gordon [6, Lemma 3.3 and Proposition 3.6]; further relations will be given below. Here "locally closed" refers to the Zariski topology.

7.1. Theorem. [Brown-Gordon] *Let V be a complex affine Poisson variety.*

(a) *Each symplectic core in V is a union of symplectic leaves, and is a smooth (nonsingular) subvariety of its closure.*

(b) *If the symplectic leaves in V are all locally closed, then they coincide with the symplectic cores.*

It is a standard result that the orbits of a connected algebraic group G acting on a variety X can be recovered from the orbit closures, as follows. Take any orbit closure C, and remove all orbit closures properly contained in C. The result will be a single G-orbit, and all G-orbits in X are obtained by this means. Yakimov has conjectured that the symplectic cores in a complex affine Poisson variety can be recovered from the closures of the symplectic leaves in a similar manner. We verify this below, with the help of the following lemma of Brown and Gordon [6, Lemma 3.5]. All topological properties are to be taken relative to the Zariski topology.

7.2. Lemma. [Brown-Gordon] *Let V be a complex affine Poisson variety, and $R = \mathcal{O}(V)$. Let L be a symplectic leaf in V, and set $K = \{f \in R \mid f = 0 \text{ on } L\}$. Then K is a Poisson-primitive ideal of R, and L is contained in the corresponding symplectic core, that is, $\mathcal{P}(\mathfrak{m}_p) = K$ for all $p \in L$.*

7.3. Lemma. *Let V be a complex affine Poisson variety, and $R = \mathcal{O}(V)$. Let K be a Poisson ideal of R, and X the closed subvariety of V determined by K. Then X is a union of symplectic cores and a union of symplectic leaves. In particular, the closure of any symplectic leaf of V is a union of symplectic leaves.*

Proof. If $p \in X$, then $\mathfrak{m}_p \supseteq K$. Since K is a Poisson ideal, it must be contained in the Poisson-primitive ideal $P = \mathcal{P}(\mathfrak{m}_p)$. Now the set $C = \{q \in V \mid \mathcal{P}(\mathfrak{m}_q) = P\}$ is the symplectic core containing p, and $C \subseteq X$ because $\mathfrak{m}_q \supseteq P \supseteq K$ for all $q \in C$. Therefore X is a union of symplectic cores. That X is a union of symplectic leaves now follows from Theorem 7.1(a),

For any symplectic leaf L of V, the ideal I of functions in R that vanish on L is a Poisson ideal by Lemma 7.2. The closed subvariety determined by I is the closure of L, and this is a union of symplectic leaves by what we have just proved. □

We can now prove that symplectic cores are obtained from symplectic leaves in the manner proposed by Yakimov; this is parts (c) and (e) of the following theorem. Here overbars denote closures.

7.4. Theorem. *Let V be a complex affine Poisson variety, and L a symplectic leaf in V.*

(a) *There is a unique symplectic core C in V containing L, and $C \subseteq \overline{L}$.*
(b) *C is the union of those symplectic leaves of V which are dense in $\overline{L}$.*
(c) *$C = \overline{L} \setminus \bigcup_M \overline{M}$ where M runs over those symplectic leaves whose closures are properly contained in $\overline{L}$.*
(d) *C is the unique symplectic core dense in $\overline{L}$.*
(e) *Each symplectic core in V is dense in the closure of every symplectic leaf it contains. Hence, it can be obtained from the closure of such a leaf as in part* (c).

Proof. Set $R = \mathcal{O}(V)$, and let K be the ideal of functions in R that vanish on L.

(a) The symplectic cores and the symplectic leaves both partition V, and the latter form a finer partition, by Theorem 7.1(a). This implies the existence and uniqueness of C.

By Lemma 7.2, K is a Poisson-primitive ideal, and the symplectic core it determines contains L. By uniqueness, this core is C, that is,

$$C = \{p \in V \mid \mathcal{P}(\mathfrak{m}_p) = K\}.$$

In particular, $\mathfrak{m}_p \supseteq K$ for all $p \in C$, from which it follows that $C \subseteq \overline{L}$.

(b) If M is a symplectic leaf which is dense in $\overline{L}$, then K equals the ideal of functions in R that vanish on M, and Lemma 7.2 implies that $M \subseteq C$. On the other hand, if M' is a symplectic leaf which is contained in but not dense in $\overline{L}$, the ideal K' of functions vanishing on M' properly contains K, whence M' is contained in a symplectic core different from C. In this case, M' is disjoint from C. Part (b) now follows, because $\overline{L}$ is a union of symplectic leaves, by Lemma 7.3.

(c) In view of Lemma 7.3, the given union $\bigcup_M \overline{M}$ equals the union of those symplectic leaves which are contained in $\overline{L}$ but not dense in $\overline{L}$. The given formula for C thus follows from part (b).

(d) Clearly C is dense in $\overline{L}$, since $L \subseteq C \subseteq \overline{L}$. If D is a different symplectic core contained in $\overline{L}$, then by (b), any symplectic leaf $N \subseteq D$ is not dense in $\overline{L}$. But $D \subseteq \overline{N}$ by (a), and thus D is not dense in $\overline{L}$.

(e) Suppose that D is a symplectic core in V, and N a symplectic leaf contained in D. By (a), D is the unique symplectic core containing N, and $D \subseteq \overline{N}$, whence D is dense in $\overline{N}$. The final statement now follows from (c), with C and L replaced by D and N. □

8. Symplectic cores versus primitive ideals for solvable Lie algebras

We now show that the concept of symplectic cores exactly overcomes the limitations of symplectic leaves with respect to the Dixmier map discussed in Section 5. Namely, the Dixmier map provides a homeomorphism from the space of symplectic cores in $\mathfrak{g}^*$ onto the primitive spectrum of $U(\mathfrak{g})$, for any solvable finite dimensional complex Lie algebra $\mathfrak{g}$. This just amounts to showing that the coadjoint orbits in $\mathfrak{g}^*$, with respect to the adjoint algebraic group of $\mathfrak{g}$, coincide with the symplectic cores. Solvability is not needed for the latter result.

All that is required to obtain the new statement about the Dixmier map is to reinterpret parts of the development of Theorem 5.4 in terms of the new concepts. This reinterpretation also shows that (for $\mathfrak{g}$ solvable) P.prim $\mathcal{O}(\mathfrak{g}^*)$ is homeomorphic to prim $U(\mathfrak{g})$. With a little extra effort, we can handle prime ideals as well, showing that P.spec $\mathcal{O}(\mathfrak{g}^*)$ is homeomorphic to spec $U(\mathfrak{g})$.

Throughout this section, $\mathfrak{g}$ will denote a finite dimensional complex Lie algebra and G its adjoint algebraic group. We do not assume $\mathfrak{g}$ solvable until Theorem 8.5. Some of the results we will need are developed in the literature in terms of $S(\mathfrak{g})$ rather than $\mathcal{O}(\mathfrak{g}^*)$. This requires use of the Poisson isomorphism $\theta : S(\mathfrak{g}) \xrightarrow{\cong} \mathcal{O}(\mathfrak{g}^*)$ of (2.6).

8.1. Actions of G and $\mathfrak{g}$. The group G acts on $\mathfrak{g}$ and $\mathfrak{g}^*$ by the adjoint and coadjoint actions, respectively, as in §5.1. In turn, these induce actions of G by $\mathbb{C}$-algebra automorphisms on $S(\mathfrak{g})$ and $\mathcal{O}(\mathfrak{g}^*)$, actions which we also refer to as *adjoint* and *coadjoint actions*. All G-actions we mention will refer to one of these four cases. Let us write $\operatorname{spec}^G S(\mathfrak{g})$ and $\operatorname{spec}^G \mathcal{O}(\mathfrak{g}^*)$ for the sets of G-stable prime ideals in $S(\mathfrak{g})$ and $\mathcal{O}(\mathfrak{g}^*)$, respectively, equipped with the relative topologies from spec $S(\mathfrak{g})$ and spec $\mathcal{O}(\mathfrak{g}^*)$.

We claim that the isomorphism θ is G-equivariant. To see this, let $\{e_1, \dots, e_n\}$ be a basis for $\mathfrak{g}$ and $\{\alpha_1, \dots, \alpha_n\}$ the corresponding dual basis for $\mathfrak{g}^*$. As in §2.6, $\mathcal{O}(\mathfrak{g}^*) = \mathbb{C}[x_1, \dots, x_n]$ where each $x_i = \theta(e_i)$. Given $\gamma \in G$, there are scalars $\gamma_{ij} \in \mathbb{C}$ such that $\gamma.e_j = \sum_i \gamma_{ij} e_i$ for all j. Consequently,

$$(\gamma.x_j)(\alpha_i) = x_j(\gamma^{-1}.\alpha_i) = (\gamma^{-1}.\alpha_i)(e_j) = \alpha_i(\gamma.e_j) = \gamma_{ij}$$

for all i, j, from which we conclude that $\gamma.x_j = \sum_i \gamma_{ij} x_i$ for all j. Therefore $\gamma.\theta(e_j) = \theta(\gamma.e_j)$ for all j, and the G-equivariance of θ follows.

For each $e \in \mathfrak{g}$, the Lie derivation $\operatorname{ad} e = [e, -]$ on $\mathfrak{g}$ extends uniquely to a derivation on $S(\mathfrak{g})$, namely the Hamiltonian $\{e, -\}$. This yields an action of $\mathfrak{g}$ on $S(\mathfrak{g})$ by derivations. We write $\operatorname{spec}^{\mathfrak{g}} S(\mathfrak{g})$ for the set of $\mathfrak{g}$-stable prime ideals of $S(\mathfrak{g})$, equipped with the relative topology from $\operatorname{spec} S(\mathfrak{g})$.

8.2. Lemma.

(a) $\operatorname{spec}^{\mathrm{G}} S(\mathfrak{g}) = \operatorname{spec}^{\mathfrak{g}} S(\mathfrak{g}) = \operatorname{P.spec} S(\mathfrak{g})$.

(b) $\operatorname{spec}^{\mathrm{G}} \mathcal{O}(\mathfrak{g}^*) = \operatorname{P.spec} \mathcal{O}(\mathfrak{g}^*)$.

(c) *θ induces a homeomorphism* $\operatorname{spec}^{\mathfrak{g}} S(\mathfrak{g}) \xrightarrow{\approx} \operatorname{P.spec} \mathcal{O}(\mathfrak{g}^*)$.

Proof. (a) Since $\mathfrak{g}$ generates the algebra $S(\mathfrak{g})$, the $\mathfrak{g}$-stable ideals of $S(\mathfrak{g})$ coincide with the Poisson ideals. Hence, $\operatorname{spec}^{\mathfrak{g}} S(\mathfrak{g}) = \operatorname{P.spec} S(\mathfrak{g})$. By [2, §13.1] or [50, §24.8.3], the $\mathfrak{g}$-stable ideals of $S(\mathfrak{g})$ coincide with the G-stable ideals. From this, we immediately obtain $\operatorname{spec}^{\mathrm{G}} S(\mathfrak{g}) = \operatorname{spec}^{\mathfrak{g}} S(\mathfrak{g})$.

(b)(c) These follow immediately from (a), because θ is both G-equivariant and a Poisson isomorphism. □

Following our previous notation for maximal ideals corresponding to points in varieties, write $\mathfrak{m}_\alpha$ for the maximal ideal of $\mathcal{O}(\mathfrak{g}^*)$ corresponding to a point $\alpha \in \mathfrak{g}^*$.

8.3. Proposition. *Let $\mathfrak{g}$ be a finite dimensional complex Lie algebra and G its adjoint algebraic group. There is a homeomorphism $\phi : \mathfrak{g}^*/G \to \operatorname{P.prim} \mathcal{O}(\mathfrak{g}^*)$ such that $\phi(G.\alpha) = \mathcal{P}(\mathfrak{m}_\alpha)$ for all $\alpha \in \mathfrak{g}^*$.*

Proof. Since $S(\mathfrak{g})$ is isomorphic to $\mathcal{O}(\mathfrak{g}^*)$, its maximal ideal space is homeomorphic to $\mathfrak{g}^*$. A coordinate-free way to express the inverse isomorphism is to send each $\alpha \in \mathfrak{g}^*$ to the ideal $\underline{m}_\alpha = \langle e - \alpha(e) \mid e \in \mathfrak{g} \rangle$ of $S(\mathfrak{g})$. Observe that $\theta(\underline{m}_\alpha) = \mathfrak{m}_\alpha$.

By [2, Lemma 13.2 and proof], there is a topological embedding

$$\tau : \mathfrak{g}^*/G \longrightarrow \operatorname{spec}^{\mathfrak{g}} S(\mathfrak{g})$$

such that $\tau(G.\alpha) = \bigcap_{\gamma \in G} \gamma.\underline{m}_\alpha$ for $\alpha \in \mathfrak{g}^*$. Thus, $\tau(G.\alpha)$ is the largest G-stable ideal of $S(\mathfrak{g})$ contained in $\underline{m}_\alpha$. Invoking [2, §13.1] or [50, §24.8.3] again, we find that $\tau(G.\alpha)$ is the largest $\mathfrak{g}$-stable ideal of $S(\mathfrak{g})$ contained in $\underline{m}_\alpha$. In particular, it now follows from [10, Lemma 3.3.2] that $\tau(G.\alpha)$ is a prime ideal. Hence, we can say that $\tau(G.\alpha)$ equals the largest member of $\operatorname{spec}^{\mathrm{G}} S(\mathfrak{g})$ contained in $\underline{m}_\alpha$. Since θ is G-equivariant, it follows that $\theta\tau(G.\alpha)$ equals the largest member of $\operatorname{spec}^{\mathrm{G}} \mathcal{O}(\mathfrak{g}^*)$ contained in $\mathfrak{m}_\alpha$. In view of Lemma 8.2(b), we conclude that $\theta\tau(G.\alpha) = \mathcal{P}(\mathfrak{m}_\alpha)$.

Combining the above with Lemma 8.2(c), we obtain a topological embedding

$$\phi : \mathfrak{g}^*/G \to \text{P.spec}\, \mathcal{O}(\mathfrak{g}^*)$$

such that $\phi(G.\alpha) = \mathcal{P}(\mathfrak{m}_\alpha)$ for $\alpha \in \mathfrak{g}^*$. Since the image of ϕ is, by definition, $\text{P.prim}\, \mathcal{O}(\mathfrak{g}^*)$, the proposition is proved. □

8.4. Corollary. *Let $\mathfrak{g}$ be a finite dimensional complex Lie algebra and G its adjoint algebraic group. The G-orbits in $\mathfrak{g}^*$ are precisely the symplectic cores.*

Proof. Injectivity and well-definedness of the homeomorphism ϕ of Proposition 8.3 say that for all $\alpha, \beta \in \mathfrak{g}^*$, we have $G.\alpha = G.\beta$ if and only if $\mathcal{P}(\mathfrak{m}_\alpha) = \mathcal{P}(\mathfrak{m}_\beta)$. Thus, α and β lie in the same G-orbit if and only if they lie in the same symplectic core. □

Corollary 8.4 allows us to phrase the Dixmier-Conze-Duflo-Rentschler-Mathieu Theorem in terms of symplectic cores:

8.5. Theorem. *Let $\mathfrak{g}$ be a solvable finite dimensional complex Lie algebra, and let X be the set of symplectic cores in $\mathfrak{g}^*$, with the quotient topology induced from $\mathfrak{g}^*$. Then the Dixmier map induces a homeomorphism $X \to \text{prim}\, U(\mathfrak{g})$.*

8.6. The extended Dixmier map. Continue to assume that $\mathfrak{g}$ is solvable. Via the embedding $\mathfrak{g}^*/G \longrightarrow \text{spec}^{\mathfrak{g}} S(\mathfrak{g})$ from [2, Lemma 13.2] used above, identify $\mathfrak{g}^*/G$ with a subspace of $\text{spec}^{\mathfrak{g}} S(\mathfrak{g})$. Borho, Gabriel, and Rentschler showed that $\overline{\text{Dx}}$ extends uniquely to a continuous map

$$\widetilde{\text{Dx}} : \text{spec}^{\mathfrak{g}} S(\mathfrak{g}) \to \text{spec}\, U(\mathfrak{g}),$$

given by the rule

$$\widetilde{\text{Dx}}(P) = \bigcap \{\text{Dx}(\alpha) \mid \alpha \in \mathfrak{g}^* \text{ and } \underline{m}_\alpha \supseteq P\}$$

for $P \in \text{spec}^{\mathfrak{g}} S(\mathfrak{g})$ [2, Satz 13.4]. They named this the *extended Dixmier map*, and proved that it is a continuous bijection [2, Satz 13.4, Kor. 15.1]. Their methods, combined with Mathieu's theorem, imply that $\widetilde{\text{Dx}}$ is a homeomorphism, as we will see shortly.

8.7. Quasi-homeomorphisms and sauber spaces. Let X and Y be topological spaces. A continuous map $\phi : X \to Y$ is a *quasi-homeomorphism* provided the induced map $F \mapsto \phi^{-1}(F)$ is an isomorphism from the lattice of closed subsets of Y onto the lattice of closed subsets of X. If X is a subspace of Y, the inclusion map $X \to Y$ is a quasi-homeomorphism if and only if $\overline{F \cap X} = F$ for all closed sets $F \subseteq Y$ [2, §1.6]. Borho, Gabriel, and Rentschler observed that the inclusion map $\text{prim}\, U(\mathfrak{g}) \to \text{spec}\, U(\mathfrak{g})$ is a quasi-homeomorphism [2, Beispiel 1.6], as is the above embedding $\mathfrak{g}^*/G \longrightarrow \text{spec}^{\mathfrak{g}} S(\mathfrak{g})$ [2, Lemma 13.2].

A *generic point* of a closed subset $F \subseteq X$ is any point $x \in F$ such that $F = \overline{\{x\}}$. The space X is *sauber* (English: *tidy*) provided every irreducible closed subset of X has precisely one generic point. As observed in [2, §13.3], the prime spectrum of any noetherian ring is sauber. We include the short argument in the lemma below, for the reader's convenience. The same argument shows that $\operatorname{spec}^{\mathfrak{g}} S(\mathfrak{g})$ is sauber. These spaces are noetherian as well, since they have Zariski topologies arising from noetherian rings.

8.8. Lemma. *Let A be a noetherian ring and R a commutative noetherian Poisson k-algebra, with* $\operatorname{char} k = 0$.

(a) *The prime spectrum* $\operatorname{spec} A$ *is a sauber noetherian space, and if A is a Jacobson ring, the inclusion map* $\operatorname{prim} A \to \operatorname{spec} A$ *is a quasi-homeomorphism.*
(b) *The Poisson prime spectrum* $\text{P.spec}\, R$ *is a sauber noetherian space, and if R is an affine k-algebra, the inclusion map* $\text{P.prim}\, R \to \text{P.spec}\, R$ *is a quasi-homeomorphism.*

Proof. (a) Suppose that $F_1 \supseteq F_2 \supseteq \cdots$ is a decreasing sequence of closed sets in $\operatorname{spec} A$. We may write each $F_j = V(I_j)$ where $I_j = \bigcap F_j$. Then $I_1 \subseteq I_2 \subseteq \cdots$ is an increasing sequence of ideals of A. Since this sequence stabilizes, so does the original sequence of closed sets. Thus, $\operatorname{spec} A$ is a noetherian space.

Let $F = V(I)$ be an arbitrary closed subset of $\operatorname{spec} A$, where I is an ideal of A. We may replace I by its prime radical, so there is no loss of generality in assuming that I is semiprime. Since A is noetherian, there are only finitely many prime ideals minimal over I, say $Q_1, \dots, Q_n$, and $I = Q_1 \cap \cdots \cap Q_n$. It follows that $F = V(Q_1) \cup \cdots \cup V(Q_n)$.

If F is irreducible, then $F = V(Q_j)$ for some j. In this case, Q_j is the unique generic point of F, proving that $\operatorname{spec} A$ is sauber.

Now assume that A is a Jacobson ring, so that all prime ideals of A are intersections of primitive ideals. It follows that

$$I = \bigcap F = \bigcap (F \cap \operatorname{prim} A),$$

from which we see that F equals the closure of $F \cap \operatorname{prim} A$ in $\operatorname{spec} A$. Thus, by [2, §1.6], the inclusion map $\operatorname{prim} A \to \operatorname{spec} A$ is a quasi-homeomorphism.

(b) The argument applied in (a) also shows that $\text{P.spec}\, R$ is a noetherian space.

As discussed in §6.1, any closed set F in $\text{P.spec}\, R$ can be written $F = V_P(I)$ for some Poisson ideal I. There are only finitely many prime ideals minimal over I, say $Q_1, \dots, Q_n$, and the Q_i are Poisson ideals by Lemma 6.2. Hence, we may replace I by $Q_1 \cap \cdots \cap Q_n$, and it follows that $F = V_P(Q_1) \cup \cdots \cup V_P(Q_n)$.

Just as in (a), if F is irreducible, $F = V_P(Q_j)$ for some j, and then Q_j is the unique generic point of F. This proves that $\text{P.spec}\, R$ is sauber.

Now assume that R is an affine k-algebra. Then R is a Jacobson ring, and it follows that every Poisson prime ideal of R is an intersection of Poisson-primitive ideals (e.g., see [13, Lemma 1.1(e)]). From this, we conclude as in (a) that the inclusion map $\text{P.prim}\, R \to \text{P.spec}\, R$ is a quasi-homeomorphism. □

8.9. Lemma. *Let $X \subseteq X'$ and $Y \subseteq Y'$ be topological spaces, such that X' and Y' are sauber and noetherian. Assume also that the inclusion maps $X \to X'$ and $Y \to Y'$ are quasi-homeomorphisms. Then any continuous map $\phi : X \to Y$ extends uniquely to a continuous map $\phi' : X' \to Y'$. Moreover, if ϕ is a homeomorphism, so is ϕ'.*

Proof. The existence and uniqueness of ϕ are proved in [2, Lemma 13.3]. The final statement follows by the usual universal property argument. □

8.10. Theorem. [Borho-Gabriel-Rentschler-Mathieu] *Let $\mathfrak{g}$ be a solvable finite dimensional complex Lie algebra. The extended Dixmier map*

$$\widetilde{\mathrm{Dx}} : \operatorname{spec}^{\mathfrak{g}} S(\mathfrak{g}) \longrightarrow \operatorname{spec} U(\mathfrak{g})$$

is a homeomorphism.

Proof. Following the proof of [2, Satz 13.4], recall that $\operatorname{spec}^{\mathfrak{g}} S(\mathfrak{g})$ and $\operatorname{spec} U(\mathfrak{g})$ are sauber noetherian spaces, and that the embedding $\mathfrak{g}^*/G \to \operatorname{spec}^{\mathfrak{g}} S(\mathfrak{g})$ and the inclusion $\operatorname{prim} U(\mathfrak{g}) \to \operatorname{spec} U(\mathfrak{g})$ are quasi-homeomorphisms. The map $\widetilde{\mathrm{Dx}}$ is defined, with the help of Lemma 8.9, to be the unique continuous map from $\operatorname{spec}^{\mathfrak{g}} S(\mathfrak{g})$ to $\operatorname{spec} U(\mathfrak{g})$ extending $\overline{\mathrm{Dx}}$. Since $\overline{\mathrm{Dx}}$ is a homeomorphism, Lemma 8.9 implies that $\widetilde{\mathrm{Dx}}$ is a homeomorphism. □

In Poisson-ideal-theoretic terms, Theorems 5.4 and 8.10 can be restated as follows.

8.11. Theorem. *Let $\mathfrak{g}$ be a solvable finite dimensional complex Lie algebra. Then there is a homeomorphism*

$$\psi : \mathrm{P.prim}\, \mathcal{O}(\mathfrak{g}^*) \longrightarrow \operatorname{prim} U(\mathfrak{g})$$

such that $\psi(\mathcal{P}(\mathfrak{m}_\alpha)) = \mathrm{Dx}(\alpha)$ for $\alpha \in \mathfrak{g}^*$, and ψ extends uniquely to a homeomorphism

$$\mathrm{P.spec}\, \mathcal{O}(\mathfrak{g}^*) \longrightarrow \operatorname{spec} U(\mathfrak{g}).$$

Proof. To obtain ψ, just compose the factorized Dixmier map $\overline{\mathrm{Dx}}$ with the inverse of the homeomorphism ϕ of Proposition 8.3. By Lemma 8.8, $\mathrm{P.spec}\, \mathcal{O}(\mathfrak{g}^*)$ and $\operatorname{spec} U(\mathfrak{g})$ are sauber noetherian spaces, and the inclusion maps $\mathrm{P.prim}\, \mathcal{O}(\mathfrak{g}^*) \to \mathrm{P.spec}\, \mathcal{O}(\mathfrak{g}^*)$ and $\operatorname{prim} U(\mathfrak{g}) \to \operatorname{spec} U(\mathfrak{g})$ are quasi-homeomorphisms. Therefore the existence and uniqueness of the desired extension of ψ follow from Lemma 8.9. □

9. Modified conjectures for quantized coordinate rings

In light of Theorems 7.1, 7.4, 8.5, and 8.11, we nominate the concept of symplectic cores as the *best algebraic approximation* for symplectic leaves. Further, we suggest that symplectic leaves should be replaced by symplectic cores in applications of the Orbit Method to algebraic problems. In particular, we revise and refine the general principle discussed in §4.4 to the following conjecture. It is, of necessity, somewhat imprecise, given the lack of a precise definition of the concept of quantized coordinate rings.

9.1. Primitive spectrum conjecture for quantized coordinate rings.

Assume that k is algebraically closed of characteristic zero, and let A be a generic quantized coordinate ring of an affine algebraic variety V over k. Then A should be a member of a flat family of k-algebras with semiclassical limit $\mathcal{O}(V)$, such that $\operatorname{prim} A$ *is homeomorphic to the space of symplectic cores in V, with respect to the semiclassical limit Poisson structure. Further, there should be compatible homeomorphisms* $\operatorname{prim} A \to \text{P.prim}\,\mathcal{O}(V)$ *and* $\operatorname{spec} A \to \text{P.spec}\,\mathcal{O}(V)$.

Each of the known types of quantized coordinate rings supports an action of an algebraic torus $H = (k^\times)^m$ (see [4, §§II.1.14–18] for a summary), which has a parallel action (by Poisson automorphisms) on the semiclassical limit (e.g., see [19, §0.2]; [14, Section 2]). We tighten the conjecture above and posit that there should exist homeomorphisms as described which are also equivariant with respect to the relevant torus actions.

9.2. Remarks

(a) The discussion of the simple example $A_q = \mathcal{O}_q(\mathbb{C}^2)$ in §4.5 indicates why Conjecture 9.1 is restricted to generic quantized coordinate rings. In particular, $\operatorname{prim} A_q$ has a generic point when q is not a root of unity, but no generic points otherwise. Since $\text{P.spec}\,\mathbb{C}[x,y]$ has a generic point, it is not homeomorphic to $\operatorname{prim} A_q$ when q is a root of unity.

(b) Each of the "standard" single parameter quantized coordinate rings is defined as a member of a one-parameter family of algebras, and it is this (flat) family to which the conjecture is meant to apply. For instance, the algebras $\mathcal{O}_q(SL_n(k))$ (with n fixed) are defined for all $q \in k^\times$ in the same way (e.g., [4, §I.2.4]), and substituting an indeterminate t for q in the definition results in a torsionfree $k[t^{\pm 1}]$-algebra A with $A/(t-q)A \cong \mathcal{O}_q(SL_n(k))$ for all $q \in k^\times$, just as with the case $n = 2$ in §§1.6, 2.2(c). The semiclassical limit is $\mathcal{O}(SL_n(k))$ with the Poisson bracket satisfying

$$\begin{aligned}
\{X_{ij}, X_{im}\} &= X_{ij}X_{im} && (j < m)\\
\{X_{ij}, X_{lj}\} &= X_{ij}X_{lj} && (i < l)\\
\{X_{ij}, X_{lm}\} &= 0 && (i < l,\ j > m)\\
\{X_{ij}, X_{lm}\} &= 2X_{im}X_{lj} && (i < l,\ j < m).
\end{aligned}$$

This Poisson structure and the above flat family should feature in the SL_n case of Conjecture 9.1, that is, for q not a root of unity, $\operatorname{prim}\mathcal{O}_q(SL_n(k))$ should be homeomorphic to the space of symplectic cores in $SL_n(k)$ and to $\text{P.prim}\,\mathcal{O}(SL_n(k))$, and $\operatorname{spec}\mathcal{O}_q(SL_n(k))$ should be homeomorphic to $\text{P.spec}\,\mathcal{O}(SL_n(k))$. Such a "standard" version of the conjecture is to be posed for $\mathcal{O}_q(M_n(k))$, $\mathcal{O}_q(GL_n(k))$, $\mathcal{O}_q(G)$, and other "standard" cases.

The situation is more involved for "nonstandard" cases, and for multiparameter families, which have to be reduced to single parameter families in order to obtain semiclassical limits. In such cases, the conjecture may be sensitive to the choice of flat family – different flat families may yield different Poisson structures in the semiclassical limit, and the conjecture may hold for some of these semiclassical limits but not for others. This phenomenon appears in an example of Vancliff [52, Example 3.14], which we discuss in Example 9.9.

(c) As discussed at the end of Example 2.6, the enveloping algebra of a finite dimensional Lie algebra $\mathfrak{g}$ is a generic member of the flat family given by the homogenization of $U(\mathfrak{g})$, and so $U(\mathfrak{g})$ should qualify as a generic quantized coordinate ring of $\mathfrak{g}^*$. The semiclassical limit of this family is the Poisson algebra $\mathcal{O}(\mathfrak{g}^*)$. For this setting, K.A. Brown has noted difficulties with Conjecture 9.1 in what one might expect to be the most canonical case, namely when $\mathfrak{g}$ is semisimple [3]. Following the Orbit Method, one would seek a bijection $\mathcal{L} \longleftrightarrow P$ between symplectic leaves in $\mathfrak{g}^*$ and primitive ideals in $U(\mathfrak{g})$ such that the Gelfand-Kirillov dimension of $U(\mathfrak{g})/P$ equals the dimension of $\mathcal{L}$. In particular, the zero-dimensional symplectic leaves of $\mathfrak{g}^*$, which are the same as the zero-dimensional symplectic cores, should match up with the maximal ideals of finite codimension in $U(\mathfrak{g})$. However, $U(\mathfrak{g})$ has infinitely many such maximal ideals, while there is only one zero-dimensional symplectic leaf in $\mathfrak{g}^*$. (The latter can be verified by using Theorem 4.2 together with the fact that the identification of $\mathfrak{g}^*$ with $\mathfrak{g}$ via the Killing form identifies the coadjoint orbits in $\mathfrak{g}^*$ with the adjoint orbits in $\mathfrak{g}$ [8, p. 12].)

Other differences are already visible in the case $\mathfrak{g} = \mathfrak{sl}_2(\mathbb{C})$. As is easily computed, all but one of the coadjoint orbits in $\mathfrak{g}^*$ are closed (compare with the adjoint orbits, computed in [8, Example 1.2.1]). It follows (using Proposition 8.3, or by direct computation) that all but one of the points of P.prim $\mathcal{O}(\mathfrak{g}^*)$ are closed. However, prim $U(\mathfrak{g})$ has infinitely many non-closed points, and therefore it is not homeomorphic to P.prim $\mathcal{O}(\mathfrak{g}^*)$.

(d) Whenever Conjecture 9.1 does hold, the space of symplectic cores in V must be homeomorphic to P.prim $\mathcal{O}(V)$. It is an open question whether the space of symplectic cores for an arbitrary affine Poisson algebra R is homeomorphic to P.prim R, but this does hold when R satisfies the *Poisson Dixmier-Moeglin equivalence*, as follows from [13, Theorem 1.5]; we excerpt the basic argument in Lemma 9.3. This equivalence requires that the Poisson-primitive ideals of R coincide with the locally closed points of P.spec R, and with those Poisson prime ideals P of R for which the Poisson center (cf. §9.6(b)) of the quotient field of R/P is algebraic over k. It holds for the semiclassical limits of many quantized coordinate rings via [13, Theorem 4.1], as shown in [14, Section 2].

(e) As in Theorem 8.11, the existence of a homeomorphism prim $A \to$ P.prim $\mathcal{O}(V)$ as in the conjecture typically implies the existence of a compatible homeomorphism spec $A \to$ P.spec $\mathcal{O}(V)$. We display this in Lemma 9.4 below for emphasis. On the other hand, a homeomorphism spec $A \to$ P.spec $\mathcal{O}(V)$ will restrict to a homeomorphism $\operatorname{prim} A \to \operatorname{P.prim} \mathcal{O}(V)$ provided $\mathcal{O}(V)$ satisfies the Poisson Dixmier-Moeglin equivalence and A satisfies the *Dixmier-Moeglin equivalence*. The latter equivalence requires that the primitive ideals of A coincide with the locally closed points of $\operatorname{spec} A$, and with those prime ideals P of A for which the center of the Goldie quotient ring of A/P is algebraic over k. It was verified for many quantized coordinate rings in [16] (see [4, Corollary II.8.5] for a summary).

9.3. Lemma. *Let R be a commutative affine Poisson k-algebra, and assume that all Poisson-primitive ideals of R are locally closed points in* $\operatorname{P.spec} R$. *Then the Zariski topology on* $\operatorname{P.prim} R$ *coincides with the quotient topology induced by the Poisson core map* $\mathcal{P}(-) : \operatorname{maxspec} R \to \operatorname{P.prim} R$. *Consequently, the space of symplectic cores in* $\operatorname{maxspec} R$ *is homeomorphic to* $\operatorname{P.prim} R$.

Proof. Observe first that the map $\mathcal{P}(-)$ is continuous. It is surjective by definition of $\operatorname{P.prim} R$.

We claim that $P = \bigcap \{\mathfrak{m} \in \operatorname{maxspec} R \mid \mathcal{P}(\mathfrak{m}) = P\}$, for any Poisson-primitive ideal P of R. Since P is locally closed in $\operatorname{P.spec} R$ (by assumption), the singleton $\{P\}$ is open in its closure $V_P(P)$, and so $\{P\} = V_P(P) \setminus V_P(J)$ for some Poisson ideal J of R. Note that $J \not\subseteq P$; hence, after replacing J by $J + P$, we may assume that $J \supsetneq P$. If $\mathfrak{m} \supseteq P$ is a maximal ideal such that $\mathcal{P}(\mathfrak{m}) \neq P$, then $\mathfrak{m} \supseteq \mathcal{P}(\mathfrak{m}) \supseteq J$. The remaining maximal ideals containing P must intersect to P by the Nullstellensatz, verifying the claim.

Now consider a set $X \subseteq \operatorname{P.prim} R$ whose inverse image, Y, under $\mathcal{P}(-)$ is closed in $\operatorname{maxspec} R$. Thus,

$$Y = \{\mathfrak{m} \in \operatorname{maxspec} R \mid \mathcal{P}(\mathfrak{m}) \in X\} = \{\mathfrak{m} \in \operatorname{maxspec} R \mid \mathfrak{m} \supseteq I\}$$

for some ideal I of R. If $P \in X$ and $\mathfrak{m} \in \operatorname{maxspec} R$ with $\mathcal{P}(\mathfrak{m}) = P$, then $\mathfrak{m} \in Y$, and so $\mathfrak{m} \supseteq I$. By the claim above, the intersection of these maximal ideals equals P, and thus $P \supseteq I$. Conversely, if $P \in \operatorname{P.prim} R$ and $P \supseteq I$, then $P = \mathcal{P}(\mathfrak{m})$ for some maximal ideal $\mathfrak{m} \supseteq I$, whence $\mathfrak{m} \in Y$ and $P \in X$. Therefore $X = \{P \in \operatorname{P.prim} R \mid P \supseteq I\}$, a closed set in $\operatorname{P.prim} R$. This proves that the topology on $\operatorname{P.prim} R$ is the quotient topology inherited from $\operatorname{maxspec} R$ via $\mathcal{P}(-)$.

The final statement of the lemma follows directly. □

9.4. Lemma. *Let A be a noetherian k-algebra and R a commutative noetherian Poisson k-algebra, with* $\operatorname{char} k = 0$.

(a) *A bijection* $\phi : \operatorname{spec} A \to \operatorname{P.spec} R$ *is a homeomorphism if and only if ϕ and ϕ^{-1} preserve inclusions.*
(b) *Assume that A is a Jacobson ring and R an affine k-algebra. Then any homeomorphism* $\operatorname{prim} A \to \operatorname{P.prim} R$ *extends uniquely to a homeomorphism* $\operatorname{spec} A \to \operatorname{P.spec} R$.
(c) *Assume that A satisfies the Dixmier-Moeglin equivalence and R the Poisson Dixmier-Moeglin equivalence. Then any homeomorphism* $\operatorname{spec} A \to \operatorname{P.spec} R$ *restricts to a homeomorphism* $\operatorname{prim} A \to \operatorname{P.prim} R$.

Proof. (a) For $P, Q \in \operatorname{spec} A$, we have $P \subseteq Q$ if and only if $Q \in \overline{\{P\}}$, and similarly in P.spec R. Hence, any homeomorphism between these spaces must preserve inclusions.

Conversely, if ϕ and ϕ^{-1} preserve inclusions, then $\phi(V(P)) = V_P(\phi(P))$ for all $P \in \operatorname{spec} A$. Since the closed sets in spec A are exactly the finite unions of $V(P)$s (recall the proof of Lemma 8.8(a)), it follows that ϕ sends closed sets to closed sets, i.e., ϕ^{-1} is continuous. Similarly, ϕ is continuous, and hence a homeomorphism.

(b) Lemmas 8.8 and 8.9.

(c) Under the assumed equivalences, prim A consists of the locally closed points in spec A, and P.prim R consists of the locally closed points in P.spec R. □

9.5. Example. Let $A_q = \mathcal{O}_q(k^2)$, where k is algebraically closed of characteristic 0 and $q \in k^\times$. View $R = \mathcal{O}(k^2)$ as the semiclassical limit of the family $(A_q)_{q \in k^\times}$, with the Poisson structure exhibited in Example 2.2(a). The torus $H = (k^\times)^2$ acts on A_q via algebra automorphisms and on R via Poisson automorphisms so that (in both cases) $(\alpha_1, \alpha_2).x_i = \alpha_i x_i$ for $(\alpha_1, \alpha_2) \in H$ and $i = 1, 2$.

Assume that q is not a root of unity. As is easily checked (e.g., [4, Example II.1.2]), the prime ideals of A_q are

- the maximal ideals $\langle x_1 - \alpha,\, x_2 \rangle$ and $\langle x_1,\, x_2 - \beta \rangle$, for $\alpha, \beta \in k$;
- (◇) the height 1 primes $\langle x_1 \rangle$ and $\langle x_2 \rangle$;
- the zero ideal.

All of these prime ideals, except for $\langle x_1 \rangle$ and $\langle x_2 \rangle$, are primitive [4, Example II.7.2]. The closed sets in spec A_q are easily found, but we shall not list them here – see [4, Example II.1.2 and Exercise II.1.C].

With very similar computations, one finds the Poisson prime and Poisson-primitive ideals in R, and a list of the closed subsets of P.spec R. In terms of notation, the answers are the same as for A_q – the list (◇) also describes the Poisson prime ideals of R, and all of these ideals, except for $\langle x_1 \rangle$ and $\langle x_2 \rangle$, are Poisson-primitive. We conclude that there exist compatible homeomorphisms prim $A_q \to$ P.prim R and spec $A_q \to$ P.spec R, sending the ideal $\langle x_1 - \alpha,\, x_2 \rangle$ of A_q to the ideal $\langle x_1 - \alpha,\, x_2 \rangle$ of R, and so on. (We say that these maps are given by "preservation of notation".) These homeomorphisms are equivariant with respect to the actions of H described above.

By inspection, all Poisson-primitive ideals of R are locally closed in P.spec R. Consequently, we conclude from Lemma 9.3 that the space of symplectic cores in maxspec $R \approx k^2$ is homeomorphic to P.prim R.

Analyzing the prime ideals in a quantized coordinate ring typically involves investigating localizations of factor algebras, which often turn out to be quantum tori. We sketch some basic procedures used to determine prime ideals in quantum tori, and similar ones for the analogous "Poisson tori".

9.6. Some computational tools

(a) A *quantum torus* over k is the localization of a quantum affine space $\mathcal{O}_{\boldsymbol{q}}(k^n)$ obtained by inverting the generators x_i, that is, an algebra

$$\mathcal{O}_{\boldsymbol{q}}((k^\times)^n) = k\langle x_1^{\pm 1}, \dots, x_n^{\pm 1} \mid x_i x_j = q_{ij} x_j x_i \text{ for all } i, j\rangle,$$

where $\boldsymbol{q} = (q_{ij})$ is a multiplicatively antisymmetric $n \times n$ matrix over k. Set $T = \mathcal{O}_{\boldsymbol{q}}((k^\times)^n)$.

Since T is a $\mathbb{Z}^n$-graded algebra, with 1-dimensional homogeneous components, its center, $Z(T)$, is spanned by central monomials [21, Lemma 1.1]. The latter are easily computed: a monomial $x_1^{m_1} x_2^{m_2} \cdots x_n^{m_n}$ is central if and only if $\prod_{j=1}^n q_{ij}^{m_j} = 1$ for all i. All ideals of T are induced from ideals of $Z(T)$ [21, Theorem 1.2]; [15, Proposition 1.4], from which it follows that contraction and extension give inverse homeomorphisms between $\operatorname{spec} T$ and $\operatorname{spec} Z(T)$ [15, Corollary 1.5(b)]. In particular, it follows from the above facts that T is a simple algebra if and only if $Z(T) = k$ [36, Proposition 1.3].

(b) Let $R = k[x_1^{\pm 1}, \dots, x_n^{\pm 1}]$ be a Laurent polynomial ring, equipped with a Poisson bracket such that $\{x_i, x_j\} = \pi_{ij} x_i x_j$ for all i, j, where (π_{ij}) is an antisymmetric $n \times n$ matrix over k. The results of part (a) all have Poisson analogs for R, as follows.

The *Poisson center* of R, denoted $Z_P(R)$, is the subalgebra consisting of those $r \in R$ for which the derivation $\{r, -\}$ vanishes. Since the Poisson bracket on R respects the $\mathbb{Z}^n$-grading, $Z_P(R)$ is spanned by the monomials it contains [21, Lemma 2.1]; [52, Lemma 1.2(a)]. A monomial $x_1^{m_1} x_2^{m_2} \cdots x_n^{m_n}$ is Poisson central if and only if $\sum_{j=1}^n \pi_{ij} m_j = 0$ for all i. All Poisson ideals of R are induced from ideals of $Z_P(R)$ [21, Theorem 2.2]; [52, Lemma 1.2(b)], from which it follows that contraction and extension give inverse homeomorphisms between $\operatorname{P.spec} R$ and $\operatorname{spec} Z_P(R)$. In particular, it follows from the above facts that R is Poisson-simple (meaning that it has no proper nonzero Poisson ideals) if and only if $Z_P(R) = k$.

9.7. Example. Let $A_q = \mathcal{O}_q(SL_2(k))$, where k is algebraically closed of characteristic 0 and $q \in k^\times$. View $R = \mathcal{O}(SL_2(k))$ as the semiclassical limit of the family $(A_q)_{q \in k^\times}$, with the Poisson structure exhibited in Example 2.2(c). The torus $H = (k^\times)^2$ again acts on A_q and R, this time so that

$$(\alpha, \beta).X_{11} = \alpha\beta X_{11} \qquad (\alpha, \beta).X_{12} = \alpha\beta^{-1} X_{12}$$
$$(\alpha, \beta).X_{21} = \alpha^{-1}\beta X_{21} \quad (\alpha, \beta).X_{22} = \alpha^{-1}\beta^{-1} X_{22}$$

for $(\alpha, \beta) \in H$.

Now restrict q to a non-root of unity. The prime ideals of A_q can be computed with the tools of §9.6(a), as outlined in [4, Exercise II.1.D]. For instance, one checks that A_q has a localization

$$A_q[X_{11}^{-1}, X_{12}^{-1}, X_{21}^{-1}] \cong k\langle x^{\pm 1}, y^{\pm 1}, z^{\pm 1} \mid xy = qyx,\ xz = qzx,\ yz = zy\rangle,$$

and that the center of the latter algebra is $k[(yz^{-1})^{\pm 1}]$. It follows that the prime ideals of A_q not containing X_{12} or X_{21} consist of $\langle 0 \rangle$ and $\langle X_{12} - \lambda X_{21} \rangle$, for $\lambda \in k^\times$.

The full list of prime ideals of A_q is as follows:

- the maximal ideals $\langle X_{11}-\lambda,\, X_{12},\, X_{21},\, X_{22}-\lambda^{-1}\rangle$, for $\lambda \in k^\times$;
- (♠) the ideal $\langle X_{12},\, X_{21}\rangle$;
- the height 1 primes $\langle X_{21}\rangle$ and $\langle X_{12}-\lambda X_{21}\rangle$, for $\lambda \in k$;
- the zero ideal.

A diagram of spec A_q, with inclusions marked, is given in [4, Diagram II.1.3].

A similar computation, using §9.6(b), yields the Poisson prime ideals of R, which can be described exactly as in (♠). This provides a natural H-equivariant bijection $\phi : \operatorname{spec} A_q \to \operatorname{P.spec} R$, given by "preservation of notation". By inspection, ϕ and ϕ^{-1} preserve inclusions, and thus, by Lemma 9.4(a), ϕ is a homeomorphism.

The algebra A_q satisfies the Dixmier-Moeglin equivalence by [4, Corollary II.8.5], and R satisfies the Poisson Dixmier-Moeglin equivalence by [13, Theorem 4.3]. Therefore Lemma 9.4(c) implies that ϕ restricts to a homeomorphism $\operatorname{prim} A_q \to \operatorname{P.prim} R$. In A_q, all prime ideals are primitive except for $\langle X_{12},\, X_{21}\rangle$ and $\langle 0\rangle$ (cf. [4, Example II.8.6]). Similarly, in R all Poisson prime ideals are Poisson-primitive except for $\langle X_{12},\, X_{21}\rangle$ and $\langle 0\rangle$. As in the previous example, we can use Lemma 9.3 to see that the space of symplectic cores in $\operatorname{maxspec} R \approx SL_2(k)$ is homeomorphic to $\operatorname{P.prim} R$.

9.8. Evidence for Conjecture 9.1. In most of the instances discussed below, k is assumed to be algebraically closed of characteristic zero.

(a) Examples 9.5 and 9.7 are the most basic instances in which the conjecture has been verified. In the same way (although with somewhat more effort), one can verify it for $\mathcal{O}_q(GL_2(k))$. In particular, most of the work required to determine the prime ideals in the generic $\mathcal{O}_q(GL_2(k))$ is done in [4, Example II.8.7].

(b) We next turn to the quantized coordinate rings $\mathcal{O}_q(G)$ and $\mathcal{O}_{q,p}(G)$ over $k = \mathbb{C}$, where G is a connected semisimple complex Lie group, $q \in k^\times$ is not a root of unity, and p is an antisymmetric bicharacter on the weight lattice of G (as in [24, §3.4]).

The Poisson structure on $\mathcal{O}(G)$ resulting from the semiclassical limit process gives G the combined structure of a *Poisson-Lie group* (e.g., see [33, Chapter 1] for the concept, and [22, §A.1] for the result). There is a known recipe for the symplectic leaves in G in case the Poisson structure arises from the standard quantization [22, Appendix A], and similarly in the multiparameter "algebraic" case [24, Theorem 1.8]. In both these cases, it follows that the symplectic leaves are Zariski locally closed (see [5, Theorem 1.9] for a more explicit statement). Hence, the symplectic leaves in G coincide with the symplectic cores (Theorem 7.1(b)).

As discussed in §4.4, Hodges and Levasseur put forward the conjecture that there should be a bijection between $\operatorname{prim} \mathcal{O}_q(G)$ and the set of symplectic leaves in G [22, §2.8, Conjecture 1]. Such bijections were developed for $G = SL_n(\mathbb{C})$ in [23], and for general G in [27] and [24]. More generally, Hodges, Levasseur, and Toro established a bijection between $\operatorname{prim} \mathcal{O}_{q,p}(G)$ and the set of symplectic leaves in G in the algebraic case. All these bijections are equivariant with respect to natural actions of a maximal torus of G.

Except for the case $G = SL_2(\mathbb{C})$ covered in Example 9.7, the topological properties of the above bijections are not known. Even when $G = SL_3(\mathbb{C})$, it is not known whether $\operatorname{prim} \mathcal{O}_q(G)$ is homeomorphic to the space of symplectic leaves (= cores) in G.

(c) The prime and primitive spectra of general multiparameter quantum affine spaces $\mathcal{O}_{\boldsymbol{q}}(k^n)$ were analyzed by Goodearl and Letzter in [17], assuming k algebraically closed together with a minor technical assumption (that either $\operatorname{char} k = 2$, or -1 is not in the subgroup $\langle q_{ij} \rangle \subseteq k^\times$). They proved that there are compatible topological quotient maps $k^n \approx \operatorname{maxspec} \mathcal{O}(k^n) \to \operatorname{prim} \mathcal{O}_{\boldsymbol{q}}(k^n)$ and $\operatorname{spec} \mathcal{O}(k^n) \to \operatorname{spec} \mathcal{O}_{\boldsymbol{q}}(k^n)$, equivariant with respect to natural actions of the torus $(k^\times)^n$ [17, Theorem 4.11]. Similar results were proved not only for quantum tori $\mathcal{O}_{\boldsymbol{q}}((k^\times)^n)$ [17, Theorem 3.11] but also for quantum affine toric varieties [17, Theorem 6.3].

Oh, Park, and Shin converted these topological quotient results into the following (assuming $\operatorname{char} k = 0$ and $-1 \notin \langle q_{ij} \rangle \subseteq k^\times$): For each $\mathcal{O}_{\boldsymbol{q}}(k^n)$, there is a Poisson structure on $\mathcal{O}(k^n)$ such that there are compatible homeomorphisms $\operatorname{P.prim} \mathcal{O}(k^n) \to \operatorname{prim} \mathcal{O}_{\boldsymbol{q}}(k^n)$ and $\operatorname{P.spec} \mathcal{O}(k^n) \to \operatorname{spec} \mathcal{O}_{\boldsymbol{q}}(k^n)$ [41, Theorem 3.5]. Goodearl and Letzter, finally, showed that such homeomorphisms could be obtained for semiclassical limit Poisson structures [18, Theorem 3.6], and extended the results to quantum affine toric varieties [18, Theorem 5.2].

All these Poisson algebra structures on $\mathcal{O}(k^n)$ satisfy the Poisson Dixmier-Moeglin equivalence [13, Example 4.6]. Hence, the space of symplectic cores in k^n is homeomorphic to $\operatorname{prim} \mathcal{O}_{\boldsymbol{q}}(k^n)$, via Lemma 9.3. The symplectic cores in k^n are algebraic, whereas this does not always hold for the symplectic leaves, as shown by Vancliff [52, Corollary 3.4]. An explicit example is computed in [18, Example 3.10].

(d) The prime and primitive spectra of the algebras $K^{P,Q}_{n,\Gamma}(k)$ introduced by Horton [25] were analyzed by Oh in [40]. These algebras are multiparameter quantizations of $\mathcal{O}(k^{2n})$, and include quantum symplectic spaces $\mathcal{O}_q(\mathfrak{sp}\, k^{2n})$, even-dimensional quantum Euclidean spaces $\mathcal{O}_q(\mathfrak{o}\, k^{2n})$, and quantum Heisenberg spaces, among others. Oh introduced Poisson algebra structures $A^{P,Q}_{n,\Gamma}(k)$ on $\mathcal{O}(k^{2n})$, and constructed compatible homeomorphisms $\operatorname{P.prim} A^{P,Q}_{n,\Gamma}(k) \to \operatorname{prim} K^{P,Q}_{n,\Gamma}(k)$ and $\operatorname{P.spec} A^{P,Q}_{n,\Gamma}(k) \to \operatorname{spec} K^{P,Q}_{n,\Gamma}(k)$, assuming the parameters involved in P, Q, Γ are suitably generic [40, Theorem 4.14].

As stated in Remark 9.2(b), a quantized coordinate ring may belong to some flat families for which Conjecture 9.1 holds and also to others for which it fails. We outline Vancliff's example [52, Example 3.14] illustrating this phenomenon.

9.9. Example

(a) Let $a_i = i - 1$ for $i = 1, 2, 3$, and set

$$R_0 = \mathbb{C}[h][(1 + a_i h)^{-1} \mid i = 1, 2, 3]$$
$$A = R_0 \langle x_1, x_2, x_3 \mid x_i x_j = r_{ij} x_j x_i \text{ for } i, j = 1, 2, 3 \rangle,$$

where $r_{ij} = (1+a_ih)(1+a_jh)^{-1}$ for all i, j. This defines a flat family of $\mathbb{C}$-algebras, whose semiclassical limit is the polynomial ring $R = \mathbb{C}[x_1, x_2, x_3]$ with the Poisson bracket satisfying

$$\{x_1, x_2\} = -x_1x_2 \quad \{x_1, x_3\} = -2x_1x_3 \quad \{x_2, x_3\} = -x_2x_3 \,.$$

It follows from [52, Corollary 3.4] that the symplectic leaves in $\mathbb{C}^3$ for this Poisson structure are algebraic; hence, they coincide with the symplectic cores (Theorem 7.1(b)). By [13, Example 4.6], R satisfies the Poisson Dixmier-Moeglin equivalence, and so Lemma 9.3 implies that the space of symplectic leaves in $\mathbb{C}^3$ is homeomorphic to P.prim R.

The Poisson-primitive ideals of R are listed in [52, Example 3.14] (where they are labelled "maximal Poisson ideals"). They consist of

($\diamondsuit$) • the maximal ideals $\langle x_1 - \alpha, x_2, x_3\rangle$, $\langle x_1, x_2 - \beta, x_3\rangle$, $\langle x_1, x_2, x_3 - \gamma\rangle$, for $\alpha, \beta, \gamma \in \mathbb{C}$;
- the height 1 primes $\langle x_1\rangle$, $\langle x_2\rangle$, $\langle x_3\rangle$, and $\langle x_1x_3 - \lambda x_2^2\rangle$, for $\lambda \in \mathbb{C}^\times$.

We can compute them by using §9.6(b) to find the Poisson prime ideals of R and then applying the Poisson Dixmier-Moeglin equivalence. For instance, the Poisson center of the localization $\mathbb{C}[x_1^{\pm 1}, x_2^{\pm 1}, x_3^{\pm 1}]$ is $\mathbb{C}[(x_1x_2^{-2}x_3)^{\pm 1}]$, from which it follows that any nonzero Poisson prime ideal of R must contain either one of the x_i or else $x_1x_3 - \lambda x_2^2$ for some $\lambda \in \mathbb{C}^\times$. The full list of Poisson prime ideals of R is

- $\langle x_1 - \alpha, x_2, x_3\rangle$, $\langle x_1, x_2 - \beta, x_3\rangle$, $\langle x_1, x_2, x_3 - \gamma\rangle$, for $\alpha, \beta, \gamma \in \mathbb{C}$;
- $\langle x_1, x_2\rangle$, $\langle x_1, x_3\rangle$, $\langle x_2, x_3\rangle$;
- $\langle x_1\rangle$, $\langle x_2\rangle$, $\langle x_3\rangle$, $\langle x_1x_3 - \lambda x_2^2\rangle$, for $\lambda \in \mathbb{C}^\times$;
- $\langle 0\rangle$.

Inspection immediately shows that the locally closed points of P.spec R are the Poisson prime ideals listed in ($\diamondsuit$).

(b) A generic member of the flat family given by A is $B_q = A/(h-q)A$, where q is a complex scalar such that $1+q$ and $1+2q$ generate a free abelian subgroup of rank 2 in $\mathbb{C}^\times$. For such q, the primitive ideals of B_q, as stated in [52, Example 3.14], consist of

- $\langle x_1 - \alpha, x_2, x_3\rangle$, $\langle x_1, x_2 - \beta, x_3\rangle$, $\langle x_1, x_2, x_3 - \gamma\rangle$, for $\alpha, \beta, \gamma \in \mathbb{C}$;
- $\langle x_1\rangle$, $\langle x_2\rangle$, $\langle x_3\rangle$, $\langle 0\rangle$.

These can be computed by finding the prime ideals using §9.6(a) and then applying the Dixmier-Moeglin equivalence, which holds because B_q is a quantum affine space [15, Corollary 2.5].

Observe that prim B_q is not homeomorphic to P.prim R. For instance, prim B_q has a generic point, while P.prim R does not.

(c) In contrast to the above, any generic B_q is a member of a flat family with a semiclassical limit R' (the algebra R, but with a different Poisson structure) such that prim $B_q \approx$ P.prim R' and spec $B_q \approx$ P.spec R', by [18, Theorem 3.6].

It would be very interesting to obtain criteria to determine which flat families yield "good" semiclassical limits relative to Conjecture 9.1. For quantum affine spaces and their Poisson analogs, one good condition appears in the work of Oh, Park, and Shin [41, Theorem 3.5] – roughly, if the scalars appearing in the defining Poisson brackets of the semiclassical limit arise from an embedding into the additive group $(k,+)$ of the subgroup of $k^{\times}$ generated by the scalars appearing in the defining commutation relations of the quantum affine space, then the prime and primitive spectra of the quantum affine space are homeomorphic to the Poisson prime and Poisson-primitive spectra of the semiclassical limit.

We close with an example of the "simplest possible" quantum group for which primitive ideals match symplectic cores but not symplectic leaves. There is no nontrivial multiparameter version of quantum SL_2, and to deal with $\mathcal{O}_{q,p}(SL_3(\mathbb{C}))$ would require investigating 36 families of primitive ideals (indexed by $S_3 \times S_3$, as in [24, Corollary 4.5]). Instead, we look at a multiparameter quantization of GL_2. It is convenient to use Takeuchi's original presentation [49].

9.10. Example. For the classification of primitive ideals, we assume only that k is algebraically closed, and we choose a generic pair of parameters $p, q \in k^{\times}$, meaning that they generate a free abelian subgroup of rank 2 in $k^{\times}$. We restrict to $k = \mathbb{C}$ and special choices of p and q when setting up a semiclassical limit and discussing symplectic leaves.

(a) Define the two-parameter quantum 2×2 matrix algebra $M_{q^{-1},p}$ as in [49]. This is the k-algebra with generators X_{11}, X_{12}, X_{21}, X_{22} and relations

$$\begin{aligned} X_{11}X_{12} &= qX_{12}X_{11} & X_{11}X_{21} &= p^{-1}X_{21}X_{11} \\ X_{21}X_{22} &= qX_{22}X_{21} & X_{12}X_{22} &= p^{-1}X_{22}X_{12} \\ X_{12}X_{21} &= (pq)^{-1}X_{21}X_{12} \quad & X_{11}X_{22} - X_{22}X_{11} &= (q-p)X_{12}X_{21}\,. \end{aligned}$$

The element $D = X_{11}X_{22} - qX_{12}X_{21}$ is the quantum determinant in $M_{q^{-1},p}$, but it is normal rather than central:

$$X_{ij}D = (pq)^{i-j}DX_{ij}$$

for all i, j [49, §2]. Since the powers of D form an Ore set, we can construct the Ore localization $A = A_{q^{-1},p} = M_{q^{-1},p}[D^{-1}]$. There is a Hopf algebra structure on A [49, §2], but we do not need that here.

For comparison with other presentations of multiparameter quantized coordinate rings, we point out that $A = \mathcal{O}_{pq^{-1},\boldsymbol{p}}(GL_2(k))$ in the notation of [12, §1.3]; [4, §I.2.4]), where $\boldsymbol{p} = \begin{bmatrix} 1 & q^{-1} \\ q & 1 \end{bmatrix}$. In particular, [4, Corollary II.6.10] applies, implying that all prime ideals of A are completely prime.

Observe that X_{12} and X_{21} are normal in A, and so we can localize with respect to their powers. Although X_{11} is not normal, its powers also form an Ore set (e.g., verify this first in $M_{q^{-1},p}$, which is an iterated skew polynomial ring over $k[X_{11}]$). Note that any ideal I of A which contains X_{11} also contains $X_{12}X_{21}$, whence $D \in I$ and $I = A$. Hence, no prime ideal of A contains X_{11}, which means that no prime ideals of A are lost in passing from A to the localization $A[X_{11}^{-1}]$.

(b) The quotient $A/\langle X_{12}, X_{21}\rangle$ is isomorphic to a commutative Laurent polynomial ring $k[x_{11}^{\pm1}, x_{22}^{\pm1}]$. Hence, we know the prime ideals of A containing $\langle X_{12}, X_{21}\rangle$. The others correspond to prime ideals in the localizations

$$\big(A/\langle X_{12}\rangle\big)[X_{21}^{-1}], \quad \big(A/\langle X_{21}\rangle\big)[X_{12}^{-1}], \quad \text{and} \quad A[X_{11}^{-1}, X_{12}^{-1}, X_{21}^{-1}].$$

We claim that these localizations are simple algebras, from which it will follow that the only prime ideals of A not containing $\langle X_{12}, X_{21}\rangle$ are $\langle X_{12}\rangle$, $\langle X_{21}\rangle$, and $\langle 0\rangle$.

First, $\big(A/\langle X_{12}\rangle\big)[X_{21}^{-1}]$ is isomorphic to the algebra

$$T_1 := k\langle x^{\pm1}, y^{\pm1}, z^{\pm1} \mid xy = p^{-1}yx, \ xz = xz, \ yz = qzy\rangle.$$

Via §9.6(a), we compute that $Z(T_1) = k$, whence T_1 is simple. Similarly, $\big(A/\langle X_{21}\rangle\big)[X_{12}^{-1}]$ is simple.

Third, observe that $X_{22} = X_{11}^{-1}(D + qX_{12}X_{21})$ in $A[X_{11}^{-1}]$, and so this algebra can be generated by $X_{11}^{\pm1}$, X_{12}, X_{21}, $D^{\pm1}$. Consequently, $A[X_{11}^{-1}, X_{12}^{-1}, X_{21}^{-1}]$ is isomorphic to the k-algebra T_3 with generators $x^{\pm1}$, $y^{\pm1}$, $z^{\pm1}$, $w^{\pm1}$ and relations

$$\begin{array}{lll} xy = qyx & xz = p^{-1}zx & xw = wx \\ yz = (pq)^{-1}zy & yw = (pq)^{-1}wy & zw = pqwz. \end{array}$$

Another application of §9.6(a) shows that T_3 is simple, establishing the claim.

Therefore, the prime ideals of A consist of

- the maximal ideals $\langle X_{11} - \lambda, X_{12}, X_{21}, X_{22} - \mu\rangle$, for $\lambda, \mu \in k^\times$;
- ($\diamondsuit$) the ideals $\langle X_{12}, X_{21}, f(X_{11}, X_{22})\rangle$, for irreducible polynomials $f(s,t) \in k[s^{\pm1}, t^{\pm1}]$;
- the ideals $\langle X_{12}, X_{21}\rangle$, $\langle X_{12}\rangle$, $\langle X_{21}\rangle$, and $\langle 0\rangle$.

(c) The torus $H = (k^\times)^4$ acts on A by k-algebra automorphisms such that

$$(\alpha_1, \alpha_2, \beta_1, \beta_2).X_{ij} = \alpha_i\beta_j X_{ij} \tag{9.10c}$$

for all i, j. Only four of the prime ideals of A are H-stable, and thus [16, Corollary 2.7(ii), Remark 5.9(i)] implies that A satisfies the Dixmier-Moeglin equivalence (cf. [4, Corollary II.8.5(c)]). Therefore, the primitive ideals of A are

- the maximal ideals $\langle X_{11} - \lambda, X_{12}, X_{21}, X_{22} - \mu\rangle$, for $\lambda, \mu \in k^\times$;
- the ideals $\langle X_{12}\rangle$, $\langle X_{21}\rangle$, and $\langle 0\rangle$.

(d) Now restrict to $k = \mathbb{C}$, choose $\alpha \in \mathbb{R} \setminus \mathbb{Q}$, assume that q is transcendental over $\mathbb{Q}(\alpha)$, and take $p = 1 + \alpha(q-1)$. The assumptions on α and q ensure that the subgroup $\langle p, q\rangle \subseteq \mathbb{C}^\times$ is free abelian of rank 2, as needed above. Our choice of p is a first-order Taylor approximation of q^α, which is convenient for extension to polynomial rings.

Choose a Laurent polynomial ring $k[z^{\pm 1}]$, set $z_\alpha = 1 + \alpha(z-1)$, and let $B = M_{z^{-1},z_\alpha}$ over $k[z^{\pm 1}, z_\alpha^{-1}]$ in the notation of [49, §2]. Thus, B is the $k[z^{\pm 1}, z_\alpha^{-1}]$-algebra given by generators X_{11}, X_{12}, X_{21}, X_{22} and relations

$$\begin{aligned} X_{11}X_{12} &= zX_{12}X_{11} & X_{11}X_{21} &= z_\alpha^{-1}X_{21}X_{11} \\ X_{21}X_{22} &= zX_{22}X_{21} & X_{12}X_{22} &= z_\alpha^{-1}X_{22}X_{12} \\ X_{12}X_{21} &= (zz_\alpha)^{-1}X_{21}X_{12} & X_{11}X_{22} - X_{22}X_{11} &= (z - z_\alpha)X_{12}X_{21}\,. \end{aligned}$$

Observe that B is an iterated skew polynomial algebra over $k[z^{\pm 1}, z_\alpha^{-1}]$, and so it is torsionfree over $k[z^{\pm 1}]$. This algebra has been arranged so that $B/(z-q)B \cong M_{q^{-1},p}$ and $B/(z-1)B \cong \mathcal{O}(M_2(k))$. In B, the quantum determinant is $D = X_{11}X_{22} - zX_{12}X_{21}$, and it is normal. We set $C = B[D^{-1}]$ and observe that C is a torsionfree $k[z^{\pm 1}]$-algebra such that $C/(z-q)C \cong A$ and $C/(z-1)C \cong R := \mathcal{O}(GL_2(k))$.

Thus, A is one of the quantizations of R in the family of algebras $C/(z-\gamma)C$. The semiclassical limit of this family is the algebra R, equipped with the Poisson bracket determined by

$$\begin{aligned} \{X_{11}, X_{12}\} &= X_{11}X_{12} & \{X_{11}, X_{21}\} &= -\alpha X_{11}X_{21} \\ \{X_{21}, X_{22}\} &= X_{21}X_{22} & \{X_{12}, X_{22}\} &= -\alpha X_{12}X_{22} \\ \{X_{12}, X_{21}\} &= -(1+\alpha)X_{12}X_{21} & \{X_{11}, X_{22}\} &= (1-\alpha)X_{12}X_{21}\,. \end{aligned}$$

To find the Poisson prime ideals of R, we can proceed in parallel with part (b) above, using §9.6(b) in place of §9.6(a). We compute that the Poisson prime ideals of R can be listed exactly as in (◇). This yields an obvious bijection ϕ : spec $A \to$ P.spec R given by "preservation of notation". Clearly ϕ and ϕ^{-1} preserve inclusions, and so ϕ is a homeomorphism by Lemma 9.4(a). (Alternatively, one can easily identify the closed sets in spec A and P.spec R and then check that ϕ and ϕ^{-1} are closed maps.)

The torus H acts on R by Poisson algebra automorphisms satisfying (9.10c), and only four Poisson prime ideals of R are stable under this action. Consequently, [13, Theorem 4.3] implies that R satisfies the Poisson Dixmier-Moeglin equivalence. Thus, the Poisson-primitive ideals of R are

(♠) • the maximal ideals $\langle X_{11}-\lambda,\ X_{12},\ X_{21},\ X_{22}-\mu\rangle$, for $\lambda, \mu \in k^\times$;
• the ideals $\langle X_{12}\rangle$, $\langle X_{21}\rangle$, and $\langle 0\rangle$,

and therefore ϕ restricts to a homeomorphism prim $A \to$ P.prim R.

(e) In view of (♠), we can now identify the symplectic cores in $GL_2(\mathbb{C}) \approx \operatorname{maxspec} R$ with respect to the Poisson structure under discussion. They are

- the singletons $\left\{\begin{bmatrix} \lambda & 0 \\ 0 & \mu \end{bmatrix}\right\}$, for $\lambda, \mu \in \mathbb{C}^\times$;
- the sets $\begin{bmatrix} \mathbb{C}^\times & \mathbb{C}^\times \\ 0 & \mathbb{C}^\times \end{bmatrix}$, $\begin{bmatrix} \mathbb{C}^\times & 0 \\ \mathbb{C}^\times & \mathbb{C}^\times \end{bmatrix}$ and $\left\{\begin{bmatrix} \lambda & \beta \\ \gamma & \mu \end{bmatrix} \in GL_2(\mathbb{C}) \;\middle|\; \beta, \gamma \neq 0\right\}$.

The space of symplectic cores in $GL_2(\mathbb{C})$ is homeomorphic to P.prim R by Lemma 9.3.

Since $\begin{bmatrix} \mathbb{C}^\times & \mathbb{C}^\times \\ 0 & \mathbb{C}^\times \end{bmatrix}$ and $\begin{bmatrix} \mathbb{C}^\times & 0 \\ \mathbb{C}^\times & \mathbb{C}^\times \end{bmatrix}$ are complex manifolds of odd dimension, they cannot be symplectic leaves. In fact, each is the union of a one-parameter family of symplectic leaves, which can be calculated as in [18, Example 3.10(v)]. For instance, the symplectic leaves contained in $\begin{bmatrix} \mathbb{C}^\times & \mathbb{C}^\times \\ 0 & \mathbb{C}^\times \end{bmatrix}$ are the surfaces

$$\left\{ \begin{bmatrix} \lambda & \beta \\ 0 & \delta\lambda^\alpha \end{bmatrix} \;\middle|\; \lambda, \beta \in \mathbb{C}^\times \right\}, \quad \text{for } \delta \in \mathbb{C}^\times.$$

Acknowledgement

We thank J. Alev, K.A. Brown, I. Gordon, S. Kolb, U. Kraehmer, S. Launois, T.H. Lenagan, E.S. Letzter, M. Lorenz, M. Martino, L. Rigal, M. Yakimov, and J.J. Zhang for many discussions on the topics of this paper.

References

[1] M. Adler, P. van Moerbeke, and P. Vanhaecke, *Algebraic Integrability, Painlevé Geometry and Lie Algebras*, Springer-Verlag, Berlin, 2004.

[2] W. Borho, P. Gabriel, and R. Rentschler, *Primideale in Einhüllenden auflösbarer Lie-Algebren*, Lecture Notes in Math. 357, Springer-Verlag, Berlin, 1973.

[3] K.A. Brown, Personal communication, Nov. 2008.

[4] K.A. Brown and K.R. Goodearl, *Lectures on Algebraic Quantum Groups*, Advanced Courses in Math. CRM Barcelona, Birkhäuser, Basel, 2002.

[5] K.A. Brown, K.R. Goodearl, and M. Yakimov, *Poisson structures on affine spaces and flag varieties. I. Matrix affine Poisson space*, Advances in Math. **206** (2006), 567–629.

[6] K.A. Brown and I. Gordon, *Poisson orders, representation theory, and symplectic reflection algebras*, J. reine angew. Math. **559** (2003), 193–216.

[7] N. Chriss and V. Ginzburg, *Representation Theory and Complex Geometry*, Birkhäuser, Boston, 1997.

[8] D.H. Collingwood and W.M. McGovern, *Nilpotent orbits in semisimple Lie algebras*, Van Nostrand Reinhold, New York, 1993.

[9] J. Dixmier, *Représentations irréductibles des algèbres de Lie résolubles*, J. Math. pures appl. **45** (1966), 1–66.

[10] ——, *Enveloping Algebras*, North-Holland, Amsterdam, 1977.

[11] D.R. Farkas and G. Letzter, *Ring theory from symplectic geometry*, J. Pure Appl. Algebra **125** (1998), 155–190.

[12] K.R. Goodearl, *Prime spectra of quantized coordinate rings*, in Interactions between Ring Theory and Representations of Algebras (Murcia 1998) (F. Van Oystaeyen and M. Saorín, Eds.), Dekker, New York, 2000, 205–237.

[13] ——, *A Dixmier-Moeglin equivalence for Poisson algebras with torus actions*, in Algebra and Its Applications (Athens, Ohio, 2005) (D.V. Huynh, S.K. Jain, and S.R. López-Permouth, Eds.), Contemp. Math. **419** (2006), 131–154.

[14] K.R. Goodearl and S. Launois, *The Dixmier-Moeglin equivalence and a Gel'fand-Kirillov problem for Poisson polynomial algebras*, Bull. Soc. Math. France (to appear), posted at arxiv.org/abs/0705.3486.

[15] K.R. Goodearl and E.S. Letzter, *Prime and primitive spectra of multiparameter quantum affine spaces*, in Trends in Ring Theory. Proc. Miskolc Conf. 1996 (V. Dlab and L. Márki, eds.), Canad. Math. Soc. Conf. Proc. Series **22** (1998), 39–58.

[16] ——, *The Dixmier-Moeglin equivalence in quantum coordinate rings and quantized Weyl algebras*, Trans. Amer. Math. Soc. **352** (2000), 1381–1403.

[17] ——, *Quantum n-space as a quotient of classical n-space*, Trans. Amer. Math. Soc. **352** (2000), 5855–5876.

[18] ——, *Semiclassical limits of quantum affine spaces*, Proc. Edinburgh Math. Soc. **52** (2009), 387–407.

[19] K.R. Goodearl and M. Yakimov, *Poisson structures on affine spaces and flag varieties. II*, Trans. Amer. Math. Soc. **361** (2009), 5753–5780.

[20] T. Hayashi, *Quantum deformations of classical groups*, Publ. RIMS **28** (1992), 57–81.

[21] T.J. Hodges, *Quantum tori and Poisson tori*, Unpublished Notes, 1994.

[22] T.J. Hodges and T. Levasseur, *Primitive ideals of $\mathbf{C}_q[SL(3)]$*, Comm. Math. Phys. **156** (1993), 581–605.

[23] ——, *Primitive ideals of $\mathbf{C}_q[SL(n)]$*, J. Algebra **168** (1994), 455–468.

[24] T.J. Hodges, T. Levasseur, and M. Toro, *Algebraic structure of multi-parameter quantum groups*, Advances in Math. **126** (1997), 52–92.

[25] K.L. Horton, *The prime and primitive spectra of multiparameter quantum symplectic and Euclidean spaces*, Communic. in Algebra **31** (2003), 2713–2743.

[26] C. Ingalls, *Quantum toric varieties*, preprint, posted at kappa.math.unb.ca/%7Ecolin/.

[27] A. Joseph, *Quantum Groups and Their Primitive Ideals*, Springer-Verlag, Berlin, 1995.

[28] A. Kamita, *Quantum deformations of certain prehomogeneous vector spaces III*, Hiroshima Math. J. **30** (2000), 79–115.

[29] A.A. Kirillov, *Unitary representations of nilpotent Lie groups*, Russian Math. Surveys **17** (1962), 57–110.

[30] ——, *Merits and demerits of the orbit method*, Bull. Amer. Math. Soc. **36** (1999), 433–488.

[31] ——, *Lectures on the Orbit Method*, Grad. Studies in Math. 64, American Math. Soc., Providence, 2004.

[32] A.U. Klimyk and K. Schmüdgen, *Quantum Groups and their Representations*, Springer-Verlag, Berlin, 1997.

[33] L.I. Korogodski and Ya.S. Soibelman, *Algebras of Functions on Quantum Groups: Part I*, Math. Surveys and Monographs 56, Amer. Math. Soc., Providence, 1998.

[34] M. Martino, *The associated variety of a Poisson prime ideal*, J. London Math. Soc. (2) **72** (2005), 110–120.

[35] O. Mathieu, *Bicontinuity of the Dixmier map*, J. Amer. Math. Soc. **4** (1991), 837–863.

[36] J.C. McConnell and J.J. Pettit, *Crossed products and multiplicative analogs of Weyl algebras*, J. London Math. Soc. (2) **38** (1988), 47–55.

[37] M. Noumi, *Macdonald's symmetric polynomials as zonal spherical functions on some quantum homogeneous spaces*, Advances in Math. **123** (1996), 16–77.

[38] S.-Q. Oh, *Catenarity in a class of iterated skew polynomial rings*, Comm. Algebra **25(1)** (1997), 37–49.

[39] ___ , *Symplectic ideals of Poisson algebras and the Poisson structure associated to quantum matrices*, Comm. Algebra **27** (1999), 2163–2180.

[40] ___ , *Quantum and Poisson structures of multi-parameter symplectic and Euclidean spaces*, J. Algebra **319** (2008), 4485–4535.

[41] S.-Q. Oh, C.-G. Park, and Y.-Y. Shin, *Quantum n-space and Poisson n-space*, Comm. Algebra **30** (2002), 4197–4209.

[42] A. Polishchuk, *Algebraic geometry of Poisson brackets*, J. Math. Sci. **94** (1997), 1413–1444.

[43] N.Yu. Reshetikhin, L.A. Takhtadzhyan, and L.D. Faddeev, *Quantization of Lie groups and Lie algebras*, Leningrad Math. J. **1** (1990), 193–225.

[44] I. Shafarevich, *Basic Algebraic Geometry*, Springer-Verlag, Berlin, 1974.

[45] Ya.S. Soibel'man, *Irreducible representations of the function algebra on the quantum group $SU(n)$, and Schubert cells*, Soviet Math. Dokl. **40** (1990), 34–38.

[46] ___ , *The algebra of functions on a compact quantum group, and its representations*, Leningrad Math. J. **2** (1991), 161–178; *Correction*, (Russian), Algebra i Analiz **2** (1990), 256.

[47] E. Strickland, *Classical invariant theory for the quantum symplectic group*, Advances in Math. **123** (1996), 78–90.

[48] M. Takeuchi, *Quantum orthogonal and symplectic groups and their embedding into quantum GL*, Proc. Japan Acad. **65** (1989), 55–58.

[49] ___ , *A two-parameter quantization of $GL(n)$ (summary)*, Proc. Japan. Acad. **66 (A)** (1990), 112–114.

[50] P. Tauvel and R.W.T. Yu, *Lie Algebras and Algebraic Groups*, Springer-Verlag, Berlin, 2005.

[51] L.L. Vaksman and Ya.S. Soibel'man, *Algebra of functions on the quantum group $SU(2)$*, Func. Anal. Applic. **22** (1988), 170–181.

[52] M. Vancliff, *Primitive and Poisson spectra of twists of polynomial rings*, Algebras and Representation Theory **2** (1999), 269–285.

[53] P. Vanhaecke, *Integrable Systems in the Realm of Algebraic Geometry*, Lecture Notes in Math. 1638, Springer-Verlag, Berlin, 1996.

[54] A. Weinstein, *The local structure of Poisson manifolds*, J. Diff. Geom. **18** (1983), 523–557.

K.R. Goodearl
Department of Mathematics
University of California
Santa Barbara, CA 93106, USA
e-mail: goodearl@math.ucsb.edu

Advances in Ring Theory
Trends in Mathematics, 205–212

On Unit-Central Rings

Dinesh Khurana, Greg Marks and Ashish K. Srivastava

Dedicated to S.K. Jain in honor of his 70th birthday.

Abstract. We establish commutativity theorems for certain classes of rings in which every invertible element is central, or, more generally, in which all invertible elements commute with one another. We prove that if R is a *semiexchange ring* (i.e., its factor ring modulo its Jacobson radical is an exchange ring) with all invertible elements central, then R is commutative. We also prove that if R is a semiexchange ring in which all invertible elements commute with one another, and R has no factor ring with two elements, then R is commutative. We offer some examples of noncommutative rings in which all invertible elements commute with one another, or are central. We close with a list of problems for further research.

Mathematics Subject Classification (2000). Primary 16U60, 16U70; Secondary 16L30.

Keywords. Unit-central, rings with commuting units, exchange rings, semiexchange rings, commutativity theorems.

1. Introduction

We say that an associative unital ring R is *unit-central* if $\mathrm{U}(R) \subseteq \mathrm{Z}(R)$, i.e., if every invertible element of the ring lies in the center. In various natural situations the unit-central condition implies full commutativity.

It is also of interest to weaken the unit-central condition and consider rings R for which $\mathrm{U}(R)$ is an abelian group. We will refer to such a ring R as having *commuting units*. Rings with commuting units have also been investigated by a number of authors (e.g., see [7], [12], [21], [22]). For a ring that is additively generated by its units (cf. [17], [18], [19], [26], [28]), having commuting units is obviously equivalent to commutativity.

Our main focus in this note will be on unit-central rings and rings with commuting units. A still wider class consists of those rings in which any two nilpotent elements commute with one another. This property proved instrumental in the study of prime rings in [5]. We will consider this condition in Theorem 2.8 below.

We will denote the Jacobson radical of a ring R by $\operatorname{rad}(R)$, the set of nilpotent elements by $\mathfrak{N}(R)$, and the right annihilator of an element a in R by $\operatorname{ann}_r^R(a)$. For any other notation not defined here, we refer the reader to [20].

We record the following construction technique for the classes of rings under consideration.

Proposition 1.1. *Let S be a ring, let M be an (S,S)-bimodule, and define $R = S \oplus M$ as an additive group, with multiplication in R defined by $(s_1, m_1)(s_2, m_2) = (s_1s_2, s_1m_2 + m_1s_2)$.*

(i) *R is unit-central if and only if S is unit-central and $sm = ms$ for all $s \in S$ and $m \in M$.*
(ii) *R has commuting units if and only if S has commuting units and $sm = ms$ for all $s \in \mathrm{U}(S)$ and $m \in M$.*

Proof. Straightforward. □

2. Commutativity theorems

We begin with a basic but useful lemma. Recall that a ring is said to be *abelian* if every idempotent element is central.

Lemma 2.1. *Let R be a ring. Then:*

(i) *If R is unit-central, then $\mathfrak{N}(R) \cup \operatorname{rad}(R) \subseteq \mathrm{Z}(R)$.*
(ii) *If R has commuting units, then for all $a, b \in \mathfrak{N}(R) \cup \operatorname{rad}(R) \cup \mathrm{U}(R)$ we have $ab = ba$.*
(iii) *If R is unit-central, then R is abelian.*
(iv) *If for all $a, b \in \mathfrak{N}(R)$ we have $ab = ba$, then R is Dedekind-finite.*

Proof. Statements (i) and (ii) are straightforward. If $e \in R$ is an idempotent in a unit-central ring, then (i) implies $eR(1-e) = \{0\}$, and (iii) follows. A Dedekind-infinite ring contains an infinite set of matrix units, whence (iv). □

Obviously neither the property of having commuting units nor the unit-central condition is Morita invariant; however, they do pass to corner rings:

Lemma 2.2. *Let R be a ring and $e \in R$ an idempotent. If R is unit-central (resp. a ring with commuting units), then the corner ring eRe is unit-central (resp. a ring with commuting units).*

Proof. Suppose R is unit-central, with $ere \in \mathrm{U}(eRe)$ and $ese \in eRe$. Then $ere + (1-e)$ is contained in $\mathrm{U}(R)$, so it commutes with ese, and hence ere and ese commute.

The proof for the "commuting units" case is analogous. □

Recall that a ring R is called an *exchange ring* if the module R_R satisfies P. Crawley and B. Jónsson's exchange property: given a set I, whenever

$$A = M \oplus N = \bigoplus_{i \in I} A_i \qquad \text{with} \qquad M \cong R$$

in the category of right R-modules, there exist submodules $A_i' \subseteq A_i$ such that

$$A = M \oplus \left(\bigoplus_{i \in I} A_i' \right).$$

R.W. Warfield Jr. showed in [29, Corollary 2] that this property is left-right symmetric. Every ring R that is *semiregular* (i.e., $R/\mathrm{rad}(R)$ is von Neumann regular and idempotents of $R/\mathrm{rad}(R)$ lift to R) is an exchange ring. Also, every *clean ring* (i.e., ring in which every element is the sum of a unit and an idempotent) is an exchange ring. For example, the endomorphism ring of an continuous module is both semiregular and clean (for the latter, see [9]). By [4, Proposition 2.6], every strongly π-regular ring is clean; by [27, Example 2.3], every π-regular ring is an exchange ring. In addition to semiregular and clean rings, the class of exchange rings includes all C^*-algebras of real rank zero and Gromov translation rings of discrete trees over von Neumann regular rings (see [2, Theorem 7.2] and [3, Theorem 2.7]).

Exchange rings can be characterized as those rings for which every pair of comaximal right ideals contain a complementary pair of idempotents, i.e., R is an exchange ring if and only if for each element $a \in R$ there exists an idempotent $e \in R$ such that $e \in aR$ and $1 - e \in (1 - a)R$. This characterization was independently discovered by K.R. Goodearl and W.K. Nicholson (see [14, p. 167] and [23, Proposition 1.1 and Theorem 2.1]), and it is very useful in practice. For example, the proof by P. Ara, K.C. O'Meara, and F. Perera that Gromov translation rings of discrete trees over von Neumann regular rings are exchange rings in [3] relied crucially on Goodearl and Nicholson's characterization.

A ring R is said to be a *semiexchange ring* if the factor ring $R/\mathrm{rad}(R)$ is an exchange ring. This common generalization of exchange rings and semilocal rings arises naturally: according to [23, Corollary 2.4], a ring is an exchange ring if and only if it is a semiexchange ring and idempotents lift modulo the Jacobson radical. The (apparently rather deep) open problem of the left-right symmetry of the quasi-duo condition has an affirmative answer for the class of semiexchange rings (see [11, Theorem 4.6]). Basic properties of semiexchange rings are developed in [10]. Of course, a semiexchange ring need not be either semilocal or an exchange ring, as can be seen by taking a direct product of a semilocal ring and an exchange ring, or an infinite direct product of semilocal rings.

A ring with commuting units can be both semilocal and an exchange ring without being commutative. On the other hand, a unit-central ring that is either semilocal or an exchange ring must be commutative, by the following theorem.

Theorem 2.3. *Every unit-central semiexchange ring is commutative.*

Proof. Let R be a unit-central semiexchange ring. To prove that R is commutative, by [20, p. 200, Ex. 12.8B] it suffices to show that $x - x^2 \in \mathrm{Z}(R)$ for every $x \in R$.

Fix $x \in R$. The exchange ring $\overline{R} = R/\mathrm{rad}(R)$ is unit-central, therefore abelian, and it is well known that abelian exchange rings are clean [23, Proposition 1.8(2)]. So $x = e + u$ for some $e, u \in R$ such that $\overline{e}$ is an idempotent and $\overline{u}$ a unit of $\overline{R}$. Then $\overline{1-2e} \in \mathrm{U}(\overline{R})$, so $1 - 2e \in \mathrm{U}(R) \subseteq \mathrm{Z}(R)$, hence $2e \in \mathrm{Z}(R)$. As $u \in \mathrm{U}(R) \subseteq \mathrm{Z}(R)$, we have $2eu \in \mathrm{Z}(R)$ and $u - u^2 \in \mathrm{Z}(R)$. Moreover, $e - e^2 \in \mathrm{rad}(R) \subseteq \mathrm{Z}(R)$. Consequently, $x - x^2 = (e - e^2) - 2eu + (u - u^2) \in \mathrm{Z}(R)$, as required. □

Remark 2.4. The classical commutativity theorems of Jacobson, Herstein, and Kaplansky made heavy use of subdirect product representations. If R is a unit-central ring, and $R/\mathrm{rad}(R)$ is a finite subdirect product of simple artinian rings, then R is semilocal, and by Theorem 2.3, R must be commutative. One might be tempted to try to extend this conclusion to the case where $R/\mathrm{rad}(R)$ is a subdirect product of an arbitrary set of simple artinian rings. Unfortunately, this generalization fails. If k is an infinite field, and $\{x_i : i \in I\}$ is an infinite set of noncommuting indeterminates, then the free algebra $R = k\langle\{x_i : i \in I\}\rangle$ is a noncommutative unit-central ring with $\mathrm{rad}(R) = (0)$, and by [1, Corollary 3], R can be represented as a subdirect product of simple artinian rings.

Example 2.5. Let

$$R = \begin{pmatrix} \mathbb{F}_2 & V \\ 0 & \mathbb{F}_2 \end{pmatrix}$$

where V is any nonzero $\mathbb{F}_2$-vector space. Then R is a noncommutative semiprimary ring with commuting units. Thus, in Lemma 2.1(iii) and Theorem 2.3 the unit-central hypothesis cannot be weakened to "commuting units."

If, however, we assume that R has no factor ring isomorphic to $\mathbb{F}_2$, then Theorem 2.3 can be extended to rings with commuting units. To prove this, we will make use of the following theorem, which occurred (with different terminology) as [21, Theorem 2.2]. (The "left suitable" condition in [21, Theorem 2.2] is equivalent to the ring being an exchange ring: see [21, Lemma 1.2] or [23, Theorem 2.1].)

Theorem 2.6 (Nicholson, Springer). *A semiprime exchange ring with commuting units is commutative.*

The following theorem strengthens a result of J. Han, [16, Theorem 2.9]. We note that Nicholson has a complementary result for semiperfect rings, [22, Corollary 1(1)].

Theorem 2.7. *Let R be a semiexchange ring with commuting units. If R has no factor ring isomorphic to $\mathbb{F}_2$, then every element of R is a sum of two units, and consequently R is commutative.*

Proof. Let $\overline{R} = R/\mathrm{rad}(R)$. As $\overline{R}$ is a semiprimitive exchange ring with commuting units, Theorem 2.6 implies $\overline{R}$ is commutative. In a commutative exchange ring with no factor ring isomorphic to $\mathbb{F}_2$, every element is the sum of two units. (See [6, Theorem 3]; the idea was already implicit in [13, Theorem 2].) Thus, every element of $\overline{R}$ is the sum of two units, whence every element of R is the sum of two units. Since R has commuting units, it is commutative. □

An element a of a ring R is a *von Neumann regular element* if there exists some $b \in R$ such that $a = aba$.

Theorem 2.8. *Let R be a ring with the property that for all $a, b \in \mathfrak{N}(R)$ we have $ab = ba$. (In particular, any ring with commuting units has this property.) Then R does not contain any nonzero nilpotent von Neumann regular element.*

Proof. Assume the contrary, that there exists some $a \in R$ such that $a = aba$ for some $b \in R$, and $a^n = 0 \neq a^{n-1}$ for some integer $n \geqslant 2$. Put $e = ab$.

The nilpotent elements a and $ea(1-e)$ commute, hence $aea(1-e) = 0$. From $a^2(1-e) = 0$ we obtain $a^2 = a^2e = a^3b$, whence $a^2 = a^{m+2}b^m$ for every $m \in \mathbb{N}$. Therefore $a^2 = 0$, which implies $a = ea(1-e)$. Hence

$$\begin{aligned} a &= \Big(ea(1-e)\Big)b\Big(ea(1-e)\Big) \\ &= \Big(ea(1-e)\Big)\Big((1-e)be\Big)\Big(ea(1-e)\Big) \\ &= \Big(ea(1-e)\Big)\Big(ea(1-e)\Big)\Big((1-e)be\Big) \\ &= 0, \end{aligned}$$

a contradiction. □

Using Theorem 2.8, we can recover the following special case of Theorem 2.6.

Corollary 2.9 (Nicholson, Springer). *Any von Neumann regular ring with commuting units is commutative.*

Proof. Let R be a von Neumann regular ring with commuting units. By Theorem 2.8, R is reduced and von Neumann regular, i.e., strongly regular. In a strongly regular ring every element is the product of a unit and a central idempotent. Hence R is commutative. □

Remark 2.10. It follows from Theorem 2.3 that any unit-central artinian ring is commutative. A unit-central noetherian ring, however, need not be commutative. For instance, the Weyl algebras over a field are noncommutative unit-central noetherian rings. For another example of this sort, let $A = k[t_1, t_2, \ldots, t_n]$ where k is a field and the t_i's are commuting indeterminates, and let σ be any nonidentity k-linear automorphism of A. Then the skew polynomial ring $R = A[x;\sigma]$ is a noncommutative unit-central noetherian ring.

We note in closing that commutativity theorems complementary to those in this section can be found in [15, §5].

3. Open problems

A ring R is called *right duo* if each right ideal of R is two-sided. We ask the following question.

Question 3.1. *Is every unit-central right duo ring commutative?*

Following Nicholson and M.F. Yousif in [24], we call a ring R *right principally injective,* or *right P-injective,* if for every $a \in R$, every right R-module homomorphism $aR \to R$ extends to a right R-module homomorphism $R \to R$. Since a right self-injective ring is an exchange ring, we know that every unit-central right self-injective ring is commutative. This suggests the following question.

Question 3.2. *Is every unit-central right principally-injective ring commutative?*

Nicholson and E. Sánchez-Campos [25] called an element a of a ring R a *right morphic element* if $R/aR \cong \text{ann}_r^R(a)$ as right R-modules. A ring R is called a *right morphic ring* if every element of R is a right morphic element. Clearly every unit and idempotent in a ring is morphic. We ask the following question.

Question 3.3. *Is every unit-central right morphic ring commutative?*

A ring R is said to have *stable range* 1 if for all $a, b \in R$ such that $aR+bR = R$, there exists $y \in R$ such that $a + by$ is a unit. As every semilocal ring has stable range 1, and Theorem 2.3 shows that every unit-central semilocal ring is commutative, we ask the following.

Question 3.4. *Is every unit-central ring with stable range* 1 *commutative?*

Acknowledgment

We thank T.Y. Lam for his valuable suggestions and advice.

References

[1] S.A. Amitsur, *The identities of PI-rings,* Proc. Amer. Math. Soc. **4** (1953), no. 1, 27–34.

[2] P. Ara, K.R. Goodearl, K.C. O'Meara, E. Pardo, *Separative cancellation for projective modules over exchange rings,* Israel J. Math. **105** (1998), 105–137.

[3] P. Ara, K.C. O'Meara, F. Perera, *Gromov translation algebras over discrete trees are exchange rings,* Trans. Amer. Math. Soc. **356** (2004), no. 5, 2067–2079.

[4] W.D. Burgess, P. Menal, *On strongly π-regular rings and homomorphisms into them,* Comm. Algebra **16** (1988), no. 8, 1701–1725.

[5] M. Chebotar, P.-H. Lee, E. R. Puczyłowski, *On prime rings with commuting nilpotent elements,* Proc. Amer. Math. Soc. **137** (2009), no. 9, 2899–2903.

[6] H. Chen, *Exchange rings with Artinian primitive factors,* Algebr. Represent. Theory **2** (1999), no. 2, 201–207.

[7] J. Cohen, K. Koh, *The group of units in a compact ring,* J. Pure Appl. Algebra **54** (1988), no. 2, 167–179.

[8] V.P. Camillo, H.P. Yu, *Exchange rings, units and idempotents,* Comm. Algebra **22** (1994), no. 12, 4737–4749.

[9] V.P. Camillo, D. Khurana, T.Y. Lam, W.K. Nicholson, Y. Zhou, *Continuous modules are clean,* J. Algebra **304** (2006), no. 1, 94–111.

[10] W. Chen, *Semiexchange rings and K_1-groups of semilocal rings,* Nanjing Daxue Xuebao Shuxue Bannian Kan **24** (2007), no. 1, 65–71.

[11] A.S. Dugas, T.Y. Lam, *Quasi-duo rings and stable range descent,* J. Pure Appl. Algebra **195** (2005), no. 3, 243–259.

[12] K.E. Eldridge, I. Fischer, *D.C.C. rings with a cyclic group of units,* Duke Math. J. **34** (1967), 243–248.

[13] J.W. Fisher, R.L. Snider, *Rings generated by their units,* J. Algebra **42** (1976), no. 2, 363–368.

[14] K.R. Goodearl, R.B. Warfield Jr., *Algebras over zero-dimensional rings,* Math. Ann. **223** (1976), no. 2, 157–168.

[15] R.N. Gupta, A. Khurana, D. Khurana, T.Y. Lam, *Rings over which the transpose of every invertible matrix is invertible,* J. Algebra, to appear.

[16] J. Han, *The structure of semiperfect rings,* J. Korean Math. Soc. **45** (2008), 425–433.

[17] M. Henriksen, *Two classes of rings generated by their units,* J. Algebra **31** (1974), 182–193.

[18] D. Khurana, A.K. Srivastava, *Right self-injective rings in which each element is sum of two units,* J. Algebra Appl. **6** (2007), no. 2, 281–286.

[19] D. Khurana, A.K. Srivastava, *Unit sum numbers of right self-injective rings,* Bull. Austral. Math. Soc. **75** (2007), no. 3, 355–360.

[20] T.Y. Lam, *A First Course in Noncommutative Rings*, Second Edition, Graduate Texts in Math. 131 (Springer-Verlag, New York, 2001).

[21] W.K. Nicholson, H.J. Springer, *Commutativity of rings with abelian or solvable units,* Proc. Amer. Math. Soc. **56** (1976), no. 1, 59–62.

[22] W.K. Nicholson, *Semiperfect rings with abelian group of units,* Pacific J. Math. **49** (1973), 191–198.

[23] W.K. Nicholson, *Lifting idempotents and exchange rings,* Trans. Amer. Math. Soc. **229** (1977), 269–278.

[24] W.K. Nicholson, M.F. Yousif, *Principally injective rings,* J. Algebra **174** (1995), no. 1, 77–93.

[25] W.K. Nicholson, E. Sánchez Campos, *Rings with the dual of the isomorphism theorem,* J. Algebra **271** (2004), 391–406.

[26] R. Raphael, *Rings which are generated by their units,* J. Algebra **28** (1974), 199–205.

[27] J. Stock, *On rings whose projective modules have the exchange property,* J. Algebra **103** (1986), no. 2, 437–453.

[28] P. Vámos, *2-Good Rings,* Q. J. Math. **56** (2005), no. 3, 417–430.

[29] R.B. Warfield Jr., *Exchange rings and decompositions of modules,* Math. Ann. **199** (1972), 31–36.

[30] D. Zelinsky, *Every linear transformation is a sum of nonsingular ones,* Proc. Amer. Math. Soc. **5** (1954), 627–630.

Dinesh Khurana
Faculty of Mathematics
Indian Institute of Science Education and Research, Mohali
MGSIPAP Complex, Sector 26
Chandigarh 160 019, India
e-mail: dkhurana@iisermohali.ac.in

Greg Marks
Department of Mathematics and Computer Science
St. Louis University
St. Louis, MO 63103–2007, USA
e-mail: marks@slu.edu

Ashish K. Srivastava
Department of Mathematics and Computer Science
St. Louis University
St. Louis, MO 63103–2007, USA
e-mail: asrivas3@slu.edu

Advances in Ring Theory
Trends in Mathematics, 213–227

Symplectic Modules and von Neumann Regular Matrices over Commutative Rings

T.Y. Lam and R.G. Swan

Abstract. A criterion is given for the existence of a symplectic structure on a finitely generated projective module P over a commutative ring R. If $P \oplus Q = R^n$, P admits a symplectic structure iff $Q = \ker(A)$ for some von Neumann regular alternating matrix $A \in \mathbb{M}_n(R)$. Every n-generated symplectic space over R arises in this way. In supplement to this, it is also shown that a matrix $A \in \mathbb{M}_n(R)$ is von Neumann regular iff each of its determinantal ideals is generated by an idempotent in R.

Mathematics Subject Classification (2000). 15A09, 15A63, 16D40, 16E50.

Keywords. Projective modules, symplectic structures, alternating and von Neumann regular matrices, determinantal ideals, idempotent ideals, Fitting invariants.

1. Introduction

Over a commutative ring R, a *symplectic module* means a finitely generated (f.g.) projective R-module P that is equipped with a bilinear pairing $B : P \times P \to R$ such that $B(v, v) = 0$ for all $v \in P$, and the mapping $v \mapsto B(v, -)$ is an R-isomorphism from P onto its R-dual P^*. While the usual R-modules (resp. bilinear R-modules) and their morphisms constitute the "linear (resp. bilinear) category" over R, the symplectic R-modules and their morphisms constitute a separate "symplectic category". The study of symplectic structures and their automorphism groups has proved to be of interest and importance in both algebra and geometry.

If P is a f.g. free R-module R^n, the symplectic structures on P are given by the alternating matrices in $\mathrm{GL}_n(R)$, where a matrix $A = (a_{ij})$ is called *alternating* if $A^T = -A$ and $a_{ii} = 0$ for all i. If n is odd, it is easy to see that this alternating condition implies that $\det(A) = 0$. On the other hand, if $n = 2m$, an obvious invertible alternating $n \times n$ matrix is given by $\begin{pmatrix} 0 & I_m \\ -I_m & 0 \end{pmatrix}$. Thus, if $R \neq 0$, symplectic structures can be found on the free module R^n iff n is even.

If P is only f.g. projective but not free, it is known that P may not support a symplectic structure. In fact, this can happen even when P is self-dual ($P \cong P^*$) and has constant rank $2m$; see, for instance, Theorem 1.4 and Corollary 6.4 in [Sw_2]. A simpler example can also be constructed as follows. Let R be a Dedekind domain having a non-principal ideal J whose square is principal, and let $P = R^{2m-1} \oplus J$ ($m \geq 1$). Since $J^* \cong J^{-1} \cong J$, P is self-dual of rank $2m$. However, P is not free (since J is not principal). According to [La_5] (comment (b) after (VII.5.8)), over a Dedekind domain, this implies that P *cannot* support a symplectic structure.

In view of the remarks above, it would thus be desirable to find, in general, some criteria for the existence of symplectic structures on a given f.g. projective module P. However, a modest search of the literature did not turn up any such criterion (except in the case of certain special modules). In this paper, we prove the following result, which offers a general matrix-theoretic criterion for P to be a symplectic module.

Theorem A. *Given a decomposition $R^n = P \oplus Q$, P admits a symplectic structure iff $Q = \ker(A)$ for some alternating von Neumann regular matrix $A \in \mathbb{M}_n(R)$. (A matrix A is said to be von Neumann regular if $A = AMA$ for some $M \in \mathbb{M}_n(R)$.)*

If one prefers to put more emphasis on A than on P, this result can be reformulated as follows; *Given an alternating matrix $A \in \mathbb{M}_n(R)$, the bilinear pairing defined by A induces a symplectic structure on the factor module $R^n/\ker(A)$ iff the matrix A is von Neumann regular.* It is of interest to note that the above formulation of the existence criteria for symplectic structures is independent of any assumptions on the rank function (on the Zariski prime spectrum) of the projective modules in question. This makes the criteria readily applicable even to projective modules of nonconstant rank.

The results above, along with some of their simplifications in the case of special ground rings R, are presented in the first half of §2. The rest of §2 gives yet other formulations of these results in case the projective module P is the syzygy module defined by a unimodular row. The results presented in §2 seemed to have escaped earlier notice as the notion of von Neumann regular matrices has not been previously brought to bear on the general study of symplectic structures. But in retrospect, this important notion (first invented by J. von Neumann in the 1930s) provided exactly the right tool to deal with the existence question addressed in the aforementioned results.

In view of the special role played by von Neumann regular matrices in this work, we offer in §3 the following criterion for a square matrix to be von Neumann regular.

Theorem B. *Over a commutative ring R, a matrix $A \in \mathbb{M}_n(R)$ is von Neumann regular iff each of its determinantal ideals is generated by an idempotent in R.*

Since we have not been able to find this result in a few standard sources in linear algebra over commutative rings (such as [Br], [HA], and [MD]), we cannot assume that it is well known. In Eisenbud's book [Ei], there is a criterion for a f.g. R-module to be projective of constant rank, in terms of the *Fitting invariants* of the module. With some additional work, Theorem B above can be deduced from this criterion. This will be done in §3, but such a proof seems somewhat circuitous as it requires the tools of Fitting invariants as well as the theory of ranks of projective modules. To compensate for this, a separate elementary and completely self-contained proof for Theorem B is offered in §3. Our proof there is largely matrix-theoretic, and again totally bypasses any consideration of projective modules and their ranks. The paper concludes with a short section (§4) on von Neumann regular matrices of small size and the explicit construction of (some of) their quasi-inverses.

Throughout this paper, R *denotes a commutative ring with* 1, and $\mathrm{U}(R)$ denotes the group of units of R. The module of $n \times n$ alternating matrices over R is denoted by $\mathbb{A}_n(R)$. All R-modules are assumed unital. Greek letters such as σ and τ are used to denote row vectors, while elements of the free module R^n are usually written as column vectors (e.g., σ^T, τ^T, where "T" means the transpose). For n-tuples u and v (rows or columns alike), $[u, v]$ denotes their dot product. For any row vector $\tau = (b_1, \ldots, b_n)$, we'll write $P(\tau) = P(b_1, \ldots, b_n)$ for the kernel of the R-homomorphism $R^n \to R$ defined by τ. This is the (first) syzygy module for the ideal in R generated by the b_i's. This module is especially important for us in case τ is *unimodular*; that is, when $\sum_i b_i R = R$.

As in the statement of Theorem A, an element A in a ring S is said to be *von Neumann regular* if $A = AMA$ for some $M \in S$. We refer to any such M as a *quasi-inverse* for A. If A has a quasi-inverse that is a unit, we say that A is *unit-regular.* If all elements $A \in S$ are von Neumann regular (resp. unit-regular), S is said to be a von Neumann regular (resp. unit-regular) ring. For other standard terminology and notations in the theory of rings and modules, we refer the reader to [La$_1$] and [La$_3$].

2. Symplectic structures on projective modules

To begin our study of the existence of symplectic structures on f.g. projective modules over a commutative ring R, we first recall a useful fact on von Neumann regular matrices, which will be used freely throughout this paper. Given a matrix $A \in \mathbb{M}_n(R)$, we have an R-endomorphism $A: R^n \to R^n$ defined by $\sigma^T \mapsto A\sigma^T$. Using the fact that R^n is a projective module, it is easy to show that A *is von Neumann regular iff* $\mathrm{im}(A)$ *is a direct summand of* R^n. The key point of the proof of this basic fact is that this condition implies that $\ker(A)$ is *also* a direct summand of R^n. The details of the proof can be found in [La$_2$: pp. 59–60].

The following notations will be fixed for the first result of this section, namely, (2.1). Let P, Q be a pair of (f.g. projective) R-modules such that $P \oplus Q = R^n$. We write E for the projection of R^n onto P with kernel Q, and let $S := \mathbb{M}_n(R)$, which we shall identify with the endomorphism ring of R^n. Using "star" to denote (as before) the formation of dual modules, we shall apply freely the canonical decomposition $(R^n)^* = P^* \oplus Q^*$, and will always identify $(R^n)^*$ with R^n.

Theorem 2.1. *With the above notations, the following statements are equivalent:*

(1) *There exists a symplectic structure on* P.

(2) $Q = \ker(A)$ *for some von Neumann regular matrix* $A \in \mathbb{A}_n(R)$.

(3) $E \in S$ *has a factorization* MA *with* $M \in S$, $A \in \mathbb{A}_n(R)$, *and* $A(Q) = 0$.

The matrix A *in* (2) *and in* (3) *is unit-regular iff the projective module* Q *is self-dual. Under* (2) *or* (3), *this is always the case if the category of f.g. projective modules over* R *satisfies the cancellation law.*

Proof. (1) $\Rightarrow$ (2). We fix a symplectic structure on P, and think of it as an isomorphism $A : P \to P^*$ such that for every $u^T \in P$, the functional Au^T vanishes on u^T. Extend A to $R^n \to R^n$ by taking $A(Q) = 0$. Then $\ker(A) = Q$, and A is von Neumann regular since $\operatorname{im}(A) = P^*$ is a direct summand in R^n. Finally, for any $u^T \in P$ and $v^T \in Q$, we have $vAu^T = [v, Au^T] = 0$ (since $Au^T \in P^*$), and hence

$$(u+v)\,A\,(u^T + v^T) = uAu^T = 0.$$

Thus, the matrix of A with respect to the standard basis on R^n is *alternating*, which proves (2).

(2) $\Rightarrow$ (3). Assuming (2), $\operatorname{im}(A) = A \cdot (P \oplus Q) = A(P)$ is a direct summand of R^n, since A is von Neumann regular. Let M be the inverse of the isomorphism $A : P \to A(P)$. Defining M to be zero on a direct complement of $A(P)$, we may view M as an R-endomorphism of R^n. Then $E = MA$ since both are zero on Q, and the identity on P. This checks (3).

(3) $\Rightarrow$ (1). For the matrices A, M in (3), we have $\ker(A) \subseteq \ker(E) = Q \subseteq \ker(A)$, so $\ker(A) = Q$. For any $u^T \in P$ and $v^T \in Q$, we have $vA = -(Av^T)^T = 0$, and hence $[v, Au^T] = vAu^T = 0$. This shows that A maps P (injectively) into P^*. We also have $uAu^T = 0$, since A is alternating. Thus, A gives a symplectic structure on P if $A : P \to P^*$ is *onto*. To check this, let $\alpha^T \in P^*$. For any $x^T \in P$, we have

$$\alpha \cdot x^T = \alpha\,(E\,x^T) = (\alpha\,M)\,A\,x^T.$$

Writing $\alpha\,M = u + v$ where $u^T \in P$ and $v^T \in Q$, and recalling that $v\,A = 0$, this gives $\alpha \cdot x^T = (u\,A)\,x^T$. But this equation holds also for all $x^T \in Q$ (since in that case both sides are zero). Thus, $\alpha = u\,A$, and hence $\alpha^T = A^T u^T = -A\,u^T$. This completes the proof of (3) $\Rightarrow$ (1), but it is useful to note a couple of additional properties of A here. First, we have $A = AE$, since they agree on P and are both zero on Q. Thus, $S \cdot E = S \cdot A$; that is, S *and* A *generate the same left ideal in the matrix ring* S. Second, A must be von Neumann regular, since $\operatorname{im}(A) = P^*$

is a direct summand of R^n. Alternatively, going back to the equation $E = MA$, we have $AMA = AE = A$.

Finally, let A be the matrix in (2) or in (3). Then

$$\ker(A) = Q, \quad \text{and} \quad \operatorname{coker}(A) = R^n/\operatorname{im}(A) = R^n/P^* \cong Q^*.$$

Since A is von Neumann regular, Ehrlich's result in [Eh] implies that A is unit-regular iff $\ker(A) \cong \operatorname{coker}(A)$; that is, iff $Q \cong Q^*$. If the category of f.g. projective R-modules satisfies the cancellation law, then under the condition (2) or (3), we have

$$P \oplus Q \cong P^* \oplus Q^* \cong P \oplus Q^* \Longrightarrow Q \cong Q^*.$$

So in this case, Q is necessarily self-dual, and A is necessarily unit-regular. □

Remark 2.2. There are many known cases where the category of f.g. projective R-modules satisfies the cancellation law. For instance, R can be any Dedekind domain, any Bézout domain, any semilocal ring, or any (commutative) von Neumann regular ring; see [La$_6$]. For each of these rings, the matrix A in (2) or (3) above will be unit-regular.

Instead of focusing on the existence of a symplectic structure on P, we may also start with a matrix $A \in \mathbb{A}_n(R)$ and ask when would A induce a symplectic structure on the factor module $R^n/\ker(A)$. From this viewpoint, we have the following variation on the theme of Theorem 2.1.

Theorem 2.3. *For a matrix $A \in \mathbb{A}_n(R)$, let $Q = \ker(A) \subseteq R^n$. The pairing*

$$(2.4) \qquad R^n/Q \times R^n/Q \to R \quad \text{defined by} \quad \langle x^T + Q, y^T + Q \rangle = xAy^T \in R$$

gives a symplectic structure on R^n/Q iff A is von Neumann regular. (In this case, A is unit-regular iff $Q \cong Q^*$.) Every n-generated symplectic module over R arises in this way, from a suitable von Neumann regular matrix $A \in \mathbb{A}_n(R)$.

Proof. First note that the pairing is well defined, since for any $v^T \in Q$, we have $Av^T = 0$ (and hence also $vA = 0$ since $A^T = -A$). If A is von Neumann regular, then Q is a direct summand in R^n, so $R^n = P \oplus Q$ for some (f.g. projective) submodule $P \subseteq R^n$. By (2) ⇒ (1) in Theorem 2.1, the pairing sending $(x^T, y^T) \in P \times P$ to $xAy^T \in R$ is a symplectic structure on P. This proves the "if" part in the theorem. For the "only if" part, recall that modules supporting symplectic structures are assumed to be f.g. projective. Thus, if R^n/Q is a symplectic module under the pairing defined in (2.4), then R^n/Q is projective and so Q is a direct summand in R^n. Fixing as before a direct complement P of Q, we see that $\operatorname{im}(A) = \{Ay^T : y^T \in P\}$ is precisely P^*, when we make the usual identification $(R^n)^* = R^n$. Since P^* is a direct summand of R^n, A is von Neumann regular, as desired. The last conclusion of the theorem follows from the proof of (1) ⇒ (2) in Theorem 2.1. □

In some special cases, Theorem 2.1 can be improved. With the same notations as in (2.1), the following result presents such an improvement.

Corollary 2.5. *If* R *is a self-injective ring or a von Neumann regular ring, and* $S = \mathbb{M}_n(R)$, *then the statements* (1)–(3) *in* Theorem 2.1 *are equivalent to*:

(2)′ $Q = \ker(A)$ *for some matrix* $A \in \mathbb{A}_n(R)$.

Proof. It suffices to show that (2)′ ⇒ (2). Assuming (2)′ holds, $\ker(A) = Q$ is a direct summand of R^n. If R (and hence R^n) is injective as an R-module, then $\operatorname{im}(A) \cong R^n/Q \cong P$ is also an injective R-module. This implies that $\operatorname{im}(A)$ is a direct summand of R^n, so $A \in S$ is von Neumann regular. (For a more general formulation, see [La$_2$: pp. 59–60].) Thus, (2) holds – for the same A in (2)′. If, instead, R is a von Neumann regular ring, then so is S by [La$_2$: Ex. 21.10B] (or by Example 3.3(B) below). In this case, *all* matrices in S are von Neumann regular, so (2) again holds. □

Next, we'll specialize Theorem 2.1 to the case where $Q \cong R$. This is the case where the projective module P in question is the syzygy module

(2.6) $$P(\tau) = \{u^T \in R^n : [u, \tau] = 0\}$$

for a unimodular row τ of length n. (In the terminology of [La$_5$: Ch. I], P is an n-generated "stably free module of type 1".) In preparation for the special version of Theorem 2.1 for such modules P, we need a special determinantal identity, as follows.

Lemma 2.7. *For any* $a \in R$ *and any rows* σ, τ *of length* n *over* R, *we have*

$$\det\left(a \cdot I_n - \sigma^T\tau\right) = a^{n-1}\left(a - [\sigma, \tau]\right).$$

Proof. This is presumably a classical result. For the convenience of the reader, we include a "modern" proof. We first observe that, for any matrix $G \in \mathrm{GL}_n(R)$, if $\sigma' := \sigma G^T$ and $\tau' := \tau G^{-1}$, then it suffices to prove the determinantal identity for σ', τ'. This follows simply by noting that

(2.8) $$a \cdot I_n - \sigma^T\tau = a \cdot I_n - G^{-1}(\sigma')^T\tau' G = G^{-1}[a \cdot I_n - (\sigma')^T\tau']G,$$

and that $[\sigma', \tau'] = \sigma'(\tau')^T = \sigma G^T (G^{-1})^T\tau^T = \sigma\tau^T = [\sigma, \tau]$.

It is sufficient to prove the determinantal identity over a field R. In this case, we can find a matrix $G \in \mathrm{GL}_n(R)$ such that $\tau' := \tau G^{-1}$ has the form $(c_1, 0, \ldots, 0)$. If $\sigma' := \sigma G^T = (b_1, \ldots, b_n)$, then $a \cdot I_n + (\sigma')^T\tau'$ is lower triangular with diagonal entries $(a - b_1c_1, a, \ldots, a)$, so its determinant is $a^{n-1}(a - b_1c_1) = a^{n-1}\left(a - [\sigma', \tau']\right)$. □

Corollary 2.9. *In the notations of* (2.7), *assume that* $a \in \mathrm{U}(R)$ *and that* $[\sigma, \tau] \in \operatorname{rad}(R)$ (*the Jacobson radical of* R). *Then* $a \cdot I_n - \sigma^T\tau \in \mathrm{GL}_n(R)$.

Proof. This follows from (2.7) since $\mathrm{U}(R) + \operatorname{rad}(R) \subseteq \mathrm{U}(R)$. □

It is now easy to reformulate Theorem 2.1 into a result for the existence of symplectic structures on syzygy modules (of even rank).

Theorem 2.10. *For two rows* σ, τ *of odd length* n *over a ring* R. *The following are equivalent*:

(3) $I_n - \sigma^T\tau$ *has a factorization* MA *with* $M \in \mathbb{M}_n(R)$, $A \in \mathbb{A}_n(R)$, *and* $A\,\sigma^T = 0$.

(4) $[\sigma,\, \tau] = 1$ *and* $P(\tau)$ *has a symplectic structure.*

The matrix A *in* (3) *is necessarily unit-regular.*

Proof. (3) $\Rightarrow$ (4). Assume (3). Since n is odd and $A \in \mathbb{A}_n(R)$, $\det(A) = 0$. Thus, from the factorization $I_n - \sigma^T\tau = MA$ in (3), we have

$$\det(I_n - \sigma^T\tau) = \det(M) \cdot \det(A) = 0.$$

On the other hand, (2.7) gives $\det(I_n - \sigma^T\tau) = 1 - [\sigma,\, \tau]$. Therefore, $[\sigma,\, \tau] = 1$ (so τ is a unimodular vector), and we have a decomposition

$$(2.11) \qquad R^n = Q \oplus P(\tau), \quad \text{where} \quad Q := R \cdot \sigma^T \cong R.$$

Let E be the projection of R^n onto $P(\tau)$ with kernel Q. Then $E = I_n - \sigma^T\tau$, since $(I_n - \sigma^T\tau)\,(u^T) = u^T - [u,\, \tau]\,\sigma^T = u^T$ for all $u^T \in P(\tau)$, and

$$(I_n - \sigma^T\tau)\,(\sigma^T) = \sigma^T - [\sigma,\, \tau]\,\sigma^T = 0.$$

Thus, by assumption, $E = MA$. Also, $A\,\sigma^T = 0$ implies that $A\,(Q) = 0$. Therefore, the condition (3) here boils down to the condition (3) in Theorem 2.1, so by that theorem, $P(\tau)$ has a symplectic structure. Finally, since $Q \cong R$ is self-dual, the last paragraph of Theorem 2.1 implies that the matrix A here is necessarily unit-regular.

(4) $\Rightarrow$ (3). Given (4), we certainly have the decomposition in (2.11), and as above, the projection E is given by $I_n - \sigma^T\tau$ in $S = \mathbb{M}_n(R)$. The assumption that $P(\tau)$ has a symplectic structure implies, by Theorem 2.1, that there is a factorization $I_n - \sigma^T\tau = MA$, where $A \in \mathbb{A}_n(R)$ vanishes on $R \cdot \sigma^T$; that is, the condition (3) in this theorem holds. □

Remark 2.12. In case R is an integral domain, we can weaken (3) into the following:

(3)′ $I_n - \sigma^T\tau \in \mathbb{M}_n(R)$ *is right-divisible by some* $A \in \mathbb{A}_n(R)$.

Indeed, if (3)′ holds, write $I_n - \sigma^T\tau = MA$ (for some $M \in \mathbb{M}_n(R)$). Since $\det(A) = 0$, $A\,w^T = 0$ for some nonzero $w^T \in R^n$. Then $E\,(w^T) = M\,(A\,w^T) = 0$ implies that $w^T \in Q$, so $w^T = r\,\sigma^T$ for some (nonzero) $r \in R$. But then

$$r \cdot A\,\sigma^T = A\,w^T = 0 \Longrightarrow A\,\sigma^T = 0$$

since R is a domain. This checks the condition (3) in Theorem 2.10..

For the syzygy module $P(\tau)$ in (2.11), there was an earlier criterion given by the second author for the existence of a symplectic structure. This criterion is expressed in terms of the notion of the "skew-completetability" of a row. For odd n, a row $\tau = (b_1, \ldots, b_n)$ is said to be *skew-completetable* if there exists a matrix $C \in \mathbb{A}_{n+1}(R) \cap \mathrm{GL}_{n+1}(R)$ with first row $(0, b_1, \ldots, b_n)$. (Of course, for this to happen, τ must be unimodular.) The second author's earlier result on the existence of symplectic structures on syzygy modules is the following, expositions on which can be found in [Kr] and [La$_5$: VII.5.28].

Theorem 2.13. *Let* $[\sigma, \tau] = 1$, *where* $\sigma = (a_1, \ldots, a_n)$, $\tau = (b_1, \ldots, b_n)$, *and* n *is odd. The syzygy module* $P(\tau)$ *has a symplectic structure iff the row* τ *is skew-completable.*

The proof of this result given in [La$_5$] was a little bit round about, since it depended on working with the skew-completability of σ instead of the skew-completability of τ. This necessitated working with the *inverse* of a certain invertible alternating matrix. As it turned out, the proof actually becomes simpler and more natural if we work directly with τ, and prove the version of the result exactly as stated above. Since this may not be well known, we give such a proof below.

Proof of Theorem 2.13. Starting with the decomposition in (2.11), we take the unit vector basis $e_1, \ldots, e_n$ on R^n, and let $R^{n+1} = R \cdot e_0 \oplus R^n$, where e_0 is a new unit vector. For the "only if" part of the theorem, fix a symplectic structure on $P(\tau)$, and extend it to a symplectic structure B on R^{n+1} by stipulating that $R \cdot e_0 \oplus Q$ is orthogonal to $P(\tau)$, and $B(e_0, \sigma^T) = 1$. Then the matrix $C = \big(B(e_i, e_j)\big)_{i,j \geq 0}$ is alternating and invertible, say with first row $(0, c_1, \ldots, c_n)$. For $i \geq 1$, we have $e_i = b_i \cdot \sigma^T + \gamma_i^T$ for some $\gamma_i^T \in P(\tau)$. Thus, for $i \geq 1$:

$$c_i = B\left(e_0,\, e_i\right) = b_i B\left(e_0,\, \sigma^T\right) + B\left(e_0,\, \gamma_i^T\right) = b_i\,.$$

This shows that C is a "skew-completion" of τ.

Conversely, assume $(0, \tau)$ is the first row of a matrix in $\mathbb{A}_{n+1}(R) \cap \mathrm{GL}_{n+1}(R)$. This matrix defines a symplectic form B on R^{n+1}. Since

$$B\left(e_0,\, \sigma^T\right) = \textstyle\sum_{i \geq 1} a_i B\left(e_0,\, e_i\right) = \sum_{i \geq 1} a_i b_i = 1,$$

the 2-space $H := R \cdot e_0 \oplus Q$ is the symplectic hyperbolic plane under the form B. We can thus decompose R^{n+1} into an orthogonal sum $H \perp H'$, where H' is the orthogonal complement of H (see [Sw$_1$: Lemma A.1], or the proof of [La$_5$: VII.5.8]). Since H' is a symplectic module and $P(\tau) \cong R^{n+1}/H \cong H'$ as R-modules, it follows that $P(\tau)$ has a symplectic structure.[1] □

The easiest and best known special case of Theorem 2.13 is where $n = 3$. In this case, the unimodular row $\tau = (b_1, b_2, b_3)$ turns out to be always skew-completable; see the last paragraph of §4. This being the case, Theorem 2.13 guarantees that $P(b_1, b_2, b_3)$ *always* has a symplectic structure (and is, in particular, self-dual). This is consistent with a result of Bass [Ba: Prop. 4.4], which implies the same conclusion for any (f.g.) stably free module of rank two.

[1] At the conclusion of this proof, it is relevant to recall that the existence criteria for symplectic structures on $P(\sigma)$ and on $P(\tau)$ are the same, since $P(\sigma) \cong P(\tau)^*$ by [La$_5$: (I.4.10)]. Thus, Theorem 2.13 actually *implies* that σ is skew-completable iff τ is skew-completable.

3. Characterization of von Neumann regular matrices

In view of (2.1) and (2.3), it is of interest to detect when a given square matrix over a commutative ring is von Neumann regular. In (3.2) below, we shall provide a necessary and sufficient condition for this which does not seem well known to matrix theorists or ring theorists. We first prove the following lemma, which prepares us for such a result.

Lemma 3.1. *Let R be any ring with Jacobson radical* $\mathrm{rad}\,(R)$. *If $A = (a_{ij}) \in \mathbb{M}_n(R)$ is a von Neumann regular matrix with $a_{ij} \in \mathrm{rad}\,(R)$ for all i, j, then $A = 0$.*

Proof. This is clear since $A \in \mathbb{M}_n\big(\mathrm{rad}\,(R)\big) = \mathrm{rad}\,\big(\mathbb{M}_n(R)\big)$ (see [La$_1$: p. 57]), and the only von Neumann regular element in the Jacobson radical of a ring is zero. □

For any matrix $A \in \mathbb{M}_n(R)$ (where R is a commutative ring), let $\mathcal{D}_i(A)$ ($1 \le i \le n$) denote the ith determinantal ideal of A, that is, the ideal in R generated by the $i \times i$ minors of A; see [No], or [Ei]. We have a descending sequence

$$\mathcal{D}_0(A) \supseteq \mathcal{D}_1(A) \supseteq \mathcal{D}_2(A) \supseteq \cdots \supseteq \mathcal{D}_n(A) = \det(A) \cdot R \supseteq (0),$$

where, by convention, $\mathcal{D}_0(A) = R$. Recall that the *McCoy rank* of A is defined to be the largest i for which $\mathcal{D}_i(A)$ is a faithful ideal (that is, $\mathrm{ann}\,(\mathcal{D}_i(A)) = 0$); see [MC: p. 159]. The following criterion for the von Neumann regularity of A is in terms of its determinantal ideals $\mathcal{D}_i(A)$'s.

Theorem 3.2. *A matrix $A = (a_{ij}) \in \mathbb{M}_n(R)$ is von Neumann regular iff each determinantal ideal $\mathcal{D}_i(A)$ ($0 \le i \le n$) is idempotent (or equivalently, each $\mathcal{D}_i(A)$ is generated by an idempotent in R : see Footnote 4 below). In this case, the McCoy rank of A is the largest integer r such that $\mathcal{D}_r(A) = R$.*

Examples 3.3.

(A) In the case where R is a connected ring, the theorem shows that A is von Neumann regular iff each $\mathcal{D}_i(A)$ is either (0) or R. Here, the McCoy rank of A is the least r such that $\mathcal{D}_i(A) = 0$ for all $i > r$.

(B) If R is a von Neumann regular ring, then all finitely generated ideals of R are generated by an idempotent (see [La$_1$: (4.23)]). In this case, Theorem 3.2 recovers the well-known fact that *any matrix ring $\mathbb{M}_n(R)$ is von Neumann regular.* (This is true even for *noncommutative* von Neumann regular rings R; see [La$_2$: Ex. 21.10B].)

(C) The matrix $A = \begin{pmatrix} 1 & 1 & 0 \\ 2 & 5 & 2 \\ -2 & -8 & -4 \end{pmatrix}$ provides a good illustration for Theorem 3.2 over the ring $R = \mathbb{Z}$. In [La$_2$: p. 60], it is shown that A is von Neumann regular. In our setting, this follows immediately from Theorem 3.2 since $\mathcal{D}_3(A) = 0$, while $\mathcal{D}_2(A) = \mathbb{Z}$ upon noting that $\det \begin{pmatrix} 1 & 0 \\ 2 & 2 \end{pmatrix} = 2$ and $\det \begin{pmatrix} 1 & 1 \\ 2 & 5 \end{pmatrix} = 3$. Of course, the McCoy rank of A here is the same as the ordinary rank of A, which is 2. But the above computation of $\mathcal{D}_2(A)$ gives more information. It implies, for instance, that A *remains* von Neumann regular if we change its third row at will, as long as we ensure that $\det(A) = 0$. More precisely, the last row can be of the form $(a, b, 2(b-a)/3)$ for any $a, b \in \mathbb{Z}$ with $a \equiv b \pmod 3$.

Proof of Theorem 3.2. Once the regularity criterion is proved, the second statement in (3.2) follows, since an idempotent-generated ideal is faithful iff it is the unit ideal.

To prove the regularity criterion, first assume A is von Neumann regular. *We claim that the ideal* $I := \mathcal{D}_1(A) = \sum a_{ij}R$ *has the property that* $I + \operatorname{ann}(I) = R$. If otherwise, there exists a maximal ideal $\mathfrak{m} \supseteq I + \operatorname{ann}(I)$. Over the localization $R_{\mathfrak{m}}$, A remains von Neumann regular. Since $a_{ij} \in \mathfrak{m}R_{\mathfrak{m}}$, Lemma 3.1 implies that $a_{ij} = 0 \in R_{\mathfrak{m}}$ for all i, j. Thus, there exists $r \in R \setminus \mathfrak{m}$ such that $ra_{ij} = 0 \in R$ for all i, j. But then $r \in \operatorname{ann}(I) \subseteq \mathfrak{m}$, a contradiction. From $I + \operatorname{ann}(I) = R$, we have an equation $e + f = 1$ where $e \in I$ and $f \in \operatorname{ann}(I)$. Multiplying this equation by e, we get $e = e^2 + ef = e^2$. For any $x \in I$, we have $x = x - fx = ex \in eR$. Thus, $I = eR$ is idempotent.[2]

Now consider any $i \in [1, n]$, and let φ be the R-homomorphism $R^n \to R^n$ defined by A. This module homomorphism induces a homomorphism on the exterior powers

$$\Lambda^i(\varphi) : \Lambda^i(R^n) \longrightarrow \Lambda^i(R^n).$$

For the unit vector basis $\{e_1, \ldots, e_n\}$ on R^n, the exterior power $\Lambda^i(R^n)$ has a basis consisting of

$$e_{k_1} \wedge \cdots \wedge e_{k_i}, \quad \text{where} \quad 1 \le k_1 < \cdots < k_i \le n.$$

With respect to this basis, $\Lambda^i(\varphi)$ has a matrix whose entries are the $i \times i$ minors of A. Fixing a matrix $B \in \mathbb{M}_n(R)$ such that $A = ABA$ and letting $\psi : R^n \to R^n$ be the homomorphism defined by B, we have (by functoriality) $\Lambda^i(\varphi) = \Lambda^i(\varphi)\Lambda^i(\psi)\Lambda^i(\varphi)$. Thus, the matrix representing $\Lambda^i(\varphi)$ is von Neumann regular. Since $\mathcal{D}_i(A)$ is just the first determinantal ideal of this matrix, it follows from the case we have settled above that $\mathcal{D}_i(A)$ is idempotent, as desired.

To prove the converse, assume that each $\mathcal{D}_i(A)$ is idempotent. To show that A is von Neumann regular, we view A as an R-homomorphism $R^n \to R^n$ and want to show that $\operatorname{im}(A)$ is a direct summand of R^n. (See the introductory remarks on von Neumann regular matrices made at the beginning of §2.) This may be checked locally (since each $\mathcal{D}_i(A)$ remains idempotent in the localizations). Thus, we may assume that R is a local ring. Then R has no nontrivial idempotents, so there exists an integer $r \in [1, n]$ such that

(3.4) $$\mathcal{D}_1(A) = \cdots = \mathcal{D}_r(A) = R \quad \text{and} \quad \mathcal{D}_{r+1}(A) = \cdots = \mathcal{D}_n(A) = 0.$$

If $r = n$, then $\mathcal{D}_n(A) = R$ implies that $\det(A) \in \mathrm{U}(R)$. In this case, A is invertible, and hence von Neumann regular. In the following, we may thus assume that $r < n$. Since R is local, $\mathcal{D}_r(A) = R$ means that some $r \times r$ minor is a

---

[2] As was pointed out by K. Goodearl, this conclusion about the first determinantal ideal could also have been gotten from an argument using the trace ideals of projective modules; see [La3: §2H].

unit. For convenience, let us assume that this is the determinant of the upper-left $r \times r$ corner B of A. Then $A = \begin{pmatrix} B & C \\ D & E \end{pmatrix}$, with $B \in \mathrm{GL}_r(R)$. Now use the fact that, for $U, V \in \mathrm{GL}_n(R)$, A is von Neumann regular iff UAV is, and that $\mathcal{D}_i(A) = \mathcal{D}_i(UAV)$ for all i (see [No]). This enables us to change A by matrix equivalence, and in particular, by elementary transformations. Using the latter (and the invertibility of B), we can bring A to the form $\mathrm{diag}\,(B, E')$. It then follows that $E' = 0$ (since $\mathcal{D}_{r+1}(A) = 0$ by (3.4)), which clearly implies that A is von Neumann regular. □

As we have mentioned in the Introduction, there is a more powerful method using which we can give a quicker proof of Theorem 3.2. This is done by exploiting the theory of *Fitting invariants* $\{\mathrm{Fitt}_j(C)\}$ of a f.g. R-module C. For an exposition on this topic, see §20 in Eisenbud's book [Ei]. For a module such as $C = \mathrm{coker}\,(A)$ where A is an $n \times n$ matrix thought of as an R-endomorphism of R^n, *the Fitting invariants* $\mathrm{Fitt}_j(C)$*'s are, up to a reindexing, precisely the determinantal ideals* $\mathcal{D}_i(A)$*'s*. A basic fact on Fitting invariants is given by the following result from [Ei: Prop. 20.8].[3]

Proposition 3.5. *A f.g. R-module C is projective of constant rank iff each* $\mathrm{Fitt}_j(C)$ *is* (0) *or* R.

From this result, we can fairly quickly deduce Theorem 3.2 as follows. Given $A \in \mathbb{M}_n(R)$, let $C = \mathrm{coker}\,(A)$. As we have mentioned before, A is von Neumann regular iff $\mathrm{im}\,(A)$ is a direct summand of R^n, and clearly, this is the case iff C is a *projective* R-module. Since C is finitely presented, this condition is equivalent to $C_{\mathfrak{p}}$ being $R_{\mathfrak{p}}$-free for each prime ideal $\mathfrak{p}$ (see [La$_5$: (I.3.4)]). By Prop. 3.5, this amounts to each $\mathrm{Fitt}_j(C_{\mathfrak{p}})$ being (0) or $R_{\mathfrak{p}}$, which can be easily translated into $\mathrm{Fitt}_j(C_{\mathfrak{p}})^2 = \mathrm{Fitt}_j(C_{\mathfrak{p}})$ (for each $\mathfrak{p}$), since $\mathrm{Fitt}_j(C_{\mathfrak{p}})$ is a f.g. ideal, and $R_{\mathfrak{p}}$ is connected.[4] Thus, A is von Neumann regular iff each $\mathrm{Fitt}_j(C)$ is idempotent; that is, iff each $\mathcal{D}_i(A)$ is idempotent. This completes the second proof of Theorem 3.2.

The proof above is certainly pretty short, but it does depend on Prop. 3.5 as well as a considerable amount of material not developed in this paper. In view of this, we believe the direct and self-contained proof given earlier in this section would still play a useful role in the literature. In addition, a careful analysis of the Fitting invariants proof above in relation to Theorem 3.2 suggests the possibility of an improved version of Prop. 3.5, which we shall discuss below.

Note that, although Prop. 3.5 is stated in terms of a (f.g.) module C over a general ring, it is basically a result of a *local* nature, since it does not address the case where the module C (in case it is projective) has possibly non-constant rank. One might thus wonder if there is a *global* version of Prop. 3.5 that would

---

[3]In Eisenbud's exposition, the base ring R was assumed to be *noetherian*. But a careful examination of the proof of [Ei: Prop. 20.8] will show that the noetherian assumption on R is not needed.

[4]Here, we used the well-known fact that a f.g. ideal is idempotent iff it is generated by an idempotent. For a proof of this, see [La$_3$: (2.43)].

give a characterization for an *arbitrary* f.g. projective module. Fortuitously, the statement of Theorem 3.2 suggests the right generalization, as follows.

Theorem 3.6. *A f.g. R-module C is projective iff each* $\mathrm{Fitt}_j(C)$ *is generated by an idempotent.*

Proof. First assume C is (f.g.) projective. Then C is also finitely presented, so each $\mathrm{Fitt}_j(C)$ is a f.g. ideal. Thus, we only need to check that, for each j, the inclusion $\mathrm{Fitt}_j(C)^2 \subseteq \mathrm{Fitt}_j(C)$ is an equality (for then $\mathrm{Fitt}_j(C)$ will be generated by an idempotent, according to the fact mentioned in Footnote 4). As in the second proof for (3.2) above, this property can be checked locally, so we may assume that R is a local ring. But then $C \cong R^k$ for some $k \geq 0$. In this case, the desired conclusion is trivial since each $\mathrm{Fitt}_j(C)$ is *surely* either R or (0).[5]

Conversely, assume that each $\mathrm{Fitt}_j(C)$ is generated by an idempotent. Then the same is true at every prime ideal $\mathfrak{p}$, so by Prop. 3.5, $C_{\mathfrak{p}}$ is free. The free rank of $C_{\mathfrak{p}}$ at every prime $\mathfrak{p}$ is determined by the sequence of Fitting invariants $\mathrm{Fitt}_j(C_{\mathfrak{p}})$. From this, we see easily that the rank function $\mathrm{rank}_C : \mathrm{Spec}\,(R) \to \mathbb{Z}$ is *locally constant.* Since C is f.g., a theorem of Bourbaki [Bo: pp. 109–111] (see also [La$_4$: Ex. 2.21]) implies that C is a (f.g.) projective R-module. □

4. Von Neumann regular matrices of small size

In this final section, we'll briefly comment on the case of von Neumann regular matrices of small size. In this case, we shall reprove the "if" part of Theorem 3.2 from a constructive point of view, and give some examples on the computation of quasi-inverses for von Neumann regular matrices.

To begin with, the case of a 1×1 matrix $A = (a)$ is basically trivial. If $aR = eR$ where $e = e^2$, then writing $e = ar$, one has $a = ea = ara$, so a is indeed von Neumann regular, with a quasi-inverse r. A more careful construction would have given an equation $a = aua$ with $u \in \mathrm{U}(R)$, so a is in fact unit-regular. (For a more general result, see [La$_2$: Ex. 12.6A].)

The case $n = 2$ can be handled directly too, as follows. If a matrix $A \in \mathbb{M}_2(R)$ has

$$\mathcal{D}_1(A) = eR \quad \text{and} \quad \mathcal{D}_2(A) = (\det A)\, R = e'R,$$

where e, e' are idempotents, there is no loss in assuming that $\det A = 0$. This is because R splits into $e'R \times (1 - e')R$, and in the component $e'R$, the projection $e'A$ of A is *invertible* (and hence von Neumann regular). Thus, it suffices to analyze the projection of A in the other component $(1 - e')\, R)$, which has determinant zero. In this case, we have the following explicit result, independently of Theorem 3.2.

[5]More precisely, $\mathrm{Fitt}_j(C)$ is R for all j in case $k = 0$; and is (0) for $j < k$ and R for $j = k$ in case $k \geq 1$.

Proposition 4.1. *Let* $A = \begin{pmatrix} a & b \\ c & d \end{pmatrix}$ *with* $ad = bc$ *and* $\mathcal{D}_1(A) = eR$, *where* $e = e^2$. *Fix an equation* $aw + bx + cy + dz = e$. *Then the matrix* $M = \begin{pmatrix} w & y \\ x & z \end{pmatrix}$ *satisfies* $A = AMA$ *(so* A *is von Neumann regular, with quasi-inverse* M*).*

Proof. This can be checked by a direct computation, using the fact that $ea = a$, $eb = b$, $ec = c$, and $ed = d$, along with $ad = bc$. □

For the case $n = 3$, let us consider the special case of *alternating* matrices; say

$$A = \begin{pmatrix} 0 & -c & b \\ c & 0 & -a \\ -b & a & 0 \end{pmatrix} \in \mathbb{A}_3(R). \tag{4.2}$$

Here, we have $\mathcal{D}_3(A) = 0$, $\mathcal{D}_1(A) = aR + bR + cR$, and quick inspection shows that $\mathcal{D}_2(A) = \mathcal{D}_1(A)^2$. (The latter is a special case of [KLS: (3.8)(2)].) Thus, we can "ignore" $\mathcal{D}_2(A)$ and $\mathcal{D}_3(A)$, and replace (3.2) by the following simpler and sharper statement.

Proposition 4.3. *The matrix* A *in* (4.2) *is von Neumann regular iff* $aR+bR+cR = eR$ *for some idempotent* $e \in R$. *(In this case,* A *is in fact unit-regular.) In particular,* A *is von Neumann regular if the row* (a, b, c) *is unimodular; in case* R *is connected and* $A \neq 0$, *the converse holds.*

Here, the "if" part of the first statement (as well as the claim in parentheses) can be checked explicitly without assuming Theorem 3.2 or Theorem 2.10. Indeed, let $aR+bR+cR = eR$ for some $e = e^2 \in R$, and let $f = 1-e$. Since $R = eR \times fR$ and A has projection 0 in fR, it is sufficient to work in the other component eR, whereby we may assume $e = 1$. Under this assumption, $\sigma = (a, b, c)$ is *unimodular.* Fix a vector $\tau = (p, q, r)$ such that $[\sigma, \tau] = ap + bq + cr = 1$. Then we have $R^3 = R \cdot \sigma^T \oplus P(\tau)$ as in (2.11). *We claim that the condition* (3) *in* Theorem 2.10 *is always satisfied.* To see this, start with the "projection matrix" in the proof of that theorem, which is

$$E = I_3 - \sigma^T \tau = \begin{pmatrix} 1 - ap & -aq & -ar \\ -bp & 1 - bq & -br \\ -cp & -cq & 1 - cr \end{pmatrix}. \tag{4.4}$$

If we let $M := \begin{pmatrix} 0 & r & -q \\ -r & 0 & p \\ q & -p & 0 \end{pmatrix} \in \mathbb{A}_3(R)$, a quick calculation shows that $MA = E$. Since obviously $A\,\sigma^T = 0$, the condition (3) in Theorem 2.10 is satisfied. Moreover,

$$AMA = A\,(I_3 - \sigma^T \tau) = A - A\,\sigma^T \tau = A. \tag{4.5}$$

Thus, *A is indeed von Neumann regular,* with a quasi-inverse M. On the other hand, we also have

$$MAM = (I_3 - \sigma^T \tau)\,M = M - \sigma^T \tau M = M \quad (\text{since } \tau M = 0), \tag{4.6}$$

so M is von Neumann regular as well, with quasi-inverse A! Finally, the fact that A is *unit-regular* (the last part of Theorem 2.10) can be directly checked by using the matrices

$$U = M + \sigma^T \sigma, \quad \text{and} \quad V = A + \tau^T \tau. \tag{4.7}$$

In view of $A,\ M \in \mathbb{A}_3(R)$ and $A\sigma^T = 0 = \tau M$, we see easily that $UV = I_3$ and $AUA = A$, so A is unit-regular with an invertible quasi-inverse U. Similarly,

$$MVM = M\,(A + \tau^T\tau)\,M = MAM = M \quad (\text{by } (4.6)), \tag{4.8}$$

so M is *also* unit-regular, with invertible quasi-inverse V.

Since we have verified condition (3) in Theorem 2.10, it follows that, in the above situation, $P(\tau)$ has a symplectic structure. Of course, in confirmation of Theorem 2.13, it is also easy to show directly that $\tau = (p, q, r)$ is *skew-completable* (as was mentioned in the last paragraph of §2). Indeed, the "bordered" alternating matrix $V = \begin{pmatrix} 0 & \tau \\ -\tau^T & -A \end{pmatrix}$ constructed from A and τ provides a natural skew-completion for τ. Here, V is invertible since its Pfaffian is $ap + bq + cr = 1$. In the study of the elementary symplectic Witt groups over commutative rings (see for instance [La$_5$: p. 320]), the (class of the) invertible alternating matrix V is known as the *Vaserstein symbol* of the unimodular row $\sigma = (a, b, c)$.

Acknowledgement

We thank Professor K. Goodearl for his valuable comments on the formulation (and proof) of Theorem B above.

References

[Ba] H. Bass: *Modules which support non-singular forms.* J. Algebra **13** (1969), 246–252.

[Bo] N. Bourbaki: *Commutative Algebra.* Hermann/Addison-Wesley, 1972.

[Br] W.C. Brown: *Matrices Over Commutative Rings.* Monographs in Pure and Applied Math., Vol. **169**, M. Dekker, Inc., New York, 1993.

[Eh] G. Ehrlich: *Unit-regular rings.* Portugal. Math. **27** (1968), 209–212.

[Ei] D. Eisenbud: *Commutative Algebra with a View Toward Algebraic Geometry.* Graduate Texts in Math., Vol. **150**, Springer-Verlag, Berlin-Heidelberg-New York, 1995.

[HA] J.A. Hermida-Alonso: *On linear algebra over commutative rings.* Handbook of Algebra (M. Hazewinkel, ed.), Vol. **3**, 3–61, North-Holland, Amsterdam, 2003.

[KLS] V. Kodiyalam, T.Y. Lam and R.G. Swan: *Determinantal ideals, Pfaffian ideals, and the principal minor theorem.* Noncommutative Rings, Group Rings, Diagram Algebras and Their Applications, Proc. International Conference, Chennai, India, December, 2006 (S.K. Jain, S. Parvathi, eds.), Contemp. Math. **456** (2008), 35–60, Amer. Math. Soc., Providence, R.I.

[Kr] M. Krusemeyer: *Completing* α^2, β, γ. Conference on Commutative Algebra, Queen's Papers in Pure and Applied Math., Vol. **42** (1975), 253–254.

[La$_1$] T.Y. Lam: *A First Course in Noncommutative Rings.* Second Edition, Graduate Texts in Math., Vol. **131**, Springer-Verlag, Berlin-Heidelberg-New York, 2001.

[La$_2$] T.Y. Lam: *Exercises in Classical Ring Theory.* Second Edition, Problem Books in Mathematics, Springer-Verlag, Berlin-Heidelberg-New York, 2003.

[La$_3$] T.Y. Lam: *Lectures on Modules and Rings.* Graduate Texts in Math., Vol. **189**, Springer-Verlag, Berlin-Heidelberg-New York, 1998.

[La$_4$] T.Y. Lam: *Exercises in Modules and Rings.* Problem Books in Mathematics, Springer-Verlag, Berlin-Heidelberg-New York, 2007.

[La$_5$] T.Y. Lam: *Serre's Problem on Projective Modules.* Monographs in Mathematics, Springer-Verlag, Berlin-Heidelberg-New York, 2006.

[La$_6$] T.Y. Lam: *A crash course on stable range, cancellation, substitution, and exchange.* J. Algebra and Its Appl. **3** (2004), 301–343.

[MC] N. McCoy: *Rings and Ideals.* Carus Mathematical Monographs, Math. Assoc. of America, 1948.

[MD] B.R. McDonald: *Linear Algebra Over Commutative Rings.* Monographs in Pure and Applied Math., Vol. **87**, M. Dekker, Inc., New York, 1984.

[No] D.G. Northcott: *Finite Free Resolutions.* Cambridge Tracts in Mathematics, Vol. **71**, Cambridge University Press, 1976.

[Sw$_1$] R.G. Swan: *Algebraic vector bundles on the* 2*-sphere.* Rocky Mountain J. Math. **23** (1993), 1443–1469.

[Sw$_2$] R. G. Swan: `http://www.math.uchicago.edu/~swan/stablyfree.pdf`.

T.Y. Lam
Department of Mathematics
University of California
Berkeley, CA 94720, USA
e-mail: `lam@math.berkeley.edu`

R.G. Swan
Department of Mathematics
University of Chicago
Chicago, IL 60637, USA
e-mail: `swan@math.uchicago.edu`

Advances in Ring Theory
Trends in Mathematics, 229–237

Extensions of Simple Modules and the Converse of Schur's Lemma

Greg Marks and Markus Schmidmeier

Abstract. The *converse of Schur's lemma* (or *CSL*) condition on a module category has been the subject of considerable study in recent years. In this note we extend that work by developing basic properties of module categories in which the CSL condition governs modules of finite length.

Mathematics Subject Classification (2000). Primary 16D90, 16G20, 16S50.
Keywords. Converse of Schur's Lemma, Gabriel quiver.

1. Introduction

Schur's Lemma states that for any ring R and any simple module M_R, the endomorphism ring $\mathrm{End}(M_R)$ is a division ring. In this note we are interested in the converse of Schur's Lemma (CSL), i.e., whether for a given module category $\mathcal{C}$, every object in $\mathcal{C}$ whose endomorphism ring is a division ring is in fact simple. If this is the case, we say that $\mathcal{C}$ *has CSL*. The case that has received almost exclusive attention in the literature (see, e.g., [1], [2], [7], [10], [9], [14]) is $\mathcal{C} = \mathcal{M}od_R$, the category of right R-modules. Here we will focus on the case $\mathcal{C} = \mathcal{FL}_R$, the category of right R-modules of finite length.

We propose to separate the study of rings R which satisfy CSL for $\mathcal{FL}_R$ from the study of rings which satisfy CSL for $\mathcal{FL}_R$ but not for Mod_R, since the two properties relate to different topics: extensions of simples versus constructions of large modules.

It turns out that the CSL property for finite length modules – and sometimes the CSL property for all modules – is controlled by the following combinatorial information:

Definition 1.1. Let R be a ring. Recall that the *right Gabriel quiver* (or *right* Ext-*quiver*) of R is the directed graph Q consisting of the following data:

- The points of Q are in bijective correspondence with the isomorphism classes of simple right R-modules.

- There is an arrow $i \longrightarrow j$ in Q whenever the corresponding simple modules S_i and S_j extend, i.e., $\mathrm{Ext}^1_R(S_i, S_j) \neq \{0\}$.

We say that the right Gabriel quiver is *totally disconnected* if there are no arrows between any two different points.

For example, the right Gabriel quiver of a semisimple ring is a disjoint union of finitely many points. The right Gabriel quiver of the discrete valuation ring $R = k[x]_{(x)}$ (for k a field) has one point and one loop.

The reader should be aware that the literature contains some variants of the definition we give here. In the classical setting where R is a finite-dimensional algebra over a field, some authors adopt the convention that in the right Gabriel quiver of R the arrow between the vertices corresponding to S_i and S_j carries as label the pair given by the dimensions of $\mathrm{Ext}^1_R(S_i, S_j)$ as a vector space over $\mathrm{End}(S_j)_R$ and $\mathrm{End}(S_i)_R$ respectively.

Theorem 1.2. *Let R be any ring. Then $\mathcal{FL}_R$ has CSL if and only if the right Gabriel quiver of R is totally disconnected.*

We will prove Theorem 1.2 in Section 2.

In general, for $\mathcal{FL}_R$ to have CSL is a considerably weaker condition than for $\mathcal{M}od_R$ to have CSL. The distinction between CSL on $\mathcal{FL}_R$ and CSL on $\mathcal{M}od_R$ is illustrated in the following examples.

Example 1.3. Let R be any commutative ring whatsoever. Then $\mathcal{FL}_R$ has CSL. To see this, suppose M is an R-module of finite length such that $\mathrm{End}(M_R)$ is a division ring. If M were not simple, then by [14, Corollary], M_R would be isomorphic to the field of fractions of $R/\mathfrak{p}$ where $\mathfrak{p} = \mathrm{ann}^R(M)$ is a prime but not maximal ideal of R, contradicting the hypothesis that M has finite length.

By contrast, in [14] it is shown that for a commutative ring R, the category $\mathcal{M}od_R$ has CSL if and only if R has Krull dimension 0.

We infer from Theorem 1.2 that the (right) Gabriel quiver of any commutative ring is totally disconnected.

Example 1.3 suggests a further reason why the (not necessarily commutative) rings R for which $\mathcal{FL}_R$ has CSL are an interesting object of study: they include all commutative rings, so this condition is a new sort of generalization of commutativity.

Example 1.4 (J.H. Cozzens [4]). Let K be an algebraically closed field of positive characteristic p, let φ: $x \mapsto x^{p^n}$ be a Frobenius automorphism on K, and let $k = K^{\langle\varphi\rangle}$ be the fixed field of φ. Then the skew Laurent polynomial ring $R = K[x, x^{-1}; \varphi]$ has, up to isomorphism, a unique simple right module S, which is injective. Thus, every finite length right R-module is semisimple, and hence $\mathcal{FL}_R$ has CSL.

Nevertheless, $\mathcal{M}od_R$ does not have CSL. It is easy to show that if R is a right or left Ore domain, then $\mathcal{M}od_R$ has CSL if and only if R is a division ring. In the present example R is a simple noetherian domain, hence an Ore domain.

Note that the category $\mathcal{FL}_R$ is equivalent to the category $\mathcal{FL}_k$, which obviously has CSL (as even $\mathcal{Mod}_k$ does).

We can generalize this example using [13, Theorem A]. Let $\mathrm{rad}(A)$ denote the Jacobson radical of a ring A. Recall that a finite-dimensional k-algebra A is called *elementary* if $A/\mathrm{rad}(A)$ is a finite direct product of copies of k.

Proposition 1.5. *Let A be a finite-dimensional elementary algebra over a finite field whose right Gabriel quiver is totally disconnected. Then there exists a noetherian ring R such that the following properties hold:*

(i) *The categories $\mathcal{FL}_A$ and $\mathcal{FL}_R$ are equivalent.*
(ii) *Both $\mathcal{FL}_A$ and $\mathcal{FL}_R$ have CSL.*
(iii) *The category $\mathcal{Mod}_A$ has CSL, but the category $\mathcal{Mod}_R$ does not.*

On the other hand, for semiprimary rings the CSL property for all modules is controlled by the right Gabriel quiver, i.e., it is controlled by the CSL property for finite length modules. Recall that a ring R is said to be *semiprimary* if the Jacobson radical $\mathrm{rad}(R)$ is nilpotent and $R/\mathrm{rad}(R)$ is a semisimple ring. Semiprimary rings figure prominently in our main object of study here: it is well known that the endomorphism ring of a finite length module is semiprimary.

Theorem 1.6. *Let R be a semiprimary ring. The following conditions are equivalent:*

(i) *$\mathcal{FL}_R$ has CSL.*
(ii) *The right Gabriel quiver of R is totally disconnected.*
(iii) *R is a finite direct product of full matrix rings over local rings.*
(iv) *$\mathcal{Mod}_R$ has CSL.*

We defer the proofs of Proposition 1.5 and Theorem 1.6 to Section 3. The literature contains results akin to Theorem 1.6, such as the following.

Theorem 1.7. *Let R be a one-sided noetherian ring or a perfect ring. Then $\mathcal{Mod}_R$ has CSL if and only if R is a finite direct product of full matrix rings over local perfect rings.*

The left noetherian case is covered by [2, Theorem 1], the right noetherian case by [5, Theorem 3.4], and the perfect case by [1, Theorem 1.2].

Example 1.8. Let k be a field of characteristic 0, and let $\mathbb{A}_1(k) = k\langle x, y\rangle/(xy - yx - 1)$ be the first Weyl algebra over k. If

$$S_1 = \mathbb{A}_1(k)/x\mathbb{A}_1(k) \qquad \text{and} \qquad S_2 = \mathbb{A}_1(k)/(x+y)\mathbb{A}_1(k),$$

then by [11, Proposition 5.6, Theorem 5.7], S_1 and S_2 are nonisomorphic simple right $\mathbb{A}_1(k)$-modules for which $\mathrm{Ext}^1_{\mathbb{A}_1(k)}(S_1, S_2) \neq \{0\}$. Therefore, the right Gabriel quiver of $\mathbb{A}_1(k)$ is not totally disconnected, so Theorem 1.2 tells us $\mathcal{FL}_{\mathbb{A}_1(k)}$ does not have CSL.

The conclusion of Example 1.8 can be extended from $\mathbb{A}_1(k)$ to certain *generalized Weyl algebras;* see [3, Theorem 1.1] for details.

Example 1.9. Let R be a right bounded Dedekind prime ring. Then $\mathcal{FL}_R$ has CSL. To prove this, first note that if R is right primitive then by [6, Theorem 4.10] it is simple artinian (so in this case even $\mathcal{Mod}_R$ has CSL). Now assume R is not right primitive. Suppose S_1 and S_2 are arbitrary nonisomorphic simple right R-modules. Then $\mathrm{ann}_r^R(S_1) = \mathfrak{m}_1$ and $\mathrm{ann}_r^R(S_2) = \mathfrak{m}_2$ are maximal ideals of R and $\mathfrak{m}_1 \neq \mathfrak{m}_2$, by [12, Theorem 3.5]. By [6, Theorem 1.2, Proposition 2.2], $\mathfrak{m}_1$ and $\mathfrak{m}_2$ are invertible ideals. We can therefore apply [8, Proposition 1] to conclude that $\mathrm{Ext}^1_{\mathbb{A}_1(k)}(S_1, S_2) = \{0\}$. Thus, by Theorem 1.2, $\mathcal{FL}_R$ has CSL.

Example 1.10. Let G be a finite group and F a field.

(i) If the characteristic of F does not divide the order of G, then FG is semisimple and hence the Gabriel quiver is totally disconnected (no arrows).

(ii) If the characteristic of F is a prime number p and G is a finite p-group, then FG is a local ring and again, the Gabriel quiver is totally disconnected (the only arrow is a loop).

(iii) In the case where the number of simples is different from the number of blocks, there is a block where two nonisomorphic simples extend, and we get a proper arrow in the Gabriel quiver.

Thus, in cases (i) and (ii), but not (iii), $\mathcal{FL}_{FG}$ has CSL.

2. CSL for finite length modules

In this section we give a proof of Theorem 1.2. First, assume that $\mathcal{FL}_R$ has CSL. As a consequence of the next lemma, the right Gabriel quiver of R must be totally disconnected.

Lemma 2.1. *Suppose $0 \to T \to M \to S \to 0$ is a non-split short exact sequence in $\mathcal{Mod}_R$ where S and T are nonisomorphic simple modules. Then* $\mathrm{End}(M_R)$ *is isomorphic to a division subring of* $\mathrm{End}(S_R)$ *and of* $\mathrm{End}(T_R)$.

Proof. Since $\mathrm{Hom}_R(M,T) = \{0\}$, $\mathrm{Hom}_R(M,M)$ embeds in $\mathrm{Hom}_R(M,S)$; since $\mathrm{Hom}_R(T,S) = \{0\}$, we can identify $\mathrm{Hom}_R(M,S)$ with $\mathrm{Hom}_R(S,S)$. This yields a ring monomorphism $\mathrm{End}(M_R) \to \mathrm{End}(S_R)$. Similarly, since $\mathrm{Hom}_R(S,M) = \{0\}$ and $\mathrm{Hom}_R(T,S) = \{0\}$, we obtain a ring monomorphism $\mathrm{End}(M_R) \to \mathrm{End}(T_R)$. Thus $\mathrm{End}(M_R)$ is isomorphic to a subring of the division rings $\mathrm{End}(S_R)$ and $\mathrm{End}(T_R)$, so $\mathrm{End}(M_R)$ is a domain. Being also semiprimary, $\mathrm{End}(M_R)$ is a division ring. □

For the converse, we assume that the right Gabriel quiver of R is totally disconnected. We first show that every finite length indecomposable module is isotypic, and then that every isotypic module is either simple or admits a nonzero nilpotent endomorphism. Note that some of the results apply both to finite length modules over an arbitrary ring and to arbitrary modules over a semiprimary ring. We will use these results again in the next section.

Definition 2.2. A module M is *isotypic* if all simple subquotients of M are isomorphic. A sequence

$$0 = M_0 \subset M_1 \subset \cdots \subset M_\ell = M$$

of submodules is called an *isotypic filtration* of M of length ℓ if for every $i = 1, \ldots, \ell$, the quotient M_i/M_{i-1} is isotypic.

Proposition 2.3. *Suppose R is a ring whose right Gabriel quiver is totally disconnected. Suppose that either*

(i) *M is an object of $\mathcal{FL}_R$, or*
(ii) *R is semiprimary, and M is an object of $\mathcal{Mod}_R$.*

Then M is a finite direct sum of isotypic modules.

Proof. Step 1: The module M has an isotypic filtration. For example, the radical filtration of M can be refined to an isotypic filtration.

Step 2: For each isotypic filtration

$$0 \subset M_1 \subset \cdots \subset M_\ell = M$$

there is an isotypic filtration $0 \subset M_1' \subset \cdots \subset M_\ell' = M$ such that $M_i \subseteq M_i'$ for each i and $\mathrm{Hom}_R(M_1', M/M_1') = \{0\}$. Zorn's Lemma can be applied to the set of all isotypic submodules of M that contain M_1; let M_1' be a maximal member of this set. We have $\mathrm{Hom}_R(M_1', M/M_1') = \{0\}$ since the socle of M/M_1' cannot contain a simple summand isomorphic to a subquotient of M_1'. For $i > 1$, put $M_i' = M_i + M_1'$. Then

$$\begin{aligned}\frac{M_i'}{M_{i-1}'} &= \frac{M_i + M_1'}{M_{i-1} + M_1'} \\ &\cong \frac{M_i}{M_i \cap (M_{i-1} + M_1')} = \frac{M_i}{M_{i-1} + (M_i \cap M_1')}\end{aligned}$$

is epimorphic image of M_i/M_{i-1} and hence isotypic.

Step 3: Let M and N be isotypic modules that both satisfy (i) *or* (ii) *of the proposition and for which* $\mathrm{Hom}_R(M, N) = \{0\}$. *Then* $\mathrm{Ext}^1_R(M, N) = \{0\}$. When M and N are semisimple, this follows from the hypothesis on the right Gabriel quiver. The general case follows by induction on the lengths of semisimple filtrations of M and N.

Step 4: The module M is a direct sum of isotypic modules. We induct on the length ℓ of the isotypic filtration of M produced in Step 2. The case $\ell = 1$ is trivial. For the induction step, let M have an isotypic filtration $0 \subset M_1' \subset \cdots \subset M_{\ell+1}' = M$. By inductive hypothesis, $M/M_1' \cong \bigoplus_i M_i''$ is a direct sum of isotypic modules M_i''. Since $\mathrm{Hom}_R(M_1', M_i'') = \{0\}$ for all i, by Step 3 we have

$$\mathrm{Ext}^1_R(M_1', M/M_1') = \bigoplus_i \mathrm{Ext}^1_R(M_1', M_i'') = \{0\}$$

Thus, the short exact sequence $0 \to M_1' \to M \to M/M_1' \to 0$ splits, and $M \cong M_1 \oplus \bigoplus_i M_i''$ is a direct sum of isotypic modules. □

Next we prove a criterion for isotypic modules to admit a nonzero nilpotent endomorphism. The argument is adapted from the proof of [1, Theorem 1.2].

Lemma 2.4. *Let R be any ring, and let M be a right R-module. Then M has no nonzero semisimple direct summand if and only if* $\mathrm{soc}(M) \subseteq \mathrm{rad}(M)$.

Proof. If M has no nonzero semisimple direct summand, then every simple submodule is superfluous, whence $\mathrm{soc}(M) \subseteq \mathrm{rad}(M)$. Conversely, if M does have a nonzero semisimple direct summand, then M has a simple direct summand, which is contained in $\mathrm{soc}(M)$ but not $\mathrm{rad}(M)$, so $\mathrm{soc}(M) \not\subseteq \mathrm{rad}(M)$. □

Proposition 2.5. *Suppose that M_R is a nonzero isotypic module that is not simple. Assume in addition that either M has finite length or R is a perfect ring. Then M has a nonzero nilpotent endomorphism.*

Proof. If M has a nonzero simple direct summand, the conclusion is clear; so assume otherwise. By Lemma 2.4, $\mathrm{soc}(M) \subseteq \mathrm{rad}(M)$. Now, M is nonzero and isotypic, and the hypotheses imply that $M/\mathrm{rad}(M)$ is semisimple; therefore, there exists a nonzero homomorphism $f_0\colon M/\mathrm{rad}(M) \to \mathrm{soc}(M)$. The composite map

$$f\colon M \xrightarrow{\pi} M/\mathrm{rad}(M) \xrightarrow{f_0} \mathrm{soc}(M) \xrightarrow{\iota} M$$

(where π is the canonical epimorphism and ι the inclusion map) is a nonzero endomorphism of M satisfying $f^2 = 0$. □

Theorem 1.2 is now established. The "only if" part follows from Lemma 2.1. The "if" part follows from Propositions 2.3 and 2.5.

3. CSL for all modules

To prove Theorem 1.6 we will show

$$\text{(i)} \Leftrightarrow \text{(ii)} \Rightarrow \text{(iii)} \Rightarrow \text{(iv)} \Rightarrow \text{(i)}.$$

By Theorem 1.2, statements (i) and (ii) are equivalent for any ring R.

(ii) $\Rightarrow$ (iii): According to Proposition 2.3, the module $R_R = \bigoplus_i P_i$ is a finite direct sum of indecomposable isotypic submodules P_i. Two such submodules are either isomorphic or have no nonzero homomorphisms between them. Thus, $R = \mathrm{End}(R_R)$ is a finite direct product of matrix rings over the local endomorphism rings of the P_i's.

(iii) $\Rightarrow$ (iv): Apply "(iv) $\Rightarrow$ (iii)" of [1, Theorem 1.2]. □

We now prove Proposition 1.5. Let A be an elementary algebra over a finite field k of p^n elements. Let K be an algebraically closed field of characteristic p and $\varphi\colon K \to K$ the Frobenius automorphism, given by $\alpha \mapsto \alpha^{p^n}$; we identify k with the fixed field of φ. Let $\Sigma = K[x, x^{-1}; \varphi]$ be the V-ring studied in [4]; we claim that the ring $R = \Sigma \otimes_k A$ has the required properties.

(i) We infer from [13, Theorem A] that the categories of $\mathcal{FL}_A$ and $\mathcal{FL}_R$ are equivalent.

(ii) Follows from (i) and (iii).

(iii) Applying Theorem 1.6 to the semiprimary ring A, we deduce that $\mathcal{Mod}_A$ has CSL. To see that R does not have CSL, note that since A is elementary, the ring homomorphism $\pi\colon A \to k$ gives rise to a surjective ring homomorphism

$$\pi \otimes 1\colon\ R = A \otimes_k \Sigma \longrightarrow k \otimes_k \Sigma.$$

Since $\mathcal{Mod}_\Sigma$ does not have CSL (as explained in Example 1.4), and Σ is isomorphic to a factor ring of R, $\mathcal{Mod}_R$ does not have CSL. □

4. Some questions

The rings in Examples 1.4 and 1.8 are both simple noetherian domains. In light of the diametrically different behavior in the two examples, we pose the following question.

Question 4.1. *For which simple noetherian domains R does $\mathcal{FL}_R$ have CSL?*

Question 4.2. *When do other subcategories of $\mathcal{Mod}_R$ have CSL? What are the conditions under which all artinian modules have CSL? All noetherian modules? Is there an example of a ring which has CSL for finite length modules, but not CSL for artinian modules?*

One may also consider categories with quasi-CSL in the following sense:

Definition 4.3. Let $\mathcal{C}$ be a category of modules, i.e., $\mathcal{C}$ is a full subcategory of $\mathcal{Mod}_R$ for some ring R. We say that an object M of $\mathcal{C}$ is *quasi-simple* if the only submodules $N \subseteq M$ such that N and M/N are objects of $\mathcal{C}$ are $N = 0$ and $N = M$. The category $\mathcal{C}$ is said to have *quasi-CSL* if the only modules with endomorphism ring a division ring are the quasi-simple ones.

Question 4.4. *Are there interesting categories with quasi-CSL?*

Example 4.5. For a given ring R, the category $\mathcal{FL}_R$ or the category $\mathcal{Mod}_R$ has quasi-CSL if and only if it has CSL.

Nevertheless, in general *quasi-CSL* and *CSL* are different conditions on a module category, as will be seen in Example 4.8 below. We preface this example with some motivating observations.

Example 4.6. Even if R has CSL, then the ring $U_2(R)$ of upper triangular 2 by 2 matrices with coefficients in R need not have CSL. Indeed, if S is a simple right R-module, then the row $\begin{pmatrix} S & S \end{pmatrix}$ is a right module over $U_2(R)$ of length 2 with endomorphism ring $\operatorname{End}(S_R)$.

In fact, the category $\mathcal{M}od_{U_2(R)}$ is just the category of all maps between right R-modules, a map $f\colon A \to B$ being a module over $U_2(R)$ via

$$(a,b)\cdot\begin{pmatrix} x & y \\ 0 & z\end{pmatrix} = (ax, f(a)y + bz).$$

Consider the full subcategory $\mathcal{S}(R)$ of $\mathcal{M}od_{U_2(R)}$ consisting of all maps which are monomorphisms ("$\mathcal{S}$=submodules").

Question 4.7. *If R has CSL, does $\mathcal{S}(R)$ have quasi-CSL?*

Categories of type $\mathcal{S}(R)$ play a role in applications of ring theory; for example, the embeddings of a subgroup in a finite abelian group, or the embeddings of a subspace in a vector spaces such that the subspace is invariant under the action of a linear operator, fall into this type of category.

Example 4.8. For Λ a commutative uniserial ring with radical generator p and radical factor field k, the category $\mathcal{S}(\Lambda)$ has quasi-CSL but not CSL, as follows. There are exactly two quasi-simple modules, $S_1 = \big(k \quad k\big)$ and $S_2 = \big(0 \quad k\big)$, up to isomorphy; both have endomorphism ring k. Then $\mathcal{S}(\Lambda)$ has quasi-CSL since any embedding $\big(A \quad B\big)$ with B a semisimple Λ-module is a direct sum of copies of S_1 and S_2. On the other hand, if B is not semisimple then multiplication by p is a nonzero nilpotent endomorphism.

References

[1] M. Alaoui, A. Haily, *Perfect rings for which the converse of Schur's lemma holds,* Publ. Mat. **45** (2001), no. 1, 219–222.

[2] M. Alaoui, A. Haily, *The converse of Schur's lemma in noetherian rings and group algebras,* Comm. Algebra **33** (2005), no. 7, 2109–2114.

[3] V.V. Bavula, *The extension group of the simple modules over the first Weyl algebra,* Bull. London Math. Soc. **32** (2000), no. 2, 182–190.

[4] J.H. Cozzens, *Homological properties of the ring of differential polynomials,* Bull. Amer. Math. Soc. **76** (1970), 75–79.

[5] M. Dombrovskaya, G. Marks, *Asymmetry in the converse of Schur's Lemma,* Comm. Algebra, to appear.

[6] D. Eisenbud, J.C. Robson, *Hereditary Noetherian prime rings,* J. Algebra **16** (1970), 86–104.

[7] C. Faith, *Indecomposable injective modules and a theorem of Kaplansky,* Comm. Algebra **30** (2002), no. 12, 5875–5889.

[8] K.R. Goodearl, R.B. Warfield Jr., *Simple modules over hereditary Noetherian prime rings,* J. Algebra **57** (1979), no. 1, 82–100.

[9] Y. Hirano, J.K. Park, *Rings for which the converse of Schur's Lemma holds,* Math. J. Okayama Univ. **33** (1991), 121–131.

[10] C. Huh, C.O. Kim, *π-regular rings satisfying the converse of Schur's lemma,* Math. J. Okayama Univ. **34** (1992), 153–156.

[11] J.C. McConnell, J.C. Robson, *Homomorphisms and extensions of modules over certain differential polynomial rings,* J. Algebra **26** (1973), 319–342.

[12] J.C. Robson, *Non-commutative Dedekind rings,* J. Algebra **9** (1968), 249–265.

[13] M. Schmidmeier, *A family of noetherian rings with their finite length modules under control,* Czechoslovak Math. J. **52(127)** (2002), no. 3, 545–552.

[14] R. Ware, J. Zelmanowitz, *Simple endomorphism rings,* Amer. Math. Monthly **77** (1970), no. 9, 987–989.

Greg Marks
Department of Mathematics and Computer Science
St. Louis University
St. Louis, MO 63103–2007, USA
e-mail: `marks@slu.edu`

Markus Schmidmeier
Mathematical Sciences
Florida Atlantic University
Boca Raton, FL 33431–0991, USA
e-mail: `markus@math.fau.edu`

Advances in Ring Theory
Trends in Mathematics, 239–255

Report on Exchange Rings

Saad H. Mohamed

Abstract. The question whether the finite (internal) exchange property implies the full (internal) exchange property, is still unsettled in general. A number of authors obtained some partial results in this direction. In this self contained expository article, we give an account of these results. Most of our representation will be concerned with the study of abstract exchange rings. The results obtained for these rings can be applied to modules.

Mathematics Subject Classification (2000). Primary 16D70, Secondary 16D50.

Keywords. Exchange property, abelian rings, square-free modules.

1. Introduction

Throughout, R is a ring with 1, and $J(R)$ its Jacobson radical. M will denote a unital right module over an unspecified ring, with endomorphisms acting on the left. S will always denote the ring of endomorphisms of M, and Δ is the ideal of S consisting of all endomorphisms with essential kernels. A *summand* of M will always mean a direct summand. If N is a summand of M, then $N = eM$ for some idempotent $e \in S$, and the endomorphism ring of N is isomorphic to the ring eSe. A ring R is *abelian* if all its idempotents are central, and a module M is an *abelian* module if S is abelian. A module M is called *square-free* if M does not contain any nonzero submodule of the form $X \oplus X$. It is known that if M is square free, then S/Δ is abelian. For definitions and properties of (quasi-) continuous modules, extending modules and the conditions (C_1), (C_2) and (C_3) we refer the reader to [9].

In their pioneering paper [3], Crawley and Jonsson defined and studied the exchange property: for a cardinal number $\aleph$, a module M is said to have the $\aleph$-*exchange property* if whenever $N \oplus M = \oplus_{i\in I} A_i$ for modules N and A_i, with $|I| \leq \aleph$, there exist submodules $A_i' \leq A_i$ such that $N \oplus M = \oplus_{i\in I} A_i' \oplus M$. The $\aleph$-exchange property is inherited by summands and finite direct sums. If M has the $\aleph$-exchange property for every finite (resp. countable) cardinal $\aleph$, then M is said to have the *finite* (*resp. countable*) *exchange property.* If M has the $\aleph$-exchange property for every cardinal $\aleph$, then M is said to have the *full exchange property.* Clearly the finite and full exchange properties coincide for finitely generated modules and indecomposable modules. It was also proved in [3] that the 2-exchange property implies the finite exchange property. The question whether the finite exchange property implies the full (or the countable) exchange property, is still open.

Modules with the finite exchange property were characterized via their endomorphism rings. Warfield [23] defined a ring R to be an exchange ring if R_R has the (finite) exchange property; (to avoid any confusion in the subsequent sections, we will call such rings *finite exchange rings*). He proved that the definition is right-left symmetric, and established that a module M has the finite exchange property if and only if its endomorphism ring S is a finite exchange ring. Nicholson [15] gave a very interesting characterization of the finite exchange property: M has the finite exchange property if and only if for each finite family $(x_i)_{i\in F}$ of elements of S with $\sum_{i\in F} x_i = 1$, there exist orthogonal idempotents $e_i \in Sx_i$ such that $\sum_{i\in F} e_i = 1$.

A major contribution to the study of the exchange property was given by Zimmerman-Huisgen and Zimmerman [25]. They proved that the $\aleph$-exchange property for a module M can be checked in a direct sum of copies of M, and then generalized Nicholson's characterization by considering summable families of endomorphisms of M. This result was utilized by a number of authors to get information regarding the exchange property of some special types of modules; we list a few:

(1) Mohamed and Müller [8] established that continuous modules have the full exchange property.
(2) Quasi-continuous modules and abelian modules do not necessarily have the finite exchange property (e.g., Z_Z). Oshiro and Rizvi [21] proved that quasi-continuous modules with the finite exchange property have the full exchange property (see also [10]).
(3) Yu [24] proved that abelian modules with the finite exchange property have the countable exchange property.
(4) Nielsen [17] strengthened the result of Yu, proving that abelian modules with the finite exchange property have the full exchange property.

The $\aleph$-exchange property, for an arbitrary cardinal $\aleph$, can be discussed for any ring equipped with an appropriate topology for which summability can be defined (cf. [11]). We consider a ring R with a left linear Hausdorff topology; this is a ring topology for which 0 has a neighborhood basis $\mathfrak{J}$ consisting of left ideals U with $\cap_{U\in\mathfrak{J}} U = 0$. With respect to this topology a family $(a_i)_{i\in I}$ is called *summable* to an element $a \in R$ if for each $U \in \mathfrak{J}$ there is a finite set $F \subseteq I$ such that $\sum_{i\in K} a_i - a \in U$ for every finite subset K with $F \subseteq K \subseteq I$. This implies that the sequence of partial sums is Cauchy in the usual sense. For simplicity, we call a family $(a_i)_{i\in I}$ *Cauchy* if the sequence of partial sums is Cauchy. If the topology is complete, then the "summable" and "Cauchy" concepts are equivalent. In this article, we call a ring R *topological* if R is equipped with a complete left linear Hausdorff topology. Such a ring is called $\aleph$-*exchange ring* if for each summable family $(a_i)_{i\in I}$ with $\sum_{i\in I} a_i = 1$ and $|I| \leq \aleph$, there exist orthogonal idempotents $e_i \in Ra_i$ such that $\sum_{i\in I} e_i = 1$. Strictly speaking, these rings should be called (right) $\aleph$-exchange rings. The right-left symmetry of such rings is not any more a valid question, as the topology is defined using left ideals only. So, there will be no confusion if we just denote these rings by $\aleph$-exchange rings. A topological ring R is called *finite (resp. countable) exchange ring* if R is an $\aleph$-exchange ring for every finite (resp. countable) cardinal $\aleph$. If R is an $\aleph$-exchange ring for every cardinal $\aleph$, then R is said to be a *full exchange ring*. If the topology is discrete, the only summable families are the finite ones, and therefore $\aleph$-exchange rings are the same as finite exchange rings. We also note that the endomorphism ring S of a module M is a topological ring with respect to the finite topology (for which the neighborhood basis of 0 consists of the annihilators of the finite subsets of M).

The motivation of studying $\aleph$-exchange rings is two fold. First it is purely ring theoretic, and can be applied to modules by considering their endomorphism rings. Second a quotient ring of an endomorphism ring S of a module M loses its connection with M, but still the $\aleph$-exchange property can be discussed for such quotient rings. This is very beneficial if the $\aleph$-exchange property can be lifted, which is indeed the case. Mohamed and Müller [11] proved that the $\aleph$-exchange ring property can be lifted modulo any closed ideal N contained in the Jacobson radical of a ring R, provided that idempotents lift modulo N. Then they proved that square-free modules with the finite exchange property have the countable exchange property. Recently Nielsen [19] proved that such modules have the full exchange property. This is a consequence of his more general result: If R is a topological finite exchange ring and $R/J(R)$ is abelian, then R is a full exchange ring. Most of the known results of the exchange property for continuous and quasi-continuous modules follow from Nielsen's theorem.

A module M is said to have the *(finite) internal exchange property*, if summands of M exchange in every (finite) direct sum decomposition of M. The 2-internal exchange property implies the finite internal exchange property [12]. The question whether the finite internal exchange property implies the (full) internal exchange property, is still open, in general. This question is answered in the affirmative for square-free modules [7]. The relation between the exchange property and the internal exchange property is discussed in [20]. Analogous to the Zimmermann's result, modules with the internal exchange property are characterized in terms of their endomorphism rings [6].

2. Finite exchange rings

Most of the material in this section can be found in Nicholson [15]. We list some results for the reader's convenience. We also give ring theoretic proofs for some basic properties of exchange rings.

The following fundamental theorem is an amalgamation of Propositions 1.1 and 1.11, and Theorem 2.1 of [15]. For a proof cf. Theorem 3.1 and Proposition 2.3 below.

2.1 Theorem. *The following are equivalent for a module M with endomorphism ring S:*

(1) *S_S has the finite exchange property;*
(2) *M has the finite exchange property;*
(3) *For each finite family $(x_i)_{i\in F}$ of elements of S with $\sum_{i\in F} x_i = 1$, there exist orthogonal idempotents $e_i \in Sx_i$ such that $\sum_{i\in F} e_i = 1$;*
(4) *${}_SS$ has the finite exchange property.*

As the 2-exchange property is equivalent to the finite exchange property, we have:

2.2 Corollary. *The following are equivalent for a ring R:*

(1) *R_R has the finite exchange property;*
(2) *For every $a \in R$, there exists $e^2 = e \in Ra$ such that $(1-e) \in R(1-a)$;*
(3) *${}_RR$ has the finite exchange property.*

Corner [2], Monk [14] and Nicholson [15] established first order ring-theoretic characterizations of finite exchange rings. The following proposition, which is a mixture of these characterizations, provides a ring-theoretic proof of the right-left symmetry of finite exchange rings (see also [16]).

2.3 Proposition. *The following are equivalent for an element a in a ring R:*

(1) *There exists $e^2 = e \in Ra$ such that $1-e \in R(1-a)$;*
(2) *There exist $r, s \in R$ such that $ras = 0$ and $ra + s(1-a) = 1$;*
(3) *There exists $\alpha \in R$ such that $(1+\alpha(1-a))(1-a\alpha) = 0$,*
(4) *There exist $r', s' \in R$ such that $r'(1-a)s' = 0$ and $ar' + (1-a)s' = 1$;*
(5) *There exists $e'^2 = e' \in aR$ such that $1-e' \in (1-a)R$.*

Proof. For any element $\alpha \in R$, define

$$r = 1+\alpha(1-a),\ s = 1-\alpha a,\ r' = 1+(1-a)\alpha,\ s' = 1-a\alpha.$$

Clearly $ra + s(1-a) = 1 = ar' + (1-a)s'$, $as = s'a$ and $(1-a)r = r'(1-a)$.

Now assume (3). Then $rs' = 0$, and so $ras = 0$ and $r'(1-a)s' = 0$. Hence (3) implies (2) and (4).

Assume (2). Then $s(1-a)s = s$, and so $s(1-a)$ is an idempotent. Then (1) follows with $e = ra$.

Assume (1). Let $e = ra$ and $1-e = s(1-a)$. We may assume that $er = r$ and $(1-e)s = s$. Hence $rar = r$ and $ras = 0$. Define $\alpha = r - s$. Then

$$1+\alpha(1-a) = 1 + r(1-a) - s(1-a) = 1 + r - e - (1-e) = r.$$

Consequently

$$(1+\alpha(1-a))(1-a\alpha) = r - ra\alpha = r - rar + ras = 0.$$

Hence (3) follows.

Similar arguments yield the equivalence of (3), (4) and (5). □

The following properties of finite exchange rings will be used frequently in this article without any further reference.

2.4 Proposition. *Let R be a finite exchange ring. Then:*

(1) *Any homomorphic image of R is a finite exchange ring;*
(2) *Idempotents lift modulo every right (left) ideal;*
(3) *$J(R)$ is the largest ideal that does not contain nonzero idempotents.*

Proof. We only give a proof for (2).Using the right-left symmetry of finit exchange rings, it is enough to consider left ideals. Let I be a left ideal in R and $a \in R$ such that $a - a^2 \in I$. By Proposition 2.3, there exists an idempotent e such that $e = ra$ and $1 - e = s(1 - a)$ for some $r, s \in R$. Then

$$e - a = e(1 - a) - (1 - e)a = ra(1 - a) - s(1 - a)a = (r - s)(a - a^2) \in I. \quad \square$$

The following basic result will be generalized in Section 3, to allow lifting of the $\aleph$-exchange property.

2.5 Theorem. ([15], Proposition 1.5) *A ring R is a finite exchange ring if and only if $R/J(R)$ is a finite exchange ring and idempotents lift modulo $J(R)$.*

We end this section by a ring theoretic proof of the known fact that the finite exchange property is inherited by summands and finite direct sums (cf. [15, Corollary 2.6]).

2.6 Corollary. *Let e be an idempotent in a ring R. Then R is a finite exchange ring if and only if eRe and $(1 - e)R(1 - e)$ are finite exchange rings.*

Proof. Throughout the proof, we will use (2) of Proposition 2.3.

"Only if": For $a \in eRe$, there exist $r, s \in R$ such that $ras = 0$ and $ra + s(1-a) = 1$. Then $(ere)a(ese) = 0$ and $(ere)a + (ese)(e - a) = e$.

"if": Let $x \in R$, and consider the element exe in the finite exchange ring eRe. There exist $\alpha, \beta \in eRe$ such that $\alpha x \beta = 0$ and $\alpha xe + \beta(1 - x)e = e$. Then $\beta(1 - x)$ and αxe are idempotents. We may assume that $\alpha = \alpha x \alpha$. Then clearly αx is an idempotent orthogonal to $\beta(1-x)$. Let $f = \alpha x + \beta(1-x)$. Then f is an idempotent with $fe = e$ and $ef = f$. Consequently $R(1 - e) = R(1 - f)$, and hence the rings $(1-e)R(1-e)$ and $(1-f)R(1-f)$ are isomorphic. Therefore the ring $(1-f)R(1-f)$ is a finite exchange ring.

Repeating the same argument with the element $(1 - f)x(1 - f)$ in the finite exchange ring $(1 - f)R(1 - f)$, we get $\gamma, \delta \in (1 - f)R(1 - f)$ such that $\gamma x \delta = 0$ and γx and $\delta(1 - x)$ are orthogonal idempotents. Write $g = \gamma x + \delta(1 - x)$. Then g is an idempotent and $(1 - g)f = 1 - g$ and $f(1 - g) = f$. Define $r = (1 - g)\alpha + \gamma$ and $s = (1 - g)\beta + \delta$. Then

$$rx + s(1 - x) = (1 - g)f + g = 1.$$

It remains to show that $rxs = 0$. First we note that $\alpha x(1 - g) = \alpha xf(1 - g) = \alpha xf = \alpha x$. Also $\alpha x \delta = \alpha xf\delta = 0$ and $\gamma x(1 - g) = \gamma xg(1 - g) = 0$. Hence

$$rxs = (1 - g)\alpha x \beta + (1 - g)\alpha x \delta + \gamma x(1 - g)\beta + \gamma x \delta = 0. \quad \square$$

3. Topological rings

Given two modules U and V, a family $(f_i)_{i\in I}$ of homomorphisms $U \to V$ is called *summable* if for each $u \in U$, $f_i(u) = 0$ for almost all $i \in I$. In that case the map $f : U \to V$ defined by $f(u) = \sum_{i\in I} f_i(u)$ is a well-defined homomorphism; we write $\sum_{i\in I} f_i$ to denote such a homomorphism. The summability just defined amounts to the convergence of the series $\sum_{i\in I} f_i$ in the finite topology of $\mathrm{Hom}(U, V)$.

Zimmerman-Huisgen and Zimmerman generalized Nicholson's result (Theorem 2.1). For the reader's convenience we include the main ideas of the proof.

3.1 Theorem. ([25, Proposition 3]). *The following are equivalent for a module M with endomorphism ring S and any cardinal $\aleph$:*

(1) *M has the $\aleph$-exchange property;*
(2) *Whenever $N \oplus M = \oplus_{i\in I} A_i$ with $A_i \cong M$ and $|I| \leq \aleph$, there exist submodules $A_i{}' \leq A_i$ such that $N \oplus M = \oplus_{i\in I} A_i{}' \oplus M$;*
(3) *For each summable family $(x_i)_{i\in I}$ in S with $\sum_{i\in I} x_i = 1$, there exist orthogonal idempotents $e_i \in Sx_i$ such that $\sum_{i\in I} e_i = 1$.*

Proof. (1) $\Rightarrow$ (2) is trivial.

(2) $\Rightarrow$ (3): Let $A = \oplus_{i\in I} A_i$ with $A_i = M$. Define $f : M \to A$ by $f(m) = (x_i(m))_{i\in I}$ and $g : A \to M$ by $g((m_i)_{i\in I}) = \sum_{i\in I} m_i$. It is clear that $gf = 1_M$, hence $A = \mathrm{Ker}\, g \oplus fM$, with $fM \cong M$. By hypothesis, $A = \oplus_{i\in I} A'_i \oplus fM$, with $A'_i \leq A_i$. Write $A_i = A'_i \oplus A''_i$. Then $fM \cong \oplus_{i\in I} A''_i$. Let $\eta : fM \to \oplus_{i\in I} A''_i$ be an isomorphism and define $e_i = g\eta^{-1}\pi_i\eta f$, where $\pi_j : \oplus_{i\in I} A''_i \to A''_j$ is the natural projection. It is easy to check that $(e_i)_{i\in I}$ is a family of orthogonal idempotents. Clearly the family $(e_i)_{i\in I}$ is summable and $\sum_{i\in I} e_i = 1$. Now let τ_i denote the projection $A_i \to A''_i$ along A'_i. Then $\pi_i\eta f = \tau_i x_i$ and consequently $e_i \in Sx_i$.

(3) $\Rightarrow$ (1): Let $A = \oplus_{i\in I} A_i = N \oplus M$. Let $p : N \oplus M \to M$ and $\nu_j : \oplus_{i\in I} A_i \to A_j$ denote the natural projections, and let $\mu_j = \nu_i|_M$. Define $x_i = p\mu_i$. Then $x_i \in S$ and the family $(x_i)^{i\in I}$ is summable to 1. By hypothesis, we can find orthogonal idempotents $e_i = s_i x_i \in Sx_i$ with $\sum_{i\in I} e_i = 1$. Define $\varphi_i : A_i \to M$ by $\varphi_i = e_i s_i p \mu_j$. Clearly $(\varphi_i)_{i\in I}$ is summable, let $\varphi = \sum_{i\in I} \varphi_i$. Noting that $\varphi_i\varphi_j = 0$ for $i \neq j$, one can check that $\ker\varphi = \oplus_{i\in I}(A_i \cap \ker\varphi_i)$. Also $\varphi_i|_M = e_i$, hence $\varphi|_M = 1|_M$. Therefore $A = \ker\varphi \oplus M$. □

Now we give some applications of this core theorem. We start by the following result which was proved by Yu [24] for endomorphism rings; the arguments are identical.

3.2 Theorem. *Let R be a topological ring. If R is an abelian finite exchange ring, then R is a countable exchange ring.*

Proof. Consider a summable family $(a_i)_{i\in I}$ with $|I| \leq \aleph_o$. For each integer n define $b_n = \sum_{i>n} a_i$. Inductively, we construct orthogonal idempotents $e_1 \in Ra_1, \ldots, e_n \in Ra_n$ and $f_n \in Rb_n$ such that $e_1 + \cdots + e_n + f_n = 1$. Write $f_n = rb_n$. Then $f_n = f_n r a_{n+1} + f_n r b_{n+1}$. As the ring $f_n R f_n$ is again a finite exchange ring by Corollary 2.6, we get $f_n = e_{n+1} + f_{n+1}$ for orthogonal idempotents $e_{n+1} \in f_nRf_nra_{n+1} \leq Ra_{n+1}$ and $f_{n+1} \in f_nRf_nrb_{n+1} \leq Rb_{n+1}$. Clearly e_{n+1} is orthogonal to $e_i(1 \leq i \leq n)$, and $e_1 + \cdots + e_n + e_{n+1} + f_{n+1} = 1$. This completes our induction process. Since $(a_i)_{i\in I}$ is summable, $b_n \to 0$ hence $f_n \to 0$, and consequently $\sum_{i\in I} e_i = 1$. □

3.3 Corollary. *An abelian module with the finite exchange property has the countable exchange property.*

Nielsen [17] extended Yu's result, using clever inductive ideas to go over limit ordinals. We include the proof for comparison with that given in Theorem 3.2.

3.4 Theorem. *An abelian module M with the finite exchange property has the full exchange property.*

Proof. Consider a summable family $(x_i)_{i\in I}$ of elements of S with $\sum_{i\in I} x_i = 1$. We may assume that I is well ordered, with first element 1 and last element τ. For any $\eta \in I$, define $y_\eta = \sum_{i>\eta} x_i$ and $z_\eta = \sum_{i\geq\eta} x_i$. Fix an element $\alpha \in I$. Suppose, by induction, that we have constructed orthogonal idempotents $e_i \in Sx_i$ and $f_i \in Sy_i(1 \leq i < \alpha)$ such that for all $\beta < \alpha$,

$$1 = \sum_{i\leq\beta} e_i + f_\beta.$$

Clearly, the family $(e_i)_{i<\alpha}$ is summable and orthogonal. Write $\varepsilon = \sum_{i<\alpha} e_i$. Then ε is an idempotent, and $(1-\varepsilon)e_i = 0$ for $i < \alpha$. It follows that $(1-\varepsilon) = f_\beta(1-\varepsilon)$ for $\beta < \alpha$. Now

$$f_\beta = r_\beta y_\beta = r_\beta \left(\sum_{\beta<i<\alpha} x_i + z_\alpha \right) = r_\beta \left(\sum_{\beta<i<\alpha} x_i \right) + r_\beta z_\alpha$$

for some element $r_\beta \in S$. As $f_\beta S f_\beta$ is an exchange ring, there exist orthogonal idempotents $h_\beta \in f_\beta S r_\beta \left(\sum_{\beta<i<\alpha} x_i\right)$ and $g_\beta \in f_\beta S r_\beta z_\alpha$ such that $f_\beta = h_\beta + g_\beta$.

Now consider an arbitrary element $m \in M$. Since $(e_i)_{i<\alpha}$ is summable, there exists $\gamma < \alpha$ such that $x_i(1-\varepsilon)(m) = 0$ for $\gamma < i < \alpha$. Then

$$\begin{aligned}(1-\varepsilon)(m) &= f_\gamma(1-\varepsilon)(m) = (h_\gamma + g_\gamma)(1-\varepsilon)(m)\\ &= g_\gamma(1-\varepsilon)(m) = r_\gamma z_\alpha(1-\varepsilon)(m).\end{aligned} \tag{1}$$

Since $g_\gamma = r_\gamma z_\alpha$ is an idempotent and S is abelian, $g_\gamma = z_\alpha r_\gamma$ (cf. [15, Proposition 1.8]), and consequently

$$(1-\varepsilon)(m) = z_\alpha r_\gamma(1-\varepsilon)(m) = z_\alpha(1-\varepsilon)r_\gamma(m) \in z_\alpha(1-\varepsilon)M. \tag{2}$$

From (1) and (2), it is clear that z_α is an automorphism of $(1-\varepsilon)M$. Hence there exists $z'_\alpha \in (1-\varepsilon)S(1-\varepsilon)$ such that $(1-\varepsilon) = z'_\alpha z_\alpha$, and so

$$(1-\varepsilon) = z'_\alpha(x_\alpha + y_\alpha) = z'_\alpha x_\alpha + z'_\alpha y_\alpha.$$

As $(1-\varepsilon)S(1-\varepsilon)$ is an exchange ring, we get orthogonal idempotents $e_\alpha \in (1-\varepsilon)Sx_\alpha$ and $f_\alpha \in (1-\varepsilon)Sy_\alpha$ such that $1-\varepsilon = e_\alpha + f_\alpha$. Clearly e_α and f_α are orthogonal to $(e_i)_{i<\alpha}$ and

$$1 = \varepsilon + e_\alpha + f_\alpha = \sum_{i<\alpha} e_i + e_\alpha + f_\alpha = \sum_{i\leq\alpha} e_i + f_\alpha.$$

This completes the induction process.

Since $(x_i)_{i\in I}$ is summable, $y_n \to 0$ hence $f_n \to 0$, and consequently

$$\sum_{i\in I} e_i = 1.$$

□

Another remarkable contribution of Nielsen [19] is the following result which generalizes Theorem 3.2. This was recently communicated to the author by Nielsen and is not published yet. So we are not including its proof, which is very technical and uses highly nontrivial arguments. We note that Theorem 3.4 follows then as a corollary; however it was a step in that direction.

3.5 Theorem. *Let R be a topological ring. If R is an abelian finite exchange ring, then R is a full exchange ring.*

Now we introduce the main theorem of [11]. It generalizes, and was inspired by, the analogous result of Nicholson (cf. Theorem 2.5). First we need the following two lemmas concerning lifting of idempotents.

3.6 Lemma. *Assume that idempotents lift modulo an ideal N contained in the Jacobson radical of a ring R. If $\bar{a}$ is an idempotent in the ring $\bar{R} = R/N$, then $\bar{a}$ lifts to an idempotent $e \in Ra$.*

Proof. There exists an idempotent $f \in R$ such that $\bar{a} = \bar{f}$. Now $a^2 - a \in N$ implies $a^p - a \in N$, and hence $a^p - f \in N$ for every positive integer p. Then $a^2 = f + n$, with $n \in N$. Write $u = 1 + fnf$. Then u is a unit in R with $u^{-1} = 1 + m$, for some $m \in N$. As $uf = fu$, we get $u^{-1}f = fu^{-1}$. Define $e = au^{-1}fa = afu^{-1}a$. Then $e \in Ra$, and

$$e^2 = au^{-1}fa^2fu^{-1}a = au^{-1}f(f+n)fu^{-1}a = au^{-1}fuu^{-1}a = e.$$

It remains to show that $e - f \in N$. We have

$$e = au^{-1}fa = a(1+m)(a^2-n)a = a^4 + k, \quad \text{with } k \in N.$$

As $a^4 - f \in N$, we get $e - f \in N$. □

3.7 Lemma. *Let g and h be idempotents in a ring R. If $gh \in J(R)$, then there exist orthogonal idempotents γ and δ such that $gR = \gamma R$ and $hR = \delta R$.*

Proof. We have $ghg \in J(R)$, and so $ghg \in J(gRg)$. Hence $g - ghg$ is a unit in the ring gRg. Let $a \in gRg$ be the inverse of $g - ghg$. Then $g = a(1-h)g$, and so $a(1-h)a = a$. Define $\gamma = a(1-h)$. Then $\gamma \in gR(1-h)$ and $\gamma^2 = \gamma$. Clearly $\gamma g = g$ and $g\gamma = \gamma$, and so $gR = \gamma R$.

As $hgh \in J(R)$, we can similarly find an idempotent $\delta \in hR(1-g)$ such that $hR = \delta R$. Clearly γ and δ are orthogonal. □

3.8 Theorem. *Assume that idempotents lift modulo a closed ideal N contained in the Jacobson radical of a topological ring R. For any cardinal $\aleph$, if R/N is an $\aleph$-exchange ring, then so is R.*

Proof. Consider a summable family $(a_i)_{i\in I}$ of elements of R with $\sum_{i\in I} a_i = 1$ and $|I| \leq \aleph$. In the quotient topology on $\bar{R} = R/N$, $(\bar{a}_i)_{i\in I}$ sums to $\bar{1}$. The hypothesis and Lemma 3.6 imply the existence of idempotents $e_i \in Ra_i$ such that the family $(\bar{e}_i)_{i\in I}$ is orthogonal in $\bar{R}$ and $\sum_{i\in I} \bar{e}_i = \bar{1}$. As the $(a_i)_{i\in I}$ are summable and the topology is linear, the family $(e_i)_{i\in I}$ is Cauchy, hence summable by completeness. Define $u = \sum_{i\in I} e_i$. Then clearly u is a unit in R, and $\sum_{i\in I} u^{-1}e_i = 1 = \sum_{i\in I} e_i u^{-1}$. It remains to show that the family $(u^{-1}e_i)_{i\in I}$ consists of orthogonal idempotents.

First we claim that $\sum_{i\in I} e_i R$ is direct. To verify our claim, it is enough to consider a finite sub-sum $e_1R + \cdots + e_nR$. Assume, by induction, that we have constructed orthogonal idempotents $f_1, \ldots, f_{n-1}$ such that $e_iR = f_iR$, $1 \leq i \leq n-1$. Write $f = f_1 + \cdots + f_{n-1}$. Hence f is an idempotent and $e_n f \in J(R)$. Then, by Lemma 3.7, we get orthogonal idempotents λ and f_n such that $fR = \lambda R$ and $e_nR = f_nR$. Clearly f_n is orthogonal to f_i, $1 \leq i \leq n-1$, and our claim is established. Now

$$e_k = \left(\sum_{i\in I} e_i u^{-1}\right) e_k = \sum_{i\in I} e_i u^{-1} e_k = e_k u^{-1} e_k + \sum_{i\neq k} e_i u^{-1} e_k.$$

Using that $\sum_{i\in I} e_iR$ is direct, we get $e_k u^{-1} e_k = e_k$ and $e_i u^{-1} e_k = 0$ for $i \neq k$. Hence $(u^{-1}e_i)_{i\in I}$ is a family of orthogonal idempotents, as desired. □

3.9 Corollary. *Let R be a topological ring, and let N be a closed ideal contained in $J(R)$. If R is a finite exchange ring and R/N is abelian, then R is a full exchange ring.*

Proof. Idempotents lift modulo N by Proposition 2.4. The result then follows by Theorems 3.5 and 3.8. □

Remark. Nielsen ([19, Lemma 3]) pointed out that if R is a finite exchange ring and R/N is abelian for an ideal $N \leq J(R)$, then $R/J(R)$ is abelian. Hence N can be replaced by $J(R)$ in the above theorem.

As noted in the introduction, the ring S of endomorphisms of a module M is a topological ring with respect to the finite topology. And if M has the finite exchange property, then the ideals Δ and $J(S)$ are closed and $\Delta \leq J(S)$ ([11, Lemma 11]). Also if M is any square free module, then S/Δ has no nonzero nilpotent elements, hence abelian ([9, Lemma 3.4]).

Then Corollary 3.9 gives rise to:

3.10 Theorem. ([19, Theorem 9]). *Square free modules with the finite exchange property have the full exchange property.*

The exchange property was established for injective modules by Warfield [22], for quasi-injective modules by Fuchs [4], and for continuous modules by Mohamed and Müller [8]. Also it is proved in [9, Theorem 2.37] that a quasi-continuous module is a direct sum of a quasi-injective module and a square-free module. Hence, to investigate the exchange property for quasi-continuous modules, it is enough to consider quasi-continuous square-free modules. Based on this, Oshiro and Rizvi [21] (see also [10]), proved that quasi-continuous modules with the finite exchange property have the full exchange property. The proofs, though ingenious, are quite lengthy and involved, and depend on other properties of quasi-continuous modules ((C_1) and (C_3)). Now all the relevant results concerning continuous and quasi-continuous modules follow from Theorem 3.10.

4. Internal exchange property

A summand X of a module Y is said to *exchange* in a decomposition $Y = \oplus_{i\in I} Y_i$, if $Y = \oplus_{i\in I} Y_i' \oplus X$, with $Y_i' \leq Y_i$. If this is true for every summand of Y, then the decomposition $Y = \oplus_{i\in I} Y_i$ is said to be *exchangeable*. For a cardinal number $\aleph$, a module M is said to have the $\aleph$-*internal exchange property*, if any decomposition $M = \oplus_{i\in I} M_i$ with $|I| \leq \aleph$ is exchangeable. If M has the $\aleph$-internal exchange property for every finite (resp. countable) cardinal $\aleph$, them M is said to have the *finite* (*resp. countable*) *internal exchange property*. If M has the $\aleph$-internal exchange property for every cardinal $\aleph$, then M is said to have the *full internal exchange property*. The finite internal exchange property follows from the 2-internal exchange property and is inherited by summands [12].

Clearly the $\aleph$-exchange property implies the $\aleph$-internal exchange property. The following two results discuss the relation between the exchange property and internal exchange property.

4.1 Lemma. *Consider the following conditions on a module M and a cardinal number $\aleph$:*

(1) *$M^{(\aleph)}$ has the internal exchange property;*
(2) *$M^{(\aleph)}$ is exchangeable;*
(3) *Any summand isomorphic to M in $M^{(\aleph)}$ exchanges in $M^{(\aleph)}$;*
(4) *M has the $\aleph$-exchange property.*

Then (1) $\Rightarrow$ (2) $\Rightarrow$ (3) $\Leftrightarrow$ (4).

Proof. That (1) $\Rightarrow$ (2) $\Rightarrow$ (3) is obvious, and (3) $\Leftrightarrow$ (4) by Theorem 3.1. □

This leads to the following corollary, which was observed by Nielsen [20].

4.2 Corollary. *A module M has the finite exchange property if and only if $M \oplus M$ has the finite internal exchange property.*

Proof. The "if" part follows by the above lemma with $\aleph = 2$. Conversely, if M has the finite exchange property, then $M \oplus M$ has the finite exchange property, and hence $M \oplus M$ has the finite internal exchange property. $\square$

Unlike the exchange property, the finite internal exchange property may not pass to finite direct sums. As an example, Z_Z has the finite internal exchange property, but not the finite exchange property, and hence $(Z \oplus Z)_Z$ does not have the internal exchange property, by Corollary 4.2.

In dealing with exchangeable decompositions, the following lemma shows that complement summands can be left intact.

4.3 Lemma. ([12, Lemma 5]) *Let $M = N \oplus K'$, with $K' \leq K \leq M$. If K has an exchangeable decomposition $K = \oplus_{i \in I} K_i$, then $M = N \oplus (\oplus_{i \in I} K_i')$ with $K_i' \leq K_i$.*

Proof. By the modular law, $K = (N \cap K) \oplus K'$. Exchanging $N \cap K$ in the decomposition $K = \oplus_{i \in I} K_i$, we get $K = (N \cap K) \oplus (\oplus_{i \in I} K_i')$, with $K_i' \leq K_i$. Clearly $M = N + (\oplus_{i \in I} K_i')$, and $N \cap (\oplus_{i \in I} K_i') = N \cap K \cap (\oplus_{i \in I} K_i') = 0$. $\square$

4.4 Proposition. ([13]) *Let $M = A \oplus B$. Then M has the finite internal exchange property if and only if the decomposition is exchangeable, and A and B have the finite internal exchange property.*

Proof. For the nontrivial direction, consider a decomposition $M = U \oplus V$ and a summand N of M. Exchanging V in the decomposition $M = A \oplus B$, we get $M = V \oplus A' \oplus B'$, with $A = A' \oplus A''$ and $B = B' \oplus B''$. Then $U \cong A' \oplus B'$ and $V \cong A'' \oplus B''$. Hence $U = U_1 \oplus U_2$ and $V = V_1 \oplus V_2$ where $U_1 \cong A'$, $U_2 \cong B'$, $V_1 \cong A''$ and $V_2 \cong B''$. Write $X = U_1 \oplus V_1$ and $Y = U_2 \oplus V_2$. Then $X \cong A$ and $Y \cong B$. Hence X and Y have the internal exchange property and $M = X \oplus Y$ is an exchangeable decomposition. Exchanging N in this decomposition, we get $M = N \oplus X' \oplus Y'$ with $X' \leq X$ and $Y' \leq Y$. Now applying Lemma 4.3, we get

$$M = N \oplus (U_1' \oplus V_1') \oplus (U_2' \oplus V_2') = N \oplus (U_1' \oplus U_2') \oplus (V_1' \oplus V_2'),$$

with $U_1' \leq U_1$, $V_1' \leq V_1$, $U_2' \leq U_2$ and $V_2' \leq V_2$. $\square$

Analogous to Theorem 3.10, square-free modules with the finite internal exchange property have the full internal exchange property. This result was obtained independently by Mohamed [7] and Nielsen [20]. The proof of this result is purely module-theoretic and needs the following lemma.

4.5 Lemma. ([6]) *A square-free module M with the finite internal exchange property has* (C_3).

Proof. Given summands A and B of M with $A \cap B = 0$. Write $M = B \oplus C$, and exchange A in this decomposition. We get $M = A \oplus B' \oplus C'$ with $B = B' \oplus B''$ and $C = C' \oplus C''$. Then $A \cong B'' \oplus C''$. As M is square-free, $B'' = 0$ and hence $M = A \oplus B \oplus C'$. $\square$

4.6 Theorem. *A square-free module M with the finite internal exchange property has the full internal exchange property.*

Proof. Let $M = \oplus_{i\in I} M_i$ and let X be a summand of M. We may assume that I is well ordered, with a last element τ. For any $\eta \in I$, define

$$A_\eta = \oplus_{i\leq\eta} M_i,\ B_\eta = \oplus_{i>\eta} M_i,\ C_\eta = \oplus_{i\geq\eta} M_i.$$

Then $M = A_\eta \oplus B_\eta = \oplus_{i<\eta} M_i \oplus C_\eta$ and $C_\eta = M_\eta \oplus B_\eta$.

Fix an element $\alpha \in I$. Suppose, by transfinite induction, that we have constructed submodules $M_i' \leq M_i$ and $B_i' \leq B_i$ $(i < \alpha)$ such that for all $\beta < \alpha$,

$$M = X \oplus (\oplus_{i\leq\beta} M_i') \oplus B_\beta' \tag{3}$$

It follows that $X \cap (\oplus_{i<\alpha} M_i') = 0$. As M has (C_3) by Lemma 4.5, $X \oplus (\oplus_{i<\alpha} M_i')$ is a summand of M. Exchanging this summand in the decomposition $M = \oplus_{i<\alpha} M_i \oplus C_\alpha$, we get

$$M = X \oplus (\oplus_{i<\alpha} M_i') \oplus (\oplus_{i<\alpha} M_i)' \oplus C_\alpha', \tag{4}$$

with $(\oplus_{i<\alpha} M_i)' \leq \oplus_{i<\alpha} M_i$ and $C_\alpha' \leq C_\alpha$. Comparing (3) and (4), then it is obvious that $(\oplus_{i<\alpha} M_i)'$ embeds in $B_\beta' \leq B_\beta$ for every $\beta < \alpha$.

Consider an element $a \in (\oplus_{i<\alpha} M_i)'$. There exists $\gamma < \alpha$ such that $aR \in \oplus_{i\gamma} M_i \leq A_\gamma$. However aR embeds in B_γ and $A_\gamma \cap B_\gamma = 0$. As M is square-free, $aR = 0$ and consequently $(\oplus_{i<\alpha} M_i)' = 0$. Therefore

$$M = X \oplus (\oplus_{i<\alpha} M_i') \oplus C_\alpha'.$$

As C_α has the finite internal exchange property, the decomposition $C_\alpha = M_\alpha \oplus B_\alpha$ is exchangeable. Thus we get by Lemma 4.3, $M = X \oplus (\oplus_{i<\alpha} M_i') \oplus M_\alpha' \oplus B_\alpha'$, with $M_\alpha' \leq M_\alpha$ and $B_\alpha' \leq B_\alpha$. Hence

$$M = X \oplus (\oplus_{i\leq\alpha} M_i') \oplus B_\alpha'.$$

This completes our induction process. Therefore

$$M = X \oplus (\oplus_{i\leq\tau} M_i') \oplus B_\tau'.$$

Since $B_\tau = 0$, $M = X \oplus (\oplus_{i\in I} M_i')$. □

Analogous to Theorem 3.1, we have the following characterizations for modules with the internal exchange property in terms of their endomorphism rings.

4.7 Theorem. ([6, Theorem 2.1]). *The following are equivalent for a module M and a cardinal number $\aleph$:*

1. *M has the $\aleph$-internal exchange property;*
2. *For every idempotent e and every summable family of orthogonal idempotents $(f_i)_{i\in I}$ in S with $\sum_{i\in I} f_i = 1$ and $|I| \leq \aleph$, there exists an idempotent $g \in S$ such that $eS = gS$ and $gf_i(1-g) = 0$ for all $i \in I$;*
3. *For every idempotent e and every summable family of orthogonal idempotents $(f_i)_{i\in I}$ in S with $\sum_{i\in I} f_i = 1$ and $|I| \leq \aleph$, there exist orthogonal idempotents $e_i \in Sf_ie$ such that $\sum_{i\in I} e_i = e$.*

Proof. (1) ⇒ (2): We have $M = \oplus_{i\in I} f_i M$ and eM is a summand of M. Then

$$M = \oplus_{i\in I} N_i \oplus eM \quad \text{with } N_i \leq f_i M.$$

Let $g : M \to eM$ be the projection along $\oplus_{i\in I} N_i$. Then $eM = gM$, hence $eS = gS$, and $\oplus_{i\in I} N_i = (1-g)M$. One can check that $gf_i(1-g) = 0$.
(2) ⇒ (3): Define $e_i = gf_ie$. Then

$$e_ie_j = gf_i(eg)f_je = (gf_ig)f_je = gf_if_je.$$

Hence $e_i^2 = e_i$ and $e_ie_j = 0$ for $i \neq j$. Clearly $(e_i)_{i\in I}$ is summable and

$$\sum_{i\in I} e_i = \sum_{i\in I} gf_ie = g\left(\sum_{i\in I} f_i\right)e = ge = e.$$

(3) ⇒ (1): Let $M = \oplus_{i\in I} M_i$ with $|I| \leq \aleph$, and let $(f_i)_{i\in I}$ be the natural projections with respect to this decomposition. Then $(f_i)_{i\in I}$ is a summable family of orthogonal idempotents in S with $\sum_{i\in I} f_i = 1$. Consider a summand N of M. Then $N = eM$ for some idempotent $e \in S$. The hypothesis implies the existence of orthogonal idempotents $e_i = s_if_ie$ with $s_i \in S$ and $\sum_{i\in I} e_i = e$. Define $g_i = e_is_if_i$. Clearly $(g_i)_{i\in I}$ is a summable, write $g = \sum_{i\in I} g_i$. Now

$$g_ig_j = g_ie_jg_j = g_iee_jg_j = e_ie_jg_j.$$

Hence $g_i^2 = g_i$ and $g_ig_j = 0$ for $i \neq j$, and consequently g is an idempotent. Now

$$eg = e\left(\sum_{i\in I} g_i\right) = \sum_{i\in I} eg_i = \sum_{i\in I} e_ig_i = \sum_{i\in I} g_i = g;$$

$$ge = \left(\sum_{i\in I} g_i\right)e = \sum_{i\in I} g_ie = \sum_{i\in I} e_i = e.$$

It follows that $N = gM$. Also $gf_i(1-g) = g_i(1-g) = 0$, and so $(1-g)f_i(1-g) = f_i(1-g)$. Hence

$$(1-g)M = \left(\sum_{i\in I} f_i\right)(1-g)M \leq \sum_{i\in I} f_i(1-g)M = \sum_{i\in I}(1-g)f_i(1-g)M \leq (1-g)M.$$

Therefore $(1-g)M = \sum_{i\in I} f_i(1-g)M = \oplus_{i\in I} f_i(1-g)M$. Write $M_i' = f_i(1-g)M$. Then $M_i' \leq M_i$, and $M = \oplus_{i\in I} M_i' \oplus N$. □

4.8 Corollary. *The following are equivalent for a module M:*

(1) *M has the finite internal exchange property;*
(2) *For any idempotents e and f of S, there exists an idempotent $g \in S$ such that $eS = gS$ and $gf(1-g) = 0$;*
(3) *For any idempotents e and f of S, there exists an idempotent $\gamma \in Sfe$ such that $e - \gamma \in S(1-f)e$;*
(4) *S_S has the finite internal exchange property.*

Proof. Obvious. □

4.9 Proposition. *The finite internal exchange property for a ring R is right-left symmetric.*

Proof. Assume that R_R has the finite exchange property. Let e and f be idempotents in R. Applying (2) of Corollary 4.8 to the idempotents $1-e$ and f, we get an idempotent g' in R such that $(1-e)R = g'R$ and $g'f(1-g') = 0$. Define $g = 1-g'$. Then $Re = Rg$ and $(1-g)fg = 0$. Hence ${}_RR$ has the finite internal exchange property. □

5. Questions

In Theorems 3.10 and 4.6, it is proved that square-free modules with the finite (internal) exchange property have the full (internal) exchange property. As a square-free module with the finite internal exchange property has (C_3), by Lemma 4.5, and also (C_2) implies (C_3), it is interesting to investigate the following questions:

5.1 Question. *Does the finite (internal) exchange property imply the full, or countable, (internal) exchange property for a module with* (C_2) *or* (C_3)*?*

Quasi continuous modules have the internal exchange property [21]. But extending modules (that is modules with (C_1)) do not, in general, enjoy even the finite internal exchange property (e.g., the abelian group $Z \oplus Z$). So it is interesting to investigate whether the finite (internal) exchange property implies the full (internal) exchange property for extending modules?

5.2 Lemma. *An extending module M has a decomposition $M = F \oplus Q$ where F is square free and Q is essential over a square.*

Proof. Same argument as in [9, Proposition 2.35].

By Zorn's Lemma, M contains a direct sum $K = \oplus_{i\in I} S_i$ maximal such that S_i is a square. Let $S_i = X_i^2$. Then $K \cong \oplus_{i\in I} X_i^2 \cong (\oplus_{i\in I} X_i)^2$. Hence

$$K = K_1 \oplus K_2, \quad \text{with } K_1 \cong \oplus_{i\in I} X_i \cong K_2.$$

Then K is a square. Let Q be a closure of K in M. Then Q is a summand of M. Write $M = F \oplus Q$. Then maximality of K implies F is square free. □

The above lemma, along with Proposition 4.4, suggests that it is enough to consider the questions:

5.3 Question. *Let M be an extending module which is essential over a square. Does the finite (internal) exchange property for M imply the full, or countable, (internal) exchange property?*

As for exchange rings, the $\aleph$-internal exchange property can be discussed for topological rings, which may not be endomorphism rings. A topological ring R is called *$\aleph$-internal exchange ring* if R_R has the $\aleph$-internal exchange property. Such a ring R is called *finite (resp. countable) internal exchange ring* if R is an $\aleph$-internal exchange ring for every finite (resp. countable) cardinal $\aleph$. If R is an $\aleph$-internal exchange ring for every cardinal $\aleph$, then R is said to be a full internal exchange ring.

The class of internal exchange rings is very large; it contains all abelian rings. An example of a finite internal exchange ring which is not abelian is the ring of all 2×2 lower triangular matrices over Z (for a proof use (2) of Corollary 4.8; cf. [6, Example 2.5]).

A slight modification of the proof given in Theorem 3.8 yields the following:

5.4 Proposition. *Assume that idempotents lift modulo a closed ideal N contained in the Jacobson radical of a topological ring R. For any cardinal number $\aleph$, if R/N is an $\aleph$-internal exchange ring, then so is R.*

Proof. Consider an idempotent $e \in R$ and a summable family $(f_i)_{i\in I}$ of orthogonal idempotents of R with $\sum_{i\in I} f_i = 1$ and $|I| \leq \aleph$. In the quotient topology on $\bar{R} = R/N$, $(\bar{f}_i)_{i\in I}$ is a family of orthogonal idempotents which sums to $\bar{1}$. By Theorem 4.7 and Lemma 3.6 we get idempotents $e_i \in Rf_ie$ such that the family $(\bar{e}_i)_{i\in I}$ is orthogonal in $\bar{R}$ and $\sum_{i\in I} \bar{e}_i = \bar{e}$. The family $(e_i)_{i\in I}$ is summable by completeness. Define $e_\alpha = 1 - e$, $I' = I \cup \{\alpha\}$ and $u = \sum_{i\in I'} e_i$. Then clearly u is a unit in R, hence

$$\sum_{i\in I'} u^{-1}e_i = 1 = \sum_{i\in I'} e_iu^{-1}.$$

It follows that $e = \sum_{i\in I} u^{-1}e_ie = \sum_{i\in I} u^{-1}e_i$. Now, the same argument as in Theorem 3.8 yields that the family $(u^{-1}e_i)_{i\in I}$ consists of orthogonal idempotents. □

5.5 Corollary. *Assume that idempotents lift modulo a closed ideal N contained in the Jacobson radical of a topological ring R. If R/N is abelian, then R is a full internal exchange ring.*

5.6 Question. *It is interesting to investigate the following questions for a topological finite internal exchange ring R:*

(a) *Find ideals $N \leq J(R)$, if any, for which idempotents lift modulo N.*
(b) *Which ideals $N \leq J(R)$ are closed?*
(c) *Determine the ideals N, satisfying* (a) *and* (b), *for which R/N is a full* (*or countable*) *internal exchange ring.*
(d) *Let S be an endomorphism ring of a module M with the finite internal exchange property. Discuss questions* (a), (b) *and* (c) *for the ring S.*

5.7 Question. *The proof given in Theorem 4.6 is purely module-theoretic. It would be of interest to have a ring-theoretic proof.*

In connection with Proposition 4.4, we have the following questions:

5.8 Question. *Let $M = \oplus_{i\in I} M_i$ be an exchangeable decomposition with all M_i having the finite internal exchange property. Does M have the finite internal exchange property?*

Repeated applications of Proposition 4.4 gives an affirmative answer for any finite index set I. Otherwise the question is still open; it might be interesting to consider the question for indecomposable M_i's.

5.9 Question. *Let* $M = \oplus_{i\in I} M_i$ *be an exchangeable decomposition with all* M_i *having the full (countable) internal exchange property. Does* M *have the full (countable) internal exchange property? This question is still not answered, even for* $|I| = 2$.

References

[1] N. Bourbaki, Topology General, Herman (1996).

[2] A.L.S. Corner, On the exchange property in additive categories, (1973), unpublished.

[3] P. Crawley and B. Jonsson, Refinements for infinite direct decompositions of algebraic systems, Pacific J. Math. 14 (1964), 797–855.

[4] L. Fuchs, On quasi-injective modules, Annali Scoula Sup. Pisa 23 (1969), 541–546

[5] M. Harada, On the exchange property of a direct sum of indecomposable modules, Osaca J. Math. 12 (1975), 719–736.

[6] S.H. Mohamed, Internal exchange rings, Proc. "Algebra and its Applications" (Athens, 2005), Contemp. Math. 419 (2006), 213–218.

[7] ___ , Internal exchange property for square-free modules, J. Egypt. Math. Soc. 16 (2008), 1–3.

[8] S.H. Mohamed and B.J. Müller, Continuous modules have the exchange property, Proc. "Abelian group theory" (Perth, 1987), Contemp. Math., Amer. Math. Soc. 87 (1989), 285–289.

[9] ___ , Continuous and discrete modules, Cambridge Univ. Press (1990).

[10] ___ , On the exchange property of quasi-continuous modules, Proc. "Abelian groups and modules" (Padova, 1994), Kluwer Acad. Publ. (1995), 367–372.

[11] ___ , $\aleph$-exchange rings, Proc. "Abelian groups, module theory and topology" (Padova,1997) , Lecture notes in pure and applied math., Marcel Dekker 201 (1998), 311–317.

[12] ___ , Ojective modules, Comm. Algebra 30 (2002), 1817–1827.

[13] ___ , Co-ojective modules, J. Egypt. Math. Soc. 12 (2004), 83–96.

[14] G.S. Monk, A characterization of exchange rings, Proc. Amer. Math. Soc.35 (1972), 349–353.

[15] W.K. Nicholson, Lifting idempotents and exchange rings, Trans. Amer. Math. Soc. 229 (1977), 269–278.

[16] ___ , On exchange rings, Comm. Algebra 25 (1997), 1917–1918.

[17] P.P. Nielsen, Abelian exchange modules, Comm. Algebra 33 (2005), 1107–1118.

[18] ___ , Countable exchange and full exchange rings, comm. Algebra 35 (2007), 3–23.

[19] ___ , Square free modules with exchange, Preprint.

[20] ___ , Internal exchange, Preprint.

[21] K. Oshiro and S.T. Rizvi, The exchange property for quasi-continuous modules, Osaka J. Math. 33 (1996), 217–234.

[22] R.B. Warfield, Decompositions of injective modules, Pacific J. Math. 31 (1969), 263–276.

[23] ——, Exchange rings and decompositions of modules, Math. Ann.. 199 (1972), 31–36.

[24] Hua-Ping Yu, On modules for which the finite exchange property implies the countable exchange property, Comm. Algebra 22 (1994), 3887–3901.

[25] Zimmerman-Huisgen and B. Zimmermann, Classes of modules with the exchange property, J. Algebra 88 (1984), 416–432.

Saad H. Mohamed
Department of mathematics
Ain Shams University
Cairo, Egypt
e-mail: sshhmohamed@yahoo.ca

Advances in Ring Theory
Trends in Mathematics, 257–268

Filtrations in Semisimple Lie Algebras, III

D.S. Passman

Dedicated to S.K. Jain on the occasion of his retirement

Abstract. This is the third in a series of papers. The first two, by Yiftach Barnea and this author, study the maximal bounded $\mathbb{Z}$-filtrations of the finite-dimensional simple Lie algebras over the complex numbers. Those papers obtain a complete characterization for all but the five exceptional Lie algebras, namely the ones of type G_2, F_4, E_6, E_7 and E_8. Here, we fill in the missing step for the algebra G_2. The proof is computational and uses MAGMA, a computer algebra package, to handle the 7×7 matrices that occur.

Mathematics Subject Classification (2000). 17B20, 17B25, 17B70.

Keywords. Lie algebras, filtrations, G_2.

1. Preliminaries

Let L be a Lie algebra over the complex field K. A $\mathbb{Z}$-filtration $\mathcal{F} = \{F_i \mid i \in \mathbb{Z}\}$ of L is a collection of K-subspaces

$$\cdots \subseteq F_{-2} \subseteq F_{-1} \subseteq F_0 \subseteq F_1 \subseteq F_2 \subseteq \cdots$$

indexed by the integers $\mathbb{Z}$ such that $[F_i, F_j] \subseteq F_{i+j}$ for all $i, j \in \mathbb{Z}$. One usually also assumes that $\bigcup_i F_i = L$ and $\bigcap_i F_i = 0$. In particular, F_0 is a Lie subalgebra of L and each F_i is an F_0-Lie submodule of L. Furthermore, we say that the filtration is bounded if there exist integers ℓ and ℓ' with $F_\ell = 0$ and $F_{\ell'} = L$. In this case, it is clear that each F_i, with $i < 0$, is ad-nilpotent on L.

If A is any finite-dimensional Lie algebra then the Ado-Iwasawa Theorem (see [4, Chapter VI]) implies that A embeds in some $L = \mathfrak{gl}_n(K)$ and therefore we obtain a filtration of L with $F_{-1} = 0$, $F_0 = A$ and $F_1 = L$. Thus, it is clearly hopeless to try to classify all the bounded filtrations of the various $\mathfrak{gl}_n(K)$, even if only up to isomorphism. Nevertheless, there is something that can be done.

Research supported in part by NSA grant 144-LQ65.

Again, let $\mathcal{F}$ be a filtration of an arbitrary Lie algebra L. If $\mathcal{G} = \{G_i \mid i \in \mathbb{Z}\}$ is a second such filtration, we say that $\mathcal{G}$ contains $\mathcal{F}$, or $\mathcal{G}$ is larger than $\mathcal{F}$, if $G_i \supseteq F_i$ for all i. In particular, it makes sense to speak about maximal bounded filtrations, and the goal of [1, 2], the first two papers in this series, is to classify such filtrations $\mathcal{F}$ when L is a simple Lie algebra over the complex numbers.

This classification is achieved in four key steps. The first step shows that F_0, the 0-component of $\mathcal{F}$, contains a Cartan subalgebra H of L. Since each component F_i is then an ad H-submodule of L, it follows easily that these F_i are sums of certain ad H-eigenspaces, that is root spaces, L_α. Note that it is necessary to allow α to equal 0 here, with $L_0 = H$. The second step makes this statement more precise by proving that $\mathcal{F} = \mathcal{F}_\lambda$ is a dual filtration. Here λ is a functional on the real root space of L, and each F_i is given by the sum of those L_α with $\lambda(\alpha) \leq i$. It turns out that not every dual filtration is maximal, and the third step shows that $\mathcal{F}_\lambda$ is maximal if and only if λ takes on integer values on an $\mathbb{R}$-basis of roots for the root space. Finally, the fourth step precisely determines these maximal λ by better understanding the $\mathbb{R}$-bases that occur.

Paper [2] deals with the fourth step, while [1] essentially handles the first three. Indeed, all that is missing is the verification of the first step in the case of the five exceptional Lie algebras, namely those of type G_2, F_4, E_6, E_7 and E_8. In this paper, we supply the verification for the smallest exception, namely G_2. The method of proof is somewhat computational, using the precise embedding of G_2 in the Lie algebra B_3. Specifically, we have $G_2 \subseteq B_3 \subseteq \mathfrak{gl}_7(K)$, and consequently our argument requires dealing with certain 7×7 matrices. For this, we use MAGMA, a computer algebra package.

We begin with some preliminary observations. For the most part, these are fairly immediate consequences of the results in [5] and [1, Section 2]. It is first necessary to deal with matrix rings. Here, of course, the filtrations satisfy $F_i F_j \subseteq F_{i+j}$, and we allow K to be a division ring. See [5] for basic definitions.

Lemma 1.1. *Let $R = \mathbf{M}_n(K)$ be the ring of $n \times n$ matrices over the division ring K and let $S \cong \oplus\sum_{i=1}^{k} \mathbf{M}_{n_i}(K)$ be the subring of R consisting of all block diagonal matrices of the form $\operatorname{diag}(s_1, s_2, \ldots, s_k)$, where $s_i \in \mathbf{M}_{n_i}(K)$. If $\mathcal{F} = \{F_a \mid a \in \mathbb{Z}\}$ is a maximal bounded $\mathbb{Z}$-filtration of S, then there exists a maximal bounded $\mathbb{Z}$-filtration $\mathcal{G} = \{G_a \mid a \in \mathbb{Z}\}$ of R with $G_a \cap S = F_a$ for all $a \in \mathbb{Z}$.*

Proof. By [5, Theorem 3.6], $\mathcal{F} = \mathcal{F}_1 \oplus \mathcal{F}_2 \oplus \cdots \oplus \mathcal{F}_k$, where each $\mathcal{F}_i$ is a maximal bounded filtration of $\mathbf{M}_{n_i}(K)$. Furthermore, by choosing an appropriate basis, we can assume that each $\mathcal{F}_i$ is a weight filtration. In other words, if N_i denotes the set of integers that correspond to the row and column positions of $\mathbf{M}_{n_i}(K)$ in $\mathbf{M}_n(K)$, then $\mathcal{F}_i$ is determined by a weight function $\omega_i \colon N_i \to \mathbb{Z}$. But $N = \{1, 2, \ldots, n\}$ is the disjoint union of the various N_i, so we can define $\omega \colon N \to \mathbb{Z}$ to extend all of the functions ω_i. Finally, the filtration $\mathcal{G} = \{G_a \mid a \in \mathbb{Z}\}$ of R determined by ω is maximal bounded, by [5, Theorem 3.6], and the definition of weight filtration clearly implies that $G_a \cap S = F_a$ for all $a \in \mathbb{Z}$. □

In the above situation, we say that $\mathcal{G}$ covers $\mathcal{F}$. Indeed, we will use this notation in all of the various contexts below. In the remainder of this paper, K will denote an algebraically closed field of characteristic 0, essentially K is the complex numbers, and we consider the finite-dimensional simple Lie algebras over K.

Lemma 1.2. *Let L be a simple Lie algebra over K and assume that $L \subseteq \mathfrak{gl}_n(K) \subseteq \mathbf{M}_n(K) = R$. If $\mathcal{F} = \{F_a \mid a \in \mathbb{Z}\}$ is a maximal bounded $\mathbb{Z}$-filtration of L, then there exists a maximal bounded $\mathbb{Z}$-filtration $\mathcal{G} = \{G_a \mid a \in \mathbb{Z}\}$ of R such that $G_a \cap L = F_a$ for all $a \in \mathbb{Z}$.*

Proof. First assume that L acts irreducibly on the vector space $V = K^n$. Then, following [1, Section 2], we let $\mathcal{F}^R = \{\widetilde{F}_a \mid a \in \mathbb{Z}\}$ to be the family of subspaces of R that are defined by

$$\widetilde{F}_a = \sum F_{i_1} F_{i_2} \cdots F_{i_t}$$

where the sum is over all $t \geq 0$ and all subscripts with $i_1 + i_2 + \cdots + i_t \leq a$. According to [1, Lemma 2.4], $\mathcal{F}^R$ is a bounded $\mathbb{Z}$-filtration of R, and hence we can extend $\mathcal{F}^R$ to $\mathcal{G} = \{G_a \mid a \in \mathbb{Z}\}$, a maximal bounded filtration of R. Since $\mathcal{G}_L = \{G_a \cap L \mid a \in \mathbb{Z}\}$ is a bounded $\mathbb{Z}$-filtration of L containing $\mathcal{F}$, by [1, Lemma 2.1], the maximality of $\mathcal{F}$ now implies that $\mathcal{F} = \mathcal{G}_L$, as required.

For the general case, we use Weyl's Theorem [3, Theorem 6.3], which asserts that L acts completely reducibly on V. Thus, with respect to a suitable basis, $R = \mathbf{M}_n(K)$ contains the subring $S \cong \oplus \sum_{i=1}^k \mathbf{M}_{n_i}(K)$ of block diagonal matrices corresponding to the irreducible constituents of this representation of L. In other words, there exist homomorphisms $\phi_i \colon L \to \mathfrak{gl}_{n_i}(K) \subseteq \mathbf{M}_{n_i}(K)$ that are either irreducible representations of L or zero maps, and with at least one ϕ_i not zero. Now if $\phi_i \neq 0$, then $\phi_i(\mathcal{F})$ is a maximal bounded $\mathbb{Z}$-filtration of $\phi_i(L) \cong L$, so by the above, $\phi_i(\mathcal{F})$ is covered by $\mathcal{G}_i$, a maximal bounded filtration of $\mathbf{M}_{n_i}(K)$. On the other hand, if $\phi_i = 0$, then $\phi_i(L) = 0$, so $\phi_i(\mathcal{F})$ is obviously covered by any maximal bounded filtration $\mathcal{G}_i$ of $\mathbf{M}_{n_i}(K)$. Since F_a is a subdirect product of its images $\phi_i(F_a)$, it follows from [5, Theorem 3.6] that $\mathcal{G} = \mathcal{G}_1 \oplus \mathcal{G}_2 \oplus \cdots \oplus \mathcal{G}_k$ is a maximal bounded filtration of S with $\mathcal{G}_L \supseteq \mathcal{F}$. Finally, we can apply the preceding lemma to find a maximal bounded filtration $\mathcal{H} = \{H_a \mid a \in \mathbb{Z}\}$ of R that covers $\mathcal{G}$. Then $\mathcal{H}_L \supseteq \mathcal{F}$, and the maximality of $\mathcal{F}$ yields the result. □

This has two consequences of interest. First, we have

Lemma 1.3. *Let $\mathcal{F} = \{F_a \mid a \in \mathbb{Z}\}$ be a maximal bounded $\mathbb{Z}$-filtration of the simple K-Lie algebra L. If $x \in F_0$ and if $x = x_s + x_n$ is its Jordan decomposition in L, then the semisimple part x_s and the nilpotent part x_n both belong to F_0.*

Proof. Using an irreducible representation of L, we embed L in the Lie algebra $\mathfrak{gl}_n(K) \subseteq \mathbf{M}_n(K) = R$. Therefore, by the previous lemma, $\mathcal{F}$ is covered by a maximal filtration $\mathcal{G} = \{G_a \mid a \in \mathbb{Z}\}$ of R and, in particular, $F_0 = G_0 \cap L$ and $x \in F_0 \subseteq G_0$. Now, by [3, Theorem 6.4], $x = x_s + x_n$ is also the usual Jordan decomposition of x in the matrix ring R. Thus, by [3, Proposition 4.2], $x_s = p(x)$ and $x_n = q(x)$, where p and q are polynomials over K without constant terms.

Since G_0 is a subalgebra of R, it now follows that $x_s, x_n \in G_0$ and consequently $x_s, x_n \in G_0 \cap L = F_0$. $\square$

Furthermore, we have

Lemma 1.4. *Let $L \subseteq \bar{L}$ be simple Lie algebras over K. If $\mathcal{F} = \{F_a \mid a \in \mathbb{Z}\}$ is a maximal bounded $\mathbb{Z}$-filtration of L, then there exists a maximal bounded $\mathbb{Z}$-filtration $\mathcal{G} = \{G_a \mid a \in \mathbb{Z}\}$ of $\bar{L}$ with $F_a = G_a \cap L$ for all $a \in \mathbb{Z}$.*

Proof. Using an irreducible representation of $\bar{L}$, we embed $\bar{L}$ in $\mathfrak{gl}_n(K) \subseteq \mathbf{M}_n(K) = R$. Then $L \subseteq \mathfrak{gl}_n(K)$, so Lemma 1.2 implies that there exists a $\mathbb{Z}$-filtration $\mathcal{H} = \{H_a \mid a \in \mathbb{Z}\}$ of R with $H_a \cap L = F_a$ for all $a \in \mathbb{Z}$. Furthermore, by [1, Lemma 2.1], $\mathcal{H}_{\bar{L}} = \{H_a \cap \bar{L} \mid a \in \mathbb{Z}\}$ is a bounded filtration of $\bar{L}$, and this extends to a maximal bounded filtration $\mathcal{G} = \{G_a \mid a \in \mathbb{Z}\}$ of $\bar{L}$. Note that

$$G_a \cap L \supseteq (H_a \cap \bar{L}) \cap L = H_a \cap L = F_a,$$

so $\{G_a \cap L \mid a \in \mathbb{Z}\}$ is a bounded filtration of L containing $\mathcal{F}$. The maximality of $\mathcal{F}$ now implies that $G_a \cap L = F_a$ for all $a \in \mathbb{Z}$. $\square$

The following is implicit in the work of [1].

Lemma 1.5. *Let $\mathcal{F} = \{F_a \mid a \in \mathbb{Z}\}$ be a maximal bounded filtration of the simple K-Lie algebra of classical type. Then $F_0 \supseteq B$, a Borel subalgebra of L.*

Proof. By [1, Section 5], F_0 contains a Cartan subalgebra H of L, and indeed $\mathcal{F} = \mathcal{F}_\lambda$ for some suitable linear functional λ on the root space. It follows from the definition of $\mathcal{F}_\lambda$ that if α is a root, then at least one of the root spaces L_α or $L_{-\alpha}$ is contained in F_0. With this, it is easy to see that if B/F_{-1} is a Borel subalgebra of F_0/F_{-1} containing $(H + F_{-1})/F_{-1}$, then $B \subseteq F_0$ is a Borel subalgebra of L. $\square$

If L is a K-Lie algebra and if S is a solvable subalgebra, then $S \subseteq B$ where B is a Borel subalgebra of L. If B is uniquely determined by S, then we say that S is uniquely extendible in L. It is clear that if $S \subseteq T \subseteq L$, with S and T both solvable, and if S is uniquely extendible, then so is T. Our reason for introducing this concept is the simple result given below that can be used in concert with the previous two lemmas.

Lemma 1.6. *Let $L \subseteq \bar{L}$ be Lie algebras over K, and let $\overline{B}$ be a Borel subalgebra of $\bar{L}$. If $\overline{B} \cap L$ is uniquely extendible in $\bar{L}$, then $\overline{B} \cap L$ is a Borel subalgebra of L.*

Proof. Obviously, $S = \overline{B} \cap L$ is a solvable subalgebra of both L and $\bar{L}$, and hence $S \subseteq B$, where B is a suitable Borel subalgebra of L. Furthermore, B extends to $\overline{B}_1$, a Borel subalgebra of $\bar{L}$. In other words, we have $S \subseteq \overline{B}$ and $S \subseteq \overline{B}_1$ so, since S is uniquely extendible in $\bar{L}$, we conclude that $\overline{B} = \overline{B}_1 \supseteq B$ and hence $S = \overline{B}_1 \cap L \supseteq B$. Thus $S = B$, as required. $\square$

We close this section with two fairly standard results from Lie theory. We include brief proofs of each for the convenience of the reader.

To start with, we say that the subalgebra S of L is ad-nilpotent if $\operatorname{ad} S$ is nilpotent in its action on L. Certainly this implies that S is a nilpotent and hence solvable subalgebra of L, so $S \subseteq B$ for some Borel subalgebra B of L. Indeed, if L is simple, then S embeds in N, the nilradical of B. The following lemma contains a sufficient condition for such a subalgebra S to be uniquely extendible. Note that the expressions S^k and N^k below are the associative powers of S and N in the endomorphism ring of $V = K^n$, the space of n-tuples over K.

Lemma 1.7. *Let S be a Lie subalgebra of $\mathfrak{gl}_n(K)$ so that S acts on the right on the vector space $V = K^n$. If $VS^n = 0$ but $VS^{n-1} \neq 0$, then S is contained in a unique Borel subalgebra of $\mathfrak{gl}_n(K)$ and hence in a unique Borel subalgebra of any intermediate Lie algebra.*

Proof. Since $VS^n = 0$, S is nilpotent in its action on V and hence ad-nilpotent in its action on $\mathfrak{gl}_n(K)$. If $B = N + H$ is a Borel subalgebra of $\mathfrak{gl}_n(K)$ containing S, then the nilradical N contains S. It follows that $VN^i \supseteq VS^i$ for all i, and we know that $VN^n = 0$. Thus, since $VS^{n-1} \neq 0$, it is clear that $VN^i = VS^i$ for all i. But B is the set of elements of $\mathfrak{gl}_n(K)$ that stabilize the flag $V = VN^0 \supseteq VN^1 \supseteq \cdots \supseteq VN^n = 0$, so since $VN^i = VS^i$ we see that B is uniquely determined by S. □

Finally, we have

Lemma 1.8. *Let S be a solvable Lie subalgebra of $\mathfrak{gl}_n(K)$ closed under Jordan decomposition in its action on $V = K^n$. Then S is the direct sum $S = M + C$, where M is the Lie ideal consisting of nilpotent elements of S and where C is a complementary commutative space of semisimple elements.*

Proof. S is contained in a Borel subalgebra $B = N + H$ of $\mathfrak{gl}_n(K)$, where N is the Lie ideal of all nilpotent elements of B and where H is a Cartan subalgebra, a commutative semisimple complement. It follows that $M = N \cap S$ is the subspace of S consisting of all nilpotent elements of S. Furthermore, M is a Lie ideal of S with S/M abelian. The goal is to find a semisimple complementary subspace for M in S, and we proceed by induction on $\dim_K S$.

Suppose first that S has a semisimple element x not contained in its center. Then $\operatorname{ad} x$ is semisimple in its action on S, so S is the direct sum $S = S_0 + S_1$ where $S_0 = \mathfrak{C}_S(x)$ and S_1 is an $\operatorname{ad} x$-stable complement. Clearly S_0 is a Lie subalgebra of S and $\dim S_0 < \dim S$ since x is not central in S. Furthermore, if $y \in S_0$, then the nilpotent and semisimple parts of y are polynomials in y and hence also commute with x. In other words, S_0 is closed under Jordan decomposition, so by induction $S_0 = M_0 + C_0$. On the other hand, S_1 is spanned by eigenvectors of $\operatorname{ad} x$ with nonzero eigenvalues and hence each such eigenvector is contained in $[S, S] \subseteq M$. Thus $S = M + C_0$, and C_0 is the required complement of semisimple elements.

It now suffices to assume that all semisimple elements of S are central in S, and we let C denote the set of all such elements. We show that C is a subspace of S. To this end, let $x, y \in C$. Then x and y commute, so they are commuting diagonalizable elements and hence they can be simultaneously diagonalized. Thus $Kx + Ky$ consists of semisimple elements and hence is contained in C. It follows that C is a subspace and since $M + C$ contains the nilpotent and semisimple parts of all elements of S, we conclude that $S = M + C$. □

2. The Lie algebra G_2

As we mentioned, in order to complete the classification of the maximal bounded $\mathbb{Z}$-filtrations of the simple Lie algebras, we must show, in the case of the exceptional Lie algebras, that the 0-component of such filtrations contains a Cartan subalgebra. The goal of this section is to prove this for G_2, and indeed we have

Theorem 2.1. *Let $\mathcal{F} = \{F_i \mid i \in \mathbb{Z}\}$ be a maximal bounded $\mathbb{Z}$-filtration of the Lie algebra L of type G_2 over the algebraically closed field K of characteristic 0. Then F_0 contains a Cartan subalgebra of L.*

Proof. We use the precise description of G_2 as given in [3, Section 19.3]. Indeed, those few pages describe a faithful 7-dimensional representation of the Lie algebra and show that $L \subseteq \overline{L}$ where $\overline{L}$ is of type B_3. Our argument requires some matrix and vector space computations and, for this, we use MAGMA, a computer algebra package. The original version of this manuscript, containing an annotated write up of the fairly simple code we require, can be found on the author's web page

`www.math.wisc.edu/~passman/abstracts.html.`

A complete MAGMA input and output text file is also available there.

Now let $\mathcal{F} = \{F_i \mid i \in \mathbb{Z}\}$ be a maximal bounded $\mathbb{Z}$-filtration of the Lie algebra L. Then it follows from Lemma 1.4 that $\overline{L}$ has a maximal bounded $\mathbb{Z}$-filtration $\mathcal{G} = \{G_i \mid i \in \mathbb{Z}\}$ such that $F_i = G_i \cap L$ for all $i \in \mathbb{Z}$. Furthermore, by Lemma 1.5, G_0 contains a Borel subalgebra $\overline{B}$ of $\overline{L}$, and hence $F_0 = G_0 \cap L \supseteq \overline{B} \cap L$, a solvable subalgebra of L. In particular, $\overline{B} \cap L \subseteq B$, a Borel subalgebra of L. Since all Borel subalgebras of L are conjugate, we can assume that B is as described in [3, Section 19.3]. Furthermore, let $\overline{N}$ denote the nilradical of $\overline{B}$, and let N be the nilradical of B. From [3, Sections 1.2 and 19.3], we have

$$\dim N = 6, \qquad \dim B = 6 + 2 = 8, \qquad \dim L = 8 + 6 = 14$$

$$\dim \overline{N} = 9, \qquad \dim \overline{B} = 9 + 3 = 12, \qquad \dim \overline{L} = 12 + 9 = 21.$$

Our computations use the basis $\{a, b, c, d, e, f\}$ for N as described in [3]. These basis members are, in fact, all root vectors corresponding, respectively, to the roots α, β, $\alpha+\beta$, $2\alpha+\beta$, $3\alpha+\beta$, and $3\alpha+2\beta$ of G_2, where α and β are simple.

We note that the 7×7 matrices for c and d are given by

$$c = \begin{bmatrix} 0 & 0 & 0 & 0 & t & 0 & 0 \\ -t & 0 & 0 & 0 & 0 & 0 & 0 \\ 0 & 0 & 0 & 0 & 0 & 0 & 0 \\ 0 & 0 & 0 & 0 & 0 & 0 & 0 \\ 0 & 0 & 0 & 0 & 0 & 0 & 0 \\ 0 & 0 & 0 & -1 & 0 & 0 & 0 \\ 0 & 0 & 1 & 0 & 0 & 0 & 0 \end{bmatrix}$$

and

$$d = \begin{bmatrix} 0 & 0 & 0 & t & 0 & 0 & 0 \\ 0 & 0 & 0 & 0 & 0 & -1 & 0 \\ 0 & 0 & 0 & 0 & 1 & 0 & 0 \\ 0 & 0 & 0 & 0 & 0 & 0 & 0 \\ 0 & 0 & 0 & 0 & 0 & 0 & 0 \\ 0 & 0 & 0 & 0 & 0 & 0 & 0 \\ -t & 0 & 0 & 0 & 0 & 0 & 0 \end{bmatrix},$$

where $t = \sqrt{2}$. Furthermore, the nonzero Lie products of the basis elements are easily found to be

$$[a, b] = -c, \quad [a, c] = -2d, \quad [a, d] = -3e, \quad [b, e] = f, \quad [c, d] = -3f,$$

and hence we have

Lemma 2.2. *The terms of the lower central series of N are given by*

$$\begin{aligned} N^{[1]} &= Ka + Kb + Kc + Kd + Ke + Kf \\ N^{[2]} &= Kc + Kd + Ke + Kf \\ N^{[3]} &= Kd + Ke + Kf \\ N^{[4]} &= Ke + Kf \\ N^{[5]} &= Kf. \end{aligned}$$

Furthermore, N is contained in a unique Borel subalgebra of the algebra $\overline{L}$. In particular, if $N \subseteq L \cap \overline{B}$, then $L \cap \overline{B}$ is a Borel subalgebra of L.

Proof. The terms of the lower central series are trivial to compute from the above commutator relations. For the last part, we want to show that $S = N$ is uniquely extendible in $\overline{L}$. For this, we first observe that N is ad-nilpotent on $\overline{L}$. Furthermore, $\overline{L}$ admits the same 7-dimensional module V as does L. Thus, in view of Lemma 1.7, it suffices to show that $N^6 \neq 0$ in its action on V. But $f \in N^{[5]} \subseteq N^5$ and we easily check that the matrix product fa is not 0. Thus $N^6 \neq 0$, and consequently Lemma 1.6 yields the result. □

It can be shown that N contains an element having one 7×7 Jordan block in its action on V, and such regular nilpotent elements are known to be contained in unique Borel subalgebras. Next, we note that

$$\dim \overline{L} \geq \dim(L + \overline{B}) = \dim L + \dim \overline{B} - \dim(L \cap \overline{B})$$

and hence

$$\dim(L \cap \overline{B}) \geq \dim L + \dim \overline{B} - \dim \overline{L} = 14 + 12 - 21 = 5.$$

Since $L \cap \overline{B} \subseteq B$, it also follows that $\dim(L \cap \overline{B}) \leq 8$.

Now $L \cap \overline{B} = F_0 \cap \overline{B}$ and, since $\overline{B}$ is closed under taking semisimple and nilpotent parts, the same is true of $L \cap \overline{B}$ by Lemma 1.3. Furthermore, recall from [3, Theorem 6.4] that the Jordan decomposition of any element of L in its action on V and in its ad-action on L are identical. It therefore follows from Lemma 1.8 that $L \cap \overline{B} = M + C$ where $M = L \cap \overline{N} \subseteq N$ and where C is a semisimple complement. If $\dim C \geq 2$, then $\dim C = 2$ and C is a Cartan subalgebra of L contained in F_0. Thus, we can assume that either $C = 0$ or $C = Kh$ has dimension 1. Indeed, by taking a suitable conjugate if necessary, we can assume that $h \in H \subseteq B$ where H is any Cartan subalgebra of our choosing.

Note that M is properly smaller than N since, if $N \subseteq L \cap \overline{B}$, then Lemma 2.2 implies that $L \cap \overline{B} = B$ contains a Cartan subalgebra of L. Thus $\dim(L \cap \overline{B}) < \dim N + 1 = 7$, and hence there are just two cases remaining to be considered. In case 1, we have $\dim(L \cap \overline{B}) = 5$ and either $L \cap \overline{B} = M$ or $L \cap \overline{B} = M + Kh$, where $M = N \cap \overline{B}$ and where h is some nonzero element of H. On the other hand, in case 2, $\dim(L \cap \overline{B}) = 6$ and, since $L \cap \overline{B} \neq N$, we have $L \cap \overline{B} = M + Kh$, where M and h are as above. Furthermore, we can assume that H is the Cartan subalgebra which we now describe.

Following [3, Section 19.3], we note that a Cartan subalgebra of $\overline{L}$ is diagonal with basis $d_1 = e_{22} - e_{55}$, $d_2 = e_{33} - e_{66}$ and $d_3 = e_{44} - e_{77}$, where of course $\{e_{ij}\}$ is the set of matrix units in $\mathbf{M}_7(K)$. Furthermore, a Cartan subalgebra $H \subseteq B$ of L is given by all elements of the form $h = k_1 d_1 + k_2 d_2 + k_3 d_3$ with $k_1, k_2, k_3 \in K$ and $k_1 + k_2 + k_3 = 0$. In particular, if all k_i are nonzero, then $\operatorname{rank} h = 6$. Thus, up to scalar factors, there are just three nonzero members of H of rank less than 6, and these all have rank 4. Specifically, we take these three elements to be $h_1 = d_1 - d_2$, $h_2 = d_2 - d_3$ and $h_3 = d_3 - d_1$. For convenience, let us define

$$N_a = Ka + Kc + Kd + Ke + Kf = Ka + [N, N] \subseteq N,$$

and

$$N_b = Kb + Kc + Kd + Ke + Kf = Kb + [N, N] \subseteq N.$$

Then, we have

Lemma 2.3. *Let H be the Cartan subalgebra of L contained in the diagonal subspace of $\mathbf{M}_7(K)$. Then $H \subseteq B$ and, up to a scalar multiple, there are just three nonzero elements of H having rank less than 6. These elements, h_1, h_2 and h_3, all have rank 4 and satisfy $\alpha(h_1) = -1$, $\beta(h_1) = 2$, $\alpha(h_2) = 1$, $\beta(h_2) = -1$ and $\alpha(h_3) = 0$, $\beta(h_3) = -1$. Furthermore, suppose M is a Lie subalgebra of N of codimension 1. Then $M \lhd N$, $M \supseteq [N, N]$ and, if M is $\operatorname{ad} h$-stable with $h = h_1$, h_2 or h_3, then $M = N_a$ or N_b.*

Proof. The values of $\alpha(h_i)$ and $\beta(h_i)$ are easily computed from the formulas $[h_i, a] = \alpha(h_i)a$ and $[h_i, b] = \beta(h_i)b$. It remains to consider the Lie subalgebra M of codimension 1 in N. Since normalizers grow in nilpotent algebras, it follows that $M \triangleleft N$ and of course N/M is abelian. Thus $M \supseteq [N, N]$, and note that $N = Ka+Kb+[N, N]$. Finally, suppose $h = h_1$, h_2 or h_3 and that M is ad h-stable. Since $\alpha(h) \neq \beta(h)$, h has distinct eigenvalues on $Ka + Kb$, with eigenvectors a and b. Thus we conclude that the only possibilities for M are $Ka + [N, N] = N_a$ or $Kb + [N, N] = N_b$. □

It follows from the values of $\alpha(h)$ and $\beta(h)$ given above that h_1, h_2 and h_3 are not regular elements of H. Since $\alpha(h_3) = 0$, h_3 is, in some sense, the worst offender.

Now, as is well known, $L = G_2$ has no nonzero representation of degree less than 7, and it has a unique irreducible representation of degree equal to 7. Indeed, this is a consequence of Weyl's dimension formula and the fact that irreducible representations are uniquely determined by their highest weight. An explicit formula for the degrees of the irreducible representations of G_2 can be found in [3, page 140]. The unique representation of degree 7 is obviously the representation described in [3, Section 19.3], where L acts on the right on a 7-dimensional vector space V. Furthermore, since $\dim \overline{L}/L = 21 - 14 = 7$, we see that the adjoint representation of L on $\overline{L}$ has the factor module $\overline{L}/L \cong V$. In other words, we can compute certain invariants for the adjoint action of L on $\overline{L}/L$ by considering the matrix action of L on V. For convenience, let $\{v_1, v_2, v_3, v_4, v_5, v_6, v_7\}$ be the natural basis for V corresponding to the matrix representation of L. We can now handle the two cases in turn. We start with

Lemma 2.4. *Case* 1 *cannot occur.*

Proof. Suppose, by way of contradiction, that $\dim(L \cap \overline{B}) = 5$. Then

$$\dim(L + \overline{B}) = \dim L + \dim \overline{B} - \dim(L \cap \overline{B}) = 14 + 12 - 5 = 21,$$

and hence $L + \overline{B} = \overline{L}$. Furthermore, since $\overline{B}$ and $L \cap \overline{B}$ are ad$(L \cap \overline{B})$-submodules of $\overline{L}$, we conclude that

$$V \cong \frac{\overline{L}}{L} = \frac{L + \overline{B}}{L} \cong \frac{\overline{B}}{L \cap \overline{B}}$$

as $(L \cap \overline{B})$-modules.

If $M = L \cap \overline{B} \subseteq N$, then M has codimension 1 in N, and hence $M \supseteq [N, N]$ by Lemma 2.3. Note also that $M \subseteq N \subseteq \overline{N}$ and that $\overline{N} \triangleleft \overline{B}$ with $\overline{B}/\overline{N}$ being abelian of dimension 3. It follows that $\overline{N}/M$ is an M-submodule of $\overline{B}/M$ of dimension 4 and that M acts trivially on the quotient $\overline{B}/\overline{N}$. Translating this to the module V, we conclude that $VM \supseteq V[N, N]$ has dimension at most 4. But our MAGMA computations show that $\dim V[N, N] = 5$, so this possibility cannot occur.

It remains to assume that $L \cap \overline{B} = M + Kh$, where $M = N \cap \overline{B}$ and h is a nonzero element of H. Obviously, $\dim M = 4$ here, so M has codimension 2 in N. Since $M \subseteq N \subseteq \overline{N}$, in this situation we have $L \cap \overline{B} = M + Kh \subseteq \overline{N} + Kh \subseteq \overline{B}$

and $\overline{N}+Kh$ is an ideal of $\overline{B}$ of codimension 2. Hence $L\cap\overline{B}$ acts trivially on this quotient. Translating to the module V, we conclude that $\dim V(L\cap\overline{B})\le 5$. In particular, $\operatorname{rank} h\le 5$ and, as we have indicated, this implies that $h=h_1,h_2$, or h_3 up to a scalar factor. Since normalizers grow in N, we have $M\lhd M_1\lhd N$ with $\dim M_1=5$. Thus $M_1\supseteq [N,N]\supseteq Kc+Kd$ and hence, since M has codimension 1 in M_1, we have $M\cap(Kc+Kd)\neq 0$. Choose $0\neq m=xc+yd\in M\cap(Kc+Kd)$ with $x,y\in K$ and not both 0.

We now compute the dimension of $Vm+Vh_i$ for all $i=1,2,3$. To this end, note that $\operatorname{rank} h_i=4$ and indeed Vh_i has a basis consisting of those v_j that correspond to the four columns where the matrix h_i has nonzero diagonal entries. Thus, to compute the dimension of $(Vm+Vh_i)/Vh_i$, we merely form the matrix of m, delete the columns of that matrix corresponding to the basis elements of Vh_i, and determine the rank of the remaining 7×3 matrix. Indeed, to compute this rank, we can certainly delete any zero row or column. When we do this, the matrices we obtain for m, corresponding to h_1, h_2 and h_3, respectively, are

$$\begin{bmatrix} 0 & yt \\ -xt & 0 \\ 0 & -x \\ -yt & 0 \end{bmatrix}, \quad \begin{bmatrix} 0 & xt \\ -xt & 0 \\ 0 & y \\ -yt & 0 \end{bmatrix}, \quad \begin{bmatrix} -xt & 0 & -y \\ -yt & x & 0 \end{bmatrix}.$$

Note that the rank of each of these matrices is equal to 2 provided that one of x or y is nonzero. Thus $\dim(Vm+Vh_i)/Vh_i=2$, so $\dim(Vm+Vh_i)=2+\dim Vh_i=6$, and this is the required contradiction since $\dim V(L\cap\overline{B})\le 5$. □

Finally, we show that the second case cannot occur. This is surprisingly a bit more complicated. Since $L\cap\overline{B}$ has dimension 6, there are actually just a few possibilities for this subalgebra. But in this case, we know less about its action on the module V.

Lemma 2.5. *Case* 2 *cannot occur.*

Proof. Suppose, by way of contradiction, that $\dim(L\cap\overline{B})=6$. Then

$$\dim(L+\overline{B})=\dim L+\dim\overline{B}-\dim(L\cap\overline{B})=14+12-6=20,$$

and thus $L+\overline{B}$ has codimension 1 in $\overline{L}$. Furthermore, we know that

$$V\cong\frac{\overline{L}}{L}>\frac{L+\overline{B}}{L}\cong\frac{\overline{B}}{L\cap\overline{B}}$$

as $(L\cap\overline{B})$-modules. Under this isomorphism, the submodule $(L+\overline{B})/L$ corresponds to a subspace W of codimension 1 in V. Unfortunately, we will not always have a precise description of this subspace.

We are also given that $L\cap\overline{B}=M+Kh$, where $M=N\cap\overline{B}$ and where h is a nonzero element of H. Certainly, $\dim M=5$, so that M has codimension 1 in N. Since $L\cap\overline{B}=M+Kh\subseteq\overline{N}+Kh$ and $\overline{N}+Kh$ is an ideal of $\overline{B}$, it follows that $L\cap\overline{B}$ acts trivially on the 2-dimensional quotient $\overline{B}/(\overline{N}+Kh)$. Translating this into V, it follows that $\dim W(M+Kh)\le 6-2=4$. In particular, h has rank at most 4 on W, and hence h has rank at most 5 on V. We conclude, as before, that $h = h_1$, h_2 or h_3.

In addition, note that M is an $\operatorname{ad} h$-stable subalgebra of N of codimension 1, so Lemma 2.3 implies that $M = N_a = Ka + [N,N]$ or $N_b = Kb + [N,N]$. The first possibility is easy to handle. Indeed, since either choice for M is nilpotent, M must act trivially on the 1-dimensional module V/W, and hence $W \supseteq VM$. In particular, if $M = N_a$, then $W \supseteq VN_a$, and the latter subspace has dimension 6 by our computations. Thus $W = VN_a$ and $W(L \cap \overline{B}) = W(M + Kh) = WN_a + Wh_i$ for some $i = 1, 2$ or 3. However, the latter three subspaces have dimension 6, 5 and 5, respectively, and this contradicts the fact that $\dim W(M + Kh) \leq 4$.

It follows that $M \neq N_a$, and hence $M = N_b$. In this case, $W \supseteq VN_b$ and VN_b has dimension 5 with basis $\{v_1, v_3, v_4, v_5, v_6\}$. Thus $W = VN_b + Ku$, where u is a nonzero vector in $U = Kv_2 + Kv_7$. Note that each of h_1, h_2 and h_3 acts on U with eigenvectors v_2 and v_7. First, let us assume that $h = h_1$ or h_2. Then the eigenvalues of h are 1 and 0, with $v_2h_1 = v_2$, $v_7h_1 = 0$, and $v_2h_2 = 0$, $v_7h_2 = v_7$. Since VN_b is h-stable and W is h-stable, it follows that $W \cap U$ must be h-stable. In particular, we must have $W = W_1 = VN_b + Kv_2 = VN_b + Vh_1$ or $W = W_2 = VN_b + Kv_7 = VN_b + Vh_2$. In this case, computations show that $\dim W_1N_b = 5$ and $\dim W_2N_b = 5$, again contradicting the fact that $\dim W(M + Kh) \leq 4$.

It remains to assume that $M = N_b$ and $h = h_3$. Here we have $W = VN_b + Ku$, where $u = xv_2 + yv_7$ with $x, y \in K$ and not both 0. We first obtain a lower bound for the dimension of WN_b. To this end, note that $V(N_b)^2$ has dimension 2 with basis $\{v_4, v_5\}$. Furthermore, $c, d \in N_b$, so $Ku(N_b) \supseteq Kuc + Kud$. We compute $\dim(V(N_b)^2 + Kuc + Kud)/V(N_b)^2)$ by constructing a 2×7 matrix with first row equal to uc and second row equal to ud. Next we delete the fourth and fifth columns, since they correspond to the vectors v_4 and v_5 in $V(N_b)^2$, and then we find the rank of the remaining matrix. Indeed, we can also delete any zero column, and when we do so, we obtain

$$\begin{bmatrix} -xt & y & 0 \\ -yt & 0 & -x \end{bmatrix},$$

a matrix quite similar to the h_3 matrix of the previous lemma. Clearly, this matrix has rank 2 provided x or y is nonzero.

In other words, we have shown that $\dim WN_b/V(N_b)^2 \geq 2$ and hence that $\dim WN_b \geq 4$. Of course, $WN_b \subseteq VN_b$. On the other hand, since $v_2h_3 = -v_2$ and $v_7h_3 = -v_7$, we see that $u = u(-h_3) \in W(L \cap \overline{B})$. Thus $W(L \cap \overline{B})$ contains the direct sum $WN_b + Ku$ and hence $\dim W(L \cap \overline{B}) \geq 5$, again a contradiction. □

The proofs of Lemmas 2.4 and 2.5 appear, in some sense, to be dual to each other. Furthermore, once these two cases are eliminated, we know that only the earlier configurations can occur, and consequently F_0 must contain a Cartan subalgebra of L. This completes the proof of Theorem 2.1. □

Since the subspaces of V considered above all have natural bases, it is clear that the above computations can all be achieved without using a computer algebra package like MAGMA. On the other hand, it is also clear that without such a package, the experimentation required to find such a proof would have made reaching this goal somewhat problematical. Finally, the author would like to thank the referee for his comment on regular nilpotent elements and for suggesting the validity of Lemma 1.8, which made some later arguments considerably cleaner.

References

[1] Y. Barnea and D.S. Passman, *Filtrations in semisimple Lie algebras, I*, Trans. AMS, **358** (2006), 1983–2010 .

[2] Y. Barnea and D.S. Passman, *Filtrations in semisimple Lie algebras, II*, Trans. AMS, **360** (2008), 801–817.

[3] J.E. Humphreys, *Introduction to Lie Algebras and Representation Theory*, second printing, Springer-Verlag, New York, 1972.

[4] N. Jacobson, *Lie Algebras*, Wiley-Interscience, New York, 1962.

[5] D.S. Passman *Filtrations in semisimple rings*, Trans. AMS, **357** (2005), 5051–5066.

D.S. Passman
Department of Mathematics
University of Wisconsin-Madison
Madison, Wisconsin 53706, USA
e-mail: passman@math.wisc.edu

Advances in Ring Theory
Trends in Mathematics, 269–282

On the Blowing-up Rings, Arf Rings and Type Sequences

D.P. Patil

In honour of Professor S.K. Jain on the occasion of his 70th birthday

Abstract. In Section 1 we recall the definitions of the Hilbert functions, multiplicity, h-polynomial, blowing-up rings, standard basis, degree of singularity, Cohen-Macaulay type, type sequences and almost Gorenstein rings etc. In Section 2 we give summary of numerical invariants of monomial curves, especially monomial curves defined by arithmetic sequences and almost arithmetic sequences. In particular, we give an explicit formula for the type sequence (see (2.1)–(6)) and give a characterization of almost-Gorensteinness of the algebroid semigroup ring $R = K[\![\Gamma]\!]$ (over a field K) of the numerical semigroup Γ generated by an arithmetic sequence. In Section 3 we mainly study Arf rings and their type sequences. We begin with recalling definition of Arf ring and its branch sequence and give a formula (see Theorem (3.4)) for the degree of singularity of R as the sum of the lengths of quotients of the successive terms of its branch sequence as well as the sum of the first coefficients of the Hilbert-Samuel polynomials of the terms of its branch sequence. Further, we use a results proved in [3] and [7] to give (see Theorem (3.6)) a characterization of complete local Arf domains with algebraically residue field using the type sequence of R and type sequences of the rings in the branch sequence of R. Finally we prove that the type sequence of the blowing-up ring of a complete local Arf domain with algebraically residue field is the sequence obtained from the type sequence of R obtained by removing its first term. In Section 4 we give some examples of Arf rings and some of not Arf rings. In Example 4, we give necessary and sufficient conditions for the algebroid semigroup ring $R = K[\![\Gamma]\!]$ of the numerical semigroup generated by an arithmetic sequence Γ over a field K to be an Arf ring.

Mathematics Subject Classification (2000). Primary 14H20; Secondary 13H10, 13P10.

Keywords. Hilbert function of a local ring, h-polynomial, semigroup ring, standard basis of a numerical semigroup, monomial curves, blowing-up ring, Arf ring.

0. Introduction

This is an expository article on some numerical invariants of one-dimensional noetherian local rings. More precisely, let $(R, \mathfrak{m}_R)$ be a noetherian local one-dimensional analytically irreducible domain, i.e., the $\mathfrak{m}$-adic completion $\hat{R}$ of R is a domain or, equivalently, the integral closure $\overline{R}$ of R in its quotient field $\mathrm{Q}(R)$ is a discrete valuation ring and a finite R-module. We further assume that R is residually rational, i.e., R and $\overline{R}$ have the same residue field. A particular important class of rings which satisfy these assumptions are algebroid semigroup rings which are coordinate rings of algebroid monomial curves.

Various algebraic and geometric properties of the ring R are described by some numerical invariants, for example, Hilbert functions, multiplicity, h-polynomial, blowing-up rings, standard basis, degree of singularity, Cohen-Macaulay type, type sequences, Gorenstein, almost Gorenstein and Arf rings etc. Several authors have studied these numerical invariants (see for example [1], [2], [3], [4] [18]). The first term t_1 of the type sequence of R is the Cohen-Macaulay type of R and the sum $\sum_{i=1}^{n} \mathrm{t}_i$ is the degree of singularity of R. Further, the "Gorensteinness" and "almost Gorensteinness" are characterized by type sequences. It is worth noting here that if R is a semigroup ring, then the above properties correspond to the properties "symmetric" and "pseudo-symmetric" of numerical semigroups, respectively. These properties are of a special interest (see [5], [21]), since each numerical semigroup can be expressed as an intersection of numerical semigroups that are either symmetric or pseudo-symmetric. Furthermore, the property "Arf" can be described by its type sequence and each term t_i is related to the ith term in the "branch sequence" of R.

In Section 1 we recall the definitions of the Hilbert functions, multiplicity, h-polynomial, blowing-up rings, standard basis, degree of singularity, Cohen-Macaulay type, type sequences and almost Gorenstein rings etc.

In Section 2 we give summary of numerical invariants of monomial curves, especially monomial curves defined by arithmetic sequences and almost arithmetic sequences. In particular, we give an explicit formula for the type sequence (see (2.1)–(6)) and give a characterization of almost-Gorensteinness of the algebroid semigroup ring $R = K[\![\Gamma]\!]$ (over a field K) of the numerical semigroup Γ generated by an arithmetic sequence.

In Section 3 we mainly study Arf rings and their type sequences. We begin with recalling definition of Arf ring and its branch sequence and give a formula (see Theorem (3.4)) for the degree of singularity of R as the sum of the lengths of quotients of the successive terms of its branch sequence as well as the sum of the first coefficients of the Hilbert-Samuel polynomials of the terms of its branch sequence. Further, we use a results proved in [3] and [7] to give (see Theorem (3.6)) a characterization of complete local Arf domains with algebraically residue field using the type sequence of R and type sequences of the rings in the branch sequence of R. Finally we prove that (see Corollary (3.9)) the type sequence of the blowing-up ring of a complete local Arf domain with algebraically residue field is the sequence obtained from the type sequence of R obtained by removing its first term. In Section 4 we give some examples of Arf rings and some of not Arf rings. In Example 4, we give necessary and sufficient conditions for the algebroid semigroup ring $R = K[\![\Gamma]\!]$ (over a field K) of the numerical semigroup Γ generated by an arithmetic sequence to be an Arf ring.

1. Preliminaries – notation, definitions and some results

Throughout this article we make the following assumptions and notation.

Notation 1.1. Let $\mathbb{N} = \{0, 1, 2, \ldots\}$ and $\mathbb{Z} = \{0, \pm 1, \pm 2, \ldots\}$ denote the set of all natural numbers and all integers respectively. Further, for $a, b \in \mathbb{N}$, we denote $[a, b] := \{r \in \mathbb{N} \mid a \leq r \leq b\}$ and $\mathbb{N}_a := \{n \in \mathbb{N} \mid n \geq a\}$.

For convenience, some definitions and some well-known results for one-dimensional noetherian local rings are collected in 1.2 below:

Notation, Definitions and Some Results 1.2. Let $(R, \mathfrak{m})$ be a 1-dimensional Cohen-Macaulay local ring of multiplicity e and embedding dimension $\nu \geq 2$. Further, let $\mathrm{h}_R := \mathrm{h}_R(Z) = \mathrm{h}_0 + \mathrm{h}_1 Z + \cdots + \mathrm{h}_s Z^s$, $s := \deg \mathrm{h}_R$ denote the h-polynomial of R. Then:

(1) (Hilbert function and h-polynomial) Let $(R, \mathfrak{m})$ be a d-dimensional noetherian local ring of the multiplicity $e = e_0(R)$ and the embedding dimension $\nu := \mathrm{Dim}_{R/\mathfrak{m}}(\mathfrak{m}/\mathfrak{m}^2) = \ell(\mathfrak{m}/\mathfrak{m}^2) = \mu(\mathfrak{m}) :=$ the minimal number of generators of $\mathfrak{m}$. For any $\mathfrak{m}$-primary ideal $\mathfrak{a}$ in R, the numerical function $\mathrm{H}_\mathfrak{a} : \mathbb{N} \to \mathbb{N}$ defined by $n \mapsto \ell(R/\mathfrak{a}^{n+1})$ (= the length of the R-module $R/\mathfrak{a}^{n+1}$) is called the Hilbert function of $\mathfrak{a}$. It is well known that there is a polynomial $\mathrm{P}_\mathfrak{a}(X) \in \mathbb{Q}[X]$ of degree d such that $\mathrm{P}_\mathfrak{a}(n) = \mathrm{H}_\mathfrak{a}(n)$ for all sufficiently large values of n and the leading coefficient of $\mathrm{P}_\mathfrak{a}(X)$ is of the form $e_0(\mathfrak{a})/d!$ with $e_0(\mathfrak{a}) \in \mathbb{N}^+$. The uniquely determined polynomial $\mathrm{P}_\mathfrak{a}(X)$ and the positive natural number $e_0(\mathfrak{a})$ are called the Hilbert-Samuel polynomial and the multiplicity of $\mathfrak{a}$, respectively. The Hilbert function (respectively, Hilbert-Samuel polynomial, multiplicity) of $\mathfrak{m}$ is also called the Hilbert function (respectively, Hilbert-Samuel polynomial, multiplicity) of R. We simply put $\mathrm{H}_R = \mathrm{H}_\mathfrak{m}$, $\mathrm{P}_R = \mathrm{P}_\mathfrak{m}$, $e(R) = e_0(R) = e_0(\mathfrak{m})$.

The generating function of the numerical function $\mathrm{H}^0_R : n \mapsto \mathrm{Dim}_{R/\mathfrak{m}}(\mathfrak{m}^n/\mathfrak{m}^{n+1}) = \mathrm{H}_R(n) - \mathrm{H}(n-1)$ is the Poincaré series $\mathrm{P}_R(Z) := \sum_{n \in \mathbb{N}} \mathrm{H}^0_R(n) Z^n$. It is well known that there exists a polynomial $\mathrm{h}_R(Z) \in \mathbb{Z}[Z]$ such that $\mathrm{P}_R(Z) = \dfrac{\mathrm{h}_R(Z)}{(1-Z)^d}$. The polynomial $\mathrm{h}_R(Z) = \mathrm{h}_0 + \mathrm{h}_1 Z + \cdots + \mathrm{h}_s Z^s$ is called the *h*-polynomial of R; $\mathrm{h}_0 = 1$, $\mathrm{h}_1 = \mathrm{H}^0_R(1) - d = \nu - d$ is called the embedding codimension of R.

(2) (Blowing-up rings) Let us recall some definitions and results proved in [7] on blowing-up rings. These results hold more generally for semi-local 1-dimensional Cohen-Macaulay rings. Let R be a semi-local Cohen-Macaulay ring of dimension 1 and let $\mathfrak{m}$ be the (Jacobson) radical of R. Let $\overline{R}$ be the integral closure of R in its total quotient ring $\mathrm{Q}(R)$. An ideal $\mathfrak{a}$ in R is called open if it is open in the $\mathfrak{m}$-adic topology on R, or, equivalently $\mathfrak{m}^n \subseteq \mathfrak{a}$ for some $n \geq 1$, or, equivalently, the length $\ell(R/\mathfrak{a})$ is finite. For any two R-submodules M, N of $\overline{R}$, we put $(M :_R N) := \{y \in R \mid yN \subseteq M\}$.

For an open ideal $\mathfrak{a}$ in R, let $\mathrm{B}(\mathfrak{a}) := \cup_{n\in\mathbb{N}}(\mathfrak{a}^n : \mathfrak{a}^n)$. The ring $\mathrm{B}(\mathfrak{a})$ is called the blowing-up of R along $\mathfrak{a}$ or the first neighbourhood ring of $\mathfrak{a}$.

(2.a) Proposition. [7, Proposition 1.1] *For an open ideal $\mathfrak{a}$ in R, the ring $\mathrm{B}(\mathfrak{a})$ is a finitely generated R-module and $R \subseteq \mathrm{B}(\mathfrak{a}) \subseteq \overline{R}$. Moreover, if R is local and if $\mathfrak{a}$ is a $\mathfrak{m}$-primary ideal which is not principal, then $R \subsetneq \mathrm{B}(\mathfrak{a})$. In particular, if R is local and if R is not a discrete valuation ring, then $R \subsetneq \mathrm{B}(\mathfrak{m})$. Furthermore, there exists a non-zero divisor $x \in \mathfrak{a}$ such that $\mathrm{B}(\mathfrak{a}) = R[\frac{z_1}{x}, \ldots, \frac{z_r}{x}]$, where $z_1, \ldots, z_r$ is a generating set for the ideal $\mathfrak{a}$. In particular, $\mathfrak{a}\,\mathrm{B}(\mathfrak{a}) = x\,\mathrm{B}(\mathfrak{a})$.*

An open ideal $\mathfrak{a}$ in R is called stable in R if $\mathrm{B}(\mathfrak{a}) = (\mathfrak{a} : \mathfrak{a})$, or, equivalently, $\mathfrak{a}\,\mathrm{B}(\mathfrak{a}) = \mathfrak{a}$. It is clear that if $\mathfrak{a}$ is an open ideal in R, then $\mathfrak{a}^n$ is stable for some $n > 0$ and if $\mathfrak{a}^n$ is stable, then $\mathfrak{a}^m$ is stable for every $m \geq n$.

Since $\dim(R) = 1$, for any open ideal $\mathfrak{a}$ in R, for all sufficiently large values of n, we have $\mathrm{H}_{\mathfrak{a}}(n) = \ell(R/\mathfrak{a}^{n+1}) = e_0(\mathfrak{a})\binom{n+1}{1} - e_1(\mathfrak{a})$ with $e_0(\mathfrak{a}) \in \mathbb{N}^+$ and $e_1(\mathfrak{a}) \in \mathbb{Z}$ and $\mathrm{P}_{\mathfrak{a}}(X) := e_0(\mathfrak{a})\binom{X+1}{1} - e_1(\mathfrak{a})$, the positive natural number $e_0(\mathfrak{a})$ is the Hilbert- Samuel polynomial of $\mathfrak{a}$, respectively.

The next proposition shows how the integers $e_0(\mathfrak{a})$ and $e_1(\mathfrak{a})$ are connected to the blowing-up $\mathrm{B}(\mathfrak{a})$ of R along $\mathfrak{a}$:

(2.b) Proposition. [7, Theorem 1.5] $\ell(\mathrm{B}(\mathfrak{a})/\mathfrak{a}\,\mathrm{B}(\mathfrak{a})) = e_0(\mathfrak{a})$ *and* $\ell(\mathrm{B}(\mathfrak{a})/R) = e_1(\mathfrak{a}) \geq e_0(\mathfrak{a}) - \ell(R/\mathfrak{a})$. *Moreover,* $\ell(\mathrm{B}(\mathfrak{a})/\mathfrak{a}^n\,\mathrm{B}(\mathfrak{a})) = e_0(\mathfrak{a})n$ *for all* $n \in \mathbb{N}$.

If $\mathfrak{m}$ is not principal, then $\mathrm{B}(\mathfrak{m}) \neq R$ and $\mathrm{B}(\mathfrak{m})$ is a finitely generated R-module. The annihilator $\mathrm{ann}_R(\mathrm{B}(\mathfrak{m})/R) := \{y \in R \mid y\mathrm{B} \subseteq R\}$ of the R-module $\mathrm{B}(\mathfrak{m})/R$ is called the conductor of R in $\mathrm{B}(\mathfrak{m})$. The conductor of R in $\overline{R}$ is called the conductor of R. We put $\mathfrak{C} := \mathrm{ann}_R(\overline{R}/R)$.

(3) (Minimal reductions and reduction number) Let $(R, \mathfrak{m})$ be a one-dimensional local Cohen-Macaulay ring. For an $\mathfrak{m}$-primary ideal $\mathfrak{a}$ in R which is not a principal ideal, we put $\mathrm{r}(\mathfrak{a}) := \min\{n \in \mathbb{N} \mid \ell(\mathfrak{a}^n/\mathfrak{a}^{n+1}) = e_0(\mathfrak{a})\}$. The natural number $\mathrm{r}(\mathfrak{a})$ are called the reduction number of $\mathfrak{a}$ (see [11]). It is easy to see (see for example [11, Theorem 5.1]) the following equalities:

$$\begin{aligned} \mathrm{r}(\mathfrak{a}) &= \min\{n \in \mathbb{N} \mid \ell(R/\mathfrak{a}^m) = e_0(\mathfrak{a})m - e_1(\mathfrak{a}) \text{ for all } m \geq n\} \\ &= \min\{n \in \mathbb{N} \mid \mathrm{B}(\mathfrak{a}) = (\mathfrak{a}^n : \mathfrak{a}^n)\} \\ &= \min\{n \in \mathbb{N} \mid x\mathfrak{a}^n = \mathfrak{a}^{n+1} \text{ for some } x \in \mathfrak{a}\} \end{aligned}$$

Note that for the last equality we need to assume either R is reduced with infinite residue field, or R is analytically irreducible; in both these cases, for every non-zero ideal $\mathfrak{a}$ in R, there exists $x \in \mathfrak{a}$ such that xR is a minimal reduction of $\mathfrak{a}$, i.e., $x\mathfrak{a}^m = \mathfrak{a}^{m+1}$ for some $m \in \mathbb{N}$. In fact in the later case (see [2, Corollary 17]) if $x \in \mathfrak{a}$, then xR is a minimal reduction of $\mathfrak{a}$ if and only if $v(x) = \min(v(\mathfrak{a}))$, where v is the discrete valuation of the integral closure $\overline{R}$ of R in its quotient field.

(4) (Standard basis of a numerical semigroups) Let $\Gamma \subseteq \mathbb{N}$ be a numerical semigroup, i.e., Γ is closed under addition, $0 \in \Gamma$ and $\gcd(\Gamma) = 1$. Then there exists a least positive integer $c := c(\Gamma) \in \Gamma$ such that $z \in \Gamma$ for all $z \geq c$. This positive integer is called the conductor of Γ. Therefore $\Gamma \setminus \mathbb{N}_c = \{0 = \mathrm{v}_0, \mathrm{v}_1, \ldots, \mathrm{v}_{n-1}\}$, where $0 = \mathrm{v}_0 < \mathrm{v}_1 < \cdots < \mathrm{v}_{n-1} < \mathrm{v}_n := c$ are elements of Γ. The least positive integer $m := v_1 \in \Gamma$ is called the multiplicity of Γ.

The set $\mathrm{S}_m(\Gamma) := \{z \in \Gamma \mid z - m \notin \Gamma\}$ is called the *standard basis* or the *Apéry set of* Γ with respect to m. We put $\mathrm{S} := \mathrm{S}_m(\Gamma)$ and write $\mathrm{S} = \{0 = s_0, s_1, \ldots, s_{m-1}\}$ with $0 = s_0 < s_1 < \cdots < s_{m-1}$. Note that every element $h \in \Gamma$ can be written in the unique form $h = \rho m + s$ with $\rho \in \mathbb{N}$ and $s \in \mathrm{S}$. Further, note that $s_{m-1} = c - 1 + m$.

(5) (Monomial curves) Let $m_0, \ldots, m_{\nu-1}$ be positive integers and let K be a field. Then the curve $\mathcal{C}(m_0, \ldots, m_{\nu-1})$ in the ν-affine space $\mathbb{A}^\nu_K$ over K defined by the parametric equations $X_0 = T^{m_0}, \ldots, X_{\nu-1} = T^{m_{\nu-1}}$ is called an affine monomial curve over K defined by $m_0, \ldots, m_{\nu-1}$. We may assume that $\gcd(m_0, \ldots, m_{\nu-1}) = 1$. The numerical semigroup $\Gamma := \sum_{i=0}^{\nu-1} \mathbb{N} m_i$ generated by $m_0, \ldots, m_{\nu-1}$, is called the value semigroup of $\mathcal{C}(m_0, \ldots, m_{\nu-1})$. The coordinate ring of $\mathcal{C}(m_0, \ldots, m_{\nu-1})$ is the subalgebra $K[\Gamma] := K[T^z \mid z \in \Gamma] = K[T^{m_0}, \ldots, T^{m_{\nu-1}}]$ of the polynomial algebra $K[T]$ and is called the affine semigroup ring of Γ over K. The kernel $\mathfrak{P}(\mathcal{C}(m_0, \ldots, m_{\nu-1}))$ of the canonical surjective K-algebra homomorphism $\varphi : K[X_0, \ldots, X_{\nu-1}] \to K[T^{m_0}, \ldots, T^{m_{\nu-1}}]$ defined by $X_i \mapsto T^{m_i}$, $i = 0, \ldots, \nu-1$, is the defining ideal of $\mathcal{C}(m_0, \ldots, m_{\nu-1})$.

The K-subalgebra $K[\![\Gamma]\!] := K[\![T^z \mid z \in \Gamma]\!] = K[\![T^{m_0}, \ldots, T^{m_{\nu-1}}]\!]$ of the power series algebra $K[\![T]\!]$ over K is called the algebroid semigroup ring of Γ over K; it is the coordinate ring of the algebroid monomial curve $\mathcal{C}(\!(m_0, \ldots, m_{\nu-1})\!) := \operatorname{Spec}(K[\![T^{m_0}, \ldots, T^{m_{\nu-1}}]\!])$ in the algebroid ν-space $\mathbb{A}^\nu_K := \operatorname{Spec}(K[\![X_0, \ldots, X_{\nu-1}]\!])$ defined by the parametric equations $X_0 = T^{m_0}, \ldots, X_{\nu-1} = T^{m_{\nu-1}}$.

The kernel $\mathfrak{P}(\mathcal{C}(\!(m_0, \ldots, m_{\nu-1})\!))$ of the canonical surjective K-algebra homomorphism $\varphi : K[\![X_0, \ldots, X_{\nu-1}]\!] \to K[\![T^{m_0}, \ldots, T^{m_{\nu-1}}]\!]$ defined by $X_i \mapsto T^{m_i}$, $i = 0, \ldots, \nu - 1$, is the defining ideal of $\mathcal{C}(\!(m_0, \ldots, m_{\nu-1})\!)$.

(6) (Type sequences and almost Gorenstein rings) Let $(R, \mathfrak{m}_R)$ be a noetherian local one-dimensional analytically irreducible domain, i.e., the $\mathfrak{m}$-adic completion $\hat{R}$ of R is a domain or, equivalently, the integral closure $\overline{R}$ of R in its quotient field $\mathrm{Q}(R)$ is a discrete valuation ring and a finite R-module. We further assume that R is residually rational, i.e., R and $\overline{R}$ have the same residue field. A particular important class of rings which satisfy these assumptions are algebroid semigroup rings which are coordinate rings of algebroid monomial curves (see (5) above).

Let $v : \mathrm{Q}(R) \to \mathbb{Z} \cup \{\infty\}$ be the discrete valuation of $\overline{R}$ and let $\mathfrak{C} := \mathrm{ann}_R(\overline{R}/R) = \{x \in R \mid x\overline{R} \subseteq R\}$ be the conductor ideal of R in $\overline{R}$. Then the value semigroup $v(R) = \{v(x) \mid x \in R, x \neq 0\}$ is a numerical semigroup, that is, $\mathbb{N} \setminus v(R)$ is finite and therefore $v(R) = \{0 = \mathrm{v}_0, \mathrm{v}_1, \ldots, \mathrm{v}_{n-1}\} \cup \{z \in \mathbb{N} \mid z \geq c\}$, where $0 = \mathrm{v}_0 < \mathrm{v}_1 < \cdots < \mathrm{v}_{n-1} < \mathrm{v}_n := c$ are elements of $v(R)$, $n := n(R) := \ell(R/\mathfrak{C}) = \mathrm{card}\,(v(R) \setminus \mathbb{N}_c)$ and $c := c(R) := \ell(\overline{R}/\mathfrak{C})$ (here $\ell(-)$ denote the length as an R-module).

The positive integer c is also determined by the equality $\mathfrak{C} = \{x \in \mathrm{Q}(R) \mid v(x) \geq c\}$ or, equivalently $\mathfrak{C} = (\mathfrak{m}_{\overline{R}})^c$. Note that c be the conductor of the value semigroup $v(R)$. The positive integer $\delta := \delta(R) := \ell(\overline{R}/R) = \mathrm{card}\,(\mathbb{N} \setminus v(R))$ is called the degree of singularity of R. It is clear that $\delta(R)$ is the sum of n positive integers $\mathrm{t}_i(R) := \ell(\mathfrak{A}_i^{-1}/\mathfrak{A}_{i-1}^{-1})$, $i = 1, \ldots, n$, where $\mathfrak{A}_i := \{x \in R \mid v(x) \geq \mathrm{v}_i\}$ and $\mathfrak{A}_i^{-1} := (R : \mathfrak{A}_i) := \{x \in \mathrm{Q}(R) \mid x\mathfrak{A}_i \subseteq R\}$. The sequence $\mathrm{t}_1(R), \mathrm{t}_2(R), \ldots \mathrm{t}_n(R)$ is called the type sequence of R.

Several authors have studied the properties of type sequences (see [1], [4]). The term "type sequence" is chosen since the first term $\mathrm{t}_1(R) = \ell(\mathfrak{m}^{-1}/R)$ is the Cohen-Macaulay type of R. Further, we have:

(6.a) Proposition.

(1) $1 \leq \mathrm{t}_i(R) \leq \tau_R$ *for every* $i = 1, \ldots, n$ (see [8, §3, Proposition 2 and Proposition 3]) and hence (see also [4, Proposition 2.1])

(2) $\ell^*(R) \leq (\tau_R - 1)\,(\ell(R/\mathfrak{C}) - 1)$, *where* $\ell^*(R) := \tau_R \cdot \ell(R/\mathfrak{C}) - \ell(\overline{R}/R)$. *Moreover, the equality holds if and only if* $\ell(\overline{R}/R) = \tau_R + \ell(R/\mathfrak{C}) - 1$, *or equivalently* $\mathrm{t}_i(R) = 1$ *for* $i = 2, \ldots, n$.

A ring R as above is called almost Gorenstein if the type sequence of R is $\{\tau_R, 1, 1, \ldots, 1\}$, or equivalently, $\ell^*(R)$ attains its upper bound, i.e., $\ell(\overline{R}/R) = \tau_R - 1 + \ell(R/\mathfrak{C})$. It is clear that Gorenstein rings are almost Gorenstein but not conversely (see [18, (1.2)-(1)]).

(7) Type sequence of a numerical semi-group Γ can also be defined analogously: Let $c = c(\Gamma) \in \mathbb{N}$ be the conductor of Γ and let $\Gamma \setminus \mathbb{N}_c = \{0 = \mathrm{v}_0, \mathrm{v}_1, \ldots, \mathrm{v}_{n-1}\}$, where $0 = \mathrm{v}_0 < \mathrm{v}_1 < \cdots < \mathrm{v}_{n-1} < \mathrm{v}_n := c$ are elements of Γ. Further, for $i = 0, \ldots, n$, let $\Gamma_i := \{h \in \Gamma \mid h \geq \mathrm{v}_i\}$, $\Gamma(i) := \{x \in \mathbb{Z} \mid x + \Gamma_i \subseteq \Gamma\}$ and let $\mathrm{t}_i = \mathrm{card}\,(\Gamma(i) \setminus \Gamma(i-1))$. Then $\Gamma = \Gamma(0) \subseteq \Gamma(1) \subseteq \cdots \subseteq \Gamma(n-1) \subseteq \Gamma(n) = \mathbb{N}$ and the sequence t_i, $i = 1, \ldots, n$ is called the type sequence of Γ. In particular, the cardinality t_1 of the set $\mathrm{T}(\Gamma) := \Gamma(1) \setminus \Gamma$ is called the Cohen-Macaulay type of the semigroup Γ.

The type sequence of a ring R need not be same as the type sequence of the numerical semi-group $v(R)$ of R (see [4]). However, if $R = K[\![\Gamma]\!]$ is the algebroid semigroup ring of Γ over a field K, then the type sequence of R is equal to the type sequence of its semigroup $v(R) = \Gamma$.

2. Numerical invariants of certain monomial curves

We begin this section with explicit descriptions of the standard basis, degree of singularity, Cohen-Macaulay type, the defining ideal, h-polynomial, type sequence of the class of monomial curves defined by an arithmetic sequence.

2.1. (A class of monomial curves defined by an arithmetic sequence) In addition to the notation introduced in (1.2)–(5), we further fix the following notation.

Let $m, d \in \mathbb{N}$, $m \geq 2$, $d \geq 1$ be such that $\gcd(m,d) = 1$ and let p be an integer $p \geq 1$ and put $m_i := m + id$ for $i = 0, 1, \ldots, p+1$. Let $\Gamma := \sum_{i=0}^{p+1} \mathbb{N} m_i$ be the semigroup generated by the arithmetic sequence $m_0, m_1, \ldots, m_{p+1}$. Further, let $R := K[\![\Gamma]\!]$ be the algebroid semigroup ring of Γ over K, $\mathfrak{P} := \mathfrak{P}((\mathcal{C}))$ be the defining ideal of the algebroid monomial curve $\mathcal{C} := \mathcal{C}((\Gamma))$ over K and $\mathrm{G} := \mathrm{G}((\Gamma))$ be the associate graded ring of R with respect to its maximal ideal.

For any positive natural number $k \in \mathbb{N}^+$, let $q_k \in \mathbb{N}$ and $r_k \in [1, p+1]$ be the unique integers defined by the equation $k = q_k(p+1) + r_k$. We put $q := q_{m-1}$ and $r := r_{m-1} - 1$. Therefore $q \in \mathbb{N}$, $r \in [0,p]$ and $m - 2 = q(p+1) + r$.

Put $s_0 = 0$ and $s_k := m_{r_k} + q_k m_{p+1} = (1+q_k)m + (r_k + q_k(p+1))\, d$ for $k \in [1, m-1]$.

Further, we put $\mathrm{S}_1 := \{m_i + j m_{p+1} \mid i \in [1, p+1] \text{ and } j \in [0, q-1]\}$ (note that $\mathrm{S}_1 = \emptyset$ if $q = 0$) and $\mathrm{S}_2 := \{m_i + q m_{p+1} \mid i \in [1, r+1]\}$.

With the notations we have:

(1) *The standard basis* $\mathrm{S} := \mathrm{S}_m(\Gamma)$ *with respect to the multiplicity* $m = m_0$ *of* Γ *is :*

$$\mathrm{S} = \{s_k \mid k \in [0, m-1]\} = \{0\} \cup \mathrm{S}_1 \cup \mathrm{S}_2.$$

Further, $c := c(\Gamma) = (m-1)(d+q) + q + 1$ *is the conductor of* Γ.

(2) *The degree of singularity* $\delta := \delta(\Gamma)$ *of* Γ *is :*

$$\delta = ((m-1)(d+q) + (r+1)(q+1))\,/2.$$

(3) *The set* $\mathrm{T} := T(\Gamma) = \Gamma(1) \setminus \Gamma = \{m_i + q m_{p+1} - m_0 \mid i \in [1, r+1]\}$. *In particular, the Cohen-Macaulay type of* Γ *is* $\tau := \tau_\Gamma = r+1$. *Furthermore,* R *is Gorenstein if and only if* $r = 1$.

(4) *The defining ideal* $\mathfrak{P}$ *of the algebroid monomial curve* $\mathcal{C}$ *is generated by* $\binom{\nu}{2} = \binom{p+2}{2}$ *elements. Moreover,* $\mathfrak{P}$ *is a set-theoretic complete intersection, that is, there exists* $F_1, \ldots, F_{\nu-1} \in \mathrm{P} := K[X_0, \ldots, X_{\nu-1}]$ *such that*

$$\mathfrak{P} = \sqrt{\mathrm{P}F_1 + \cdots + \mathrm{P}F_{\nu-1}}\,.$$

(5) *The associated graded ring* G *of* R *is always Cohen-Macaulay and the* h*-polynomial of* R *is*

$$\mathrm{h}_R := \begin{cases} 1+\sum_{i=1}^{q}(p+1)Z^i+(r+1)Z^{q+1}, & \text{if } \ r\le p-2\,,\\ 1+\sum_{i=1}^{q+1}(p+1)Z^i+pZ^{q+2}, & \text{if } \ r=p-1\,,\\ 1+\sum_{i=1}^{q+1}(p+1)Z^i, & \text{if } \ r=p\,. \end{cases}$$

In particular, the Hilbert function of R *is non-decreasing.*

(6) *The* i*th term* $\mathrm{t}_i=\mathrm{t}_i(\Gamma)$ *of the type sequence* $(\mathrm{t}_1,\mathrm{t}_2,\dots,\mathrm{t}_n)$ *of* Γ *is*

$$t_i = \begin{cases} 1, & \text{if } \ \mathrm{v}_{i-1}\ne jm_{p+1} \ \text{ for every } j\in[0,q],\\ r+1, & \text{if } \ \mathrm{v}_{i-1}= jm_{p+1} \ \text{ for some } j\in[0,q]. \end{cases}$$

In particular, if $d=1$*, then the* i*-the term* t_i *of the type sequence* $(\mathrm{t}_1,\,\mathrm{t}_2,\,\dots,\,\mathrm{t}_n)$ *of* Γ *is*

$$t_i = \begin{cases} r+1, & \text{if } \ i=\binom{j+1}{2}(p+1)+j+1 \ \text{ for some } j\in[0,q],\\ 1, & \text{otherwise.} \end{cases}$$

Furthermore, R *is almost Gorenstein if and only if either* R *is Gorenstein, or* $m=p+2$*. Moreover, in this case we have* $\tau_R=m-1$*.*

Proof. (1), (3) and (4) are special cases of the general results proved in [17, (3.5)], [16, § 5] and [14] (see also [13]). (2) is proved in [22, § 3, Supplement 6]. (5) is proved in [9, Theorem 3.8]. (6) is proved in [20, Theorem 3.8 and Corollaries 3.8, 3.9]. □

Remark 2.2. (A class of monomial curves defined by an almost arithmetic sequence) Suppose that $\Gamma:=\sum_{i=0}^{p}\mathbb{N}m_i+\mathbb{N}n$ be the semigroup generated by an almost arithmetic sequence $m_0,m_1,\dots,m_p,n$ with $\gcd(m_0,m_1,\dots,m_p,n)=\gcd(m,d,n)=1$, i.e., $m_i=m_0+id$, $i=0,1,\dots,p$ is an arbitrary arithmetic sequence with $m=m_0\ge 2, d\ge 1, p\ge 1$ and n is an arbitrary positive integer. Results analogous to (1), (3) and (4) are proved in [17], [16] and [14]. Moreover, an explicit formula for the minimal number $\mu(\mathfrak{P})$ is given in [15]. The characterization for the Cohen-Macaulayness of G is given (in most cases) in [10] and the explicit computation of the h-polynomial is done in [18]. So far no explicit formulas (in terms of the generating set of Γ) for the degree singularity and the terms in the type sequence are known even if Γ is generated by an almost arithmetic sequence.

Remark 2.3. A well-known question: *whether or not monomial curves in the affine* ν*-space* $\mathbb{A}_K^{\nu}$ *defined by an arbitrary sequence of positive integers* $m_0,\dots,m_{\nu-1}$ *are set-theoretic complete intersections* is still open in general even in the case of embedding dimension $\nu=4$. For an arbitrary numerical semigroup Γ the characterization for the Cohen-Macaulayness of G and the behaviour of the Hilbert function H_R are not known in general.

3. Blowing-up rings and Arf rings

In this section let us recall the definition of an *Arf ring* studied by Lipman in [7].

Let R be a semi-local Cohen-Macaulay ring of dimension 1 and let $\mathfrak{m}$ be the (Jacobson) radical of R. Let $\overline{R}$ be the integral closure of R in its total quotient ring $\mathrm{Q}(R)$.

An element $z \in R$ is said to be integral over the ideal $\mathfrak{a}$ in R if z satisfies an integral equation $z^n + a_1 z^{n-1} + \cdots + a_n = 0$ with $a_j \in \mathfrak{a}^j$ for all $j = 1, \ldots, n$. The set $\overline{\mathfrak{a}}$ of all elements in R which are integral over $\mathfrak{a}$ is an ideal in R and is called the integral closure of $\mathfrak{a}$ in R. An ideal $\mathfrak{a}$ in R is said to be integrally closed in R if $\mathfrak{a} = \overline{\mathfrak{a}}$. A semi-local Cohen-Macaulay ring R of dimension 1 is called an Arf ring if every integrally closed open ideal in R is stable, or, equivalently (see [7, Theorem 2.2]), if A is any local ring infinitely near to R, then A has maximal embedding dimension, i.e., $\mathrm{embdim}(A) = e_0(A)$. In particular, if a local ring R is Arf, then R has maximal embedding dimension.

The next proposition gives necessary and sufficient conditions for the equality (see (1.2)-(2)-(2.b)) $e_0(\mathfrak{a}) - e_1(\mathfrak{a}) = \ell(R/\mathfrak{a})$.

Proposition 3.1. *Let $(R, \mathfrak{m})$ be a one-dimensional local Cohen-Macaulay ring and let $\mathfrak{a}$ be a $\mathfrak{m}$-primary ideal. Then the following statements are equivalent:*

(i) $e_0(\mathfrak{a}) - e_1(\mathfrak{a}) = \ell(R/\mathfrak{a})$.
(ii) $\mathrm{B}(\mathfrak{a}) = (\mathfrak{a} : \mathfrak{a})$, *i.e,* $\mathfrak{a}$ *is stable.*
(iii) *There exists* $z \in \mathfrak{a}$ *such that* $z\mathfrak{a} = \mathfrak{a}^2$.
(iv) $\ell(\mathfrak{a}/\mathfrak{a}^2) = e_0(\mathfrak{a})$.
(v) $\mathrm{r}(\mathfrak{a}) \leq 1$, *i.e.,* $\ell(\mathfrak{a}^n/\mathfrak{a}^{n+1}) = e_0(\mathfrak{a})$ *for all* $n \geq 1$.
(vi) $\mathrm{H}_{\mathfrak{a}}(n) = \mathrm{P}_{\mathfrak{a}}(n)$ *for all* $n \in \mathbb{N}$.

In particular, the maximal ideal $\mathfrak{m}$ *is stable* $\iff$ $\mathrm{embdim}(R) = e_0(R)$ $\iff$ $e_0(R) - e_1(R) = 1$.

Proof. Most of these equivalences are proved in [11, Theorem 5.1]. □

The next proposition provides a link between *stability of* $\mathfrak{m}$, *conductor of* $\mathrm{B}(\mathfrak{m})$ *over* R and the *type* τ_R *of* R.

Proposition 3.2. *Let $(R, \mathfrak{m})$ be a one-dimensional Cohen-Macaulay local ring. The following statements are equivalent:*

(i) $\ell(\mathfrak{m}/\mathfrak{m}^2) = e_0(R)$, *i.e.,* $\mathrm{embdim}(R) = e_0(R)$.
(ii) $\mathrm{ann}_R(\mathrm{B}(\mathfrak{m})/R) = (R : \mathrm{B}(\mathfrak{m})) = \mathfrak{m}$.
(iii) $\tau_R = e_0(R) - 1$.

Moreover, in this case $\tau_R = e_1(R)$.

Proof. The equivalence of (i) and (ii) is proved in [12, Corollary 2].
(i) $\iff$ (iii): In view of the equivalence of (i), (iii) and (iv) in (3.3) (for $\mathfrak{a} = \mathfrak{m}$), it is enough to prove that: $\tau_R = e_0(R) - 1 \iff x\mathfrak{m} = \mathfrak{m}^2$ for some $x \in \mathfrak{m}$. Let $x \in \mathfrak{m}$ be a minimal reduction of $\mathfrak{m}$. Then, since R is Cohen-Macaulay,

$\ell(R/xR) = e_0(R)$ and from $xR \subseteq \cdots \subseteq (xR : \mathfrak{m}) \subseteq \cdots \subseteq \mathfrak{m} \subsetneq R$ we have $\tau_R = \ell((R:\mathfrak{m})/R) = \ell((xR:\mathfrak{m})/xR) \leq \ell(R/xR) - 1 = e_0(R) - 1$. Moreover, the equality $\tau_R = e_0(R)-1 \iff \ell((xR:\mathfrak{m})/xR) = \ell(R/xR)-1 \iff \ell(R/(xR:\mathfrak{m})) = 1 = \ell(R/\mathfrak{m}) \iff (xR:\mathfrak{m}) = \mathfrak{m} \iff x\mathfrak{m} = \mathfrak{m}^2$. □

(3.3) Branch and multiplicity sequences. Let $(R, \mathfrak{m})$ be a semilocal Cohen-Macaulay ring of dimension 1. Then, since the integral closure $\overline{R}$ of R in the quotient field of R is a finite R-module, there exists a finite sequence

$$R = R_0 \subsetneq R_1 \subsetneq \cdots \subsetneq R_{m-1} \subsetneq R_m = \overline{R} \tag{3.3.1}$$

of one-dimensional noetherian semilocal rings such that for each $i = 1, \ldots, m$, the ring R_i is obtained from the ring R_{i-1} by blowing up the radical $\mathfrak{m}_{i-1}$ of R_{i-1}. Furthermore, for each maximal ideal $\mathfrak{n}$ of $\overline{R}$, every local ring $R_i' := (R_i)_{\mathfrak{n}\cap R_i}$ is called *infinitely near to* R. For each $i = 1, \ldots, m$, the multiplicity and the residue field of the local ring R_i' are denoted by $e(R_i')$ and k_i respectively. The sequence $R_0', R_1', \ldots, R_m'$ is called the *branch sequence of* R *along* $\mathfrak{n}$ and the sequence of pairs $((e_0(R_i'), [k_i : k_0])$, $i = 0, \ldots, m$ is called the *multiplicity sequence* of R, where for each $0 \leq i \leq m$, k_i denotes the residue field of R_i' and $[k_i : k_0]$ denotes the degree of the field extension $k_i|k_0$ (see [7, pages 661, 669]. In particular, if R is analytically irreducible, residually rational and $R \neq \overline{R}$, then each R_i in (3.3.1) is also analytically irreducible, residually rational; if $\mathfrak{m}_i$ is the maximal ideal of R_i, then the ring R_i is obtained from R_{i-1} by blowing up $\mathfrak{m}_{i-1}$. Further, $R_i = R_i'$ for each $i = 0, \ldots, m$, since $\overline{R}$ is local (see [6, Theorem 4]) and $\mathfrak{n}$ is the only maximal ideal in $\overline{R}$.

In the next result we give a formula for the degree of singularity in terms of the branch sequence:

Proposition 3.4. *Let $(R, \mathfrak{m})$ be a one-dimensional noetherian local analytically irreducible, residually rational domain and let $R = R_0 \subsetneq R_1 \subsetneq \cdots \subsetneq R_{m-1} \subsetneq R_m = \overline{R}$ be the branch sequence of R. For each $i = 0, \ldots, m$, let $\Gamma_i, \delta_i, e_0(R_i)$ and $e_1(R_i)$ be the value-semigroup, degree of singularity, multiplicity and the first coefficient of the Hilbert-Samuel polynomial of R_i respectively. Then:*

$$\delta(R) = \delta_0 = \sum_{i=1}^{m} \text{card}\,(\Gamma_i \setminus \Gamma_{i-1}) = \sum_{i=1}^{m} \ell(R_i/R_{i-1}) = \sum_{i=1}^{m} e_1(R_{i-1}).$$

Proof. Note that $\Gamma_0 \subsetneq \Gamma_1 \subsetneq \cdots \subsetneq \Gamma_{m-1} \subsetneq \Gamma_m = \mathbb{N}$ and $\delta_i = \text{card}\,(\mathbb{N} \setminus \Gamma_i)$ for each $i = 0, \ldots, m$. Therefore $\delta_0 = \sum_{i=1}^{m} \text{card}\,(\Gamma_i \setminus \Gamma_{i-1}) = \sum_{i=1}^{m} \ell(R_i/R_{i-1}) = \sum_{i=1}^{m} e_1(R_{i-1})$ by (1.2)-(2)-(2.b). □

Now recall the following characterization of complete local Arf rings in terms of its value semigroup given in [7] which will be used in the proof of the Theorem (3.6).

Proposition 3.5. ([7, Theorem 2.2 and Corollary 3.8]) *Let $(R, \mathfrak{m})$ be a one-dimensional noetherian local analytically irreducible ring and let $R = R_0 \subsetneq R_1 \subsetneq$*

$\cdots \subsetneq R_{m-1} \subsetneq R_m = \overline{R}$ *be the branch sequence of* R. *Then* R *is an Arf ring if and only if* $\operatorname{embdim}(R_i) = e_0(R_i)$ *for each* $i = 0, \ldots, m$. *Moreover, if* R *is complete with algebraically residue field* k, *then* R *is an Arf ring if and only if the value semi-group* $v(R)$ *of* R *is* $\{0, e_0(R_0), e_0(R_0)+e_0(R_1), \ldots, e_0(R_0)+\cdots+ e_0(R_{m-2})\} \cup \mathbb{N}_c$, *where* $c = e_0(R_0) + \cdots + e_0(R_{m-2}) + e_0(R_{m-1})$.

Now we can give the characterization for completer local Arf domains with algebraically closed residue field using the type sequences of R and the terms in its branch sequence.

Theorem 3.6. *Let* $(R, \mathfrak{m})$ *be a complete local domain with algebraically closed residue field* k. *Let* $R = R_0 \subsetneq R_1 \subsetneq \cdots \subsetneq R_{m-1} \subsetneq R_m = \overline{R}$ *be the branch sequence of* R. *For each* $j = 0, \ldots, m-1$, *let* $\mathfrak{C}_j$ *be the conductor of* $\overline{R}$ *over* R_j, *and let* $n_j = n(R_j)$, $c_j = \ell(\overline{R}/\mathfrak{C}_j)$ *and* $\mathrm{t}_i(R_j)$ *be the ith term in the type sequence of* R_j. *Then:* R *is an Arf ring if and only if for each* $j = 0, \ldots, m-1$ *and* $i = 1, \ldots, n_j$, *we have* $n_j = m - j$ *and* $\mathrm{t}_i(R_j) = e_0(R_{j+i-1}) - 1 = \mathrm{t}_{i+1}(R_{j-1})$.

For the proof of this theorem we use of the following result proved in [3] (see also [4]) which shows how the property Arf is described by its type sequence.

Proposition 3.7. [3, Theorem 1.7-(5)] *Let* $(R, \mathfrak{m})$ *be a one-dimensional noetherian local analytically irreducible, residually rational domain. Let* v *be the discrete valuation of* $\overline{R}$ *and let* $v(R) = \{0 = \mathrm{v}_0, \mathrm{v}_1, \ldots, \mathrm{v}_{n-1}\} \cup \mathbb{N}_c$ *be the value semi-group of* R, *where* $0 = \mathrm{v}_0 < \mathrm{v}_1 < \cdots < \mathrm{v}_{n-1} < \mathrm{v}_n = c$, $\mathfrak{C}$ *is the conductor of* $\overline{R}$ *over* R, $n := n(R) = \ell(R/\mathfrak{C})$ *and* $c = c(R) := \ell(\overline{R}/\mathfrak{C})$. *If* R *is an Arf ring, then* $\mathrm{t}_i = \mathrm{v}_i - \mathrm{v}_{i-1} - 1$ *is the ith term in the type sequence of* R.

Proof of (3.6) : ($\Rightarrow$): By the assumptions on R and (3.5), for each $j = 0, \ldots, m-1$ we have R_j is an Arf complete domain with integral closure $\overline{R}$, the same residue field k, $R_j \subsetneq R_{j+1} \subsetneq \cdots \subsetneq R_{m-1} \subsetneq R_m = \overline{R}$ is the branch sequence of R_j and the value semigroup $v(R_j)$ is $\{0, \mathrm{v}_{1,j}, \mathrm{v}_{2,j}, \ldots, \mathrm{v}_{m-j-1,j}\} \cup \mathbb{N}_{c_j}$, where $\mathrm{v}_{i,j} = e_0(R_j) + \cdots e_0(R_{j+i-1})$, $i = 1, \ldots, m-j-1$ and $c_j = e_0(R_j) + \cdots + e_0(R_{m-1})$. Therefore we have $n_j = n(R_j) = (m-j-1) + 1 = m - j$. Further, for each $j = 0, \ldots, m-1$, if $\{\mathrm{t}_i(R_j) \mid 1 \le i \le m-j\}$ is the type sequence of R_j, then by (3.7) we have $\mathrm{t}_i(R_j) = \mathrm{v}_{i,j} - \mathrm{v}_{i-1,j} - 1 = e_0(R_{j+i-1}) - 1 = \mathrm{v}_{i+1,j-1} - \mathrm{v}_{i,j-1} - 1 = \mathrm{t}_{i+1}(R_{j-1})$ for every $1 \le i \le m - j$.
($\Leftarrow$): For each $j = 0, \ldots, m-1$, by assumption, we have $\tau_{R_j} = \mathrm{t}_1(R_j) = e_0(R_j) - 1$. Therefore $\operatorname{emdim}(R_j) = e_0(R_j)$ by (3.1) and (3.2) and hence R is an Arf ring by the first part of (3.6). □

In particular, for the ready reference we note the following formulas for the ith term t_i in the type sequence of R, in terms of the types, the multiplicities and the lengths arising from the terms of the branch sequence of R.

Corollary 3.8. *Let* $(R, \mathfrak{m})$ *be an Arf complete local domain with algebraically closed residue field* k *and let* $R = R_0 \subsetneq R_1 \subsetneq \cdots \subsetneq R_{m-1} \subsetneq R_m = \overline{R}$ *be the branch sequence of* R. *Then :* $m = n = n(R)$ *and for each* $i = 1, \ldots, n$, *the ith term* t_i *in the type sequence of* R *is given by :* $\mathrm{t}_i = \tau(R_{i-1}) = e_0(R_{i-1}) - 1 = \ell(R_i/R_{i-1})$.

Corollary 3.9. *Let* $(R, \mathfrak{m})$ *be an Arf complete local domain with algebraically closed residue field* k *and let* $\mathrm{B} = \mathrm{B}(\mathfrak{m})$ *be the blowing up of* R *along* $\mathfrak{m}$. *If* $\mathrm{t}_1, \ldots, \mathrm{t}_n$ *is the type sequence of* R, *then* $\mathrm{t}_2, \ldots, \mathrm{t}_n$ *is the type sequence of* B.

4. Examples

In this section we give some examples of Arf rings and some of not Arf rings. In the following examples R denote the algebroid semi-group ring $K[\![\Gamma]\!]$ of the numerical semi-group Γ over a field K. Note that in this case each term R_j in the branch sequence of R is also semigroup ring; in fact, if Γ is generated by $n_1, n_2, \ldots, n_p$ with $n_1 < n_2 < \cdots < n_p$, then $R_1 = K[\![\Gamma_1]\!]$, where $\Gamma_1 = v(R_1)$ is generated by $n_1, n_2 - n_1, \ldots, n_p - n_1$; by repeating this argument we get the result for R_j, $j \geq 2$.

Example 1. (see [19, Lemma 3.11-(1)] for details) Let $e, p \in \mathbb{N}$ with $e \geq 3, p \geq 1$ and Γ be the semigroup generated by $e, pe+1, pe+2, \ldots, pe+(e-1)$. Then there are exactly $p+1$ terms in the branch sequence of R and $\operatorname{embdim}(R_j) = e = e_0(R_j)$ for every $j = 0, \ldots, p$. Therefore R is an Arf ring by (3.5).

Example 2. (see [19, Lemma 3.11-(2)] for details) Let $e, p, a \in \mathbb{N}$ with $e \geq 3$, $p \geq 2$, $1 \leq a \leq e-1$ and Γ be the semigroup generated by $e, pe-a, pe-a+1, \ldots, pe-a+(a-1)$. Then there are exactly $p+1$ terms in the branch sequence of R and $\operatorname{embdim}(R_j) = e = e_0(R_j)$ for every $j = 0, \ldots, p-2$ and $\operatorname{embdim}(R_{p-1}) = e - a = e_0(R_{p-1})$. Therefore R is an Arf ring by (3.5).

Example 3. (see [19, Example 4.1] for details) Let $e, r, r' \in \mathbb{N}$ with $e \geq 3, 1 \leq r$, $1 \leq r'$, $r + r' \leq e - 1$ and let Γ be the semi-group generated by the sequence $e, e+r, e+r+r', e+r+r'+1, \ldots, 2e+r+r'-1$. The type sequence of R is

$$\begin{cases} e-1, & \text{if } r' = r = 1, \\ e-1, r-1, & \text{if } r' = 1 \text{ and } r \geq 2, \\ e-1, r'-1, r-1, & \text{if } r' = r, \\ e-1, r-1, r'-1, & \text{if } r' < r, \\ e-2, r, r'-1, & \text{if } r < r'. \end{cases}$$

In particular, R is almost Gorenstein if and only if $(r', r) \in \{(1,2), (2,2), (2,1)\}$. Further, R is an Arf ring in cases (i), (ii), (iii) and R is not Arf in the case (iv).

Example 4. Let $m, d, p \in \mathbb{N}$, $m \geq 2$, $p \geq 1, d \geq 1$, $\gcd(m, d) = 1$, Γ be the semigroup generated by an arithmetic sequence $m, m+d, \ldots, m+pd$ and let $R = K[\![\Gamma]\!]$. Let B be the blowing-up of R along the maximal ideal of R. Then $\mathrm{B} = K[\![\Gamma']\!]$, where Γ' is the semigroup generated by m, d, and so $\operatorname{embdim}(\mathrm{B}) = 2$. Further, by (3.5):

(i) If $d = 1$, then R is Arf if and only if $\operatorname{embdim}(R) = m$ (in fact, in this case, $\mathrm{B} = K[\![T]\!]$).

(ii) If $d = 2$ or $m = 2$, then for every $j \geq 2$ the jth term in the branch sequence of R is $R_j = K[\![\Gamma_j]\!]$, where Γ_j is the semigroup generated by $2, 2n+1$ for some integer $n \geq 1$ and so $\mathrm{embdim}(R_j) = e_0(R_j)$ for every $j \geq 1$. Therefore, R is an Arf ring if and only if $\mathrm{embdim}(R) = m$; in particular, if $m = 2$, then R is an Arf ring.

(iii) If $d \geq 3$ and $m \geq 3$, then $e_0(\mathrm{B}) \geq 3$, $\mathrm{embdim}(\mathrm{B}) = 2 < 3 \leq \min\{m, d\} = e_0(\mathrm{B})$ and hence R is not an Arf ring.

Acknowledgment

This paper was presented at the "International Conference on Algebra and its Applications" in the honour of Professor S.K. Jain on the occasion of his 70th birthday held at the Center of Ring theory and its Applications, Department of Mathematics, Ohio University, USA during June 18–21, 2008. Participation in this conference was financially supported by IISc from GARP-PTF-International Conferences fund.

References

[1] V. Barucci, D.E. Dobbs and M. Fontana, *Maximality properties in numerical semigroups and applications to one-dimensional analytically irreducible local domains*, Memoirs Amer. Math. Soc. **125** (1994), no. 598.

[2] V. Barucci and R. Fröberg, *One Dimensional Almost Gorenstein Rings*, Journal of Algebra **188** (1997), 418–442.

[3] M. D'Anna, *Canonical Module and One Dimensional Analytically Irreducible Arf Domains*, Lecture Notes in Pure and Applied Math., Marcel Dekker **185** (1997).

[4] M. D'Anna and D. Delfino, *Integrally closed ideals and type sequences in one dimensional local rings*, Rocky Mountain Jour. of Math. **27** (1997), no. 4, 1065–1073.

[5] R. Fröberg, C. Gottlieb and R. Häggkvist, *On numerical semigroups*, Semigroup Forum **35** (1987), no. 1, 63–83.

[6] D. Katz, *On the number of minimal primes in the completion of a local domain*, Rocky Mountain Jour. Math. **16** (1986), 575–578.

[7] J. Lipman, *Stable ideals and Arf rings*, Amer. Jour. Math. **93** (1993), 649–685.

[8] T. Matsuoka, *On the degree of singularity of one-dimensional analytically irreducible noetherian rings*, Journal Math. Kyoto Univ. **11** (1971), no. 3, 485–494.

[9] S. Molinelli and G. Tamone, *On the Hilbert function of certain rings of monomial curves*, J. Pure and Applied Algebra **101**, no. 2, (1995), 191–206.

[10] S. Molinelli, D.P. Patil and G.Tamone, *On the Cohen-Macaulayness of the associated graded ring of certain monomial curves*, Beiträge zur Algebra und Geometrie – Contributions to Algebra and Geometry **39**, no. 2 (1998), 433–446.

[11] A. Ooishi, *Genera and arithmetic genera of commutative rings*, Hiroshima Math. J. **17** (1987), 47–77.

[12] A. Ooishi, *On the conductor of the blowing-up of a one-dimensional Gorenstein local ring*, Journal of Pure and Applied Algebra **76** (1991), 111–117.

[13] D.P. Patil, *Generators for the module of derivations and the relation ideals of certain curves*, Ph.D. Thesis, Tata Institute of Fundamental Research, Bombay, 1989.

[14] D.P. Patil, *Certain monomial curves are set-theoretic complete intersections*, Manuscripta Math **68** (1990), no. 4, 399–404.

[15] D.P. Patil, *Minimal set of generators for the defining ideals of certain monomial curves*, Manuscripta Math **80** (1993), no. 3, 239–248.

[16] D.P. Patil and I. Sengupta, *Minimal set of generators for the derivation module of certain monomial curves*, Communications in Algebra **27** (1999), no. 11, 5619–5631.

[17] D.P. Patil and B. Singh, *Generators for the module of derivations and the relation ideals of certain curves*, Manuscripta Math **68** (1990), no. 3, 327–335.

[18] D.P. Patil and G. Tamone, *On the* h*-polynomial of certain monomial curves*, Rocky Mountain J. Math. **34**, no. 1 (2004), 289–307.

[19] D.P. Patil and G. Tamone, *On the type sequences and Arf rings*, Annales Academiae Paedagogicae Cracoviensis, Studia Mathematica, VI **45** (2007), 35–50.

[20] D.P. Patil and G. Tamone, *On the type sequences of some one-dimensional rings*, Universitatis Iagellonicae Acta Mathematica, Fasciculus XLV (2007), 119–130.

[21] J.C. Rosales and M.B. Branco, *Numerical semigroups that can be expressed as an intersection of symmetric numerical semigroups*, Journal of Pure and Applied Algebra **171** (2002), no. 2-3, 303–314.

[22] G. Scheja and U. Storch, *Regular Sequences and Resultants*, Research Notes in Mathematics **8** A.K. Peters, Natick, Massachusetts 2001.

D.P. Patil
Department of Mathematics
Indian Institute of Science
Bangalore 560 012, India
e-mail: patil@math.iisc.ernet.in

Advances in Ring Theory
Trends in Mathematics, 283–302

A Guide to Supertropical Algebra

Zur Izhakian and Louis Rowen

Abstract. This paper describes a new algebraic structure to enrich the algebraic theory underlying "tropical geometry," an area of mathematics that has developed considerably over the last ten years, with applications to combinatorics, polynomials (Newton's polytope), linear algebra, and algebraic geometry.

Mathematics Subject Classification (2000). 13B22, 11C, 11S, 16D25, 11S, 12D, 15A09, 15A03, 15A15, 65F15.

Keywords. Semiring theory, supertropical structures, polynomial algebra, matrix algebra, determinant, characteristic polynomial, eigenvector, eigenvalue, Hamilton-Cayley theorem, Vandermonde matrix, resultant.

1. Introduction

Our object in this paper is to describe a new algebraic structure to enrich the algebraic theory underlying "tropical geometry," an area of mathematics that has developed considerably over the last ten years, [1, 8, 24, 25, 27, 28, 29, 30, 33], with applications to combinatorics, polynomials (Newton's polytope), linear algebra, and algebraic geometry; cf. [12] and [26]. A survey of tropical geometry and its applications can be found in [23], but we review some of the basics here, for the convenience of the reader not versed in tropical geometry.

1.1. Brief overview of tropical geometry

Given a complex variety $W = \{(z_1, \dots, z_n) : z_i \in \mathbb{C}\} \subset \mathbb{C}^{(n)}$, and any small $t > 0$, one can define its **amoeba**

$$\mathcal{A}(W) = \{(\log_t |z_1|, \dots, \log_t |z_n|) : (z_1, \dots, z_n) \in W\} \subset \mathbb{R}_0^{(n)},$$

The first author has been supported by the Chateaubriand scientific post-doctorate fellowships, Ministry of Science, French Government, 2007–2008.
The second author has been supported in part by the Israel Science Foundation, grant 1178/06.

where $\mathbb{R}_0 = \mathbb{R} \cup \{-\infty\}$. Note that $\log_t |z_1 z_2| = \log_t |z_1| + \log_t |z_2|$. Furthermore, we recall for any s, t that that

$$\log_t z = \log_s z \log_t s,$$

so changing t merely rescales the amoeba, which becomes more compressed as t decreases. As explained in [8, §1.1], the limiting case $t \to 0$ degenerates to a polyhedral complex in $\mathbb{R}_0^{(n)}$, where now $\mathbb{R}_0$ is given the structure of the **max-plus algebra** $\mathcal{M} = (\mathbb{R}_0, \oplus, \odot)$, for which the new addition $\oplus$ is defined as the maximum, and multiplication $\odot$ is taken to be the original addition in $\mathbb{R}$.

Any polynomial f over $\mathcal{M}$ in the commuting indeterminates $\lambda_1, \dots, \lambda_n$ defines a graph in $\mathcal{M}^{(n+1)}$, whose points are

$$(a_1, \dots, a_n, f(a_1, \dots, a_n)), \qquad a_i \in \mathcal{M}.$$

But because addition is taken to be the maximum, the polynomial defines a piecewise linear function $\mathcal{M}^{(n)} \to \mathcal{M}$, whose graph is a collection of n-dimensional planar sections. For example, the graph of a polynomial in one indeterminate is a sequence of line segments and two rays; the graph of a polynomial in two indeterminates consists of "slices" of planes.

An affine **tropical hypersurface** is defined as the domain of non-differentiability, also called also the **corner locus**, of the piecewise linear function determined by a polynomial. (Tropical varieties are defined more generally as weighted rational polyhedral complexes satisfying a certain "balancing condition.")

Example 1.1. Graph of the tropical polynomial $0 \odot \lambda^2 \ \oplus\ 3 \odot \lambda \ \oplus\ 2$:

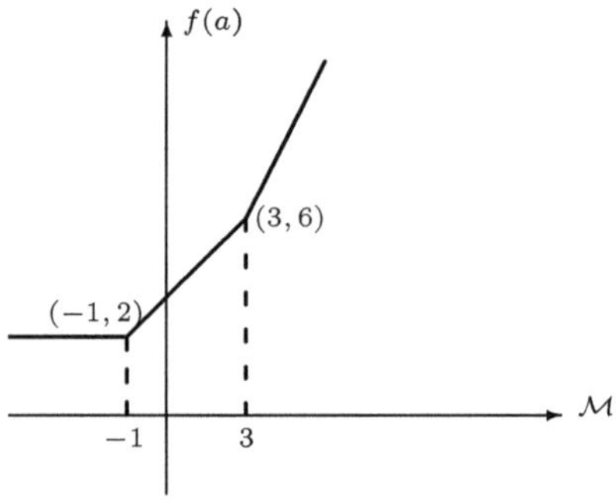

Here the tropical hypersurface is comprised merely of the two points $\{-1, 3\}$ in the real line, the first coordinates of the respective vertices $\{(-1, 2), (3, 6)\}$ where the graph is not differentiable.

The tropical structure can also be described as the target of a certain field with non-Archimedean valuation, as explained in [8, §1.2]. Passing from the original algebraic variety to this "tropical variety" preserves various topological and geometric invariants involving intersections. From this perspective, "curves" are much easier to study in the tropical framework than in customary algebraic geometry, and tropical geometry has been used to simplify proofs of deep results from algebraic geometry [4].

1.2. The semiring structure of the max-plus algebra

The two operations $\oplus$ and $\odot$ endow the max-plus algebra $\mathcal{M}$ with the structure of an (associative) semiring, in which $a \oplus a = a$ for every element $a \in \mathcal{M}$. In the sequel, for elements $a, b \in \mathcal{M}$, we write $a + b$ for $a \oplus b$ and ab for $a \odot b$.

Recall that a **semiring** $(R, +, \cdot, 0_R, 1_R)$ is a set R endowed with two binary operations $+$ and $\cdot$, such that:

1. $(R, +, 0_R)$ is an Abelian monoid with "zero element" 0_R;
2. $(R, \cdot\, , 1_R)$ is a monoid with "unit element" 1_R;
3. Multiplication distributes over addition;
4. $0_R \cdot a = a \cdot 0_R = 0_R$, $\forall a \in R$.

We use [11] as a standard reference on semiring theory. For any semiring R, one can define the semiring $R\lfloor\lambda\rfloor$ of **polynomials**, where polynomial addition and multiplication are defined in the familiar way:

$$\Big(\sum_i \alpha_i \lambda^i\Big)\Big(\sum_j \beta_j \lambda^j\Big) = \sum_k \Big(\sum_{i+j=k} \alpha_i \beta_{k-j}\Big)\lambda^k.$$

(The reason for the unusual notation $\lfloor\ \rfloor$ will become clear in Example 2.4 below.) Since the polynomial semiring was defined over an arbitrary semiring, we can define inductively

$$R\lfloor\lambda_1, \dots, \lambda_n\rfloor = R\lfloor\lambda_1, \dots, \lambda_{n-1}\rfloor\lfloor\lambda_n\rfloor.$$

A semiring with additive inverses is a ring, and many of the basic definitions (including homomorphisms, sub-semirings, ideals, prime ideals, and modules) mimic those from ring theory.

Unfortunately, the max-plus algebra $\mathcal{M}$ has no additive inverse (even if one formally adjoins a zero element, often denoted $-\infty$). In fact, every element of $\mathcal{M}$ is additively idempotent, and in any semiring, any additive idempotent having an additive inverse is 0. Indeed, if $a + b = 0$ then

$$0 = a + b = (a + a) + b = a + 0 = a.$$

Thus, the theory of cosets is useless here. To facilitate algebraic computations, Izhakian [13] introduced the **extended tropical arithmetic** $\mathbb{T}$, to be described explicitly in Example 2.10(a) below. This survey indicates how any max-plus semiring is "covered" by a certain semiring, which we call a **supertropical semiring**, that has a more manageable structure theory than the max-plus semiring, and in whose language many algebraic concepts related to tropical geometry can be described much more naturally. We will discuss the algebraic structure theory of supertropical semirings, including polynomials and their roots, cf. [15, 20], matrix theory, cf. [17], [19], and valuation theory, cf. [14].

Conceptually, the relation of $\mathbb{T}$ to the max-plus algebra $\mathcal{M}$ is similar to that of the complex numbers $\mathbb{C}$ to $\mathbb{R}$. Although one can manage with $\mathbb{R}$ and reformulate properties of $\mathbb{C}$ in terms of $\mathbb{R}$, it quickly becomes much easier to work with $\mathbb{C}$, and, moreover, imaginary numbers have their own meaning in applications; likewise, supertropical semirings have new elements, called **ghost elements**, which may well have their own intrinsic meaning, as we shall discuss at the end of this overview.

2. Basic notions

Instead of the special case $(\mathbb{R}, +)$, we start with an arbitrary monoid $\mathcal{G}$ (totally) ordered under $\leq$. Although our main example is $(\mathbb{R}, +)$, we write the monoid $\mathcal{G}$ in multiplicative notation, since it is to be a multiplicative submonoid of our supertropical semiring. Polynomials over a semiring R are written with coefficients in R. For example, taking $R = \mathbb{R}_0$, where the unit element 1_R is 0 and the neutral element 0_R is $-\infty$, the monomial λ means $1_R\lambda$, which as just noted is $0 \odot \lambda$ in this setting, one has the equality

$$(\lambda \oplus 3) \odot (\lambda \oplus 4) \quad = \quad \lambda \oplus \lambda \ \oplus \ 4 \odot \lambda \ \oplus \ 7.$$

This notation is rather cumbersome, so from now on, even when giving examples in a semiring built over $\mathbb{R}$, we utilize the more familiar semiring notation $\cdot$ and $+$ in place of $\odot$ and $\oplus$, and our equation becomes

$$(\lambda + 3)(\lambda + 4) = \lambda^2 + 4\lambda + 7.$$

Recall that a monoid $\mathcal{G}$ is **ordered** if $ab \leq ac$ and $ba \leq ca$ for all $b \leq c$ and a in $\mathcal{G}$. Any ordered monoid $\mathcal{G}$ can be viewed as a semiring $\mathcal{G}_0 := \mathcal{G} \cup \{0\}$, which is $\mathcal{G}$ with a formal element 0 adjoined, declaring that $a > 0$ for any $a \in \mathcal{G}$. The semigroup multiplication in $\mathcal{G}_0$ is the original monoid operation of $\mathcal{G}$ (with $0a = a0 = 0$ for all $a \in \mathcal{G}_0$), and the semigroup sum in $\mathcal{G}_0$ is taken to be the maximum in $\mathcal{G}$ (with $0 + a = a + 0 = a$ for all $a \in \mathcal{G}_0$). Thus, 0 is the zero element in the semiring $\mathcal{G}_0$, which we call the **associated semiring** of the ordered monoid.

We say that a pair of elements (a, b) in a semiring is (additively) **bipotent** if $a + b \in \{a, b\}$; the semiring is **bipotent** if every pair of elements is bipotent. A **semidomain** is a semiring in which $a, b \neq 0$ implies $ab \neq 0$. Thus, $\mathcal{G}_0$ as constructed above is a bipotent semidomain.

Conversely, any semidomain R yields a multiplicative monoid $\mathcal{G} = R \setminus 0$, together with a partial order on $\mathcal{G}$ given by

$$a \leq b \quad \text{iff} \quad a + b = b. \tag{2.1}$$

Clearly, two elements are related under this order iff they are additively bipotent in R. Thus, the semidomain R is bipotent iff our order on $\mathcal{G}$ is a total order; in this case, R can be identified with the associated semiring of $\mathcal{G}_0$.

In this way, the category of ordered monoids (where morphisms are required to respect the order) is isomorphic to the category of bipotent semidomains (whose morphisms are the semiring homomorphisms). So far we have merely changed languages. Here is the abstract algebraic formulation of Izhakian's main idea.

Definition 2.1. A **cover** of a semiring $\mathcal{G}_0$ is a semiring R with a semiring projection $\nu : R \to \mathcal{G}_0$; in other words, ν is an onto semiring homomorphism satisfying $\nu^2 = \nu$.

A **supertropical semiring** is a semiring $R = (R, \mathcal{G}_0, \nu)$ having an ideal $\mathcal{G}_0$, called the **ghost ideal**, which itself is an idempotent semiring (but whose unit element differs from that of R unless $\mathcal{G}_0 = R$), such that R is a cover of $\mathcal{G}_0$ (with respect to the projection ν), satisfying the following conditions, where we write a^ν for $\nu(a)$:

(i) (**Bipotence**) If $a^\nu \neq b^\nu$, then the pair (a, b) is bipotent;
(ii) (**Supertropicality**) $a + b = a^\nu$ whenever $a^\nu = b^\nu$.

A special case of supertropicality is

$$a + a = a^\nu, \quad \forall a \in R. \tag{2.2}$$

In particular, $1_R^\nu = 1_R + 1_R$ is a multiplicative idempotent of R, which also serves as the unit element of $\mathcal{G}_0$.

Definition 2.2. For any supertropical semiring $R = (R, \mathcal{G}_0, \nu)$, we define the ν-**topology** to have a base of open sets of the form

$$W_{\alpha,\beta} = \{a \in R : \alpha^\nu < a^\nu < \beta^\nu\}$$

and

$$W_{\alpha,\beta;\mathcal{T}} = \{a \in \mathcal{T} : \alpha^\nu < a^\nu < \beta^\nu\}, \quad \alpha^\nu, \beta^\nu \in \mathcal{G}_0.$$

This topology enables us to employ density arguments. Note that points in $\mathcal{G}$ are closed, but not points in $\mathcal{T}$.

2.1. Semirings with ghosts

In order to deal with polynomials and matrices, we need to describe the set-up in somewhat greater generality.

Definition 2.3. A **semiring with ghosts** $(R, \mathcal{G}_0, \nu)$ is a semiring R together with a semiring ideal $\mathcal{G}_0$, which we call the **ghost ideal**, and an idempotent semiring map

$$\nu : R \to \mathcal{G}_0,$$

called the **ghost map**, such that Equation (2.2) holds.

The ghost ideal $\mathcal{G}_0$ is equipped with the natural partial order given in Equation (2.1).

The following examples fit into this broader context.

Example 2.4. Suppose $(R, \mathcal{G}_0, \nu)$ is a supertropical semiring.

1. $\mathrm{Fun}(R^{(n)}, R)$ denotes the set of functions from $R^{(n)}$ to R; here $f \in \mathrm{Fun}(R^{(n)}, R)$ is **ghost** if $f(r_1, \dots, r_n) \in \mathcal{G}_0$ for every $r_1, \dots, r_n \in R$.
2. The sub-semiring with ghosts $\mathrm{CFun}(R^{(n)}, R)$ of $\mathrm{Fun}(R^{(n)}, R)$ consists of the continuous functions (with respect to the ν-topology).

3. There is a natural semiring homomorphism
$$\Phi : R\lfloor\lambda_1,\dots,\lambda_n\rfloor \to \mathrm{CFun}(R^{(n)}, R),$$
and we write $R[\lambda_1,\dots,\lambda_n]$ for the image of Φ; in this semiring, each element is the image of a set of polynomials that correspond to the same function. For example, $\lambda^2+\lambda+7$ (which we recall means $0\odot\lambda\odot\lambda \oplus 0\odot\lambda \oplus 7$) and λ^2+7 are the same function over $\mathbb{R}_0$. The ghost ideal of $R[\lambda_1,\dots,\lambda_n]$ is $\mathcal{G}_0[\lambda_1,\dots,\lambda_n]$.
4. Likewise, for $\Lambda = \{\lambda_1,\dots,\lambda_n\}$, one can define the Laurent polynomial semiring $R\lfloor\Lambda,\Lambda^{-1}\rfloor$ and its image $R[\Lambda,\Lambda^{-1}]$ in $\mathrm{CFun}(R^{(n)}, R)$.
5. We define the matrix semiring with ghosts $M_n(R)$; its ghost ideal is $M_n(\mathcal{G}_0)$.

None of these are supertropical semirings. Bipotence fails for polynomials: $(2\lambda+1)+(\lambda+2) = 2\lambda+2$. Likewise, bipotence fails for matrices.

Remark 2.5. The semiring $\mathrm{Fun}(R^{(n)}, R)$ satisfies the following amazing property, called the **Frobenius Property**:
$$\Big(\sum f_i\Big)^m = \sum f_i^m \tag{2.3}$$
for *every* natural number m. For example,
$$(f+g)^2 \;=\; f^2+g^2+fg+gf \;=\; f^2+g^2+(fg)^\nu \;=\; f^2+g^2. \tag{2.4}$$
since, for any $\mathbf{a} = (a_1,\dots,a_n) \in R^{(n)}$,
$$(f(\mathbf{a})g(\mathbf{a}))^\nu \;\le\; \max\{\nu(f(\mathbf{a})^2), \nu(g(\mathbf{a})^2)\}.$$

A suggestive way of viewing (2.3) is to note that for any m there is a semiring endomorphism $\mathrm{Fun}(R^{(n)}, R) \to \mathrm{Fun}(R^{(n)}, R)$ given by $f \mapsto f^m$, reminiscent of the Frobenius automorphism in classical algebra. But here the Frobenius property holds for **every** m. This plays an important role in our theory.

Here is a fundamental relation, called **ghost surpasses**, which replaces equality in many theorems taken from classical algebra.

Definition 2.6. $a \models_{\mathcal{G}} b$ in a semiring with ghosts $(R, \mathcal{G}_0, \nu)$, if $a = b + c$ for some element $c \in \mathcal{G}_0$.

Example 2.7. Any ring R of characteristic 2 is a semiring with ghosts, taking $\mathcal{G}_0 = \{0_R\}$, where $a^\nu = 0_R$ for all a. Note in this case that ghost surpasses is the same as equality, and Equation (2.2) is equivalent to $1_R + 1_R = 1_R^\nu = 0_R$. In this way, ring theory of characteristic 2 is subsumed in the theory of semirings with ghosts.

2.2. Supertropical domains and semifields

We finally are ready to introduce the fundamental objects of supertropical algebra.

Definition 2.8. A **supertropical domain** is a commutative supertropical semiring $(R, \mathcal{G}_0, \nu)$ for which the following extra properties hold (where $\mathcal{G} = \mathcal{G}_0 \setminus \{0\}$):

(i) The set $\mathcal{T} = R \setminus \mathcal{G}_0$ is a monoid, called the monoid of **tangible elements**;
(ii) The restriction $\nu_{\mathcal{T}} : \mathcal{T} \to \mathcal{G}$ is onto; in other words, $\mathcal{G}$ is a monoid, and every element of $\mathcal{G}$ has the form a^ν for some $a \in \mathcal{T}$.

A supertropical domain $(R, \mathcal{G}_0, \nu)$ is called a **supertropical semifield** if every tangible element $\neq 0_R$ is invertible.

Note in a supertropical domain that if $a \models_{\mathcal{G}} b$ with a tangible, then $a = b$. On the other hand, if $a + b \models_{\mathcal{G}} 0_R$ with b tangible, then $a \models_{\mathcal{G}} b$. These properties help explain the importance of tangible elements as well as the ghost surpassing relation.

Example 2.9. Given a homomorphism of Abelian monoids $\nu : \mathcal{T} \to \mathcal{G}$, where the monoid $\mathcal{G}$ is ordered, one can define the semiring

$$R := \mathcal{T} \cup \mathcal{G}_0,$$

where $\mathcal{G}_0 = \mathcal{G} \cup \{0\}$, with respect to the following operations (writing a^ν for $\nu(a)$):

$$ab = \begin{cases} \text{product in } \mathcal{T} & \text{for } a, b \in \mathcal{T}; \\ \text{product in } \mathcal{G} & \text{for } a, b \in \mathcal{G}; \\ a^\nu b & \text{for } a \in \mathcal{T},\ b \in \mathcal{G}; \\ ab^\nu & \text{for } a \in \mathcal{G},\ b \in \mathcal{T}; \\ 0 & \text{for } a = 0 \text{ or } b = 0, \end{cases} \qquad \begin{aligned} a + b &= \begin{cases} a & \text{for } a^\nu > b^\nu; \\ b & \text{for } a^\nu < b^\nu; \\ a^\nu & \text{for } a^\nu = b^\nu, \end{cases} \\ a + 0 &= 0 + a = a. \end{aligned}$$

Note that $ab \in \mathcal{G}_0$ if $a \in \mathcal{G}_0$ or $b \in \mathcal{G}_0$, so $\mathcal{G}_0$ is an ideal of R, and clearly 1_R^ν is the multiplicative unit of $\mathcal{G}_0$, implying $(R, \mathcal{G}_0, \nu)$ is a supertropical domain. $(R, \mathcal{G}_0, \nu)$ is a supertropical semifield, iff ν is onto and $\mathcal{G}$ is a group.

Example 2.10. Here are the motivating examples of supertropical semifields, using Remark 2.9:

(a) $\mathcal{T} = (\mathbb{R}, +)$ and $\mathcal{G} = (\mathbb{R}, +)$, with ν the identity map (Izhakian's original example of the extended tropical arithmetic $\mathbb{T}$);
(b) $\mathcal{T} = F^\times$ (F a field) and $\mathcal{G}$ is an ordered group, with ν a valuation. (Here we forget the original addition on the field F!) As a special case, one could take the field of locally convergent Puiseux series, which plays a key role in tropical geometry.

From this vantage point, a supertropical domain could be viewed as a generalization of a valuation, thereby giving rise to the "supervaluation theory" described in [14].

The innovation in this structure is that the ideal $\mathcal{G}_0 = R^\nu$ is to be treated much the same way, via the ghost surpassing relation, as one would customary treat the 0 element in classical commutative algebra. This provides R with a rich algebraic structure, in which much of the theory of real commutative algebra can be formulated.

3. Polynomials and their roots

Classical results about polynomials over fields that we use as signposts for the supertropical theory:

- The fundamental theorem of algebra;
- Hilbert's Nullstellensatz;
- Unique factorization into irreducibles;
- Properties of the resultant of two polynomials in a single indeterminate;
- The Euclidean algorithm;
- Finite generation of ideals in polynomial rings.

What happens in the supertropical case? Polynomials should play a role in any reasonable algebraic theory. Our initial excursion into supertropical algebra is to try to build a foundation for algebraic geometry by means of polynomials and their roots. A nonzero polynomial over the max-plus algebra cannot have any zeroes in the classical sense! But here is an alternate definition.

An n-tuple $\mathbf{a} = (a_1, \dots, a_n) \in R^{(n)}$ is called a **(supertropical) root** of a polynomial $f \in R[\lambda_1, \dots, \lambda_n]$ if $f(\mathbf{a}) \in \mathcal{G}_0$, which can be written also as $f(\mathbf{a}) \underset{\mathcal{G}}{\models} 0_R$.

We are interested mainly in the set of tangible roots, called the **tangible root set** of f; this is readily seen to be a tropical hypersurface.

Example 3.1. The tangible roots of the polynomial $f = \lambda_1 + \lambda_2 + 0$ over $\mathbb{R}_0$ are:

$$\begin{cases} (0, a) \text{ for } a < 0; \\ (a, 0) \text{ for } a < 0; \\ (a, a) \text{ for } a > 0. \end{cases}$$

The "curve" of tangible roots of f is comprised of three rays, all emanating from $(0, 0)$.

3.1. Polynomials in one indeterminate over a supertropical semifield

Classically, every polynomial has a root in the algebraic closure. The tropical version: Given any supertropical semifield R, we define its **divisible closure**

$$\bar{R} = \left\{ \frac{a}{m} : \ a \in R, \ m \in \mathbb{N} \right\}.$$

which is a supertropical semifield when we extend ν to $\bar{R}$ by putting $\nu(\frac{a}{m}) = \frac{a^\nu}{m}$. For example, $\mathbb{Q}$ is divisibly closed, but $\mathbb{Z}$ is not. One sees easily that any supertropical polynomial has a tangible root in the divisible closure; cf. [15, Proposition 4.7].

There are two kinds of tangible roots of a polynomial $f = \sum_{\mathbf{j} \in J} h_{\mathbf{j}}$, where $h_{\mathbf{j}}$ are distinct monomials. Given $\mathbf{a} \in \mathcal{T}^{(n)}$, we let $c_{\mathbf{j}} = h_{\mathbf{j}}(\mathbf{a})$, and $S(\mathbf{a}) = \{c_{\mathbf{j}}^\nu : \mathbf{j} \in J\}$:

Corner root. At least two of the $c_{\mathbf{j}}$'s are ν-maximal (and thus equal) in $S(\mathbf{a})$. In this case,

$$f(\mathbf{a}) = c_{\mathbf{j}}^\nu \in \mathcal{G}_0.$$

Cluster root. There is a unique $\mathbf{j}$ for which $c_{\mathbf{j}}^{\nu}$ is maximal in $S(\mathbf{a})$. Then $f(\mathbf{a}) = c_{\mathbf{j}}$; this will necessarily be ghost when the coefficient of the monomial $h_{\mathbf{j}}$ is ghost.

Example 3.2. The set of tangible roots of the polynomial $f = \lambda^4 + 3^{\nu}\lambda^3 + 5^{\nu}\lambda^2 + 6\lambda + 6$ over $\mathbb{T}$ is

$$\{0\} \cup \{a : 1^{\nu} \le a^{\nu} \le 3^{\nu}\}.$$

The tangible corner roots are 0,1,2, and 3, as seen by noting in each evaluation that two terms match:

$$\begin{aligned} f(0) &= 0^4 + 3^{\nu}0^3 + 5^{\nu}0^2 + 6 \cdot 0 + 6 = 0 + 3^{\nu} + 5^{\nu} + 6 + 6 = 6^{\nu};\\ f(1) &= 1^4 + 3^{\nu}1^3 + 5^{\nu}1^2 + 6 \cdot 1 + 6 = 4 + 6^{\nu} + 7^{\nu} + 7 + 6 = 7^{\nu};\\ f(2) &= 2^4 + 3^{\nu}2^3 + 5^{\nu}2^2 + 6 \cdot 2 + 6 = 8 + 9^{\nu} + 9^{\nu} + 8 + 6 = 9^{\nu};\\ f(3) &= 3^4 + 3^{\nu}3^3 + 5^{\nu}3^2 + 6 \cdot 3 + 6 = 12 + 12^{\nu} + 11^{\nu} + 9 + 6 = 12^{\nu}. \end{aligned} \tag{3.1}$$

For any other number a between 1 and 3, a single monomial with ghost coefficient dominates in the evaluation $f(a)$, and thus a is a cluster root. For example, $f(1.5) = 8^{\nu}$ and $f(2.5) = 10.5^{\nu}$.

We turn to factorization of polynomials, *which is always done as functions.* Here are some results for polynomials in one indeterminate over a divisibly closed supertropical semifield.

It is not hard to see that any polynomial f with tangible coefficients can be factored as the product

$$\prod_j (\lambda + a_j)^{i_j},$$

where the a_j range over the tangible roots of f. In particular, all irreducible polynomials with tangible coefficients are linear.

For example, over $\mathbb{T}$ (and also over $\mathcal{M}$),

$$\lambda^2 + 5\lambda + 7 = (\lambda + 5)(\lambda + 2);$$

$$\lambda^2 + 8 = (\lambda + 4)^2.$$

An example of an irreducible quadratic polynomial:

$$\lambda^2 + 5^{\nu}\lambda + 7.$$

In general, any polynomial factors (as functions) as the product of linear and quadratic polynomials. But unique factorization of polynomials can fail.

Example 3.3.

$$\begin{aligned} \lambda^4 + 4^{\nu}\lambda^3 + 6^{\nu}\lambda^2 + 5^{\nu}\lambda + 3 &= (\lambda^2 + 4^{\nu}\lambda + 2)(\lambda^2 + 2^{\nu}\lambda + 1)\\ &= (\lambda^2 + 4^{\nu}\lambda + 2)(\lambda + 2)(\lambda + (-1))\\ &= (\lambda^2 + 4^{\nu}\lambda + 3)(\lambda^2 + 2^{\nu}\lambda + 0). \end{aligned}$$

Nevertheless, there is a version of unique factorization in one indeterminate.

Theorem 3.4. [15, Theorem 7.41] *The factorization of a polynomial in one indeterminate into linear and quadratic irreducible factors, which is minimal in the number of ghost terms, is unique.*

In Example 3.3, it comes from the second line: $(\lambda^2+4^\nu\lambda+2)(\lambda+2)(\lambda+(-1))$. Another interpretation of this factorization is in terms of the roots of a polynomial:

Proposition 3.5. [15, Proposition 7.47] *Suppose f is a polynomial of degree t, whose tangible root set is the closed interval $[\alpha_1, \alpha_t]$, and $\alpha_1, \dots, \alpha_t$ are corner roots of f, arranged in ascending ν-value. Then*

$$f = (\lambda^2 + \alpha_t^\nu \lambda + \alpha_t \alpha_1) \prod_{k=2}^{t-1} (\lambda + \alpha_k),$$

and this is the (unique) factorization minimal in ghosts, having only one ghost term.

The notion of "relatively prime" for polynomials is quite delicate. Given a polynomial f, we write $\underline{\deg}(f)$ for the degree of the lowest order monomial of f. For example, $\underline{\deg}(\lambda^3 + 2\lambda^2 + \lambda^\nu) = 1$.

Definition 3.6. Two polynomials f and g of respective degrees m and n are **relatively prime** if there do not exist tangible polynomials p and q (not both 0) with $\deg(p) < n$ and $\deg(q) < m$, such that $pf + qg$ is ghost with $\deg(pf) = \deg(qg)$ and $\underline{\deg}(pf) = \underline{\deg}(qg)$.

Theorem 3.7. [20, Theorem 3.14] *Two non-constant monic polynomials f and g in $F[\lambda]$ are relatively prime iff f and g do not have a common tangible root.*

3.2. Factorization of polynomials in several indeterminates

In two indeterminates, we have a worse violation of unique factorization:

$$\begin{aligned}(0 + \lambda_1 + \lambda_2)(\lambda_1 + \lambda_2 + \lambda_1\lambda_2) &= \lambda_1 + \lambda_2 + \lambda_1^2 + \lambda_2^2 + \nu(\lambda_1\lambda_2) + \lambda_1^2\lambda_2 + \lambda_2^2\lambda_1 \\ &= (0 + \lambda_1)(0 + \lambda_2)(\lambda_1 + \lambda_2) \\ &= (0 + \lambda_1 + \lambda_2 + \lambda_1\lambda_2)(\lambda_1 + \lambda_2).\end{aligned}$$

But this is an instance of the following important phenomenon (in arbitrarily many indeterminates):

Theorem 3.8. [15, Theorem 7.53] *Suppose $f = \sum_{i=1}^m f_i$ is written as a sum of monomials, for $m \geq 2$. Then f divides $\prod_{j\neq i}(f_i + f_j)$. More precisely,*

$$\prod_{i<j}(f_i + f_j) = g_1 \cdots g_{m-1} \tag{3.2}$$

where $g_1 = f = \sum_i f_i$, $g_2 = \sum_{i<j} f_i f_j$, $\dots$, and $g_{m-1} = \sum_i \prod_{j\neq i} f_j$.

It follows that the root set of any polynomial is contained as a subvariety of a union of hyperplanes. Algebraically, it implies that every prime ideal of the polynomial semiring contains a binomial (sum of two monomials).

3.3. A tropical version of the Hilbert Nullstellensatz

As in commutative ring theory, we define the **radical** $\sqrt{A}$ of a subset $A \subset R$ as

$$\sqrt{A} = \{a \in R : a^k \in A \text{ for some } k\}.$$

An ideal A of R is **radical** if $A = \sqrt{A}$. The situation here is actually nicer than the classical one – for any supertropical semiring R, the radical of any sub-semiring of $\mathrm{Fun}(R^{(n)}, R)$ is also a sub-semiring, because of the Frobenius property (Remark 2.5).

The connection between supertropical algebra and geometry is more subtle than in the classical case. Given a polynomial $f = \sum_j h_j$ (written as a sum of monomials), for each tangible monomial h_j we define the jth **tangible component** D_j of f to be the set

$$D_j := \{\mathbf{a} \in (\mathcal{T} \cup \{0\})^{(n)} : f(\mathbf{a}) = h_j(\mathbf{a})\}.$$

Thus, the union of the tangible components of f is the complement of its tangible root set. Assuming F is a supertropical semifield, radical semiring ideals of $F[\lambda_1, \dots, \lambda_n]$ correspond to the tangible components, as follows:

For any tangible component D of f, we write $f \preceq_D g$ if g has a tangible component containing D; $f \in_{\text{ir-com}} S$ for $S \subseteq F[\lambda_1, \dots, \lambda_n]$, if, for every tangible component $D = D_j$ of f, there is some tangible $g \in S$ (depending on D_j) with $f \preceq_D g$. The supertropical version of the Nullstellensatz is:

Theorem 3.9. [15, Theorem 6.17] *Suppose* $A \triangleleft F[\lambda_1, \dots, \lambda_n]$ *and* $f \in R[\lambda_1, \dots, \lambda_n]$. *Then* $f \in_{\text{ir-com}} A$ *iff* $f \in \sqrt{A + \mathcal{G}_0[\lambda_1, \dots, \lambda_n]}$.

Generation of ideals is tricker: For example, the ideal of $\mathbb{T}[\lambda]$ generated by $\{\lambda + n : n \in \mathbb{N}\}$ is not finitely generated.

4. Supertropical matrix theory

We turn to matrix theory, again referring to the classical literature for inspiration. Let $|A|$ denote the determinant of the matrix A. Classical results in linear algebra (for matrices A and B) include:

- $|AB| = |A||B|$;
- $|A| = 0$ iff the rows of A are linearly dependent;
- $A\,\mathrm{adj}(A) = |A|I$;
- $f_A(A) = 0$, where $f_A = |\lambda I - A|$;
- The roots of f_A are eigenvalues of A;
- If f_A is separable, then $|A|$ is diagonalizable;
- Cramer's rule enables one to solve the matrix equation $Ax = v$.

4.1. Review of matrix theory over the max-plus algebra

There is a considerable literature for matrices over the max-plus algebra, especially relating to the asymptotic behavior of matrices with entries < 1, which are used in the theories of discrete event systems, stochastic processes, and dynamic programming. We review some previously known results:

An important combinatoric tool is the **weighted digraph** $G_A = (V, E)$ of an $n \times n$ matrix $A = (a_{i,j})$. This graph has vertex set $V = \{1, \dots, n\}$, and an edge (i, j) from i to j (given **weight** $a_{i,j}$) whenever $a_{i,j} \neq 0_R$. The matrix A is said to be **irreducible** if G_A is strongly connected. Each summand of the determinant corresponds to a multicycle of length n.

The **maximal cycle mean** $\rho_{\max}(A)$, in G_A is the maximum ratio $w(C)/\ell(C)$ over all the cycles C in G_A, where $w(C)$ and $\ell(C)$ denote respectively the weight and the length of the cycle C in G_A.

1. For any matrix A, $\rho_{\max}(A)$ is the maximal eigenvalue of A and is associated with the eigenvector $v_{\rho_{\max}}$ of $\rho_{\max}$.
2. When A is irreducible, $\rho_{\max}(A) > 0$ is the only eigenvalue of A.
3. $\rho_{\max}(A)$ is the maximal root of the characteristic polynomial of A.

The **Kleene star** A^* of a matrix $A \in M_n(\mathbb{R})$ is

$$A^* := I + A + A^2 + A^3 + \dots,$$

Given a matrix $A = (a_{ij})$, where each cycle of G_A has weight < 1, then:

1. $A^* = \sum_{i=0}^{n-1} A^i$,
2. $A^* u$ is the unique solution of the system $x = Ax + u$, where $u, x \in \mathbb{R}^{(n)}$.

A combinatoric version of the Cayley-Hamilton theorem was proved by Straubing [31]. Definitions of matrix rank have been put forward by Gondran-Minoux, and by Develin, Santos and Sturmfels, Barvinok (also called Schein) and Kapranov ranks [2, 5, 6, 9, 10, 21, 22], e.g., the largest k such that the matrix has a $k \times k$ nonsingular submatrix. Also, Sturmfels [32] has utilized a version of the resultant in combinatorics.

4.2. Supertropical determinants

These various notions can be unified and strengthened by means of the supertropical structure. Supertropical matrix theory starts with the determinant. Since we cannot take negatives, we simply drop the minus signs (which is reasonable since $+1 = -1$ in characteristic 2). Our main tool is the permanent $|A|$, which can be defined for any matrix A over any commutative semiring. To emphasize its parallel to the determinant, we call the permanent the **tropical determinant**. Although the permanent is not multiplicative over arbitrary commutative semirings, it is multiplicative in this theory, in a certain sense.

Let us develop the supertropical determinant in detail. Assume that $R = (R, \mathcal{G}_0, \nu)$ is a supertropical domain. Take $V = R^{(n)}$, with the standard basis $(e_1, \dots, e_n)$. V has the **ghost subspace** $\mathcal{H}_0 = \mathcal{G}_0^{(n)}$. As before, we define the relation

$v \underset{\mathcal{H}}{\models} w$ to denote $v = w + y$ for $y \in \mathcal{H}_0$. Let us define the function $\Phi_\gamma : V^{(n)} \to R$ by the following formula, where $v_i = (v_{i,1}, \dots, v_{i,n})$:

$$\Phi_\gamma(v_1, \dots, v_n) = \gamma \sum_{\pi \in S_n} v_{1,\pi(1)} \cdots v_{n,\pi(n)}, \tag{4.1}$$

where $\gamma \in R$ is fixed.

Theorem 4.1. *Φ_γ satisfies the following properties:*

1. *Φ_γ is linear in each tangible component; i.e.,*
$$\begin{aligned}\Phi_\gamma(v_1, \dots, \alpha_i v_i + \alpha_i' v_i', \dots, v_n) &= \alpha_i \Phi_\gamma(v_1, \dots, v_i, \dots, v_n) \\ &\quad + \alpha_i' \Phi_\gamma(v_1, \dots, v_i', \dots, v_n).\end{aligned}$$
2. *$\Phi_\gamma(v_1, \dots, v_n) \in \mathcal{G}_0$ if $v_i = v_j$ with $i \neq j$.*
3. *$\Phi_\gamma(v_1, \dots, v_n) = 0_R$ if some $v_i = 0_V$.*
4. *$\Phi_\gamma(v_{\pi(1)}, \dots, v_{\pi(n)}) = \Phi_\gamma(v_1, \dots, v_n)$, for all permutations π.*
5. *$\Phi_\gamma(e_1, \dots, e_n) = \gamma$.*

Furthermore, Φ_γ is unique up to ghost surpassing, in the sense that if Φ_γ' is another function satisfying the same properties (1)–(5) and $\Phi_\gamma(v_1, \dots, v_n)$ is tangible, then

$$\Phi_\gamma'(v_1, \dots, v_n) \underset{\mathcal{G}}{\models} \Phi_\gamma(v_1, \dots, v_n).$$

Define the **tropical determinant** as Formula (4.1) (normalized):

$$|(a_{i,j})| = \sum_{\pi \in S_n} a_{\pi(1),1} \cdots a_{\pi(n),n}. \tag{4.2}$$

Thus, there are two ways for $|(a_{i,j})|$ to be ghost: Two dominant summands in the right side of (4.2) are ν-matched, or the dominant summand is ghost. The former possibility matches the tropical version of a singular matrix over the max-plus algebra, but one also needs the latter possibility to develop the matrix theory along the lines of classical matrix algebra.

Definition 4.2. A matrix A is **nonsingular** if $|A|$ is tangible; A is **singular** when $|A| \in \mathcal{G}_0$. Likewise, the **rank** of a matrix is the largest k such that A has a nonsingular $k \times k$ submatrix.

Theorem 4.3. [17, Theorem 3.5] $|AB| \underset{\mathcal{G}}{\models} |A|\,|B|$, *for any $n \times n$ matrices over a supertropical semiring. In particular, $|AB| = |A|\;|B|$ whenever AB is nonsingular.*

Proof. Take $\Phi_{|B|}(A) = |AB|$ in Theorem 4.1. This satisfies (1)–(5), for $\gamma = |B|$, and thus must be $|A|\;|B|$ except when $|AB|$ is ghost. □

Given Theorem 4.3, one might expect a Zariski topology-type argument to imply that $|AB|^\nu = |A|^\nu\,|B|^\nu$ for all matrices A and B, presumably seen by modifying A and B slightly to get tangible tropical determinants. But in [17, Example 6.11] it is seen that $A = \begin{pmatrix} 0 & 0 \\ 1 & 2 \end{pmatrix}$ (over $\mathbb{T}$) satisfies $|A|^\nu = 2^\nu$ whereas

$|A^2|^\nu = 5^\nu$. This might seem to discredit the Zariski density approach, but note that if $\tilde{A} = \begin{pmatrix} a_0 & a_0' \\ a_1 & a_2 \end{pmatrix}$ and $\tilde{A} = \begin{pmatrix} b_0 & b_0' \\ b_1 & b_2 \end{pmatrix}$ where a_0, a_0', b_0, b_0' are "close" to 0 and a_1, b_1 are "close" to 1 and a_2, b_2 are "close" to 2, then $\tilde{A}\tilde{B} = \begin{pmatrix} a_0' b_1 & a_0' b_2 \\ a_2 b_1 & a_2 b_2 \end{pmatrix}$, whose tropical determinant is $(a_0' a_2 b_1 b_2)^\nu$, which is still a ghost! In other words, if $\tilde{A}, \tilde{B}$ are "close" to A, then $\tilde{A}\tilde{B}$ remains singular, and the Zariski density argument is inapplicable! This can be understood generically as follows, for the 2×2 matrix $A = \begin{pmatrix} a_{1,1} & a_{1,2} \\ a_{2,1} & a_{2,2} \end{pmatrix}$: In [17, Example 3.11] it is seen that

$$|A^2| = (a_{1,1}^2 + a_{2,2}^2) a_{1,2} a_{2,1}^\nu + |A|^2 .$$

Thus, the familiar product rule for determinants only holds for A when the ghost term $(a_{1,1}^2 + a_{2,2}^2) a_{1,2} a_{2,1}^\nu$ is inessential.

4.3. Adjoints

Definition 4.4. The **minor** $A'_{i,j}$ is obtained by deleting the i row and j column of A. The **adjoint** matrix $\mathrm{adj}(A)$ is the transpose of the matrix $(a'_{i,j})$, where $a'_{i,j} = |A'_{i,j}|$.

Some easy calculations:

- $|A| = \sum_{j=1}^n a_{i,j}\, a'_{i,j}, \quad \forall i.$
- $\sum_{j=1}^n a_{i,j}\, a'_{k,j} \in \mathcal{G}_0, \quad \sum_{j=1}^n a'_{j,i}\, a_{j,k} \in \mathcal{G}_0, \quad \forall k \neq i.$

Theorem 4.5. [17, Theorem 4.9]

1. $|A \ \mathrm{adj}(A)| = |A|^n$,
2. $|\mathrm{adj}(A)| = |A|^{n-1}$.

The proof is a direct consequence of a special case of a celebrated theorem of Birkhoff and von Neumann which states that any directed graph where all in-degrees and out-degrees are equal to k is a disjoint union of k multicycles; one can quote instead the graph-theoretic version of Hall's marriage theorem.

Definition 4.6. A **quasi-identity matrix** is a nonsingular matrix $I_{\mathcal{G}}$ which is a multiplicatively idempotent matrix of tropical determinant 1_R, equal to the identity matrix on the diagonal and ghost off the diagonal.

The identity matrix is a quasi-identity matrix, but there are many other examples.

Theorem 4.7. *For any nonsingular matrix A over a supertropical semifield,*

$$A \ \mathrm{adj}(A) = |A| \, I_A,$$

for a suitable quasi-identity matrix I_A.

Likewise $\mathrm{adj}(A)A = |A|\, I'_A$, *for a suitable quasi-identity matrix I'_A (perhaps different from I_A).*

4.4. The Hamilton-Cayley theorem

One also has a supertropical version of the Hamilton-Cayley theorem: We say that the matrix A **satisfies** the polynomial $f \in R[\lambda]$ if $f(A) \in M_n(\mathcal{G}_0)$.

Theorem 4.8. [17, Theorem 5.2] *Any matrix A satisfies its characteristic polynomial $f_A = |\lambda I + A|$.*

(This actually follows from the result of Straubing [31] quoted earlier.)

Definition 4.9. A vector v is a **supertropical eigenvector** of A, with **supertropical eigenvalue** $\beta \in \mathcal{T}$, if $Av \underset{\mathcal{H}}{\models} \beta v$.

Theorem 4.10. [17, Theorem 7.10] *Every (tangible) root of the polynomial f_A is a supertropical eigenvalue of A.*

4.5. Tropical dependence

Definition 4.11. A subset $W \subset R^{(n)}$ is **tropically dependent** if there is a finite sum $\sum \alpha_i w_i \underset{\mathcal{H}}{\models} (0)$, with each α_i tangible or 0_R, but not all of them 0_R; otherwise W is called **tropically independent**.

Theorem 4.12. [17, Theorem 6.5] *Suppose R is a supertropical domain. Vectors $v_1, \dots, v_n \in R^{(n)}$ are tropically dependent, iff the matrix whose rows are $v_1, \dots, v_n$ is singular.*

More generally, the following numbers are equal (and could be viewed as the rank of a matrix A over a supertropical domain), cf. [18]:

(a) The number of tropically independent rows of A;
(b) The number of tropically independent columns of A;
(c) The largest size of a nonsingular square submatrix of A.

4.6. Solving supertropical equations

One basic application of matrices in classical algebra is to solve the matrix equation $Ax = v$, for a given nonsingular matrix A and vector v. This cannot be done in general over the max-plus algebra, but one does have the following result:

Theorem 4.13. [19, Theorems 3.3 and 3.5] *The matrix equation $Ax \underset{\mathcal{H}}{\models} v$, for $|A|$ invertible and v a tangible vector, has a tangible solution given by an analog of Cramer's rule, which is the unique largest solution (in terms of ν-values).*

This solution gives the solution for $Ax = v$ when it exists.

Example 4.14. Define the **Vandermonde matrix** A to be the $n \times n$ matrix $(a_{i,j})$, where $a_{i,j} = a_i^{j-1}$ and $a_i^0 = 1$. Then $|A| = \prod_{i \neq j}(a_i + a_j)$; cf. [16]. Thus, if the a_i are distinct and tangible, A is nonsingular, and the Vandermonde matrix is an important tool in solving equations.

Nevertheless, we saw above that the Vandermonde matrix $A = \begin{pmatrix} 0 & 0 \\ 1 & 2 \end{pmatrix}$ has the poor behavior that $A^2 = \begin{pmatrix} 1 & 2 \\ 3 & 4 \end{pmatrix}$ is singular.

4.7. The resultant of two polynomials

The tropical resultant has already been studied by Sturmfels [32, 33], Dickenstein, Feichtner, and Sturmfels [7], and Tabera [34], but our purely algebraic approach is quite different, leading to a characterization of relatively prime polynomials and a supertropical version of Bézout's Theorem. Given two polynomials $f(\lambda)$ and $g(\lambda)$ over a supertropical domain R, one can define the resultant $\mathfrak{R}(f, g)$ just as in the classical definition, where one uses the tropical determinant instead of the usual determinant. Then one has

Theorem 4.15. [20, Theorem 3.14] $\mathfrak{R}(f, g) \in \mathcal{T}$ *iff* f, g *do not have a common tangible root in* R.

This leads to a version of Bézout's theorem, [20, Theorem 5.1].

5. The structure theory of semirings with tangibles and ghosts

One of the pillars of the theory of field extensions is the formal construction of a field extension $K = F[\lambda]/\langle f \rangle$ of F, in which f formally has a root (the coset $\overline{\lambda}$ of λ); moreover, for any field extension L of F containing a root a of f, there is an injection $K \to L$ sending $\overline{\lambda} \mapsto a$.

We would like a parallel result in supertropical algebra. However, as stated at the outset, semirings do not in general have a natural theory of quotient structures (modulo a given ideal), and this poses one of the main challenges to the structure theory.

In order to circumvent this difficulty, we introduce a quotient structure reminiscent of the Rees quotient from semigroup theory.

Definition 5.1. Suppose $R = (R, \mathcal{G}_0, \nu)$ is a semiring with ghosts. Given an ideal A of R, we define R/A to be $(R, A + \mathcal{G}_0, \nu)$, the *same* semiring R but we now enlarge the ghost ideal to be $A + \mathcal{G}_0$.

When $A \supset B$ are ideals of R, we can identify R/B with $(R/B)/(A/B)$, a rather trivial version of Noether's second isomorphism theorem.

Remark 5.2. If P is a prime ideal of R containing $\mathcal{G}_0$, then R/P becomes a supertropical domain, since by definition of prime ideal, its set of tangible elements $R \setminus P$ is a monoid.

Example 5.3. Suppose we want to adjoin the square root of 3 (which is really $\frac{3}{2}$) to the supertropical semiring $F = (\mathbb{Z}, \mathbb{Z}^\nu, \nu)$, the cover of the max-plus algebra of $\mathbb{Z}$. We could take $F[\lambda]/I = (F[\lambda], I, \nu)$ where $I = F[\lambda](\lambda^2 + 3) + \mathbb{Z}^\nu$, and note that $\lambda^2 + 3 \in I$ implies that the element λ can now be viewed as a root of the polynomial $\lambda^2 + 3$.

Instead, one could simply adjoin $\frac{3}{2}$ to $\mathbb{Z}$ (taken in the divisible closure $\mathbb{Q}$ of $\mathbb{Z}$), so we would like some version of Noether's First Isomorphism Theorem to identify the two constructions.

Definition 5.4. A **supertropical homomorphism** $\varphi : (R, \mathcal{G}_0, \nu) \to (R', \mathcal{G}'_0, \nu')$ of semirings with ghosts is a semiring homomorphism such that $\varphi(a^\nu) = \varphi(a)^{\nu'}$ for all $a \in R$.

In particular, $\varphi(1_R^\nu) = \varphi(1_R)^{\nu'} = 1_{R'}^{\nu'}$, so $\varphi(\mathcal{G}_0) \subseteq \mathcal{G}'_0$.

Definition 5.5. The **ghost kernel** g-ker φ of a supertropical homomorphism $\varphi : (R, \mathcal{G}_0, \nu) \to (R', \mathcal{G}'_0, \nu')$ is $\varphi^{-1}(\mathcal{G}'_0)$. The homomorphism φ is a **ghost injection** if g-ker $\varphi \subseteq \mathcal{G}_0$.

Ghost injections have the following application to supertropical domains:

Remark 5.6. If φ is a ghost injection and $a, b \in \mathcal{T}$ such that $\varphi(a)^{\nu'} = \varphi(b)^{\nu'}$, then $a^\nu = b^\nu$. (Indeed,

$$\varphi(a+b) = \varphi(a) + \varphi(b) = \varphi(a)^{\nu'},$$

implying $a + b \in \mathcal{G}_0$, so $a^\nu = b^\nu$.)

We are now ready for our version of Noether's First Isomorphism Theorem, which essentially holds by definition:

Remark 5.7. Suppose $\varphi : R \to W$ is a supertropical homomorphism, with ghost kernel A. Then φ induces a natural ghost injection $R/A \to W$.

Although immediate, this result is quite useful in the structure theory.

Example 5.8. Suppose $f \in F[\lambda]$ and a is a root of f in some semiring with ghosts R containing F. Let $I = F[\lambda]f$. Then there is a supertropical homomorphism from $F[\lambda]/I$ to R, induced by the substitution homomorphism $g(\lambda) \mapsto g(a)$. (Indeed, since $f(a)$ is ghost, the ghost kernel contains I.)

6. Conclusions and directions for further research

6.1. The role of ghosts

"Ghost elements" were introduced into supertropical domains in order to facilitate calculations of roots of polynomials and supertropical determinants. But they seem to have their own significance, as a kind of "noise." The fact that the product law holds for tangible determinants but not for ghost determinants suggests some sort of "uncertainty" arising from ghosts, and it would be interesting to see whether this uncertainty is compatible with other notions of uncertainty arising in physics and statistics. Also, the breakdown of the product law for determinants may be related to pathological situations in game theory and economics. A careful analysis of matrix multiplication involving singular matrices should lead to insight in these matters.

6.2. Structure theory

The structure theory of supertropical algebras is still in its early development, and one can "tropicalize" many algebraic notions in order to apply their techniques to supertropical algebra. For example, there is the natural notion of a supertropical module $(V, \mathcal{H}, \mu)$ over a supertropical domain $(R, \mathcal{G}_0, \nu)$, namely a semiring module V supplied with a ghost submodule $\mathcal{H}$, together with a ghost projection μ compatible with the ghost map ν of R.

Finally, in view of Example 2.7, and since many theorems of supertropical algebra parallel the theory of algebras of characteristic 2, there should be a theory of semirings with ghosts that encompasses both theories. M. Akian, S. Gaubert, and A. Guterman [3] prove some pretty results in this direction.

6.3. Linear algebra

Having the basic properties of matrices in hand, one should go on to develop various analogs of vector spaces. In the supertropical situation, a vector subspace of a vector space can have the same dimension, leading to some interesting situations. The authors have launched such an investigation together with Knebusch, including supertropical vector spaces and their bases, and supertropical inner products.

6.4. Category theory

Having established the basic supertropical algebraic notions, one could study supertropical categories, and develop the appropriate homology and cohomology theories. One major advance would be to put all of this in the general framework of category theory, both as a source of examples and as a way of tapping general categorical results.

6.5. Multiple ghost layers

Another promising direction of research is to refine the ghost ideal, so as to obtain different layers of ghosts, which thereby should enable one to consider the multiplicity of a root. This approach, although considerably more intricate, leads to a cleaner theory, including unique factorization (with one exceptional class) in the polynomial semiring $F[\lambda]$.

References

[1] O. Aharony, A. Hanany, and B. Kol, Webs of (p,q) 5-branes, five dimensional field theories and grid diagrams, *J. High Energy Phys*, pages 1126–6708, 1998, preprint at http://arxiv.org/abs/hep-th/9710116v2.

[2] M. Akian, R. Bapat, and S. Gaubert, Min-plus methods in eigenvalue perturbation theory and generalised Lidskii-Vishik-Ljusternik theorem, preprint at arXiv:math.math.SP/0402090, 2005.

[3] M. Akian, S. Gaubert, and A. Guterman. Linear independence over tropical semirings and beyond. Contemp. Math., to appear.

[4] F. Ardila and S. Billey, Flag arrangements and triangulations of products of simplices, preprint at math/0605598, 2006.

[5] P. Butkovic, Max-algebra: the linear algebra of combinatorics? *Lin. Alg. and Appl.*, pages 313–335, 2003.

[6] M. Develin, F. Santos, and B. Sturmfels, On the rank of a tropical matrix, *Combinatorial and computational geometry, Math. Sci. Res. Inst. Publ., Cambridge Univ. Press, Cambridge*, 53:213–242, 2005, (preprint at arXiv:math.CO/0312114).

[7] A. Dickenstein, E.M. Feichtner, and B. Sturmfels. Tropical discriminants. *J. Amer. Math. Soc*, (20):1111–1133., 2007. (Preprint at arXiv:math/0510126.)

[8] A. Gathmann, Tropical algebraic geometry, Jahresbericht der DMV, 108:3–32, 2006. (Preprint at arXiv:math.AG/0601322.)

[9] S. Gaubert and R. Katz, Rational semimodules over the max-plus semiring and geometric approach of discrete event systems, *Kybernetika-Volume* 40, 2:153, 2004.

[10] S. Gaubert and M. Plus, Methods and applications of (max, +) linear algebra, In Proc. 14th Symposium on Theoretical Aspects of Computer Science, LNCS Springer-Verlag, Proceedings of STACS '97, 1997.

[11] J. Golan, *The theory of semirings with applications in mathematics and theoretical computer science*, volume 54, Longman Sci & Tech., 1992.

[12] I. Itenberg, G. Mikhalkin, and E. Shustin, *Tropical algebraic geometry*, volume 35, Birkhäuser, 2007, Oberwolfach seminars.

[13] Z. Izhakian, Tropical arithmetic and algebra of tropical matrices. *Communications in Algebra*, 37(4):1–24, 2009. (Preprint at arXiv:math.AG/0505458.)

[14] Z. Izhakian, M. Knebusch, and L. Rowen, Supertropical semirings and supervaluations, preprint, 2009.

[15] Z. Izhakian and L. Rowen, Supertropical algebra, preprint at arXiv:0806.1175, 2007.

[16] Z. Izhakian and L. Rowen, Completions, reversals, and duality for tropical varieties, preprint at arXiv:0806.1175, 2008.

[17] Z. Izhakian and L. Rowen, Supertropical matrix algebra, to appear in Israel J. Math., preprint at arXiv:0806.1178, 2008.

[18] Z. Izhakian and L. Rowen. The tropical rank of a matrix. *Communications in Algebra*, 37(11):3912–3927, 2009. (Preprint at arXiv:math.ac/060420.)

[19] Z. Izhakian and L. Rowen. Supertropical matrix algebra II: Solving tropical equations. Preprint at arXiv:0902.2159, 2009.

[20] Z. Izhakian and L. Rowen, Supertropical resultants, preprint at arXiv:0902.2155, 2009.

[21] K.H. Kim, *Boolean Matrix Theory and Applications*, volume 70 of *Monographs and Textbooks in Pure and Applied*, Marcel Dekker, New York, 1982.

[22] K.H. Kim and F.W. Roush. Kapranov rank vs. tropical rank, *Proc. Amer. Math. Soc*, 134:2487–2494, 2006. (Preprint at arXiv:math.CO/0503044.)

[23] G. Litvinov, The Maslov dequantization, idempotent and tropical mathematics: a very brief introduction, *J. of Math. Sciences*, 140(3):1072–3374, 2007. (Preprint at arXiv:math.GM/0501038.)

[24] G. Mikhalkin, Enumerative tropical algebraic geometry in $\mathbb{R}^2$, *J. Amer. Math. Soc*, 18:313–377, 2005.

[25] J.-E. Pin, Tropical semirings, *Cambridge Univ. Press, Cambridge*, 11:50–69, 1998, Publ. Neton Inst. 11, Cambridge Univ.

[26] J. Richter-Gebert, B. Sturmfels, and T. Theobald, First steps in tropical geometry, *Idempotent mathematics and mathematical physics*, pages 289–317, 2005, Contemp. Math., Amer. Math. Soc., Providence, RI, 377.

[27] E. Shustin, A tropical approach to enumerative geometry, *Algebra i Analiz*, 17(2): 170–214, 2005.

[28] D. Speyer and B. Sturmfels, Tropical Grassmannians, *Adv. Geom.*, 4(3):389–411, 2004.

[29] D. Speyer and L. Williams, The tropical totally positive Grassmannian, *Journal of Algebraic Combinatorics*, 22(2):189–210, September 2005.

[30] D.E. Speyer and B. Sturmfels, Tropical mathematics, *Math. Mag.*, 82:163–173, 2009. (Preprint at arXiv:math.CO/0408099.)

[31] H. Straubing A combinatorial proof of the Cayley-Hamilton Theorem Discrete Math. 43 (2-3): 273-279, 1983.

[32] B. Sturmfels. On the Newton polytope of the resultant. *Journal of Algebraic Combinatorics*, 3:207–236, 1994.

[33] B. Sturmfels, *Solving Systems of Polynomial Equations* (*Cbms Regional Conference Series in Mathematics*), American Mathematical Society, October 2002.

[34] L.F. Tabera. Tropical resultants for curves and stable intersection, 2009. arXiv: 0805.1305v1.

Zur Izhakian and Louis Rowen
Department of Mathematics
Bar-Ilan University
Ramat-Gan 52900, Israel
e-mail: `zzur@math.biu.ac.il`
`rowen@macs.biu.ac.il`

Advances in Ring Theory
Trends in Mathematics, 303–326

Projective Modules, Idempotent Ideals and Intersection Theorems

Patrick F. Smith

For S.K. Jain on the occasion of his 70th birthday

Abstract. We investigate the relationship between projective modules and idempotent ideals for group rings, polynomial rings and more general rings, giving a survey of known results, proving some new results and raising a number of questions. In particular, it is proved that if R is any ring, X a projective right R-module and A an ideal of R such that the R-module X/XA can be generated by a set of elements of cardinality $\aleph$, for some infinite cardinal $\aleph$, then X/XB can be generated by a set of elements of cardinality $\aleph$, where B is the unique maximal idempotent ideal of R contained in A. A recurring theme is that of "intersection theorems" which give information about intersections of powers of ideals of the ring.

Mathematics Subject Classification (2000). Primary: 16D25, 16D40. Secondary: 16D30, 16P40, 16S30, 16S34, 16S36.

Keywords. Projective module, idempotent ideal, Noetherian ring, semiprime ring, group ring, polynomial ring.

1. Projective modules

All rings are associative with identity element and all modules are unital. Let R be a ring. For any R-module M we denote $Hom_R(M, R)$ by M^*. Recall that an ideal A of R is called *idempotent* if $A = A^2$. An idempotent ideal A of R will be called *trivial* if $A = R$ or $A = 0$, otherwise A is called *non-trivial*. Clearly if e is an idempotent element of R and A is the ideal ReR then A is an idempotent ideal of R. However, it is easy to give examples of rings R for which the only idempotent elements are 0, 1 but R contains non-trivial idempotent ideals and we mention some examples later. We are interested in the relationship between idempotent ideals of R and projective R-modules. Recall the *Dual Basis Lemma* (see, for example, [2, p. 203] or [43, 18.6]).

Lemma 1.1. *Let R be any ring. A right R-module X is projective if and only if there exist an index set Λ, elements $x_\lambda\,(\lambda \in \Lambda)$ of X and $\varphi_\lambda\,(\lambda \in \Lambda)$ of X^* such that for each $x \in X$ there exists at most a finite number of elements $\lambda \in \Lambda$ such that $\varphi_\lambda(x) \neq 0$ and $x = \sum_{\lambda\in\Lambda} x_\lambda \varphi_\lambda(x)$.*

Next we mention a famous theorem of Kaplansky [17] (see, for example, [2, Corollary 26.2] or [43, 8.10]).

Theorem 1.2. *Let R be any ring. Then every projective (right or left) R-module is a direct sum of countably generated submodules.*

As a consequence Kaplansky [17] proved the following result.

Corollary 1.3. *Let R be a ring with Jacobson radical J such that the ring R/J is a division ring. Then every projective (right or left) R-module is free.*

Let R be a general ring and again let J denote the Jacobson radical of R. Corollary 1.3 has been generalized in a number of different ways. For example, Beck [4] proved that if X is a projective right R-module such that X/XJ is a free (R/J)-module then X is a free R-module. This is a consequence of the fact that if F is a free right R-module such that $F = FJ + F_1$, for some submodule F_1 of F, then there exists a direct summand F_2 of F such that $F_2 \subseteq F_1$ and $F_2 \cong F$. Beck's Theorem was improved by Příhoda [28] who proved that if X and Y are projective right R-modules and $X/XJ \cong Y/YJ$ then $X \cong Y$. Recall that a projective R-module is called a *generator* in case there exists a positive integer n such that the projective R-module $X^{(n)}$ contains a non-zero free direct summand. In another direction, Akasaki [1] proved that if R/J is a finite direct product of division rings and every projective R-module is a generator then every projective R-module is free. This not only generalized Corollary 1.3 but also a theorem of Hinohara [11] who proved the result in case R is commutative. Hinohara [12] extended his theorem to what he called "weakly Noetherian" commutative rings.

We also ought to mention work of Warfield [41] here. An R-module M is said to have the *exchange property* if, for any R-module L such that $L = M \oplus N = \oplus_{i\in I} L_i$, for some submodules N and $L_i\,(i \in I)$ of L, then there exist submodules $H_i \subseteq L_i\,(i \in I)$ such that $L = M \oplus (\oplus_{i\in I} H_i)$. The ring R is called an *exchange ring* if the module R_R has the exchange property and in this case the module ${}_RR$ also has the exchange property. The ring R is an exchange ring provided for each r in R there exists an idempotent e in R such that $rR + J = eR + J$. If R is an exchange ring then every projective right R-module is isomorphic to a direct sum of right ideals each generated by an idempotent. See [41] for more details.

2. Idempotent ideals

Let R be a ring and let M be a right R-module. Following [43, p. 154] the *trace ideal* of M, denoted by $Tr(M)$, is defined by

$$Tr(M) = \sum_{\varphi \in M^*} \varphi(M).$$

It is well known and easy to check that $Tr(M)$ is a two-sided ideal of R. Now suppose that X is a projective right R-module. By Lemma 1.1 there exist an index set Λ, elements x_λ $(\lambda \in \Lambda)$ of X and φ_λ $(\lambda \in \Lambda)$ of X^* such that for each $x \in X$ there exists at most a finite number of elements $\lambda \in \Lambda$ such that $\varphi_\lambda(x) \neq 0$ and $x = \sum_{\lambda \in \Lambda} x_\lambda \varphi_\lambda(x)$. Let $\theta \in X^*$. For each $x \in X$,

$$\theta(x) = \theta\left(\sum_{\lambda \in \Lambda} x_\lambda \varphi_\lambda(x)\right) = \sum_{\lambda \in \Lambda} \theta(x_\lambda)\varphi_\lambda(x) \in Tr(X)^2.$$

It follows that $Tr(X)$ is an idempotent ideal of R. This and other properties of trace ideals we state in the next result. The proofs are very easy and are omitted.

Lemma 2.1. *Let R be any ring and let X and X_i $(i \in I)$ be projective right R-modules. Then*

(i) *$Tr(X)$ is an idempotent ideal of R.*
(ii) *$X = XTr(X)$.*
(iii) *$Tr(X) = \cap\{A : A$ is an ideal of R such that $X = XA\}$.*
(iv) *$Tr(X) = R$ if and only there exists a positive integer n such that $X^{(n)} \cong R \oplus Y$ for some R-module Y.*
(v) *$Tr(eR) = ReR$ for every idempotent element e in R.*
(vi) *$Tr(\oplus_{i \in I} X_i) = \sum_{i \in I} Tr(X_i)$.*

Lemma 2.1 has the following consequence (see [3, Proposition 2.4]).

Corollary 2.2. *Let R be any ring and let X be a projective right R-module. Then $Tr(X) = R$ if and only if the countable direct sum $X \oplus X \oplus \ldots$ is free.*

Let R be any ring and let X be any non-zero projective right R-module with trace ideal T. Note that $X = XT$ so that the (R/T)-module X/XT is zero. Given an infinite cardinal $\aleph$, in [3] Bass defines X to be *uniformly $\aleph$-big* provided X has a generating set of cardinality $\aleph$ (and in this case we shall call X *$\aleph$-generated*) but there does not exist a proper ideal A of R such that X/XA has a generating set of cardinality less than $\aleph$. Then X is *uniformly big* if X is uniformly $\aleph$-big for some infinite cardinal $\aleph$. Clearly (non-zero) uniformly big projective modules are generators. As usual $\aleph_0$ will denote the cardinality of the natural numbers. Bass [3, Theorems 2.2 and 3.1] proves the next result.

Theorem 2.3. *Let R be a ring with Jacobson radical J .*

(i) *If R/J is right or left Noetherian then every uniformly $\aleph$-big projective right R-module is free, for every uncountable cardinal $\aleph$.*

(ii) *If R/J is right Noetherian then every uniformly $\aleph_0$-big projective right R-module is free.*

Bass [3] defines a ring R to be *right p-connected* provided $Tr(X) = R$ for every non-zero projective right R-module X, in other words if every non-zero projective right R-module is a generator. He proves the following consequence of Theorems 1.2 and 2.3 (see [3, Corollary 3.4]).

Corollary 2.4. *Let R be a right p-connected ring with Jacobson radical J such that the ring R/J is right Noetherian. Then every infinite direct sum of projective right R-modules is free. In particular, every non-countably generated projective right R-module is free.*

Note that Corollary 2.2 shows that every ring R such that non-finitely generated (or even non-countably generated) projective modules are free is right p-connected. Note also that if R is a ring with Jacobson radical J such that R/J is a right Noetherian ring then an $\aleph_0$-generated projective right R-module X is free if and only if X/XA is not finitely generated for every proper ideal A of R. This raises the following obvious question.

Question 2.5. *Let R be a ring with Jacobson radical J such that the ring R/J is left Noetherian. Is every uniformly $\aleph_0$-big projective right R-module free?*

The relationship that exists between projective modules and idempotent ideals was further illustrated by Whitehead [42] who gives a necessary and sufficient condition for an idempotent ideal to be the trace ideal of a countably generated projective module. This gives as a consequence the following result.

Theorem 2.6. *Let A be an idempotent ideal of a ring R such that the left ideal A is finitely generated. Then there exists a countably generated projective right R-module X such that $A = Tr(X)$.*

Corollary 2.7. *Let R be a left Noetherian ring. Then R is right p-connected if and only if R has no non-trivial idempotent ideals.*

Note that Corollary 2.7 is not true for rings R which are not left Noetherian. In Section 5 we shall give an example of a commutative ring R such that every projective R-module is free, and therefore R is p-connected, but R contains a non-trivial idempotent ideal.

3. Projective modules and idempotent ideals

In this section we investigate further the relationship between projective modules and idempotent ideals. Let R be a ring and let A be any ideal of R. Consider the descending chain of ideals

$$A = A^1 \supseteq A^2 \supseteq \cdots \supseteq A^\alpha \supseteq A^{\alpha+1} \supseteq \ldots$$

of R, where, for each ordinal $\alpha \geq 1$, $A^{\alpha+1} = A^\alpha A$ and $A^\alpha = \bigcap_{1 \leq \beta < \alpha} A^\beta$ when α is a limit ordinal. Because R is a set there exists an ordinal $\rho \geq 1$ such that $A^\rho = A^{\rho+1}$ and in this case we write $\kappa(A) = A^\rho$. Jacobson [13, Theorem 11] proves that if R is a right Noetherian ring with Jacobson radical J then $\kappa(J) = 0$. We can then define a descending chain of ideals

$$A = \kappa^0(A) \supseteq \kappa^1(A) \supseteq \cdots \supseteq \kappa^\alpha(A) \supseteq \kappa^{\alpha+1}(A) \supseteq \ldots,$$

where, for each ordinal $\alpha \geq 0$, $\kappa^{\alpha+1}(A) = \kappa(\kappa^\alpha(A))$ and $\kappa^\alpha(A) = \bigcap_{0 \leq \beta < \alpha} \kappa^\beta(A)$ for every limit ordinal α. In particular, note that $\kappa^1(A) = \kappa(A)$. Because R is a set there exists an ordinal $\nu \geq 0$ such that $\kappa^\nu(A) = \kappa^{\nu+1}(A)$. In this case, clearly $\kappa^\nu(A)$ is an idempotent ideal of R contained in A. Let $\operatorname{id}(A)$ denote the sum of all idempotent ideals of R contained in A. Clearly $\operatorname{id}(A)$ is an idempotent ideal of R and $\operatorname{id}(A) \subseteq \kappa^\alpha(A)$ for every ordinal $\alpha \geq 0$. We have proved the following result.

Lemma 3.1. *Let A be an ideal of an arbitrary ring R. Then there exists an ordinal $\nu \geq 0$ such that $\kappa^\nu(A) = \operatorname{id}(A)$.*

Let A be any ideal of a ring R and let $B = \operatorname{id}(A)$. Note that B is the unique largest idempotent ideal of R contained in A. Let C be any ideal of R such that $B \subseteq C \subseteq A$ and C/B is an idempotent ideal of the ring R/B. Then

$$C \subseteq C^2 + B = C^2 + B^2 \subseteq C^2,$$

and hence $C = C^2$ and $B = C$. We have proved that $\operatorname{id}(A/\operatorname{id}(A)) = 0$.

We now define, for any ideal A of a ring R, $\delta(A) = \bigcap_{i=1}^\infty A^i$ and a descending chain

$$A = \delta^0(A) \supseteq \delta^1(A) \supseteq \cdots \supseteq \delta^\alpha(A) \supseteq \delta^{\alpha+1}(A) \supseteq \ldots,$$

where, for each ordinal $\alpha \geq 0$, $\delta^{\alpha+1}(A) = \delta(\delta^\alpha(A))$ and $\delta^\alpha(A) = \bigcap_{0 \leq \beta < \alpha} \delta^\beta(A)$ for every limit ordinal α.

Lemma 3.2. *With the above notation, if A is a proper ideal of a ring R then $\kappa^\alpha(A) \subseteq \delta^\alpha(A)$ for every ordinal $\alpha \geq 0$.*

Proof. By transfinite induction on $\alpha \geq 0$. □

Note that the above remarks show that, for any ideal A of a ring R, there exists an ordinal $\mu \geq 0$ such that $\delta^\mu(A) = \operatorname{id}(A)$, and, of course, in this case $\kappa^\mu(A) = \operatorname{id}(A)$. Thus the sequence $\{\kappa^\alpha(A)\}$ "converges" to $\operatorname{id}(A)$ faster than the sequence $\{\delta^\alpha(A)\}$. In [14], Jategaonkar shows that, for each ordinal $\alpha \geq 1$, there exists a principal right ideal ring with unique maximal ideal J such that R/J is a division ring and $\kappa(J) = 0$ but $\delta^\alpha(J) \neq 0$.

Let R be a ring with Jacobson radical J. Krause and Lenagan [20] prove that if R is a ring with right Krull dimension α, for some ordinal $\alpha \geq 0$, then $\delta^\alpha(J)$ is nilpotent and hence $\delta^{\alpha+1}(J) = 0$. Herstein [9] gives an example of a right Noetherian PI ring with $\delta(J) \neq 0$ and he and Small [10] show that, for each right Noetherian PI ring R, $\delta^m(J) = 0$ for some positive integer m. Cauchon [6] proved that $\delta^2(J) = 0$ for every right Noetherian PI ring R and Jategaonkar [15] proved that $\delta(J) = 0$ for every right and left FBN ring (and in particular every right and left Noetherian PI ring). (For the definition and basic properties of Krull dimension, PI rings and FBN rings see [23].) Now we ask the following question:

Question 3.3. *Does there exist a right and left Noetherian ring R and an ideal A of R such that $\kappa^n(A) = 0$, for some positive integer n, but $\delta^m(A) \neq 0$ for every positive integer m?*

In view of Bass' work referred to in Section 1 above we are interested in the following situation: X is a projective right module over a ring R and A is an ideal of R such that X/XA has a generating set of cardinality $\aleph$ (i.e., X/XA is $\aleph$-generated). We first deal with the situation when X/XA is finitely generated.

Theorem 3.4. *Let A be an ideal of a ring R and let X be a projective right R-module such that X/XA is a finitely generated module. Then $X/X\kappa(A)$ is finitely generated.*

Proof. Let $B = \kappa(A)$. Let Y be a finitely generated submodule of X such that $X = Y + XA$. Let F be a free right R-module with basis $\{f_\lambda : \lambda \in \Lambda\}$, for some index set Λ, such that $F = X \oplus X'$, for some submodule X' of F. There exists a finite subset Λ_1 of Λ such that $Y \subseteq F_1$, where $F_1 = \sum_{\lambda\in\Lambda_1} f_\lambda R$. Let $\Lambda_2 = \Lambda \setminus \Lambda_1$ and let $F_2 = \sum_{\lambda\in\Lambda_2} f_\lambda R$. Note that

$$X = Y + XA \subseteq F_1 + (F_1 + F_2)A = F_1 \oplus F_2 A.$$

Now we claim that

$$X \subseteq F_1 \oplus F_2 A^\alpha,$$

for every ordinal $\alpha \geq 1$. Suppose that this statement is false and let α be the least ordinal such that $X \nsubseteq F_1 \oplus F_2 A^\alpha$. Then $\alpha \geq 2$. If $\alpha - 1$ exists then $X \subseteq F_1 \oplus F_2 A^{\alpha-1}$ and hence

$$X = Y + XA \subseteq Y + (F_1 \oplus F_2 A^{\alpha-1})A \subseteq F_1 \oplus F_2 A^\alpha.$$

Thus α is a limit ordinal and

$$X \subseteq F_1 \oplus F_2 A^\beta,$$

for all ordinals $1 \leq \beta < \alpha$. It follows that

$$X \subseteq \cap_{1\leq\beta<\alpha}(F_1 \oplus F_2 A^\beta) = F_1 \oplus [\cap_{1\leq\beta<\alpha}(F_2 A^\beta)] = F_1 \oplus F_2 A^\alpha,$$

a contradiction. Thus $X \subseteq F_1 \oplus F_2 A^\alpha$ for all ordinals $\alpha \geq 1$. In particular, $X \subseteq F_1 \oplus F_2 B$. It follows that

$$X \subseteq F_1 \oplus F_2 B \subseteq F_1 + (X + X')B \subseteq F_1 + XB + X'B,$$

and hence $X = \pi(X) = \pi(F_1) + XB$, where $\pi : F \to X$ is the canonical projection. Because, F_1 is finitely generated, we deduce that $\pi(F_1)$ is finitely generated and hence so too is X/XB. □

Note that the above proof can easily be adapted to prove that if A an ideal of R such that, for some projective right R-module X, X/XA is $\aleph$-generated, for some infinite cardinal $\aleph$, then $X/X\kappa(A)$ is also $\aleph$-generated.

Corollary 3.5. *Let R be a right Noetherian ring with Jacobson radical J such that every projective right (R/J)-module is finitely generated or free. Then every projective right R-module is finitely generated or free.*

Proof. Let X be any projective right R-module. Then X/XJ is a projective right (R/J)-module. If X/XJ is free then X is a free R-module by Beck's Theorem mentioned above (see [4]). Suppose that X/XJ is finitely generated. By Theorem 3.4 and [13, Theorem 11], X is finitely generated. □

Now we ask:

Question 3.6. *Let R be a ring with Jacobson radical J such that every projective (R/J)-module is finitely generated or free. Is every projective R-module finitely generated or free?*

Note also the following immediate corollary of Theorem 3.4.

Corollary 3.7. *Let A be an ideal of a ring R and let X be a projective right R-module such that X/XA is a finitely generated module. Then $X/X\kappa^n(A)$ is finitely generated for every positive integer n.*

This corollary immediately raises the following question.

Question 3.8. *Let A be an ideal of a ring R and let X be a projective right R-module such that X/XA is a finitely generated module. Is $X/X\kappa^\omega(A)$ also finitely generated, where ω is the first infinite ordinal?*

The difficulty that arises in trying to answer Question 3.8 is that the generating sets for the modules $X/X\kappa^n(A)$ may increase in size as the positive integer n increases. For example in the proof of Theorem 3.4, $\pi(F_1)$ may require more generators than Y. We shall show that this problem disappears for infinite generating sets. We first prove a lemma.

Lemma 3.9. *Let R be an arbitrary ring, let F be a free right R-module with basis $\{f_\lambda : \lambda \in \Lambda\}$ and let X be a direct summand of F with $\pi : F \to X$ the canonical projection. Let Y be a countably generated submodule of X. Then there exist a countably generated submodule Z of X with $Y \subseteq Z$ and a countable subset Ω of Λ such that if $G = \sum_{\lambda\in\Omega} f_\lambda R$ then $Z \subseteq G$ and $\pi(G) = Z$.*

Proof. Suppose that S is a countable generating set for Y. For each $s \in S$ there exists a finite subset Λ_s of Λ such that $s \in \sum_{\lambda\in\Lambda_s} f_\lambda R$. Let $\Lambda_Y = \cup_{s\in S} \Lambda_s$ and let $F_Y = \sum_{\lambda\in\Lambda_Y} f_\lambda R$. Then Λ_Y is a countable subset of Λ and $Y \subseteq F_Y$. Note that $Y \subseteq \pi(F_Y) = \sum_{\lambda\in\Lambda_Y} \pi(f_\lambda)R$.

Let $Y_1 = \sum_{\lambda\in\Lambda_Y} \pi(f_\lambda)R$. As before, starting with Y_1 we can find a countable subset Λ_{Y_1} of Λ such that $Y_1 \subseteq F_{Y_1} = \sum_{\lambda\in\Lambda_{Y_1}} f_\lambda R$. Let $Y_2 = \sum_{\lambda\in\Lambda_{Y_1}} \pi(f_\lambda)R$. Repeat this process to obtain countably generated submodules $Y \subseteq Y_1 \subseteq Y_2 \subseteq \cdots$ of X and countable subsets $\Lambda_Y \subseteq \Lambda_{Y_1} \subseteq \Lambda_{Y_2} \subseteq \cdots$ of Λ such that $Y_i \subseteq \sum_{\lambda\in\Lambda_{Y_i}} f_\lambda R$ and $Y_{i+1} = \sum_{\lambda\in\Lambda_{Y_i}} \pi(f_\lambda)R$, for all positive integers i. Let $Z = \cup_{i\geq 1} Y_i$, $\Omega =$

$\cup_{i\geq 1}\Lambda_{Y_i}$ and let $G = \sum_{\lambda\in\Omega} f_\lambda R$. Then Ω is a countable subset of Λ. Clearly $Z \subseteq G$. Let $\lambda \in \Omega$. Then $\pi(f_\lambda) \in Y_i \subseteq Z$, for some $i \geq 1$. Thus $\pi(G) \subseteq Z$. But $Z \subseteq G$ implies that $Z = \pi(Z) \subseteq \pi(G)$, so that $\pi(G) = Z$. □

For convenience, we have chosen to state and prove Lemma 3.9 for a countably generated submodule Y but it is clear that the same proof would prove the corresponding result for any $\aleph$-generated submodule Y, for any infinite cardinal $\aleph$. More precisely, if, in Lemma 3.9, Y is $\aleph$-generated then so too is Z and Ω has cardinality at most $\aleph$. But we have the additional information that if Y is finitely generated then Ω is countable. This brings us to the main new theorem.

Theorem 3.10. *Let A be an ideal of a ring R and let X be a projective right R-module such that X/XA is $\aleph$-generated, for some infinite cardinal $\aleph$. Then $X/X\,\mathrm{id}(A)$ is $\aleph$-generated.*

Proof. Let $B = \mathrm{id}(A)$. Let Y be an $\aleph$-generated submodule of X such that $X = Y + XA$. Let F be a free right R-module with basis $\{f_\lambda : \lambda \in \Lambda\}$, for some index set Λ, such that $F = X \oplus X'$, for some submodule X' of F. Let $\pi : F \to X$ denote the canonical projection. By Lemma 3.9 there exist an $\aleph$-generated submodule Z of X with $Y \subseteq Z$ and a subset Ω of Λ of cardinality at most $\aleph$ such that if $G = \sum_{\lambda\in\Omega} f_\lambda R$ then $Z \subseteq G$ and $\pi(G) = Z$. Let $\Omega' = \Lambda \setminus \Omega$ and let $G' = \sum_{\lambda\in\Omega'} f_\lambda R$. Note that

$$X = Y + XA \subseteq Z + (G + G')A \subseteq G \oplus G'A.$$

Now we claim that

$$X \subseteq G \oplus G'\kappa^\alpha(A),$$

for every ordinal $\alpha \geq 0$. Suppose that this is not the case. Let α be the least ordinal such that $X \not\subseteq G \oplus G'\kappa^\alpha(A)$. Clearly $\alpha \geq 1$. If $\alpha - 1$ exists then $X \subseteq G \oplus G'C$ where $C = \kappa^{\alpha-1}(A)$. Next we prove by induction that

$$X \subseteq G \oplus G'C^\beta,$$

for every ordinal $\beta \geq 1$. Suppose that this statement is false and let β be the least ordinal such that $X \not\subseteq G \oplus G'C^\beta$. Clearly $\beta \geq 2$. If $\beta - 1$ exists then $X \subseteq G \oplus G'C^{\beta-1}$. Note that

$$X \subseteq G + G'C = G + (X + X')C = G + XC + X'C,$$

so that $X = \pi(X) \subseteq \pi(G) + XC = Z + XC$. Thus

$$X \subseteq Z + XC \subseteq Z + (G \oplus G'C^{\beta-1})C \subseteq Z + GC + G'C^\beta \subseteq G \oplus G'C^\beta,$$

a contradiction. Thus β is a limit ordinal and

$$X \subseteq G \oplus G'C^\gamma,$$

for all ordinals $1 \leq \gamma < \beta$. It follows that

$$X \subseteq \cap_{1\leq\gamma<\beta}(G \oplus G'C^\gamma) = G \oplus [\cap_{1\leq\gamma<\beta}(G'C^\gamma)] = G \oplus G'C^\beta,$$

a contradiction. Thus $X \subseteq G \oplus G'C^\beta$ for all ordinals $\beta \geq 1$.

In particular,

$$X \subseteq G \oplus G'\kappa(C) = G \oplus G'\kappa^{\alpha}(A),$$

another contradiction. Thus α is a limit ordinal. It follows that $X \subseteq G \oplus G'\kappa^{\mu}(A)$ for every ordinal $0 \leq \mu < \alpha$. Now we have

$$X \subseteq \cap_{0 \leq \mu < \alpha}[G \oplus G'\kappa^{\mu}(A)] = G \oplus [\cap_{0 \leq \mu < \alpha}(G'\kappa^{\mu}(A))] = G \oplus G'\kappa^{\alpha}(A),$$

a contradiction. Thus $X \subseteq G \oplus G'\kappa^{\alpha}(A)$ for all ordinals $\alpha \geq 0$. In particular,

$$X \subseteq G \oplus G' \operatorname{id}(A) \subseteq G + X \operatorname{id}(A) + X' \operatorname{id}(A),$$

so that $X = \pi(X) \subseteq \pi(G) + X \operatorname{id}(A) = Z + X \operatorname{id}(A)$ and hence $X = Z + X \operatorname{id}(A)$. Thus $X/X \operatorname{id}(A)$ is $\aleph$-generated. □

Theorem 3.10 has many consequences and we mention some of these next.

Corollary 3.11. *Let A be an ideal of a ring R and let X be a projective right R-module such that X/XA is finitely generated. Then $X/X \operatorname{id}(A)$ is countably generated.*

Corollary 3.12. *Let R be any ring and let $\aleph$ be any cardinal with $\aleph > \aleph_0$. Then an $\aleph$-generated projective right R-module X is uniformly $\aleph$-big if and only if X/XA is not $\aleph'$-generated for any proper idempotent ideal A of R and cardinal $\aleph' < \aleph$.*

Proof. The necessity is clear. Conversely, suppose that X is not uniformly $\aleph$-big. There exists a cardinal $\aleph' < \aleph$ and a proper ideal B of R such that X/XB is $\aleph'$-generated. By Theorem 3.10 $X/X \operatorname{id}(B)$ is $\aleph'$-generated and $\operatorname{id}(B)$ is an idempotent ideal of R. □

Compare the next result with Corollary 2.4.

Corollary 3.13. *Let R be a ring with no non-trivial idempotent ideals. Then every projective (right or left) R-module is countably generated or uniformly big.*

Proof. By Corollary 3.12. □

4. Shallow rings

Let R be any ring. Recall that, for any ideal A of R, $\operatorname{id}(A)$ denotes the unique maximal idempotent ideal contained in A. Following [36], given a non-negative integer n, the ring R will be called *right n-shallow* provided $\kappa^n(A) = \operatorname{id}(A)$ for every ideal A of R. Next, the ring R is called *right shallow* if for each ideal A of R there exists a positive integer m such that $\kappa^m(A) = \operatorname{id}(A)$. For example, Jacobson's Theorem mentioned above implies that if R is a right Noetherian ring with Jacobson radical J and J is a maximal ideal of R then R is right 1-shallow (in fact, $\kappa(A) = 0$ for every proper ideal A of R). More generally, it is proved in [36, Theorem 3.4] that if R is a right Noetherian ring with Jacobson radical J such that J is an intersection of n distinct maximal ideals, for some positive integer n, then R is right ($2n$-1)-shallow. If R is a commutative Noetherian ring with n minimal prime ideals, where n is a positive integer, then R is (right) $(n+1)$-shallow (see [36, Theorem 2.6]).

A ring R is called *right* $\aleph_0$*-hereditary* provided every countably generated right ideal is projective. Bergman [5, Corollary 8.3] has shown that if A is an ideal of a left $\aleph_0$-hereditary ring then $\kappa(A) = \cap_{i=1}^{\infty} A^i$. If R is also right $\aleph_0$-hereditary then $\kappa(A)$ is an idempotent ideal (see [5, Corollary 5.3]). Thus right and left $\aleph_0$-hereditary rings are right and left 1-shallow. Moreover, right hereditary rings are right 1-shallow by [36, Proposition 4.4]. Other examples of right shallow rings include right Artinian rings (clearly), principal right ideal rings (see [36, Theorem 2.8]) and rings with right Krull dimension 1 (see [36, Proposition 2.9]).

In order to find examples of right shallow rings we now prove the following result.

Theorem 4.1. *Let R be a semiprime ring such that R contains only a finite number of minimal prime ideals and whenever $P \subseteq Q$ are prime ideals of R there exists a positive integer m such that $\kappa^m(Q) \subseteq P$ (respectively, $\delta^m(Q) \subseteq P$). Let A be any ideal of R. Then there exists a positive integer n such that $\kappa^n(A)$ (respectively, $\delta^n(A)$) is generated by a central idempotent of R.*

Proof. We prove the result for κ, the proof for δ being similar. Let $P_1, \ldots, P_k$ denote the minimal prime ideals of R, for some positive integer k. Note that $P_1 \cap \cdots \cap P_k = 0$. Let A be any proper ideal of R. Clearly $A \neq R$ implies that $A + P_i \neq R$ for some $1 \leq i \leq k$. Without loss of generality, $i = 1$. There exists a maximal ideal M of R such that $A + P_1 \subseteq M$. By hypothesis, $\kappa^t(A) \subseteq \kappa^t(M) \subseteq P_1$ for some positive integer t. Let $A_1 = \kappa^t(A)$ and note that $A_1 \subseteq P_1$. Let $B = \cap_{i=2}^{n} P_i$. If $R = A_1 + P_i \, (2 \leq i \leq n)$ then $R = A_1 + B$ and $A_1 \cap B = 0$ so that A_1 is generated by a central idempotent. Otherwise we can suppose without loss of generality that $R \neq A_1 + P_2$. Then, as before, there exists a positive integer s such that $\kappa^s(A_1) \subseteq P_2$, so that $\kappa^{s+t}(A) \subseteq P_1 \cap P_2$. Repeating this argument there exist a positive integer n and an integer $1 \leq j \leq k$ such that $\kappa^n(A) \subseteq P_1 \cap \cdots \cap P_j$ and $R = \kappa^n(A) + (P_{j+1} \cap \cdots \cap P_k)$. Thus $R = \kappa^n(A) \oplus (P_{j+1} \cap \cdots \cap P_k)$ and hence $\kappa^n(A)$ is generated by a central idempotent element of R. □

Note that in Theorem 4.1 one only needs that Q be a maximal ideal of R. To see how Theorem 4.1 is useful in providing examples of right shallow rings we prove the following corollary.

Corollary 4.2. *Let R be a ring which satisfies the ascending chain condition on (two-sided) ideals such that whenever $P \subseteq Q$ are prime ideals of R there exists a positive integer m such that $\kappa^m(Q) \subseteq P$. Then R is a right shallow ring.*

Proof. It is well known that R contains only a finite number of minimal prime ideals and that if $P_1, \ldots, P_k$ are the minimal prime ideals of R, for some positive integer k, then the ideal $N = P_1 \cap \cdots \cap P_k$ is nilpotent. Let $\overline{R}$ denote the semiprime ring R/N. Let A be any proper ideal of R. Let $\overline{A}$ denote the ideal $(A + N)/N$ of $\overline{R}$. By Theorem 4.1 there exists a positive integer n such that $\kappa^n(\overline{A})$ is generated by a central idempotent element $\overline{e} = e + N$ of $\overline{R}$. Because $e + N \in \overline{A} = (A + N)/N$ we can suppose without loss of generality that $e \in A$. Because idempotents can be lifted modulo the nilpotent ideal N, we can suppose without loss of generality that $e = e^2$. It is easy to check that $(\kappa^n(A) + N)/N \subseteq \kappa^n(\overline{A})$ and hence we have the following:

$$ReR \subseteq \kappa^n(A) \subseteq ReR + N.$$

Suppose that $N^t = 0$ for some positive integer t. Then $(\kappa^n(A))^t \subseteq (ReR + N)^t \subseteq ReR + N^t = ReR$, so that $\kappa^{n+1}(A) = ReR = \mathrm{id}(A)$. It follows that R is a right shallow ring. □

An ideal A of a general ring R has the *right Artin-Rees property* or simply the *right AR property* provided for each right ideal L of R there exists a positive integer n such that $L \cap A^n \subseteq LA$. For a discussion of the AR property see [23, Section 4.2] or [39]. Perhaps the most familiar ideals with the right AR property are those found by Nouazé and Gabriel [25]. For any ring S, let $C(S)$ denote the centre of S. Following [23, 4.1.13], given a positive integer k, elements $c_1, c_2, \ldots, c_k$ of R form a *centralizing sequence of elements* of R provided

$$c_1 \in C(R) \text{ and } c_i \in C(R/(c_1R + \cdots + c_{i-1}R)) \text{ for all } 2 \leq i \leq k.$$

Nouazé and Gabriel [25, 2.7] proved that if R is a right Noetherian ring and A an ideal of R such that A is generated by a centralizing sequence of elements of R then A satisfies the right AR property. This fact was generalized by McConnell (see [23, 4.2.7]). By a *right Ore set* in a general ring R we mean a multiplicatively closed subset T of R such that $1 \in T$ and, for all $r \in R$ and $t \in T$, there exist $r_1 \in R$ and $t_1 \in T$ such that $rt_1 = tr_1$. Recall that an element c of an arbitrary ring R is called *regular* if $rc \neq 0$ and $cr \neq 0$ for every non-zero element r of R. Let U be a non-empty subset of a ring R. Then $\mathbf{r}_R(U)$ will denote the right annihilator and $\mathbf{l}_R(U)$ the left annihilator of U in R. The next result is taken partially from [37, Theorem 1.1] (see also [23, 4.2.9]).

Theorem 4.3. *Let R be a right Noetherian ring and let A be an ideal of R. Consider the following statements.*

(i) *A has the right AR property.*
(ii) *$\{1 - a : a \in A\}$ is a right Ore set.*
(iii) *$\kappa(A) = \{r \in R : r(1 - a) = 0$ for some $a \in A\}$.*
(iv) *$E \cap \kappa(A) \subseteq EA$ for every right ideal E of R.*

Then (i) ⇒ (ii) ⇒ (iii) ⇔ (iv).

Proof. (i) ⇒ (ii) Let $r \in R$, $a \in A$. For each positive integer n let $b_n = r(1 - a^n) - (1 - a^n)r = a^n r - ra^n \in A^n$. Let $E = b_1R + b_2R + \ldots$. Because R is right Noetherian, there exists a positive integer k such that $E = b_1R + \cdots + b_kR$. Moreover, by hypothesis, $E \cap A^m \subseteq EA$ for some positive integer m. Let $n = max\{k, m\}$. Then $b_n \in E \cap A^m \subseteq EA = b_1A + \cdots + b_kA$. Thus there exist $a_i \in A\,(1 \leq i \leq k)$ such that $b_n = b_1a_1 + \ldots b_ka_k$ and this gives

$$r(1 - a^n) - (1 - a^n)r = [r(1 - a) - (1 - a)r]a_1 + \cdots + [r(1 - a^k) - (1 - a^k)r]a_k.$$

Rearranging the terms we obtain $r(1-c) = (1-a)r_1$, where

$$c = a^n + (1-a)a_1 + \cdots + (1-a^k)a_k,$$

and

$$r_1 = (1 + a + \cdots + a^{n-1})r - ra_1 - \cdots - (1 + a + \cdots + a^{k-1})ra_k.$$

Note that $c \in A$. Now (ii) follows.

(ii) $\Rightarrow$ (iii) Let S denote the set of elements $1-a$ where $a \in A$. Let $B = \{r \in R : rs = 0 \text{ for some } s \in S\}$. Then it can be shown using (ii) that B is an ideal of R. Clearly $B \subseteq \kappa(A)$. Let $\overline{R}$ denote the ring R/B and for each element r in R let $\overline{r}$ denote the coset $r + B$. Let $r \in R$ and $s \in S$. Suppose that $rs \in B$. Then there exists $s' \in S$ such that $r(ss') = rss' = 0$, so that $r \in B$. Now suppose that $sr \in B$. Consider the ascending chain $\mathbf{r}_{\overline{R}}(\overline{s}) \subseteq \mathbf{r}_{\overline{R}}(\overline{s}^2) \subseteq \cdots$ of right ideals of $\overline{R}$. Because R is right Noetherian, there exists a positive integer k such that $\mathbf{r}_{\overline{R}}(\overline{s}^k) = \mathbf{r}_{\overline{R}}(\overline{s}^{k+1})$. By (ii) there exists $r_1 \in R$ and $s_1 \in S$ such that $s^k r_1 = rs_1$. Now $sr \in B$ implies that $s^{k+1}r_1 \in B$ and hence $s^k r_1 \in B$. Thus $rs_1 \in B$ and it follows that $r \in B$. We deduce that $\overline{S} = \{\overline{s} : s \in S\}$ is a right Ore set of regular elements of $\overline{R}$ and we can form the partial right quotient ring $Q = \overline{R}_{\overline{S}}$. It is easy to check that the set C of Q consisting of all elements $\overline{r}_{\overline{s}}$, where $r \in A$ and $s \in S$, is an ideal of Q and C is contained in the Jacobson radical J of Q. Because Q is a right Noetherian ring, $\kappa(C) \subseteq \kappa(J) = 0$. Again it is an easy check to show that $(A^\alpha + B)/B \subseteq C^\alpha$ for every ordinal $\alpha \geq 0$. It follows that $\kappa(A) \subseteq B$, as required.

(iii) $\Rightarrow$ (iv) Let $e \in E \cap \kappa(A)$. Then $e(1-a) = 0$ for some $a \in A$ and hence $e = ea \in EA$.

(iv) $\Rightarrow$ (iii) Let $r \in \kappa(A)$. Then (iv) gives $rR \subseteq rR \cap \kappa(A) \subseteq rRA = rA$ and hence $r(1-b) = 0$ for some $b \in A$. Moreover, if $s \in R$ and $s(1-c) = 0$ for some $c \in A$ then $s = sc \in \kappa(A)$. □

Corollary 4.4. *Let R be a prime right Noetherian ring and let A be a proper ideal of R such that $\{1 - a : a \in A\}$ is a right Ore set. Then $\kappa(A) = 0$.*

Proof. By Goldie's Theorem R is a right order in a simple right Artinian ring (see, for example, [23, Theorem 2.3.6]). Thus R satisfies the descending chain condition on right annihilators and hence also the ascending chain condition on left annihilators. There exists $c \in A$ such that $\mathbf{l}_R(1-c)$ is maximal in the collection of left annihilators of the form $\mathbf{l}_R(1-a)$, where $a \in A$. Let $b \in A$ and let $r \in R$ such that $r(1-b) = 0$. Then $(1-b)u = (1-c)(1-d)$ for some $u \in R$ and $d \in A$. But $\mathbf{l}_R(1-c) \subseteq \mathbf{l}_R(1-c)(1-d)$ so that $\mathbf{l}_R(1-c) = \mathbf{l}_R(1-c)(1-d)$ by the choice of c. It follows that $r(1-c) = 0$. By Theorem 4.3, $\kappa(A)(1-c) = 0$. Since $\kappa(A)$ is an ideal of the prime ring R and $1 - c \neq 0$ it follows that $\kappa(A) = 0$, as required. □

Corollary 4.5. *Let R be a right Noetherian ring such that whenever $P \subseteq Q$ are prime ideals of R there exists a positive integer m such that $\{1 - (c + P) : c \in \kappa^m(Q/P)\}$ is a right Ore set in the ring R/P. Then R is a right shallow ring.*

Proof. By Corollaries 4.2 and 4.4. □

We now return to consider projective modules. Let R be a right shallow ring and let X be a projective right R-module such that X/XA is finitely generated for some ideal A of R. There exists a positive integer n such that $\kappa^n(A) = \mathrm{id}(A)$. By Corollary 3.7 the module $X/X\,\mathrm{id}(A)$ is also finitely generated. Thus a countably (but not finitely) generated projective right R-module X is $\aleph_0$-big if and only if there does not exist a proper idempotent ideal I of R such that X/XI is finitely generated.

Theorem 4.6. *Let R be a right shallow ring with no non-trivial idempotent ideals such that the ring R/J is right Noetherian, where J is the Jacobson radical of R. Then every projective right R-module is finitely generated or free.*

We conclude this section with the following two related questions.

Question 4.7. *let R be a right and left Noetherian right shallow ring. Is R left shallow?*

Question 4.8. *Let R be a ring such that every projective right R-module is finitely generated or free. Is every projective left R-module finitely generated or free?*

5. Group rings

For the definition and basic properties of group rings see [27]. Let S be a ring, G a group and let R denote the group ring SG. Every element of R is a finite sum of the form $s_1g_1 + \cdots + s_ng_n$, for some positive integer n and elements $s_i \in S, g_i \in G\,(1 \le i \le n)$. Let $\epsilon : R \to S$ denote the *augmentation map* defined by

$$\epsilon(s_1g_1 + \cdots + s_ng_n) = s_1 + \cdots + s_n,$$

for all positive integers n and elements $s_i \in S, g_i \in G\,(1 \le i \le n)$. Then ϵ is an epimorphism whose kernel is the *augmentation ideal* ωG given by

$$\omega G = \left\{ \sum_{1\le i\le n} s_ig_i \,:\, \sum_{1\le i\le n} s_i = 0 \right\} = \sum_{g\in G}(g-1)R = \sum_{g\in G} R(g-1).$$

For any subgroup H of G we set $\omega H = \sum_{h\in H}(h-1)R$. Note that ωH is a right ideal of R. In case H is a normal subgroup of G, $\omega H = \sum_{h\in H} R(h-1)$, ωH is a two-sided ideal of R and ωH is the kernel of the canonical epimorphism $\epsilon_H : R \to S(G/H)$ defined by

$$\epsilon(s_1g_1 + \cdots + s_ng_n) = s_1(g_1H) + \cdots + s_n(g_nH),$$

for all positive integers n and elements $s_i \in S, g_i \in G\,(1 \le i \le n)$.

An idempotent element e of a ring S will be called *non-trivial* if $e \neq 0$ and $e \neq 1$. We shall call the ring S *connected* if it has no non-trivial idempotent elements. We can think of S as a subring of the group ring SG. Thus every idempotent element of S is an idempotent element of SG. It can turn out that even if S is connected, the group ring $R = SG$ can have idempotent elements. For example, suppose that S is any ring and that there exists an element $x \in G$ such that x

has order n for some integer $n \geq 2$, where n is a unit in S. Consider the element $e = n^{-1}(1+x+\cdots+x^{n-1})$ of R. It is easy to check that e is a non-trivial idempotent element of R distinct from 0 and 1. Moreover, $\epsilon(e) = 1$ so that $e - 1 \in \omega G$ and $R(e-1)R$ is a non-trivial idempotent ideal of R. Non-trivial idempotent ideals of R can arise in other ways. Let G be a finite non-soluble group such that the order of each non-trivial element of G is not a unit in S. Because G is finite and non-soluble, there exists a normal subgroup N of G such that $N = N'$, where N' denotes the derived subgroup of N. Then N is generated by the set of commutators $x^{-1}y^{-1}xy$, with $x, y \in N$, and it is easy to check that ωN is generated as a right (or left) ideal by the elements $x^{-1}y^{-1}xy - 1$ of R. But, for all $x, y \in N$,

$$x^{-1}y^{-1}xy-1 = x^{-1}y^{-1}(xy-yx) = x^{-1}y^{-1}[(x-1)(y-1)-(y-1)(x-1)] \in (\omega N)^2.$$

Thus $\omega N = (\omega N)^2$ and ωN is a non-trivial idempotent ideal of R.

There are even more possibilities for manufacturing non-trivial idempotent ideals in group rings. Let K be a field of characteristic p, for some prime p, and let G denote the Prüfer p-group. Then G is an Abelian (and hence soluble) group generated by elements $x_1, x_2, \ldots$ such that $x_1^p = 1$ and $x_{i+1}^p = x_i$ for every positive integer i. It follows that in the group ring KG, $(x_1-1)^p = 0$ and $x_i - 1 = (x_{i+1}-1)^p$, for every positive integer i. Thus ωG is a nil non-trivial idempotent ideal of the group ring KG. Note that the ring KG is a commutative ring with unique maximal ideal ωG. By Corollary 1.3 every projective KG-module is free and hence KG is p-connected. But ωG is a non-trivial idempotent ideal of KG. Compare Corollary 2.7. Moreover the ring KG is clearly connected but contains a non-trivial idempotent ideal.

At this point we want to mention another famous theorem of Kaplansky [18] (see [27, Theorem 2.1.8]):

Theorem 5.1. *The group ring $\mathbb{Z}G$ is connected for every group G.*

Note that one consequence of Theorem 5.1 is that if G is any finite non-soluble group then the group ring $\mathbb{Z}G$ is connected but contains a non-trivial idempotent ideal. Before proceeding we recall some well-known results. The first is the following version of Krull's Intersection Theorem (see, for example, [19, Theorem 77]).

Lemma 5.2. *Let S be a commutative Noetherian domain. Then $\delta(A) = \cap_{i=1}^{\infty} A^i = 0$ for every proper ideal A of R.*

The second well-known result we want to quote is the following one (see [27, Lemma 3.1.6]).

Lemma 5.3. *Let S be any ring, let N be a normal subgroup of a group G such that N is a finite p-group, for some prime p, and let R be the group ring SG. Then there exists a positive integer k such that $(\omega N)^k \subseteq Rp$.*

Corollary 5.4. *Let S be a commutative Noetherian domain, let N be normal subgroup of a group G such that N is a finite p-group, for some prime p which is not a unit in S, and let R be the group ring SG. Then $\delta(\omega N) = 0$.*

Proof. By Lemma 5.3, we have that $\delta(\omega N) = \cap_{i=1}^{\infty}(\omega N)^i \subseteq \cap_{i=1}^{\infty} Rp^i$. Because p is not a unit in the commutative Noetherian domain S, Lemma 5.2 gives that $\cap_{i=1}^{\infty} Rp^i = 0$. The result follows. □

Suppose that S is a commutative Noetherian domain of characteristic p, for some prime p, G is a finite p-group and $R = SG$. By Lemma 5.3, there exists a positive integer k such that $(\omega G)^k = 0$. Let P be a maximal ideal of the ring R. Then $\omega G \subseteq P$. The ring $R/\omega G \cong S$ and hence $\cap_{i=1}^{\infty}(P/\omega G)^i = 0$ by Lemma 5.2. Thus $\delta(P) \subseteq \omega G$ and hence $\delta^2(P) = 0$. It follows that $\delta^2(A) = 0$ for every proper ideal A of R. On the other hand, recall that if H is a finite group which is not a p-group then SH is not connected and contains non-trivial idempotent ideals.

In what follows we shall consider commutative Noetherian domains S of characteristic 0. In this case we shall identify the ring $\mathbb{Z}$ with the subring $S' = \{n1 : n \in \mathbb{Z}\}$ of S. The next theorem was proved by Swan [40, Theorem 7] in 1963.

Theorem 5.5. *Let S be a Dedekind domain of characteristic 0 and let G be a finite soluble group such that no prime divisor of the order of G is a unit in S. Then every projective module over the group ring SG is either finitely generated or free.*

By using one of the main results of [40], Roggenkamp [29, Theorem 3] proved the next result.

Theorem 5.6. *Let S be a Dedekind domain of characteristic 0 and let G be a finite soluble group such that no prime divisor of the order of G is a unit in S. Then the group ring SG has no non-trivial idempotent ideals.*

We now prove a theorem which gives both Theorems 5.5 and 5.6 as corollaries. As we shall see the proof uses Roggenkamp's Theorem which in turn used Swan's Theorem so we are not really saying anything new but merely viewing these results in a new light!

Theorem 5.7. *Let S be a Dedekind domain of characteristic 0 and let G be a finite soluble group such that no prime divisor of the order of G is a unit in S. Let A be any proper ideal of $R = SG$. Then there exists a positive integer n such that $\delta^n(A) = 0$.*

Proof. We prove the result by induction on the order $|G|$ of G. If $|G| = 1$ then $R \cong S$ and hence R is a commutative Noetherian domain. Then $\delta(A) = 0$ by Lemma 5.2. Suppose that $|G| > 1$. Let N be a minimal normal subgroup of G. Then N is a p-group for some prime p. Let $B = \delta(A)$. Suppose that $R \neq B + \omega N$. Then $(B + \omega N)/\omega N$ is a proper ideal of the ring $R/\omega N \cong S(G/N)$. By induction, there exists a positive integer m such that $\delta^m((B + \omega N)/N) = 0$ and hence $\delta^{m+1}(A) = \delta^m(B) \subseteq \omega N$. Corollary 5.4 shows that $\delta^{m+2}(A) = 0$.

Now suppose that $R = B + \omega N$. By Lemma 5.3 it follows that $R = B + Rp$. Let $r \in R$ such that $1 - rp \in B$. If $r' \in R$ and $r'(1 - rp) = 0$ then

$$r' = r'rp \in \cap_{i=1}^{\infty} Rp^i = R(\cap_{i=1}^{\infty} Sp^i) = 0,$$

by Lemma 5.2. Similarly, $(1 - rp)r' = 0$ implies that $r' = 0$. Thus $1 - rp$ is a regular element of R. If Q is the quotient field of S then the ring QG is right and left Artinian and hence every regular element of QG is a unit. It follows that $R(1-rp) \cap S \neq 0$. Let $C = B \cap S \neq 0$. The ring S/C is Artinian and hence the ring $R/RC \cong (S/C)G$ is right Artinian. It follows that R/B is a right Artinian ring and hence $A^k = A^{k+1}$ for some positive integer k. Thus A^k is an idempotent ideal of R. By Theorem 5.6, $A^k = 0$. But R is a semiprime ring by Passman's Theorem (see [27, Theorem 4.2.13]). Thus $A = 0$ and hence $\delta(A) = 0$. □

Note that Theorem 5.5 follows from Theorem 5.7 by Theorem 4.6. Note also that the first paragraph of the proof of Theorem 5.7 gives the following result.

Theorem 5.8. *Let S be a commutative Noetherian domain and let G be a finite soluble group such that no prime divisor of the order of G is a unit in S. Then in the group ring SG there exists a positive integer n such that $\delta^n(\omega G) = 0$.*

Corollary 5.9. *Let S be a commutative Noetherian domain and let G be a finite soluble group such that no prime divisor of the order of G is a unit in S. Then there exists a positive integer n such that $\delta^n(A) = 0$ for every ideal A of SG such that $SG \neq A + \omega G$.*

Proof. By Theorem 5.8 there exists a positive integer k such that $\delta^k(\omega G) = 0$. Let $R = SG$. Let $\epsilon : R \to S$ denote the canonical epimorphism. Let A be any ideal of R such that $R \neq A + \omega G$. Then $\epsilon(A)$ is a proper ideal of S and hence, by Lemma 5.2, $\delta(\epsilon(A)) = 0$. It follows that $\delta(A) \subseteq \omega G$. Hence $\delta^{k+1}(A) = 0$. □

Note that Gruenberg [7] proves that if $\mathbb{Z}$ is the ring of integers and G a finite group then, in the integral group ring $\mathbb{Z}G$, $\delta(\omega G) = 0$ if and only if G is a finite p-group for some prime p. This brings us to the following question.

Question 5.10. *Let S be a commutative Noetherian domain of characteristic 0 and let G be a finite soluble group such that p is not a unit in S for each prime divisor p of the order of G. Does there exist a proper ideal A of the group ring SG such that $\delta^n(A) \neq 0$ (or even $\kappa^n(A) \neq 0$) for every positive integer n?*

Recall that a group G is called *metanilpotent* provided there exists a normal subgroup N of G such that the groups N and G/N are both nilpotent. We have the following partial answer to Question 5.10.

Theorem 5.11. *Let S be a commutative Noetherian domain of characteristic 0 and let G be a finite metanilpotent group such that no prime divisor of the order of G is a unit in S. Then for each proper ideal A of the group ring SG there exists a positive integer n such that $\delta^n(A) = 0$.*

Proof. Note first that the ring R is right and left Noetherian. By Passman's Theorem (see [27, Theorem 4.2.13]) the ring R is semiprime. Let N be a normal subgroup of G such that N and G/N are both nilpotent groups. We shall prove the result by induction on the order $|N|$ of the subgroup N. Suppose first that

$|N| = 1$. In this case the group G is finite nilpotent. By [35, Theorems 2.1 and 3.2] (see [34] for the terminology) every ideal of R has the right AR property. It follows that if $Q \subseteq P$ are prime ideals of R then $\delta(P) \subseteq Q$ by [33, Theorem 1.2]. Now apply Theorem 4.1 to obtain that for every proper ideal A of R there exists a positive integer n such that $\delta^n(A) = 0$.

Now suppose that $|N| > 1$. Let P be a maximal ideal of R. Suppose that $|N| \in P$. Then there exists a prime divisor p of $|N|$ such that $Rp \subseteq P$. Let H be the Sylow p-subgroup of N. Because N is nilpotent, H is a normal subgroup of G. By Lemma 5.3 $(\omega H)^k \subseteq P$, for some positive integer k, and hence $\omega H \subseteq P$. Note that $R/\omega H \cong S(G/H)$. By induction on the order of N, there exists a positive integer n such that $\delta^n(P/\omega H) = 0$ and hence $\delta^n(P) \subseteq \omega H$. By Corollary 5.4, $\delta^{n+1}(P) \subseteq \cap_{i=1}^{\infty}(\omega H)^i = 0$. Suppose that $|N| \notin P$. By [38, Lemma 3.4], P has the right AR property.

Suppose that the result is false and let A be any proper ideal of R such that $\delta^n(A) \neq 0$ for every positive integer n. Let $P_1, \ldots, P_t$ denote the minimal prime ideals of R and note that $0 = P_1 \cap \cdots \cap P_t$ by the above remarks. By the above proof we know that, for every positive integer k, every maximal ideal of R containing $\delta^k(A)$ has the right AR property. By [33, Theorem 1.2] and the proof of Theorem 4.1, there exists a positive integer t such that $\delta^t(A)$ is generated by a central idempotent element e in R.

With the above notation, $e + \omega H$ is a central idempotent of the ring $R/\omega H$ and is not a unit. By induction, $e \in ReR \subseteq \omega H$. By Corollary 5.4, $e = 0$. Thus $\delta^t(A) = 0$, a contradiction. □

Note that the proof of Corollary 5.9 shows that if S and G are as in Corollary 5.9 and A is an ideal of the group ring SG such that $SG \neq \delta^t(A) + \omega G$, for some positive integer t, then $\delta^n(A) = 0$ for some positive integer n. In fact, the proof shows rather more, namely if $SG \neq \kappa^t(A) + \omega G$, for some positive integer t, then $\kappa^n(A) = 0$ for some positive integer n. This raises the following question which is a special case of Question 3.3.

Question 5.12. *With the notation of Theorem* 5.8 *does there exist an ideal A of the group ring SG such that $\kappa^n(A) = 0$ for some positive integer n but $\delta^m(A) \neq 0$ for every positive integer m?*

6. Group rings of infinite groups

Now we consider group rings of certain infinite groups. A group G is called *polycyclic* provided there exist a finite chain

$$G = G_0 \supseteq G_1 \supseteq \cdots \supseteq G_n = 1,$$

where, for each $1 \leq i \leq n$, G_i is a normal subgroup of G_{i-1} such that the factor group G_{i-1}/G_i is cyclic. Note that a group G is polycyclic if and only if G is a soluble group such that every subgroup is finitely generated. Next a group G is called *polycyclic-by-finite* if G contains a polycyclic normal subgroup N such that the factor group G/N is finite. Polycyclic-by-finite groups are interesting for us because of the following theorem of Hall [8, Theorem 1] (or see [27, Theorem 10.2.7]):

Theorem 6.1. *Let S be a right Noetherian ring and let G be a polycyclic-by-finite group. Then the group ring SG is right Noetherian.*

In what follows we need a mechanism to reduce from an infinite group G to a finite factor group of G and this is provided by a theorem of Roseblade. By a *capital* of a commutative ring S is meant a field S/P, for some maximal ideal P of S. A commutative ring S is called a *Hilbert ring* provided every prime ideal of S is an intersection of maximal ideals, i.e., the Jacobson radical of every prime homomorphic image of S is zero. A field is called *absolute* provided every non-zero element is a root of unity. With this terminology Roseblade [31, Corollary A] (see also [27, Theorem 12.3.7]) proved:

Theorem 6.2. *Let S be a commutative Noetherian Hilbert ring all of whose capitals are absolute, let G be a polycyclic-by-finite group and let R be the group ring SG. Then every simple right R-module is finite dimensional over a capital of S.*

Theorem 6.2 has the following consequence (see [30] or [27, Corollary 12.3.9]):

Corollary 6.3. *With the notation of Theorem 6.2, for every maximal ideal M of R there exists a normal subgroup N of finite index in G such that $\omega N \subseteq M$.*

We want to apply Roseblades's Theorem to certain group ring situations. To do this we again need information about intersections of powers of the augmentation ideal of certain group rings. Note that Jennings [16] proved that if G is a finitely generated torsion-free nilpotent group then $\delta(\omega G) = 0$ in the group ring $\mathbb{Z}G$. For related results see [26]. We could use Jennings result in the proof of the next result but choose not to do so in order that the presentation be more self-contained.

Lemma 6.4. *Let S be a commutative Noetherian domain of characteristic 0 and let G be a polycyclic group such that no prime p is a unit in S. Then in the group ring $R = SG$, $\delta^n(\omega G) = 0$ for some positive integer n.*

Proof. Let $G = G_0 \supseteq G_1 \supseteq \cdots \supseteq G_t = 1$ denote the derived series of G. We shall prove the result by induction on t. If $t = 0$ then $\omega G = 0$. Suppose that $t \geq 1$. Let $N = G_{t-1}$. Then N is a finitely generated Abelian subgroup of G. By induction on t we know that in the ring $R/\omega N \cong S(G/N)$, $\delta^m(\omega G/\omega N) = 0$ for some positive integer m. It follows that $\delta^m(\omega G) \subseteq \omega N$. Let T denote the torsion subgroup of N. Then T is a finite subgroup of N and a finite normal subgroup of G. We now complete the proof by induction on $|T|$.

Suppose first that $|T| = 1$. Then N is a finitely generated torsion-free Abelian group. The ring SN is a commutative Noetherian domain. Let $A = \sum_{x\in N}(x-1)SN$, the augmentation ideal of SN. For any $g \in G$ and $x \in N$, $g(x-1) = (gxg^{-1}-1)g$. It follows that $\omega N = AR = RA$. By Lemma 5.2 $\delta(A) = 0$

and hence $\delta(\omega N) = \delta(A)R = 0$. In this case, $\delta^{m+1}(\omega G) = 0$. Now suppose that T is non-trivial. Let p be any prime divisor of the order of T and let H be the Sylow p-subgroup of T. By induction on $|T|$, there exists a positive integer n such that $\delta^n(\omega G) \subseteq \omega H$ and hence $\delta^{n+1}(\omega G) = 0$ by Corollary 5.4. The result follows. □

Theorem 6.5. *Let S be a Dedekind domain of characteristic 0 such that S has zero Jacobson radical and such that no prime is a unit in S. Suppose further that every capital of S is absolute. Let G be a polycyclic group and let $R = SG$. Then for each proper ideal A of R there exists a positive integer n such that $\delta^n(A) = 0$.*

Proof. Let A be any proper ideal of R. Let M be a maximal ideal of R such that $A \subseteq M$. By Corollary 6.3 there exists a normal subgroup N of finite index in G such that $\omega N \subseteq M$. Then $M/\omega N$ is a proper ideal of the ring $R/\omega N \cong S(G/N)$. By Theorem 5.7 there exists a positive integer k such that $\delta^k(M/\omega N) = 0$ and hence

$$\delta^k(A) \subseteq \delta^k(M) \subseteq \omega N.$$

By Lemma 6.4,

$$\delta^{k+1}(A) \subseteq \delta(\omega N) \subseteq \delta(\omega G) = 0,$$

and the result is proved. □

Note that if S and G are as in Theorem 6.5, in particular if S is the ring $\mathbb{Z}$ of integers, then $R = SG$ has no non-trivial idempotent ideals and every projective right R-module is finitely generated or free. We have the following result for polycyclic groups that are metanilpotent.

Theorem 6.6. *Let S be a commutative Noetherian domain and let G be a finitely generated group having a finite normal subgroup N such that both N and G/N are nilpotent and such that no prime divisor of the order of an element of finite order in G is a unit in S. Then for each proper ideal A of R there exists a positive integer n such that $\delta^n(A) = 0$.*

Proof. Adapt the proof of Theorem 5.11. □

These results lead to the following obvious question.

Question 6.7. *Let R be the group ring SG of a commutative Noetherian domain S over a polycyclic-by-finite group G. Is R right (and left) shallow?*

We end this section with two recent theorems. The first is due to Linnell, Puninski and the author [21].

Theorem 6.8. *The following statements are equivalent for a polycyclic-by-finite group G.*

(i) *G is polycyclic.*
(ii) *The integral group ring $\mathbb{Z}G$ has no non-trivial idempotent ideals.*
(iii) *Ever projective right $(\mathbb{Z}G)$-module is finitely generated or free.*

The second and related theorem is due to McGovern, Puninski and Rothmaler [24].

Theorem 6.9. *Let G be a polycyclic-by-finite group. Then G is polycyclic if and only if every projective right $(\mathbb{Z}G)$-module is a direct sum of finitely generated submodules.*

7. Ideals with a centralizing sequence of generators

Let R be any ring and let A be any ideal of R. We shall say that A has a *centralizing sequence of generators* if A can be generated by a centralizing sequence of elements of R and in this case, following [23, 4.1.13], we call A *polycentral.* For example, in [22] McConnell proved that if U is the universal enveloping algebra of a finite dimensional Lie algebra $\mathfrak{g}$ over a field, then every ideal of U is polycentral if and only if $\mathfrak{g}$ is nilpotent (see also [37]). Corresponding theorems for group rings can be found in [30] and [32]. In particular, it is proved in [37] that if K is a field and G a finitely generated group such that there exists a finite normal subgroup N of G such that G/N is nilpotent and the order of N is a unit in K then every ideal of the group algebra KG is polycentral. In this section we show that every ring such that every ideal is polycentral is right and left shallow. We begin with a simple result whose proof is standard and therefore omitted.

Lemma 7.1. *Let R be a ring such that every ideal is polycentral. Then R satisfies the ascending chain condition* (*acc*) *on* (*two-sided*) *ideals.*

Next we shall give more examples of rings for which every ideal is polycentral. We shall call a ring R *hypercentral* provided for all ideals $A \subset B$ of R the non-zero ideal B/A of the ring R/A contains a non-zero central element of R/A, i.e., there exists an element $b \in B \setminus A$ such that $rb - br \in A$ for all $r \in R$. We shall be interested in polynomial rings $S[x]$ in an indeterminate x over a ring S.

Lemma 7.2. *Let S be any hypercentral ring. Then the polynomial ring $R = S[x]$ is also hypercentral.*

Proof. Let $A \subset B$ be any ideals of R. Let k be the non-negative integer which is the least degree of an element which belongs to B but not A. Let A_k and B_k denote the sets consisting of the zero element and the leading coefficients of elements of degree k in A and B, respectively. Then $A_k \subset B_k$ are ideals of S. By hypothesis, there exists $b \in B_k \setminus A_k$ such that $sb - bs \in A_k$ for all $s \in S$. Let c be any element of B of degree k with leading coefficient b. Note that $c \notin A$. For any $s \in S$ there exists $d \in A$ such that $sc - cs$ and d have the same leading coefficient and hence $sc - cs - d = 0$, by the choice of k. Thus $sc - cs \in A$. Because $xc = cx$, we conclude that $rc - cr \in A$ for all $r \in R$. It follows that R is hypercentral. □

Proposition 7.3. *Let S be any ring such that every ideal is polycental and let R denote the polynomial ring $S[x_1, \ldots, x_n]$ over S in commuting indeterminates $x_1, \ldots, x_n$, for some positive integer n. Then every ideal of R is polycentral.*

Proof. By Lemma 7.1 the ring S satisfies acc on ideals. By adapting the proof of the Hilbert Basis Theorem, it follows that the ring R satisfies the acc on ideals. By Lemma 7.2 the ring R is hypercentral. The result follows. □

Corollary 7.4. *Let S be any simple ring and let R denote the polynomial ring $S[x_1, \ldots, x_n]$ over S in commuting indeterminates $x_1, \ldots, x_n$, for some positive integer n. Then every ideal of R is polycentral.*

Proof. By Proposition 7.3. □

In a similar way it can be proved that if S is any ring such that every ideal is polycentral and R denotes the ring $S[[x_1, \ldots, x_n]]$ of formal power series over S in commuting indeterminates $x_1, \ldots, x_n$, for some positive integer n, then every ideal of R is polycentral. In particular, if S is a simple ring then every ideal of the ring $S[[x_1, \ldots, x_n]]$ is polycentral.

Now let R and R' be any rings and let M be a left R'-, right R-bimodule. By an *essential* sub-bimodule of ${}_{R'}M_R$ we mean a sub-bimodule L of M such that $K \cap L \neq 0$ for every non-zero sub-bimodule K of M.

Lemma 7.5. *Let R and R' be rings and let M be any left R'-, right R-bimodule such that M satisfies acc on sub-bimodules. Let L be an essential sub-bimodule of M and let A be a polycentral ideal of R such that $LA = 0$. Then $MA^n = 0$ for some positive integer n.*

Proof. Suppose that the result if false. Let k be the least positive integer such that there exist rings S' and S and a left S'-, right S-bimodule N, an essential sub-bimodule K of N and a polycentral ideal B of S with a centralizing sequence of generators $b_1, \ldots, b_k$ such that N satisfies acc on sub-bimodules and $KB = 0$ but $NB^n \neq 0$ for every positive integer n. Suppose that $k = 1$. Let $b = b_1$. For each positive integer i, let $H_i = \{m \in N : mb^i = 0\}$. Clearly H_i is a sub-bimodule of N for each $i \geq 1$. Moreover, $H_1 \subseteq H_2 \subseteq \ldots$. By hypothesis, there exists a positive integer t such that $H_t = H_{t+1}$. It is easy to check that $K \cap Nb^t = 0$. But Nb^t is a sub-bimodule of N. Hence $Nb^t = 0$ so that $NB^t = 0$, a contradiction. Thus $k \geq 2$.

By the above proof $Nb_1^t = 0$ for some positive integer t. Suppose that $t = 1$. Let $\overline{S}$ denote the ring S/Sb_1 and note that N is a left S'-, right $\overline{S}$-bimodule with essential sub-bimodule K such that $K(B/Sb_1) = 0$. By the choice of k, there exists a positive integer n such that $N(B/Sb_1)^n = 0$ and hence $NB^n = 0$, a contradiction. Thus $t \geq 2$. Consider the left S'-, right $\overline{S}$-bimodule G, where $G = \{m \in N : mb_1 = 0\}$. Note that G satisfies acc on sub-bimodules and G contains an essential sub-bimodule $K \cap G$ with $(K \cap G)(B/Sb_1) = 0$. By the choice of k, there exists a positive integer n_1 such that $GB^{n_1} = 0$. Now consider the left S'-, right S-bimodule Nb_1 which satisfies acc on sub-bimodules and which contains an essential sub-bimodule $K \cap Nb_1$ such that $(K \cap Nb_1)B = 0$. By induction on t, there exists positive integer n_2 such that $(Nb_1)B^{n_2} = 0$, so that $NB^{n_2} \subseteq G$ and hence $NB^n = 0$, where $n = n_1 + n_2$, a contradiction. The result follows. □

Theorem 7.6. *Let R be a ring which satisfies acc on ideals and let A be a polycentral ideal of R. Then for each ideal B of R there exists a positive integer n such that $B \cap A^n \subseteq BA$,*

Proof. Let B be any ideal of R. Consider the collection $\mathcal{S}$ of ideals C of R such that $B \cap C = BA$. Note that BA belongs to $\mathcal{S}$. Let D be a maximal member of $\mathcal{S}$. Consider the left R-, right R-bimodule $M = R/D$. Clearly M satisfies acc on sub-bimodules. Moreover, if $(B + D) \cap E = D$ for some ideal E of R then $B \cap E \subseteq B \cap D = BA$ so that $B \cap E = BA$ and hence $E = D$. It follows that $(B + D)/D$ is an essential sub-bimodule of ${}_RM_R$ with $[(B + D)/D]A = 0$. By Lemma 7.5 there exists a positive integer n such that $MA^n = 0$ and hence $A^n \subseteq D$. Thus $B \cap A^n \subseteq BA$, as required. □

Corollary 7.7. *Let R be a prime ring such that every ideal is polycentral. Then $\delta(A) = \cap_{i=1}^{\infty} A^i = 0$, for every proper ideal A of R.*

Proof. Let A be any proper ideal of R and let $B = \cap_{i=1}^{\infty} A^i$. Suppose that $B \neq 0$ and let c be a non-zero central element of R such that $c \in B$. By Lemma 7.1, the ring R satisfies acc on ideals. Now Theorem 7.6 gives that $cR \cap A^n \subseteq cRA = cA$, for some positive integer n. Then $c \in cA$ and hence $c(1-a) = 0$ and $cR(1-a) = 0$. Because R is prime, we have $1 - a = 0$ and $A = R$, a contradiction. □

In particular, note that prime rings such that every ideal is polycentral do not contain a non-trivial idempotent ideal. Applying Corollary 7.4, we deduce that if S is a simple ring then the polynomial ring $S[x_1, \ldots, x_n]$ and the formal power series ring $S[[x_1, \ldots, x_n]]$ do not contain non-trivial idempotent ideals. This raises the following question.

Question 7.8. *Does there exist a ring S which does not contain any non-trivial idempotent ideal but the polynomial ring $S[x]$ does contain a non-trivial idempotent ideal?*

Of course, there is a corresponding question to Question 7.8 for the ring $S[[x]]$ of formal power series. Combining Corollary 7.7 with Corollary 4.2 we see that every ring for which every ideal is polycentral is right and left shallow. Finally we return to consider projective modules.

Theorem 7.9. *Let S be a simple right Noetherian ring and let R be the polynomial ring $S[x_1, \ldots, x_n]$ over S in the commuting indeterminates $x_1, \ldots, x_n$. Then every projective right R-module is finitely generated or free.*

References

[1] T. Akasaki, Projective R-modules with chain conditions on R/J, Proc. Japan. Acad. 46 (1970), 94–97.

[2] F.W. Anderson and K.R. Fuller, *Rings and Categories of Modules* (Springer-Verlag, New York 1973).

[3] H. Bass, Big projective modules are free, Illinois J. Math. 7 (1963), 24–31.

[4] I. Beck, Projective and free modules, Math. Z. 129 (1972), 231–234.

[5] G.M. Bergman, Infinite multiplication of ideals in $\aleph_0$-hereditary rings, J. Algebra 24 (1973), 56–70.

[6] G. Cauchon, Anneaux semi-premiers noethériens à identités polynômiales, Bull. Soc. Math. France 104 (1976), 99–111.

[7] K.W. Gruenberg, The residual nilpotency of certain presentations of finite groups, Archiv. der Math. 13 (1962), 408–417.

[8] P. Hall, Finiteness conditions for soluble groups, Proc. London Math. Soc. (3) 4 (1954), 419–436.

[9] I.N. Herstein, A counter-example in Noetherian rings, Proc. Nat. Acad. Sc. 54 (1965), 1036–1037.

[10] I.N. Herstein and L.W. Small, The intersection of the powers of the radical in Noetherian P.I. rings, Israel J. Math. 16 (1973), 176–180.

[11] Y. Hinohara, Projective modules over semilocal rings, Tôhoku Math. J. (2) 14 (1962), 205–211.

[12] Y. Hinohara, Projective modules over weakly noetherian rings, J. Math. Soc. Japan, 15 (1963), 75–88.

[13] N. Jacobson, The radical and semisimplicity for arbitrary rings, Amer. J. Math. 67 (1945), 300–320.

[14] A.V. Jategaonkar, A counter-example in Ring Theory and Homological Algebra, J. Algebra 12 (1969), 418–440.

[15] A.V. Jategaonkar, Jacobson's conjecture and modules over fully bounded Noetherian rings, J. Algebra 30 (1974), 103–121.

[16] S.A. Jennings, The group ring of a class of infinite nilpotent groups, Canad. J. Math. 7 (1955), 169–187.

[17] I. Kaplansky, Projective modules, Ann. Math. 68 (1958), 372–377.

[18] I. Kaplansky, *Fields and Rings* (University of Chicago Press, Chicago 1969).

[19] I. Kaplansky, *Commutative Rings* (Allyn and Bacon, Boston 1970).

[20] G. Krause and T.H. Lenagan, Transfinite powers of the Jacobson radical, Comm. Algebra 7 (1979), 1–8.

[21] P.A. Linnell, G. Puninski and P.F. Smith, Idempotent ideals and nonfinitely generated projective modules over integral group rings of polycyclic-by-finite groups, J. Algebra 305 (2006), 845–858.

[22] J.C. McConnell. The Intersection Theorem for a class of non-commutative rings, Proc. London Math. Soc. (3) 17 (1967), 487–498.

[23] J.C. McConnell and J.C. Robson, *Noncommutative Noetherian Rings* (Wiley-Interscience, Chichester 1987).

[24] W.W. McGovern, G. Puninski and P. Rothmaler, When every projective module is a direct sum of finitely generated modules, J. Algebra 315 (2007), 454–481.

[25] Y. Nouazé and P. Gabriel, Ideaux premiers de l'algèbre enveloppante d'une algèbre de Lie nilpotents, J. Algebra 6 (1967), 77–99.

[26] M.M. Parmenter and S.K. Sehgal, Idempotent elements and ideals in group rings and the Intersection Theorem, Arch. Math. 24 (1973), 586–600.

[27] D.S. Passman, *The Algebraic Structure of Group Rings* (Wiley-Interscience, New York 1977).

[28] P. Příhoda, Projective modules are determined by their radical factors, J. Pure Appl. Algebra 210 (2007), 827–835.

[29] K.W. Roggenkamp, Integral group rings of solvable finite groups have no idempotent ideals, Archiv Math. (Basel) 25 (1974), 125–128.

[30] J.E. Roseblade, The integral group rings of hypercentral groups, Bull. London Math. Soc. 3 (1971), 351–355.

[31] J.E. Roseblade, Group rings of polycyclic groups, J. Pure Appl. Algebra 3 (1973), 307–328.

[32] J.E. Roseblade and P.F. Smith, A note on hypercentral group rings, J. London Math. Soc. (2) 13 (1976), 13–190.

[33] P.F. Smith, On the Intersection Theorem, Proc. London Math. Soc. (3) 21 (1970), 385–398.

[34] P.F. Smith, Localization and the AR property, Proc. London Math. Soc. (3), 22 (1971), 39–68.

[35] P.F. Smith, Localization in group rings, Proc. London Math. Soc. (3) 22 (1971), 69–90.

[36] P.F. Smith, A note on projective modules, Proc. Roy. Soc. Edin. 75A (1975/76), 23–32.

[37] P.F. Smith, On non-commutative regular local rings, Glasgow Math. J. 17 (1976), 98–102.

[38] P.F. Smith, A note on idempotent ideals in group rings, Archiv der Math. 27 (1976), 22–27.

[39] P.F. Smith, The Artin-Rees property, in *Séminaire Dubreil-Malliavin (1981)*, ed. M.-P. Malliavin, Lecture Notes in Mathematics 924 (Springer-Verlag, Berlin 1981), 197–240.

[40] R.G. Swan, The Grothendieck ring of a finite group, Topology 2 (1963), 85–110.

[41] R.B. Warfield, Jr., Exchange rings and decompositions of modules, Math. Ann. 199 (1972), 31–36.

[42] J.M. Whitehead, Projective modules and their trace ideals, Comm. Algebra 8 (1980), 1873–1901.

[43] R. Wisbauer, *Foundations of Ring and Module Theory* (Gordon and Breach, Philadelphia 1991).

Patrick F. Smith
Department of Mathematics
University of Glasgow
Glasgow G12 8QW, Scotland, UK
e-mail: pfs@maths.gla.ac.uk

Advances in Ring Theory
Trends in Mathematics, 327–334

On Ef-extending Modules and Rings with Chain Conditions

Le Van Thuyet and Truong Cong Quynh

Abstract. A module M is called ef-extending if every closed submodule which contains essentially a finitely generated submodule is a direct summand of M. A ring R is called right ef-extending if R_R is an ef-extending module. In this paper, we obtain some properties of a ring R for which $R \oplus R$ is ef-extending as a right R-module. Then we study the structure of rings for which the direct sum of any two ef-extending right R-modules is ef-extending.

Mathematics Subject Classification (2000). 16D50, 16L60, 16D80.

Keywords. Ef-extending rings, extending (or CS) rings, PF rings, QF rings.

1. Introduction

Throughout the paper, R represents an associative ring with identity $1 \neq 0$ and all modules are unitary R-modules. We write M_R (resp., ${}_RM$) to denote that M is a right (resp., left) R-module. For a right R-module M, $\sigma[M]$ denotes the full subcategory of Mod-R whose objects are submodules of M-generated modules. Unless otherwise mentioned, by a module we will mean a right R-module.

We recall the concepts and notations will be used in this paper. We denote the Jacobson radical of a ring R by J and the injective hull of M by $E(M)$. If A is a submodule of M (resp., proper submodule), we denote by $A \leq M$ (resp., $A < M$).

A submodule K of M is *essential* in M if $K \cap L \neq 0$ for every non-zero submodule L of M. In this case, M is called an *essential extension* of K and we write $K \leq^e M$. A submodule C of M is *closed* in M if C has no proper essential extension in M. A module M is called *uniform* if $M \neq 0$ and every non-zero submodule of M is essential in M. A module M is called *to have finite uniform dimension* if M does not contain an infinite direct sum of non-zero submodules.

The work was supported by the NAFOSTED of Vietnam.

A module M is called *small* if M is small in $E(M)$. If M is not a small module, we say that M is *non-small.*

We consider the following conditions on a module M_R:

C1: Every submodule of M is essential in a direct summand of M.

C2: Every submodule of M that is isomorphic to a direct summand of M is itself a direct summand of M.

C3: $M_1 \oplus M_2$ is a direct summand of M for any two direct summands M_1, M_2 of M with $M_1 \cap M_2 = 0$.

Module M_R is called *extending* (or CS) (resp., *continuous, quasi-continuous)* if it satisfies C1 (resp., both C1 and C2, both C1 and C3). R is called right extending (resp., continuous) if R_R is an extending (resp., continuous) module. A module M is called *uniform-extending* if every uniform submodule is essential in a direct summand of M.

Recall that a module M is called *ef-extending* if every closed submodule which contains essentially a finitely generated submodule is a direct summand of M. A ring R is called right ef-extending if R_R is an ef-extending module (see [13]). Ef-extending modules were studied in Thuyet and Wisbauer [13], Chien and Thuyet [3], Quynh and Thuyet [11, 12]. Some characterizations of ef-extending modules and rings were obtained. It is well known that a right extending ring is right ef-extending, but the converse is not true in general (see [12]). Some characterizations of QF-rings via ef-extending properties have been studied in [11]. In this paper, we characterize the QF-ring via ef-extending modules and prove that R is QF if and only if $R_R^{(\mathbb{N})}$ is an ef-extending module with finite uniform dimension and R_R is pseudo-injective. Moreover, we also study the structure of rings for which the direct sum of two ef-extending modules is an ef-extending module.

2. Results

It is well known that the direct sum of two extending modules may not be an extending module, in general. The same situation happens for ef-extending modules. In fact, let p be a prime number. Then $\mathbb{Z}$-modules $\mathbb{Z}/p\mathbb{Z}$, $\mathbb{Z}/p^3\mathbb{Z}$ are ef-extending. Since $(1+p\mathbb{Z}, p+p^3\mathbb{Z})\mathbb{Z}$ is a closed submodule of $M = \mathbb{Z}/p\mathbb{Z} \oplus \mathbb{Z}/p^3\mathbb{Z}$ which contains essentially a finitely generated submodule. But it is not a direct summand of M, so it follows that M is not ef-extending.

We first consider rings for which the direct sum of two ef-extending modules is an ef-extending module.

By using the technique of proving [5, Lemma 7.3], we have:

Lemma 2.1. *Let A and B be uniform modules with local endomorphism rings such that $M = A \oplus B$ is ef-extending. Let C be a submodule of A and let $f : C \to B$ be a homomorphism. Then the followings hold:*

1. *If f cannot be extended to a homomorphism from A to B, then f is a monomorphism and B is embedded in A.*

2. *If any monomorphism $B \to A$ is an isomorphism, then B is A-injective.*
3. *If B is not embedded in A, then B is A-injective.*

Proof. (1). Suppose f cannot be extended to a homomorphism from A to B. Let

$$U = \{x - f(x) | x \in C\} \leq A \oplus B.$$

Then $U \cong C$ is a uniform submodule of M and clearly $U \cap B = 0$. Hence there is a direct summand U^* of M such that $U \leq^e U^*$. By the Krull-Schmidt-Azumaya Theorem ([2, Corollary 12.7]), we have $M = A \oplus U^*$ or $M = U \oplus B$. Suppose that $M = B \oplus U^*$. Let $\pi : B \oplus U^* \to B$ be the projection. Then it is easy to see that $\pi|_A$ extends $f : C \to B$, a contradiction. Thus $M = A \oplus U^*$ which implies that$f(x) \neq 0$ for $x \neq 0$, i.e., f is a monomorphism. Since $U^* \cap B = 0$, clearly B is embedded in A.

(2). As in the proof of (1), let $f : C \to B$ be any homomorphism, with $C \leq A$ and suppose that $M = A \oplus U^*$, $\psi : A \oplus U^* \to A$ be the projection. Then clearly $\psi|_B$ is a monomorphism (because U is essential in U^*), hence an isomorphism by the hypothesis. It follows easily that $M = B \oplus U^*$, so that, as in (1), f can be extended to a homomorphism from A to B. It follows that B is A-injective.

(3) can be obtained easily by (1). □

Lemma 2.2. *Let M be a module such that* Soc (M) *is finitely generated and essential in M. Then M is an extending module if and only if M is an ef-extending module.*

Proof. It is obvious. □

Lemma 2.3. *The following statements are equivalent for a module M:*

1. *The direct sum of any two uniform modules is ef-extending.*
2. *Any uniform self-injective module has length at most* 2.
3. *Any direct sum of uniform modules is extending.*

Proof. (1) $\Rightarrow$ (2). Consider any uniform injective module U. Suppose $x \in$ Rad U and T is a maximal nonzero submodule of xR. Then U and xR/T have local endomorphism rings and $U \oplus xR/T$ is ef-extending by assumption. Hence the map $f : xR \to xR/T$ can be extended to $\bar{f} : U \to xR/T$ by Lemma 2.1. However $xR \leq$ Rad $U \leq$ Ker $\bar{f}$ which yields a contradiction. We conclude that Rad U is semisimple and hence simple.

Assume that K_1, K_2 are two distinct maximal submodules of U. Then any monomorphism $f : K_i \to K_j$ is onto for $i, j \in \{1, 2\}$, since f extends to a monomorphism of U which has to be an automorphism. So the endomorphism rings of K_1 and K_2 are local. Now $K_i \oplus K_j$ is extending for $i, j \in \{1, 2\}$ and hence K_1 is both K_2-injective and K_1-injective by Lemma 2.1. Since $K_1 + K_2 = U$, this implies that K_1 is U-injective and hence is a direct summand of U, a contradiction.

(2) $\Rightarrow$ (3). By [5, 13.1].

(3) $\Rightarrow$ (1). Obvious. □

Consider the following property for a ring R:

(W): *The direct sum of any two ef-extending right R-modules is ef-extending.*

In [7], Er has proved that if R has finite uniform dimension and the direct sum of any two extending right R-modules is extending, then R is right Artinian. We prove the following:

Proposition 2.4. *Assume that R has property (W) and $E(R_R) = \bigoplus_{i\in I} E_i$ where E_i is indecomposable for all $i \in I$. Then R is a right Artinian ring whose uniform right R-modules have length at most two.*

Proof. Since $E(R_R) = \bigoplus_{i\in I} E_i$ where E_i is indecomposable for all $i \in I$, E_i is an uniform submodule for all $i \in I$. Let $\{V_j\}_{j\in J}$ be any nonempty family of injective hulls of simple modules. Let $V = \oplus_{j\in J} V_j$. Then the module

$$M = E(R_R) \oplus V$$

is extending by Lemma 2.3, hence quasi-injective by [5, Lemma 8.10]. Therefore V is a quasi-injective module which is $E(R_R)$-injective. Thus V is injective. This implies that R is right Noetherian.

As R is right Noetherian, every injective module is a direct sum of uniform modules. Since by Lemma 2.1, each uniform module has length at most two, every injective module is a direct sum of injective hulls of simple modules. This proves that R is right Artinian. □

Corollary 2.5. *Assume that R has finite uniform dimension and property (W). Then R is a right Artinian ring whose uniform right R-modules have length at most two.*

Lemma 2.6. *Let M be an ef-extending, C3 module. If R satisfies ACC on right ideals of the form $r(m)$, $m \in M$, then M is a direct sum of uniform submodules.*

Proof. Same argument of [5, 8.2], M contains a maximal local direct summand $N = \bigoplus_{i\in I} N_i$. By [5, 8.1], N is closed in M. Let H be a complement of N in M. That means $N \oplus H \leq^e M$. Assume that $H \neq 0$. Let $r(y)$ be maximal in $\{x \in M \setminus \{0\} | xR \cap N = 0\}$. Since M is ef-extending, $yR \leq^e K \leq^{\oplus} M$. Repeat above proving, we have K is uniform. Since M has C3 and so $N \oplus K$ is a local direct summand of M. This is a contradiction. Therefore $H = 0$ and so $N \leq^e M$. It implies that $N = M$. □

Hence we have a characterization of a right Artinian ring via ef-extending property:

Proposition 2.7. *The following statements are equivalent for a ring R:*

1. *R has finite uniform dimension and the direct sum of any two ef-extending, C3 right R-modules is ef-extending.*
2. *R has finite uniform dimension and direct sum of any two uniform right R-modules is ef-extending.*

3. *R has finite uniform dimension and direct sum of any two uniform right R-modules is extending.*
4. *R is a right Artinian ring whose uniform right R-modules have length at most two.*

Proof. (1) $\Rightarrow$ (2). Since every uniform module has C3, it follows that direct sum of any two uniform modules is ef-extending by (1).

(3) $\Leftrightarrow$ (2). By Lemma 2.3.

(4) $\Leftrightarrow$ (2). By [7, Theorem 1].

(4) $\Rightarrow$ (1). By Lemma 2.6 and [8, Lemma 6]. □

Corollary 2.8. *The following statements are equivalent for a ring R:*

1. *R has finite uniform dimension and the direct sum of any two ef-extending, C3 right R-modules is ef-extending.*
2. *R has finite uniform dimension and the direct sum of any two extending right R-modules is extending.*

Proof. By Proposition 2.7 and [7, Theorem 1]. □

Next we have some results about an ef-extending, pseudo-injective module. A module M is called *pseudo-N-injective* (*resp., essentially pseudo-N-injective*) if, for any submodule A (resp., essential submodule A) of N, every monomorphism $f : A \to M$ can be extended to a homomorphism $\bar{f} : N \to M$.

Proposition 2.9. *Assume that R satisfies ACC on right ideals of the form $r(m)$, $m \in M$. Then M is quasi-injective if and only if M is pseudo-injective and ef-extending.*

Proof. Let M be a pseudo-injective, ef-extending module. Since M is pseudo-injective, M has C2 by [4, Theorem 2.6]. Thus M is a direct sum of uniform modules by Lemma 2.6. Then M is quasi-injective by [1, Lemma 3.5]. □

Corollary 2.10 ([4], [1, Theorem 3.6])**.** *Let R be a right Noetherian ring. Then M is quasi-injective if and only if M is pseudo-injective and extending.*

Corollary 2.11. *The following statements are equivalent for a ring R:*

1. *R is QF.*
2. *R is a right pseudo-injective, right ef-extending ring satisfying ACC on right annihilators.*

If the right R-module $R_R^{(\mathbb{N})}$ is ef-extending, then we have:

Theorem 2.12. *The following statements are equivalent for a ring R:*

1. *R is QF.*
2. *$R_R^{(\mathbb{N})}$ is an ef-extending module and R_R is an essentially pseudo-injective module with finite uniform dimension.*
3. *$R_R^{(\mathbb{N})}$ is an ef-extending module and R_R is a pseudo-injective module with finite uniform dimension.*

Proof. (2) $\Leftrightarrow$ (3). By [1, Theorem 3.2].

(1) $\Rightarrow$ (3) is clear.

(3) $\Rightarrow$ (1). Since R has finite uniform dimension and R is right ef-extending, $R = e_1R \oplus e_2R \oplus \cdots \oplus e_nR$ for some uniform right ideals e_iR of R. Therefore R is right self-injective by [1, Lemma 3.5]. It implies that e_iR is injective for every $i = 1, 2, \ldots, n$. On the other hand, $R_R^{(\mathbb{N})} = (e_1R \oplus \cdots \oplus e_nR)^{(\mathbb{N})} = \bigoplus_{\omega'} \varepsilon_i R$ is uniform-extending, for some countable set ω' and $\varepsilon_i \in \{e_1, e_2, \ldots, e_n\}$ for each $i \in \omega'$. By [5, Corollary 8.10], $R_R^{(\mathbb{N})}$ is injective. By a well-known result of Faith ([6]), R has ACC on right annihilators and hence R is QF.

□

Corollary 2.13 ([1, Theorem 3.7]). *The following statements are equivalent for a ring R:*

1. *R is QF.*
2. *$R_R^{(\mathbb{N})}$ is an extending module and R_R is an essentially pseudo-injective module with finite uniform dimension.*

We characterize a right PF ring via the ef-extending property of $(R \oplus R)_R$.

Theorem 2.14. *The following statements are equivalent for a ring R:*

1. *R is right PF.*
2. *$(R \oplus R)_R$ is ef-extending, left Kasch and $S_r \leq^e R_R$.*
3. *$(R \oplus R)_R$ is ef-extending, left Kasch and $J \leq Z_l$.*

Proof. (1) $\Rightarrow$ (2), (3) are clear.

(2) $\Rightarrow$ (1). By the same argument of [11, Theorem 2.7], we claim that R is semiperfect. Let T be a maximal left ideal of R. Since R is left Kasch, $r(T) \neq 0$. There exists $0 \neq a \in r(T)$ or $T \leq l(a)$ which yields $T = l(a)$ by maximality of T and so $r(T) = rl(a)$. Since R is right ef-extending, $aR \leq^e eR$ for some $e^2 = e \in R$. On the other hand, $aR \leq rl(a) \leq eR$ and then $rl(a) \leq^e eR$. Hence $r(T) \leq^e eR$. It implies that R is semiperfect by [10, Lemma 4.1]. Thus $R = e_1R \oplus \cdots \oplus e_nR$, where $\{e_i\}_{i=1}^n$ is the complete set of orthogonal local idempotents. For every $i \neq j$ $(i, j \in \{1, 2, \ldots, n\})$, let $f : e_iR \longrightarrow e_jR$ be a monomorphism. We have $e_iR \cong f(e_iR) \leq e_jR$. Since R satisfies the right C2 (because R is left Kasch), then $f(e_iR)$ is a direct summand of e_jR or $f(e_iR) = e_jR$ (because e_jR is indecomposable). Hence f is an isomorphism. Since R is right ef-extending, every uniform right ideal of R is essential in direct summand of R_R. Therefore for every $i_0 \in \{1, 2, \ldots, n\}$, $\bigoplus_{\{1,2,\ldots,n\}\setminus\{i_0\}} e_iR$ is $e_{i_0}R$-injective by [5, Corollary 8.9]. Since e_iR is also ef-extending, indecomposable and so e_iR is quasi-continuous. By [9, Theorem 2.13], R is right quasi-continuous. Thus R is right continuous.

By Utumi's Theorem (see [10, Theorem 1.26]), $J = Z_r$. By [10, Example 7.18], $(R \oplus R)_R$ satisfies the C2 and so $(R \oplus R)_R$ is continuous. Therefore R is right self-injective by [10, Theorem 1.35]. Thus R is right PF.

(3) $\Rightarrow$ (1). By (2) $\Rightarrow$ (1) and [15, Theorem 2]. □

Acknowledgment

This paper was written during a stay of the first author at the Department of Mathematics of Ohio University, Athens, Ohio, USA as a visiting professor. We would like to thank Professor Dinh Van Huynh for many useful suggestions and comments. We would also like to thank the Department of Mathematics and their members for their warm hospitality.

References

[1] A. Al-Ahmadi, N. Er, S.K. Jain, *Modules which are invariant under monomorphisms of their injective hulls*, J. Aust. Math. Soc. **79** (2005), 349–360.

[2] F.W. Anderson and K.R. Fuller, *Rings and Categories of Modules*, Springer-Verlag: New York, 1974.

[3] N. Chien and L.V. Thuyet, *On ef-extending modules*, Southeast Asian Bull. Math., 26(2003), 909–916.

[4] H.Q. Dinh, *A note on pseudo-injective modules*, Comm. Algebra **33** (2005), 361–369.

[5] N.V. Dung, D.V. Huynh, P.F. Smith and R. Wisbauer, *Extending Modules*, Pitman, 1996.

[6] C. Faith, *Rings with ascending condition on annihilators,* Nagoya Math. J. 27(1966), 179–191.

[7] N. Er, *Artinian Rings Characterized by Direct Sums of CS modules*, *Comm. Algebra* **32** (2004) 4821–4833.

[8] D.V. Huynh and B.J. Müller, *Rings over which direct sums of CS modules are CS*, Advances in Ring theory, Eds. S.K. Jain and S.T. Rizvi, Birkhäuser Verlag, Boston-Basel-Berlin 1997, pp. 151–159.

[9] S.H. Mohammed and B.J. Müller, *Continuous and Discrete Modules*, London Math. Soc. LN 147: Cambridge Univ.Press., 1990.

[10] W.K. Nicholson and M.F. Yousif. *Quasi-Frobenius Rings*, Cambridge Univ. Press. 2003.

[11] T.C. Quynh and L.V. Thuyet, *Some properties of ef-extending rings*, to appear in Math. J. Okayama Univ.

[12] L.V. Thuyet and T.C. Quynh, *On general injective rings with chain conditions*, Algebra Colloquium, **16**(2)(2009), 243–252.

[13] L.V. Thuyet and R. Wisbauer, *Extending property for finitely generated submodules*, Vietnam J. Math., 25(1997), 65–73.

[14] R. Wisbauer, *Foundations of Module and Ring Theory.* Gordon and Breach: Reading, 1991.

[15] M.F. Yousif, *CS rings and Nakayama permutations*, Comm. Algebra **25** (1997) 3787–3795.

Le Van Thuyet
Department of Mathematics
Hue University
3 LeLoi
Hue city, Vietnam
e-mail: lvthuyet@hueuni.edu.vn
lvthuyethue@gmail.com

Truong Cong Quynh
Department of Mathematics
Danang University
459 Ton Duc Thang
DaNang city, Vietnam
e-mail: tcquynh@dce.udn.vn

Advances in Ring Theory
Trends in Mathematics, 335–345

On Clean Group Rings

Yiqiang Zhou

Dedicated to S.K. Jain on the occasion of his 70th birthday

Abstract. A ring with unity is called clean (uniquely clean) if each of its elements is (uniquely) the sum of an idempotent and a unit. When is a group ring clean? The question seems to be difficult in general. For example, it is even unknown when the group ring of a cyclic group of order 2 is clean. After reviewing the known results, we present several new partial answers to the question. We also give a different proof of the main result on uniquely clean group rings obtained by Chen, Nicholson and Zhou [6].

Mathematics Subject Classification (2000). Primary 16U99, 16S34.

Keywords. Clean ring, group ring, uniquely clean ring.

All rings here are associative rings with unity. Let R be a ring and let G be a group. We denote by RG the group ring of G over R. Let us recall some concepts and notation needed. We write $J(R)$ and $U(R)$ for the Jacobson radical and the set of units of R, respectively. An element of a ring is called *clean* if it is the sum of an idempotent and a unit, and the ring is called *clean* if each of its elements is clean. A ring is called *uniquely clean* if each of its elements can be uniquely written as the sum of a unit and an idempotent. A ring R is (*von Neumann*) *regular* if $a \in aRa$ for all $a \in R$. A ring R such that $a \in aU(R)a$ for all $a \in R$ is called *unit-regular*. A ring whose idempotents are central is called *abelian*, and an abelian regular ring is called *strongly regular*. A ring R is called *strongly π-regular* if $aR \supseteq a^2R \supseteq \cdots$ terminates for every $a \in R$, or equivalently $Ra \supseteq Ra^2 \supseteq \cdots$ terminates for every $a \in R$. As usual, we write C_n and S_n for the cyclic group of order n and the symmetric group of degree n, respectively. A group G is called *locally finite* if every finitely generated subgroup of G is finite. Let p be a prime number. A group G is called a *p-group* if the order of each element of G is a power of p. A group G is said to be an *elementary p-group* if all non-identity elements of G are of order p. It is well known that a finite abelian, elementary p-group is a direct product of finitely many copies of C_p.

1. A brief review

When is a group ring RG clean? This question was first considered by Han and Nicholson [10] who observed the following facts: (1) If R is semiperfect then RC_2 is clean; (2) if R is a Boolean ring and G is a locally finite group, then RG is clean; (3) $\mathbb{Z}_{(7)}C_3$ is not clean where $\mathbb{Z}_{(7)}$ is the localization of $\mathbb{Z}$ at the prime ideal generated by 7. This example gave a negative answer to the question of J.K.Park whether the group ring RG is clean if R is a clean ring and G is a finite group with $|G|$ a unit of R. The authors in [10] also raised the question whether RG is clean when R is a commutative regular ring and G is a locally finite group. This question has a positive answer by a result in [7] that if R is strongly regular or commutative strongly π-regular and if G is a locally finite group then RG is clean (strongly π-regular, indeed). It is a result of [4] that if R is right pure-injective and if G is locally finite then RG is clean. There is a discussion about the uniquely clean group rings in [6], where it was shown that, for a locally finite group G, RG is uniquely clean if and only if R is uniquely clean and G is a 2-group. McGovern [11] discussed the cleanness of RG when R is a commutative clean ring and G is an abelian group. In particular, he proved that if R is a commutative clean ring and G is an abelian elementary 2-group, then RG is clean.

In general, the question when RG is clean seems to be difficult. It is even unknown when RC_2 is clean. If RG is clean, then R must be clean (being an image of RG). But, for a commutative ring R and an abelian group G, RG being clean implies that G is locally finite (see [11, Proposition 2.7] or [7, Proposition 7]). Because some of the important examples of clean rings are semiperfect rings (see [2]), unit-regular rings (see [3]), strongly π-regular rings (see [1]), and abelian clean rings, the following questions seem worthy of consideration: If G is a locally finite group and if R is a semiperfect or unit-regular or strongly π-regular or abelian clean ring, then when is RG clean? We present several new partial answers to these questions. In the last part of the paper, we give a different proof of the main result obtained in [6].

For a group ring RG, the ring homomorphism $\epsilon : RG \to R$, $\Sigma r_g g \mapsto \Sigma r_g$, is called the *augmentation mapping* of RG and its kernel, denoted by $\Delta(RG)$, is $\Delta(RG) = \left\{\sum_{g\in G} a_g(g-1) : 1 \neq g \in G, a_g \in R\right\}$. For $r = \sum r_g g \in RG$, the support of r is the set $\{g \in G : r_g \neq 0\}$. The center of the group G is denoted by $Z(G)$. The ring of $n \times n$ matrices over R is denoted by $\mathbb{M}_n(R)$.

2. A sufficient condition

We start with a well-known result due to Connell.

Lemma 1. [8, Proposition 9] *If R is a ring and G is a locally finite group then $J(R) = J(RG) \cap R$. In particular, $J(R)(RG) \subseteq J(RG)$.*

Lemma 2. *Let p be a prime with $p \in J(R)$. If G is a locally finite p-group, then $\triangle(RG) \subseteq J(RG)$.*

Proof. We first suppose G is finite and prove the claim by induction on $|G|$. Take $z \in Z(G)$ with order $o(z) = p$ and let $\langle z \rangle$ be the subgroup of G generated by z. Then the group $\overline{G} = G/\langle z \rangle$ has smaller order, so $J(R\overline{G}) \supseteq \Delta(R\overline{G})$ by induction hypothesis. The mapping $\varphi : RG \to R\overline{G}$, $\Sigma r_g g \mapsto \Sigma r_g \bar{g}$, is an onto ring homomorphism whose kernel is $(RG)(1-z)$. Since $z^p = 1$, $(1-z)^p \in p(RG) \subseteq J(R)(RG) \subseteq J(RG)$ by Lemma 1. Since $1-z$ is central in RG, it follows that $1-z \in J(RG)$, so $(RG)(1-z) \subseteq J(RG)$. Therefore, $J(R\overline{G}) = \varphi(J(RG))$. Thus, for any $h \in G$, $\varphi(1-h) = 1-\bar{h} \in \Delta(R\overline{G}) \subseteq \varphi(J(RG))$. This shows that $1-h \in (RG)(1-z) + J(RG) = J(RG)$. So $\Delta(RG) \subseteq J(RG)$.

Now for the general case let $r \in \Delta(RG)$. If H is the subgroup of G generated by the support of r, then $r \in \Delta(RH)$ and, since H is a finite p-group, $\Delta(RH) \subseteq J(RH)$ as proved above. Thus $r \in J(RH)$ is quasi-regular. Since r is arbitrary in $\Delta(RG)$, $\Delta(RG)$ is a quasi-regular ideal. So $\Delta(RG) \subseteq J(RG)$. □

Lemma 3. *Let p be a prime number with $p \in J(R)$. Let G be a locally finite group with $G = KH$ where K is a normal p-subgroup of G and H is a subgroup of G. If RH is clean then RG is clean.*

Proof. For $g \in G$, there exist $k \in K$ and $h \in H$ such that $g = kh = (k-1)h + h \in \sum_{k\in K}(1-k)(RG) + RH$. So $RG = \sum_{k\in K}(1-k)(RG) + RH$. By Lemma 2, $\Delta(RK) \subseteq J(RK)$. Since G is locally finite, G/K is locally finite, so $J(RK) \subseteq J(RG)$ by [15, Lemma 4.1]. Hence $\Delta(RK) \subseteq J(RG)$ and this shows that $\sum_{k\in K}(1-k)(RG) \subseteq \Delta(RK)(RG) \subseteq J(RG)$. Hence one obtains

$$RG = J(RG) + RH. \tag{2.1}$$

By [8, Proposition 9], $RH \cap J(RG) \subseteq J(RH)$. But, since $RH/[RH \cap J(RG)] \cong RG/J(RG)$ is semiprimitive, $J(RH) = RH \cap J(RG)$. Therefore, $RH/J(RH) \cong RG/J(RG)$. Suppose RH is clean. Then $RH/J(RH)$ is clean and hence so is $RG/J(RG)$. Now to show that RG is clean, it suffices to show that idempotents of $RG/J(RG)$ can be lifted to idempotents of RG by [10, Proposition 6]. Let $x^2 - x \in J(RG)$ where $x \in RG$. By (2.1), write $x = y + z$ with $y \in J(RG)$ and $z \in RH$. Then $z^2 - z \in RH \cap J(RG) = J(RH)$. Since RH is clean, there exists $e^2 = e \in RH \subseteq RG$ such that $z - e \in J(RH)$ (by [10, Proposition 6]). So $x - e = -y + (z-e) \in J(RG)$. □

Theorem 4. *Let p be a prime with $p \in J(R)$. If R is a clean ring and G is a locally finite p-group, then RG is clean.*

Proof. We may assume that G is finite, and the claim follows by Lemma 3 with $H = \{1\}$. □

By [10], $\mathbb{Z}_{(7)}C_3$ is not clean. This shows that the assumption that $p \in J(R)$ in Theorem 4 is essential.

Example 5. *Let $p > 2$ be a prime number and D_p be the Dihedral group of order $2p$. Let R be a clean ring with $p \in J(R)$. Then RD_p is clean.*

Proof. We have $D_p = \langle a, b : a^p = 1, b^2 = 1, ba = a^{-1}b \rangle = \langle a \rangle \langle b \rangle$, where $\langle a \rangle$ is a normal p-subgroup of D_p. Since $p \in J(R)$, $p-1 \in U(R)$, so 2 is a unit of R. Hence $R\langle b \rangle \cong RC_2$ is clean by Lemma 6 below. So RD_p is clean by Lemma 3. □

Lemma 6. *If R is a clean ring with $2 \in U(R)$ and if G is an abelian elementary 2-group, then RG is clean.*

Proof. We may assume that G is a finite group. Then G is a direct product of n copies of C_2 for some $n \geq 1$. Since $2 \in U(R)$, $RC_2 \cong R \oplus R$. Because 2 is a unit of RC_2, we have $R(C_2 \times C_2) \cong (RC_2)(C_2) \cong RC_2 \bigoplus RC_2 \cong R \oplus R \oplus R \oplus R$. A similar argument shows that RG is isomorphic to the direct sum of $2n$ copies of R. So the claim follows. □

3. Unit-regular and strongly π-regular rings

An element $a \in R$ is *π-regular* if for some $n > 0$, $a^n \in a^n R a^n$. The ring R is called *π-regular* if each of its elements is π-regular.

Lemma 7. *If a nonzero integer k is π-regular in a ring R, then there exists a direct sum decomposition $R = R_1 \oplus R_2$ such that $k \in U(R_1)$ and k is nilpotent in R_2.*

Proof. There exists $n \geq 1$ such that $k^n \in k^n R k^n = k^{2n} R$. Write $k^n = k^{2n} a$ with $a \in R$ and let $e = k^n a$. Then $e^2 = e$ and $k^n = k^n e$. So $eR = k^n R = Rk^n = Re$. Hence e is central and so $R = eR \oplus (1-e)R$. Since $e = k^n a = k^n \cdot eae = eae \cdot k^n$, k^n is a unit of eR; so k is a unit of eR. From $k^n = k^{2n} a$, it follows that $(k(1-e))^n = k^n(1-e) = k^n - k^n e = k^n - k^n = 0$. Thus, k is nilpotent in $(1-e)R$. □

Theorem 8. *Let R be a clean ring and let G be an abelian elementary 2-group. If $2 \in R$ is π-regular, then RG is clean.*

Proof. By Lemma 7, $R = R_1 \oplus R_2$ where $2 \in J(R_1)$ and $2 \in U(R_2)$, so $RG \cong R_1 G \bigoplus R_2 G$. Since R_1, R_2 are still clean, $R_1 G$ is clean by Theorem 4 and $R_2 G$ is clean by Lemma 6. So RG is clean. □

Lemma 9. [5, Lemma 4.4] *If $2 = 0$ in a ring R then $RS_3 \cong RC_2 \bigoplus \mathbb{M}_2(R)$.*

Theorem 10. *Let R be a clean ring and let $G = H \times K$ where H is an abelian elementary 2-group and K is the direct product of finitely many copies of S_3. If $2, 3 \in R$ are π-regular, then RG is clean.*

Proof. We have $RG \cong (RK)H$. By Theorem 8, RK being clean will imply that $(RK)H$ is clean. So we only need to show that RK is clean. Thus it suffices to show that RS_3 is clean, because we can easily finish the proof by a simple induction process. By Lemma 7,

$$R = R_1 \oplus R_2,$$

where 2 is a unit of R_1 and 2 is nilpotent in R_2. Thus, $R_2/2R_2$ is a clean ring and $2 = 0$ in $R_2/2R_2$; so $(R_2/2R_2)S_3$ is clean by Lemma 9 and Theorem 4. Because $(2R_2)S_3$ is a nilpotent ideal of $R_2 S_3$ and because $(R_2 S_3)/(2R_2)S_3 \cong (R_2/2R_2)S_3$,

R_2S_3 is clean by [10, Proposition 6]. Because $RS_3 \cong R_1S_3 \bigoplus R_2S_3$, it remains to show that R_1S_3 is clean. Again by Lemma 7,

$$R_1 = R_{11} \oplus R_{12}$$

where $3 \in U(R_{11})$ and 3 is nilpotent in R_{12}. Thus, $6 \in U(R_{11})$, so

$$R_{11}S_3 \cong R_{11} \bigoplus R_{11} \bigoplus \mathbb{M}_2(R_{11})$$

by [5, Lemma 4.7]. Hence $R_{11}S_3$ is clean. Let $g = (123) \in S_3$ and let $\langle g\rangle$ be the subgroup of S_3 generated by g. Then

$$R_{12}S_3/(R_{12}S_3)(1-g) \cong R_{12}(S_3/\langle g\rangle) \cong R_{12}C_2.$$

Since 2 is a unit in R_{12}, $R_{12}C_2$ is clean by Lemma 6, so $R_{12}S_3/(R_{12}S_3)(1-g)$ is clean. Since $(R_{12}S_3)(1-g) = (1-g)(R_{12}S_3)$,

$$[(R_{12}S_3)(1-g)]^3 \subseteq (R_{12}S_3)(1-g)^3 = (R_{12}S_3)[3(g^2-g)] \subseteq 3(R_{12}S_3).$$

Because 3 is a nilpotent element in R_{12}, $3(R_{12}S_3)$ is a nilpotent ideal of $R_{12}S_3$; so $(R_{12}S_3)(1-g)$ is nilpotent in $R_{12}S_3$. Thus, idempotents of $(R_{12}S_3)/(R_{12}S_3)(1-g)$ can be lifted to idempotents of $R_{12}S_3$. So, by [10, Proposition 6], $R_{12}S_3$ is clean. Hence $R_1S_3 \cong R_{11}S_3 \bigoplus R_{12}S_3$ is clean. □

Let R be a simple ring. If n is an integer, then either $nR = 0$ or $nR = R$. So either $n = 0$ or $n \in U(R)$. In either case, n is π-regular in R.

Corollary 11. *Let R be a clean ring and let M_i $(i = 1, \ldots, n)$ be maximal ideals of R. If $G = H \times K$ where H is an abelian elementary 2-group and K is the direct product of finitely many copies of S_3, then $\left(R/\cap_{i=1}^n M_i\right)G$ is a clean ring.*

Proof. Since R/M_i is a simple ring, every integer is π-regular in R/M_i. So, by Theorem 10, $\left(R/M_i\right)G$ is clean for each i. By Chinese Remainder Theorem,

$$R/\cap_{i=1}^n M_i \cong R/M_1 \bigoplus \cdots \bigoplus R/M_n.$$

So

$$\left(R/\cap_{i=1}^n M_i\right)G \cong \left(R/M_1\right)G \bigoplus \cdots \bigoplus \left(R/M_n\right)G$$

is clean. □

Corollary 12. *Let R be a π-regular clean ring and let $G = H \times K$ where H is an abelian elementary 2-group and K is the direct product of finitely many copies of S_3. Then RG is a clean ring.*

Corollary 13. *Let R be a unit-regular or strongly π-regular ring and let $G = H \times K$ where H is an abelian elementary 2-group and K is the direct product of finitely many copies of S_3. Then RG is a clean ring.*

4. Abelian clean rings

Lemma 14. *If R is a local ring and G is an abelian elementary 2-group, then RG is clean.*

Proof. If $2 \in J(R)$, then RG is clean by Theorem 4. If $2 \notin J(R)$, then $2 \in U(R)$ because R is local; so RG is clean by Lemma 6. □

It is worth noting that C_2 is the only cyclic group G for which RG is clean for every local ring R (see [7, Proposition 12]).

Lemma 15. [15, Lemma 6.1] *If R is a semiperfect ring and G is the direct product of finitely many copies of S_3, then RG is semiperfect.*

Corollary 16. *Let R be a semiperfect ring and let $G = H \times K$ where H is an abelian elementary 2-group and K is the direct product of finitely many copies of S_3. Then RG is clean.*

Proof. By Lemma 15, RK is a semiperfect ring. For each local idempotent e of RK, $(e(RK)e)H$ is clean by Lemma 14. So $(RK)H$ is clean by [10, Proposition 2]. Hence $RG \cong (RK)H$ is clean. □

Proposition 17. *For a ring R and for a locally finite group G, RG is clean if and only if SG is clean for every indecomposable image S of R.*

Proof. ($\Rightarrow$). For an image S of R, SG is an image of RG. So the implication follows.

($\Leftarrow$). If I is an ideal of R and if $a_i \in R$ and $g_i \in G$ ($i = 1, \ldots, n$), we abuse the notation "$\sum \bar{a}_i g_i \in (R/I)G$" to mean that $\sum (a_i + I) g_i \in (R/I)G$. Suppose that RG is not clean. Then there exists a finite subset F of G such that $\sum_{g \in F} a_g g$ is not clean in RG, where each $a_g \in R$. Thus,

$$\mathcal{F} := \Big\{ I \lhd R : \sum_{g \in F} \bar{a}_g g \text{ is not clean in } (R/I)G \Big\}$$

is not empty. For a chain $\{I_\lambda\}$ of elements of $\mathcal{F}$, let $I = \cup_\lambda I_\lambda$. Then I is an ideal of R. Assume that $\sum_{g \in F} \bar{a}_g g$ is clean in $(R/I)G$. Because G is a locally finite group, there exists a finite subgroup H of G with $F \subseteq H$ such that

$$\sum_{g \in H} \bar{a}_g g = \sum_{g \in H} \bar{e}_g g + \sum_{g \in H} \bar{u}_g g \tag{4.1}$$

where $a_g = 0$ for all $g \in H \backslash F$, $\sum_{g \in H} \bar{e}_g g$ is an idempotent in $(R/I)H$ and $\sum_{g \in H} \bar{u}_g g$ is a unit in $(R/I)H$ with inverse $\sum_{g \in H} \bar{v}_g g$. Write

$$H = \{1 = g_1, g_2, \ldots, g_n\}.$$

Thus, the following hold in R/I for $k = 1, \ldots, n$,

$$\begin{cases} \bar{a}_{g_k} = \bar{e}_{g_k} + \bar{u}_{g_k} \\ \bar{e}_{g_k} = \sum_{g_i g_j = g_k} \bar{e}_{g_i} \bar{e}_{g_j} \\ \sum_{g_i g_j = g_k} \bar{u}_{g_i} \bar{v}_{g_j} = \delta_{1k} \bar{1} = \sum_{g_i g_j = g_k} \bar{v}_{g_i} \bar{u}_{g_j} \end{cases} \tag{4.2}$$

where $\delta_{11} = 1$ and $\delta_{1k} = 0$ for $k \neq 1$. It follows that all the following elements (for $k = 1, \dots, n$) are in I:

$$a_{g_k} - e_{g_k} - u_{g_k},\ e_{g_k} - \sum_{g_i g_j = g_k} e_{g_i} e_{g_j},\ \delta_{1k} - \sum_{g_i g_j = g_k} u_{g_i} v_{g_j},\ \delta_{1k} - \sum_{g_i g_j = g_k} v_{g_i} u_{g_j}.$$

Because $\{I_\lambda\}$ is a chain, there exists some I_λ such that all these elements are in I_λ. Hence (4.2) holds in R/I_λ and (4.1) holds in $(R/I_\lambda)G$. So, $\sum_{g \in F} \bar{a}_g g$ is clean in $(R/I_\lambda)G$. This contradiction shows that I is in $\mathcal{F}$. By Zorn's Lemma, $\mathcal{F}$ contains a maximal element, say I. It now suffices to show that R/I is indecomposable.

Assume that R/I is decomposable. Then there exist ideals $K_j \supsetneq I$ of R $(j = 1, 2)$ such that

$$R/I \cong R/K_1 \bigoplus R/K_2, \quad \text{via } r + I \mapsto (r + K_1, r + K_2).$$

Then

$$(R/I)G \cong (R/K_1 \bigoplus R/K_2)G \cong (R/K_1)G \bigoplus (R/K_2)G,$$

where the composition of the two isomorphisms is

$$\sum (r_g + I)g \mapsto \Big(\sum (r_g + K_1)g, \sum (r_g + K_2)g\Big).$$

By the maximality of I in $\mathcal{F}$, $\sum_{g \in F}(a_g + K_j)g$ is clean in $(R/K_j)G$ for $j = 1, 2$. Hence

$$\Big(\sum_{g \in F} (a_g + K_1)g, \sum_{g \in F} (a_g + K_2)g\Big)$$

is a clean element of $(R/K_1)G \bigoplus (R/K_2)G$; so $\sum_{g \in F} \bar{a}_g g$ is clean in $(R/I)G$. This is a contradiction. □

By [7, Theorem 3.1], if R is a commutative strongly π-regular ring and if G is a locally finite group then RG is a clean ring. The next corollary generalizes this. It is also interesting to compare the next result with Corollary 12.

Corollary 18. *If R is a commutative π-regular clean ring and if G is a locally finite group, then RG is clean.*

Proof. Let R/I be an indecomposable image of R. By Proposition 17, it suffices to show that $(R/I)G$ is clean. Since every ideal of R/I not contained in $J(R/I)$ contains a nonzero idempotent, it follows that R/I is local. However, R/I is π-regular and the Jacobson radical of a π-regular ring is nil. So $J(R/I)$ is nil. Note that a local ring is strongly π-regular if and only if its Jacobson radical is nil. So R/I is a commutative strongly π-regular ring. Hence $(R/I)G$ is clean by [7, Theorem 3.1]. □

Corollary 19. *If R is a commutative local ring such that $J(R)$ is nil and if G is a locally finite group, then RG is clean.*

In [11], McGovern proved that if R is a commutative clean ring and if G is an abelian elementary 2-group then RG is clean. McGovern's proof is technical, relying on a result of Johnstone [9] that a commutative ring A is clean if and only if $\mathrm{Max}(A)$ (the set of maximal ideals of A equipped with the hull-kernel topology) is zero-dimensional and every prime ideal of A is contained in a maximal ideal.

Corollary 20. *Let R be an abelian ring and let $G = H \times K$ where H is an abelian elementary 2-group and K is the direct product of finitely many copies of S_3. Then RG is clean if and only if R is clean.*

Proof. ($\Rightarrow$). This is obvious.

($\Leftarrow$). Let R be an abelian clean ring and let R/I be an indecomposable image of R. Since idempotents lift modulo I, R being abelian implies that R/I is abelian; so R/I is an abelian, indecomposable clean ring. Since every left ideal of R/I not contained in $J(R/I)$ contains a nonzero idempotent, it follows that R/I is local. So $(R/I)G$ is clean by Corollary 16. □

A ring R is called *semi-abelian* if there exist orthogonal idempotents e_i ($i = 1, \dots, n$) such that $1 = e_1 + \cdots + e_n$ and each e_iRe_i is an abelian ring. Examples of semi-abelian rings include abelian rings and semiperfect rings.

Lemma 21. *A ring R is semi-abelian and clean if and only if there exist orthogonal idempotents e_i ($i = 1, \dots, n$) such that $1 = e_1 + \cdots + e_n$ and each e_iRe_i is an abelian clean ring.*

Proof. ($\Leftarrow$). This is by [10, Theorem, p. 2590].

($\Rightarrow$). Since R is semi-abelian, there exist orthogonal idempotents e_i ($i = 1, \dots, n$) such that $1 = e_1 + \cdots + e_n$ and each e_iRe_i is an abelian ring. Since R is clean, it is an exchange ring (i.e., for each $a \in R$, there exists $e^2 = e \in R$ such that $e - a \in R(a - a^2)$) by [13, Proposition 1.8]. Thus each e_iRe_i is also an exchange ring by [13, Proposition 1.10]. Since e_iRe_i is abelian, it is clean by [13, Proposition 1.8]. □

Corollary 22. *Let R be a semi-abelian ring and let $G = H \times K$ where H is an abelian elementary 2-group and K is the direct product of finitely many copies of S_3. Then RG is clean if and only if R is clean.*

Proof. Let R be a semi-abelian ring and let G be given as in the corollary. If R is clean, then, by Lemma 21, there exist orthogonal idempotents e_i ($i = 1, \dots, n$) such that $1 = e_1 + \cdots + e_n$ and each e_iRe_i is an abelian clean ring. Thus, for each i, $e_i(RG)e_i \cong (e_iRe_i)G$ is clean by Corollary 20. So RG is clean by [10, Theorem]. □

We do not know if R being clean always implies that RC_2 is clean. The next proposition may be useful in considering this question.

Proposition 23. *Let R be a ring and let $a+bg \in RC_2$. Then $a+bg \in U(RC_2)$ if and only if $a \pm b \in U(R)$. In this case,*

$$(a+bg)^{-1} = (a-b)^{-1}a(a+b)^{-1} - (a-b)^{-1}b(a+b)^{-1}g.$$

Proof. ($\Leftarrow$). From

$$\begin{cases} a(a-b)^{-1} - b(a-b)^{-1} = 1 \\ (a-b)^{-1}a - (a-b)^{-1}b = 1, \end{cases}$$

one obtains

$$\begin{aligned} a(a-b)^{-1}&a(a+b)^{-1} - b(a-b)^{-1}b(a+b)^{-1} \\ &= a\big[1+(a-b)^{-1}b\big](a+b)^{-1} + \big[1-a(a-b)^{-1}\big]b(a+b)^{-1} \\ &= \big[a + a(a-b)^{-1}b + b - a(a-b)^{-1}b\big](a+b)^{-1} \\ &= 1 \end{aligned}$$

and

$$\begin{aligned} b(a-b)^{-1}&a(a+b)^{-1} - a(a-b)^{-1}b(a+b)^{-1} \\ &= b\big[1+(a-b)^{-1}b\big](a+b)^{-1} - \big[1+b(a-b)^{-1}\big]b(a+b)^{-1} \\ &= \big[b + b(a-b)^{-1}b - b - b(a-b)^{-1}b\big](a+b)^{-1} \\ &= 0. \end{aligned}$$

So

$$\begin{aligned} (a+bg)&\big[(a-b)^{-1}a(a+b)^{-1} - (a-b)^{-1}b(a+b)^{-1}g\big] \\ &= \big[a(a-b)^{-1}a(a+b)^{-1} - b(a-b)^{-1}b(a+b)^{-1}\big] \\ &\qquad + \big[b(a-b)^{-1}a(a+b)^{-1} - a(a-b)^{-1}b(a+b)^{-1}\big]g \\ &= 1. \end{aligned}$$

Similarly,

$$\big[(a-b)^{-1}a(a+b)^{-1} - (a-b)^{-1}b(a+b)^{-1}g\big](a+bg) = 1.$$

($\Rightarrow$). Note that, $a+bg \mapsto a+b$ and $a+bg \mapsto a-b$ give two ring homomorphisms from RC_2 onto R, so the claim follows. □

5. Uniquely clean group rings

The following theorem is the main result proved in [6]. Here we give a different proof. A lemma is needed.

Lemma 24. *Let R be a ring and let G be a locally finite group. Then RG is uniquely clean if and only if SG is uniquely clean for every indecomposable image S of R.*

Proof. The proof (with a slight modification) of Proposition 17 works well. □

Theorem 25. [6] *Let R be a ring and let G be a group. If RG is uniquely clean, then R is uniquely clean and G is a 2-group; the converse holds if G is locally finite.*

Proof. Suppose that RG is a uniquely clean ring. Then, by [14, Theorems 20 and 22], R is uniquely clean and $RG/J(RG)$ is Boolean. The Booleanness of $RG/J(RG)$ implies that $1 - g \in J(RG)$ for all $g \in G$, so $\triangle(RG) \subseteq J(RG)$. Thus, G is a p-group and $p \in J(R)$ for some prime p by [8, Proposition 15(i)]. But, $R/J(R)$ is Boolean (by [14, Theorem 20]), so $p = 2$.

For the converse, we suppose that R is a uniquely clean ring and G is a locally finite 2-group. Let S be an indecomposable image of R. Then S is uniquely clean by [14, Theorem 22], so S is abelian by [14, Lemma 4]. Thus, S is a uniquely clean with only trivial idempotents, so $S/J(S) \cong \mathbb{Z}_2$ by [14, Theorem 15]. By [12, Theorem, p.138], SG is a local ring. But, from $S/J(S) \cong \mathbb{Z}_2$, we see that $\mathbb{Z}_2$ is an image of SG; this implies that $SG/J(SG) \cong \mathbb{Z}_2$. So SG is uniquely clean by [14, Theorem 15]. It is now by Lemma 24 that RG is uniquely clean. □

Acknowledgment

The research of the author was supported by Natural Sciences and Engineering Research Council of Canada, Grant OGP0194196. The author thanks Drs. Goodaire and Parmenter for helpful conversations.

References

[1] W.D. Burgess and P. Menal, On strongly π-regular rings and homomorphisms into them, *Comm. Algebra* **16**(1988), 1701–1725.

[2] V.P. Camillo and H.-P. Yu, Exchange rings, units and idempotents, *Comm. Algebra* **22**(1994), 4737–4749.

[3] V.P. Camillo and D. Khurana, A characterization of unit-regular rings, *Comm. Algebra* **29**(2001), 2293–2295.

[4] V.P. Camillo, D. Khurana, T.Y. Lam, W.K. Nicholson and Y. Zhou, Continuous modules are clean, *J. Algebra* **304**(1)(2006), 94–111.

[5] J. Chen, Y. Li and Y. Zhou, Morphic group rings, *J. Pure Appl. Algebra* **205**(3)(2006), 621–639.

[6] J. Chen, W.K. Nicholson, and Y. Zhou, Group rings in which every element is uniquely the sum of a unit and an idempotent, *J. Algebra* **306**(2)(2006), 453–460.

[7] J. Chen and Y. Zhou, Strongly clean power series rings, *Proc. Edinb. Math. Soc.* (2) **50**(1)(2007), 73–85.

[8] I.G. Connell, On the group ring, *Canadian J. Math.* **15** (1963), 650–685.

[9] P.T. Johnstone, Stone Spaces, Cambridge Stud. Adv. Math., **3**, Cambridge Univ. Press, 1982.

[10] J. Han and W.K. Nicholson, Extensions of clean rings, *Comm. Algebra* **29**(2001), 2589–2595.

[11] W.Wm. McGovern, A characterization of commutative clean rings, *Int. J. Math. Game Theory Algebra* **15**(4)(2006), 403–413.

[12] W.K. Nicholson, Local group rings, Canadian Math. Bull. **15**(1972), 137–138.

[13] W.K. Nicholson, Lifting idempotents and exchange rings, *Trans. Amer. Math. Soc.* **229**(1977), 269–278.

[14] W.K. Nicholson and Y. Zhou, Rings in which elements are uniquely the sum of an idempotent and a unit, *Glasgow Math. J.* **46**(2004), 227–236.

[15] S.M. Woods, Some results on semiperfect group rings, *Canadian J. Math.* **26**(1974), 121–129.

Yiqiang Zhou
Department of Mathematics and Statistics
Memorial University of Newfoundland
St. John's, NL A1C 5S7, Canada
e-mail: zhou@mun.ca

MIX
Papier aus verantwortungsvollen Quellen
Paper from responsible sources
FSC® C105338

If you have any concerns about our products,
you can contact us on
ProductSafety@springernature.com

In case Publisher is established outside the EU,
the EU authorized representative is:
Springer Nature Customer Service Center GmbH
Europaplatz 3, 69115 Heidelberg, Germany

Printed by Libri Plureos GmbH
in Hamburg, Germany